Enzymatic Conversion of Biomass for Fuels Production

ACS SYMPOSIUM SERIES 566

Enzymatic Conversion of Biomass for Fuels Production

Michael E. Himmel, EDITOR
National Renewable Energy Laboratory

John O. Baker, EDITOR
National Renewable Energy Laboratory

Ralph P. Overend, EDITOR
National Renewable Energy Laboratory

Developed from a symposium sponsored
by the Division of Cellulose, Paper and Textile
at the 205th National Meeting
of the American Chemical Society,
Denver, Colorado,
March 28–April 2, 1993

American Chemical Society, Washington, DC 1994

Library of Congress Cataloging-in-Publication Data

Enzymatic conversion of biomass for fuels production / Michael E. Himmel, editor; John O. Baker, editor; Ralph P. Overend, editor.

p. cm.—(ACS symposium series, ISSN 0097–6156; 566)

"Developed from a symposium sponsored by the Division of Cellulose, Paper, and Textile at the 205th national meeting of the American Chemical Society, Denver, Colorado, March 28–April 2, 1993."

Includes bibliographical references and indexes.

ISBN 0–8412–2956–2

1. Biomass energy—Congresses. 2. Enzymes—Industrial applications—Congresses.

I. Himmel, Michael E. II. Baker, John O., 1944– . III. Overend, R. P. IV. American Chemical Society. Cellulose, Paper, and Textile Division. V. Series.

TP360.E66 1994
662′.88—dc20 94–30879
CIP

The paper used in this publication meets the minimum requirements of American National Standard for Information Sciences—Permanence of Paper for Printed Library Materials, ANSI Z39.48–1984. ∞

PRINTED IN THE UNITED STATES OF AMERICA

1994 Advisory Board

ACS Symposium Series

M. Joan Comstock, *Series Editor*

Foreword

THE ACS SYMPOSIUM SERIES was first published in 1974 to provide a mechanism for publishing symposia quickly in book form. The purpose of this series is to publish comprehensive books developed from symposia, which are usually "snapshots in time" of the current research being done on a topic, plus some review material on the topic. For this reason, it is necessary that the papers be published as quickly as possible.

Before a symposium-based book is put under contract, the proposed table of contents is reviewed for appropriateness to the topic and for comprehensiveness of the collection. Some papers are excluded at this point, and others are added to round out the scope of the volume. In addition, a draft of each paper is peer-reviewed prior to final acceptance or rejection. This anonymous review process is supervised by the organizer(s) of the symposium, who become the editor(s) of the book. The authors then revise their papers according to the recommendations of both the reviewers and the editors, prepare camera-ready copy, and submit the final papers to the editors, who check that all necessary revisions have been made.

As a rule, only original research papers and original review papers are included in the volumes. Verbatim reproductions of previously published papers are not accepted.

M. Joan Comstock
Series Editor

Contents

INDEXES

Preface

THE PROCESS OF CONVERTING LOW-VALUE lignocellulosic biomass to ethanol via fermentation depends keenly on the development of economical biocatalysts to achieve effective depolymerization of the carbohydrate content of biomass to sugars. This process shares several features in common with contemporary, large-scale uses of cellulases and other carbohydrases (i.e., food and horticultural processing), including the requirement for cost-effective and reliable enzyme production. However, unlike enzymes used in the food industry, cellulases used in ethanologenic processes need not be derived from GRAS (generally recognized as safe) microorganisms. Still, proper stewardship of environmental concerns must be considered.

Cellulase enzymology research, supported by the Department of Energy's Biofuels Program, investigates key aspects of reducing the substantial cost of cellulase enzymes used in industrial processes. The driving force for this effort is the extraordinarily high ratio of enzyme required to fully depolymerize cellulose (that is, one kilogram of cellulase for 50 kilograms of cellulose). Because native biomass is resistant to enzymatic action, chemical–mechanical methods of conditioning the substrate must be employed prior to saccharification. These methods of "pretreatment" require optimization for different substrates and economies of conversion to ethanol. Furthermore, the ideal cellulase complex used in biomass processing must be carefully chosen to ensure high activities on the actual substrate used (such as dilute acid pretreated woods, waste paper, and agricultural residues). The key question to be considered for economic biomass conversion is, "Which pretreatment and cellulase system should be chosen for this task?"

Enzymatic Conversion of Biomass for Fuels Production features 24 chapters written by many of the leading international authorities in the fields of cellulase biochemistry, biofuel cell development, and ethanologenic fermentation. The chapters in this book highlight the diversity of approaches used to enhance our understanding of the action of cellulases on insoluble substrates, as well as efforts to pretreat and ferment the sugars thus released to ethanol. Also reported in this book are recent studies important to the production of biodiesel fuels and electrolytic fuel cells. We believe the chapters in this book bring, individually and collectively, a unique perspective to the field of biomass conversion.

Acknowledgments

We are grateful to the Biotechnology Secretariat of the ACS for sponsoring the symposium upon which this book is based. We also wish to thank the organizers and moderators of the three sessions around which the symposium and book were constructed: John O. Baker, Session I; Paul G. Roessler, Session II; and Michael E. Himmel, Session III. Their contribution was essential to the success of the symposium and book. We also thank Liz Yanda for her expert assistance in production of the final versions of all 24 chapters. Our editor at ACS Books was Barbara E. Pralle. Her guidance and efficient processing of the manuscripts were key to the timely production of the book. Finally, we thank the scientists contributing chapters to this book. In all cases their high-quality work and caring attitude have enhanced our understanding of this intriguing field.

In concluding this preface, we offer the following excerpt from college lecture notes written during the 1940s. Despite the extremely eventful five decades that lie between the writing of the notes and their inclusion in this preface, we believe that this excerpt fully captures the fundamental world view and spirit underlying the efforts described in this volume.

> As you huddle around the campfire on a dark and chilly night with a group of friends and quietly watch the tongues of flame play over the burning wood, have you ever tried to visualize what is happening before you? Here are the pieces of wood which seem so real. They have form, color, weight, hardness for a moment; and then the next moment they disappear from view. Whither is this substance going? The chemist tells us that all the elements which were originally taken from the atmosphere as gases are now returning to the atmosphere as gases. All the materials which originally came from the soil are now returning to the soil as ashes. So most of the substance of the wood is being diffused into space in a colorless, intangible, transparent form. But the leaves of the tree are able to recapture these elements again and convert them into recognizable forms once more. But whence all this light and heat? The leaves in the forming of sugar out of carbon dioxide and water capture the sun's energy and convert it into potential energy stored in the wood. When the wood is burned it combines with oxygen, and the energy captured from the sunlight is again released. So the light and warmth of the campfire is really "canned" sunshine. It is kin to the rainbow and sunset and glimmerings of distant stars. Thus the process of photosynthesis carried on in the leaves and

the process of oxidation comprise a cycle in which the carbon and oxygen of nature continually move, ever storing and releasing heat in the process.

—Edward N. Himmel, Professor of Botany from 1909 to 1958, North Central College, Naperville, IL.

MICHAEL E. HIMMEL
Applied Biological Sciences Branch

JOHN O. BAKER
Applied Biological Sciences Branch

RALPH P. OVEREND
Alternative Fuels Division

National Renewable Energy Laboratory
1617 Cole Boulevard
Golden, CO 80401–3393

June 16, 1994

Chapter 1

Bioconversion for Production of Renewable Transportation Fuels in the United States

A Strategic Perspective

John J. Sheehan

National Renewable Energy Laboratory, 1617 Cole Boulevard, Golden, CO 80401–3393

Ideally, the scope of this introductory chapter would encompass all of the major issues that face bioconversion technology for fuel production in the United States, both technical and social. This is, obviously, a far more ambitious scope than can actually be covered in the space allotted here. For many of the social and political issues presented here, I make no claims of expertise. My perspective is presented strictly as that -- my own perspective on issues facing bioconversion technology development. I will be satisfied if I am able to provide a flavor for some of the bigger issues and to give the technical audience of this book an appreciation for the broad range of opportunities and challenges that face the research community in this area.

What is a "Strategic Perspective?"

I approach the topic of bioconversion for fuels from the vantage point of strategic planning or strategic management. I use the terms strategic planning and strategic management interchangeably. Strategic planning means many things to many people. In fact, for every expert on strategic planning, you will find a pet definition. For my purposes, strategic planning is, in the broadest sense, a way of thinking that forces us to "see the forest for the trees" in everyday decisions and actions. This very broad definition blurs the line between individual planning steps and day-to-day management. And well it should, since certain aspects of strategic planning never stop and are not performed as an isolated effort, separate from day-to-day activities.

Steiner *(1)* views strategic planning from several different perspectives. These include:

- Philosophy (or, as I have said, a way of thinking)
- "Futurity" of current decisions
- Process

0097–6156/94/0566–0001$13.76/0

The essence of strategic planning is the systematic identification of opportunities and threats that lie in the future to provide a basis for making current decisions that take advantage of opportunities and avoid threats. It is an attempt to design a desired future and to find a way to make it happen. Strategic planning embodies a process of continually evaluating your situation, defining aims, setting strategies, and starting the process all over again. This cyclic aspect is critical because, as we all know, the only thing that is truly certain is uncertainty. It is a misconception to think that strategic planning can provide a fixed blueprint for the future. By extension, my intent is not to provide such a blueprint for bioconversion technology.

Figure 1 lays out the basic process of strategic management, which consists of three stages: a situation audit, a planning stage, and an implementation stage. This is a simplified model that can be applied to any organization or activity. The focus of this article is on the situation audit, which brings together all elements that can affect an organization. From an "internal" perspective, it forces an organization to look closely at itself to determine its strengths and weaknesses. Externally, it forces an organization to look outside itself at all the forces at play that could impact its success. The latter is particularly important, as most organizations tend to become myopic and lose sight of the important factors which at first glance do not seem immediately relevant to the tasks at hand.

Strategic planning concepts have traditionally been applied to the operation of private companies. This is changing. Public entities at all levels (local, state and federal) are increasingly seeing the need to apply these principles to their own situation *(2)*. The translation of strategic planning to the public sector is difficult because many of the goals and issues are less quantifiable than they would be for a profit-making organization. In this paper, I have, perhaps, taken this approach one step further by applying strategic thinking to technology development -- in this case, bioconversion for fuels. Strategic planning has been applied to similarly broad efforts involving social initiatives (*3*). It is simply a matter of defining the "internal" and "external" aspects of this technology effort.

The reasons for not using a strategic approach to technology development are many. They are not really very different from the reasons any organization will give to avoid strategic planning. These include:

- Unquantifiable external forces (e.g. political, etc.)
- Unpredictable nature of research and development
- Increasingly turbulent fiscal environments and shifting cultural perception

The first item is perhaps the most readily defended. It is very difficult to put a quantitative framework around the external forces at play for something like renewable fuels technology development. Still, this does not mean that it is not worth trying. The more qualitative such an analysis is, the more prone to error it will be. Though it is difficult to substantiate this claim, I agree with Steiner's supposition *(1)* that the more times we try to "simulate the future" the better we get. In the words of poet Piet Hein *(4)*:

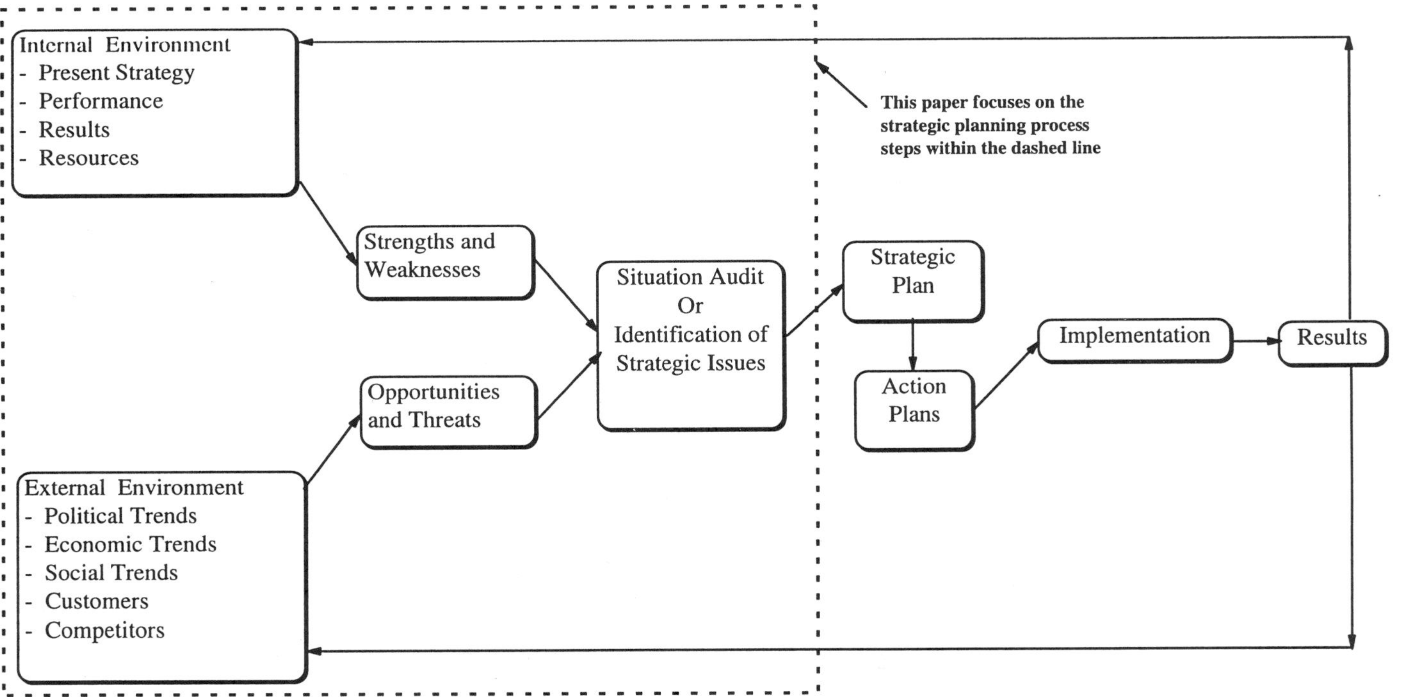

Figure 1: Strategic planning model.

"The road to wisdom -- it is plain and simple to express:

Err
and err
and err again
but less
and less
and less."

The unpredictable nature of research emphasizes the need to continually reassess strategic plans. The same is true when considering the turbulent environment surrounding research on renewable fuels. Strategic planning is a tool that allows us to respond more sensibly to sudden change, and sometimes, to even anticipate it.

With all of this as background, let me restate my goals in this chapter:

- I intend to provide a situation audit that touches on the factors external and internal to the effort of developing technology for biological conversion of renewable biomass to transportation fuels that could impact its ultimate success.

- External issues addressed include: political, social, environmental (that is, ecological), and economic factors.

- Internal factors addressed are limited to a brief status on technology development for three examples covered in this book.

Situation Audit for Bioconversion of Fuels: External Factors

Ask a group of three or more thoughtful people what important issues need to be considered in understanding the external forces that may influence development of bioconversion technology, and you will no doubt end up with dozens of possible concerns. I limit myself to the following areas in this analysis:

- Environmental (ecological) issues
- Energy trends and national security
- Public opinion on environmental and energy issues
- Public policy debate
- Legislative and policy trends

None of these issues is exhaustively addressed. Each in itself could entail volumes of discussion.

The Environment. No issue impacts our daily lives and those of future generations more than the questions being raised about the effects of humankind's activities on the health of our planet. Concern about the global environment came to a head in June 1992, with the United Nations Conference on Environment and Development in Rio de Janeiro, more commonly known as the "Earth Summit." Vice President Al Gore characterized this meeting as a "...critical turning point in the long struggle to increase international awareness of the true nature of the global environmental crisis" *(5)*. Bioconversion technology could play a key role in mitigating potential environmental problems on the horizon. But, what do we really know about these issues? To be sure, there is little we know with great certainty. These are questions which scientists continue to debate and will for years to come. To illustrate these uncertainties, I will consider two of the more serious environmental problems: global climate change and urban air pollution.

Global Climate Change. "The Greenhouse Effect" -- the concept that atmospheric gases such as carbon dioxide and methane can trap heat from sunlight -- is a well-established phenomenon. Venus and Mars are well documented planetary examples of this effect *(6,7)*. The basic theory that certain gases in the atmosphere could absorb radiation, trapping heat like a greenhouse, was introduced by the French mathematician Joseph Fourier. In the 1850's, carbon dioxide was identified as one of the atmospheric gases with high absorptive capacity. In 1896, the Swedish chemist and Nobel prize winner Svante Arrhenius proposed that carbon dioxide produced from burning fossil fuels might result in global warming. The concept fell into obscurity for a long time, however, because of doubts that there would be any appreciable build-up of carbon dioxide in the atmosphere. Most felt that ocean uptake of carbon dioxide would more than compensate for increased anthropogenic carbon dioxide emissions *(8)*. The question was raised again in 1939 by G. S. Callendar, in light of data showing a 60-year warming trend, and yet a third time in 1950 by Gilbert Plass. Concern dwindled quickly because, at about this time, the average global temperature stopped rising.

In 1957 (the International Geophysical Year), Revelle *(7)* presented data that first showed that uptake of anthropogenic carbon dioxide could not possibly occur as rapidly as had been assumed in the past. His findings spurred the first comprehensive effort to collect data on atmospheric carbon dioxide and temperature levels. Data show a dramatic rise in carbon dioxide levels from around 290 ppm in 1860 to almost 350 ppm in 1988. (Note that pre-1958 data was estimated using ice core drillings containing trapped atmospheric gas bubbles *(9)*.)

Temperature rise over the same period is less readily measured. Temperature data reconstructed for the period from 1860 to 1988 are shown in Figure 2. While carbon dioxide levels have shown a strong and steady increase over this time period, the same is not true for temperature. The net temperature increase over the period from 1860 to 1988 is somewhere between 0.5 and 0.7 degrees overall *(6,9)*. But, global temperature may have actually declined between 1940 and 1970. This period of decline was then followed by a decade of record-breaking high temperatures in the 1980s *(6)*.

Herein lies the controversy. The issue is not whether carbon dioxide *can* cause global warming or whether carbon dioxide levels have accumulated in the last century. The greatest uncertainty involves predicting how the Earth's surface temperature will respond as a result of a given increase in carbon dioxide. Research efforts today continue to focus on developing predictive models for surface temperature and on studying ice core and other sources of historical data to ascertain what, if any, clear relationship can be gleaned from past temperature and carbon dioxide levels.

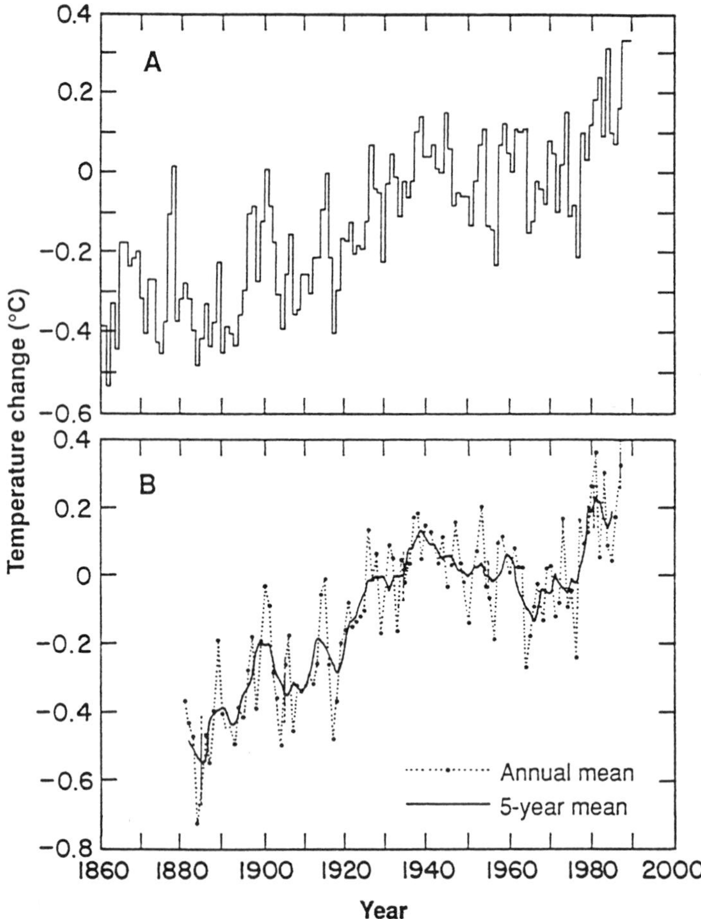

Figure 2: Estimates of global temperature history. A. Based on data from Climate Research Unit; B. Based on data from Goddard Institute for Space Studies. Reproduced with permission from ref. 6. Copyright 1989 American Association for the Advancement of Science (AAAS).

Early ice core data from Antarctica showed a strong correlation between temperature swings and atmospheric greenhouse gas levels. However, the temperature increases observed were five to 14 times higher than expected for measured carbon dioxide changes. One possible interpretation of this data is that global warming may be the causative fact for increased carbon dioxide levels, and not the reverse. Or, this data may indicate that positive feedback mechanisms for warming can occur *(9)*.

More recent data from ice cores in Greenland have revealed even stranger behavior. During the 100,000-year-long ice age that ended 15,000 years ago, it appears that rather abrupt and dramatic warming trends occurred every few thousand years. These warmer periods would continue for only a few years before reverting back to typical ice age conditions. Such observations fly in the face of previously held assumptions that climate change was strictly a gradual process *(10)*. Another implication from these observations is that there may be climatic "switches" that trigger such dramatic changes. There is evidence that ocean circulation patterns may be related to these swings in temperature, especially in the North Atlantic region. Still, the possibility remains, that if the Earth has one such trigger for dramatic change, it may have others that we do not yet know about.

Many Global Climatic Models (GCMs) for predicting temperature-carbon dioxide relationships have been developed over the past few decades. These models are difficult to verify quantitatively based on historical data because of the complexity of the systems being modelled *(6,8)*. Oreske *(5)* has made this point even more strongly, stating that, based on a fundamental understanding of their nature, such numerical models can *never* be verified. An extension of this point is that, even if a model shows good agreement with past and present observations, there is no guarantee that such a model will perform equally well when used to predict the future *(11)*. Why? Because, in complex systems such as those being modeled to predict climate change, quite often more information is available on the dependent variable than is known about the nature of the independent variables. Thus, these models require constant recalibration based on the data and conditions used to confirm the model. As indicated previously in looking at ice core data, it is not even clear whether carbon dioxide levels should be classified as dependent or independent variables. This kind of brutal honesty about the inherently uncertain nature of modeling in the earth sciences will likely not be well received by many researchers, but it is a critical issue that needs to be clearly understood by both the research community and the public at large.

Another major issue with global climate models is that they cannot predict behavior at regional levels. At best, they may predict long term, average global responses. It is an accepted fact that global temperature changes will not be felt uniformly in regions around the world *(6,9,12)*. These regional variations are critical to predictions made about droughts, rises in sea level, and other dire economic and ecological predictions associated with global warming *(6,9,11-14)*.

In 1989, model predictions for temperature rise associated with a doubling in carbon dioxide concentrations over the next century ranged from 2 to 6 degrees centigrade *(6)*. Recently, Michaels *(12)* has pointed out that many researchers have reduced their estimates of temperature rise as more factors are introduced into the models. However, many of these factors involve localized effects such as cloud cover and urban aerosol levels, and these models cannot accurately account for these effects.

Add to this the problem of trying to use equilibrium (or steady-state) based models to predict behavior under conditions which are highly transient and far from steady state, and it becomes clear that reliable predictive models that truly account for regional phenomenon may be decades away *(6,13)*.

In 1989, Schneider *(6)* suggested that increased research might actually confuse rather than clarify the uncertainties about global climate change. He appears to be right. It seems the more we understand about global climate and the complex interactions that are involved in it, the less we are able to make reasonable predictions about the future impacts of greenhouse gases. Examples of this paradox abound in the scientific press. Interpretation of prehistoric data from ice cores continues to draw controversy in the research community *(15)*. The recent eruption of Mt. Pinatubo is an example of a natural global experiment that has confounded researchers. The resulting increase in atmospheric particulates from the volcano should have, by all accounts, caused global cooling. Instead, recent research indicates that it actually caused warming trends in North America and Eurasia *(16)*. Finally, recent data on the concentration of carbon dioxide and methane indicates that their accumulation has slowed abruptly since 1991. At the same time atmospheric oxygen levels have increased. Explanations for this unexpected slowdown in accumulation range from increased carbon uptake by plants to the effects of Mt. Pinatubo, but these are purely speculative *(17)*.

Today, discussions about global warming dominate environmental policy *(5)*. What can we truly say about this potentially disastrous problem? We know that greenhouse warming is a real physical phenomenon. We also know that greenhouse gas levels over the past century have climbed at rates never seen before in the geophysical record. We also know that there is an 80 to 90% heuristic likelihood that the 0.5-degree centigrade rise in global temperature rise over this time period is "not a wholly natural climatic effect" *(12)*. That is, humankind's industrial activities over the last century have had a demonstrated impact on global temperature. What remains to be seen is what future effects will be seen if fossil fuel-driven activities persist or increase beyond current levels. There is a finite risk that unabated greenhouse gas emissions will force unprecedented climatic changes with potentially disastrous repercussions. However, despite all of the data that has been accumulated over the past 35 years, we cannot characterize the situation any better than Roger Revelle and Hans Suess stated it in their landmark 1957 paper on greenhouse gases *(*18):

> "Human beings are carrying out a large-scale geophysical experiment of a kind that could not have happened in the past nor be produced in the future. Within a few centuries, we are returning to the atmosphere and the oceans the concentrated organic carbon stored in sedimentary rocks over hundreds of millions of years."

Urban Air Pollution. The problems of urban air pollution have been recognized for a long time. Nowhere in the U.S. has this been more apparent than in Los Angeles. Southern California's air quality is the worst in the United States, and yet residents of the area have been fighting the problem for over 40 years *(19)*. Much progress has been made, but the experiences of cities like Los Angeles demonstrate the need for a better understanding of the factors influencing pollutant levels in urban areas *(20)*.

The urban atmosphere is essentially a giant chemical reactor in which pollutants such as hydrocarbons, sulfur oxides (SO_x), nitrogen oxides (NO_x), carbon monoxide (CO), and ozone (O_3) react together to create a variety of products. The presence of *both* nitrogen oxides and hydrocarbons in urban environments contributes to the increased level of ozone production in cities, though in an indirect way. The only significant reaction producing ozone is between atomic and molecular oxygen:

$$O + O_2 + M \rightleftharpoons O_3 + M \tag{1}$$

where M is any third body (molecular oxygen or nitrogen) that removes the energy of the reaction and stabilizes ozone formations. At lower altitudes, the following reactions occur:

$$NO_2 + h\nu \rightleftharpoons NO + O \tag{2}$$
$$NO + O_3 \rightleftharpoons NO_2 + O_2 \tag{3}$$

Where NO is nitric oxide, NO_2 is nitrogen dioxide and $h\nu$ is a photon of energy. The net effect of these three reactions is no ozone formation because reaction (3) occurs rapidly. But, of course, the distinguishing feature of urban atmospheric chemistry relative to other areas is the high level of anthropogenic hydrocarbons formed. The presence of hydrocarbons allows a fourth reaction to occur which consumes NO, thus shifting the equilibrium in favor of ozone production:

$$RO_2 + NO \rightleftharpoons NO_2 + RO \tag{4}$$

where R is from a hydrocarbon. But, the formation of RO_2 requires the presence of hydroxyl radicals, which are generated by photodissociation of ozone itself, aldehydes, and/or nitrous acids. These reactions demonstrate the highly interconnected relation that exists among these pollutants. The role of hydrocarbons in urban atmospheric chemistry is tremendously complex. The nature and extent of reactions involved with the various alkanes, alkenes, aldehydes, alcohols, and aromatics is poorly understood. Aromatics represent the most important reactive organic in urban atmosphere, but their role in urban atmospheric chemistry is highly uncertain.

Added to this complex light-catalyzed chemistry are two other phenomena: complex transport processes that affect the chemistry, and the effects of cycling between daytime and nighttime chemistry, in which photo-catalyzed reactions cease. Complex physical and chemical modeling is used to try to estimate the effect of NO_x and hydrocarbon levels on maximum ozone levels for a particular region.

Today, the effects of the various ozone precursors on ozone production is not well understood. There is general agreement that reductions in hydrocarbons will lead to decreases in ozone. On the contrary, the role of NO_x on ozone production is less clear, and its effect depends on the level of organics present in the atmosphere. At high concentrations of organics, NO_x levels control ozone formation. However, at low organics levels, a decrease in NO_x may actually lead to increased ozone production *(19)*.

Visibility problems in cities like Los Angeles are due to the presence of aerosols -- particles of two microns or larger. These aerosols contain sulfates, nitrates, ammonium, trace metals, carbonaceous material, and water. Carbon (soot) is generally responsible for the largest fraction of aerosols. These aerosols probably exacerbate the many health problems that have been encountered in high-smog areas because these particles are easily respired *(19,20)*.

Much progress has been made in reducing SO_x and CO levels. Meanwhile, ozone remains today the most difficult pollutant to reduce *(19)*. As with global climate issues, the problems of understanding and dealing with urban ozone are fraught with uncertainty. Nevertheless, developing a more detailed picture of ozone chemistry is critical to developing strategies for reducing ozone. The answers to these questions will determine the role which can and should be played by alternative fuels such as ethanol, methanol, biodiesel and biogas.

Energy Trends and National Security. World resources of fossil fuels are quite large. By far the largest available resource is coal, and about one-fourth of the total coal reserves are in the United States. To put this in perspective, world crude oil resources would last 70 years at current rates. Natural gas would last 127 years; worldwide coal reserves would last 224 years *(21)*. Fossil fuels will thus likely play a major role in fuel usage for the foreseeable future, this being especially true for coal. The International Panel on Climate Change, which has developed baseline projections for worldwide fuel use, estimates that two-thirds of the total world wide fuel consumption will be based on coal by the year 2050. Also, fuel consumption will shift more to developing nations, who will account for 40% of fuel use by the end of the next century *(22)*.

The U.S. Department of Energy's (DOE's) Energy Information Administration provides the following picture of the U.S. energy situation *(23)*:

- *Rising U.S. Oil Import Dependency.* Between 1980 and 1990, net oil imports in the United States increased from 30% to 40% of the total oil supplied in the United States *(24)*. Projections for imports in 2010 show imported oil supplies increasing to between 52% and 72%. The cost to the U.S. economy for imported oil is projected to rise dramatically from a current cost of $43 billion to $117-140 billion in 2010.

- *Increased Contribution of Electricity to Total U.S. Energy Market.* New demand will initially be met using natural-gas-fired plants, but after 2005, coal-fired plants will be built to meet increased demand.

- *Natural Gas Usage.* Natural gas prices may almost quadruple by the year 2010. These projections are seriously affected by uncertainty in known U.S. supply and rate of technology development for exploration.

- *Coal Usage Increases at Highest Rate of All Fossil Fuels.* Demand will come from electricity generation in the United States and increased U.S. exporting of coal. Coal will remain the mainstay of U.S. baseline electricity generation, accounting for almost half of electricity generation by 2010.

From the perspective of national security, the United States faces a serious crisis in terms of its vulnerability to imports. As the numbers from DOE suggest, the United States has, since 1980, entered into a period of unprecedented dependence on foreign oil imports. The transportation sector accounts for 60% of the total petroleum demand, and is essentially completely dependent on petroleum as its main source of fuel. Declining domestic reserves and continued demand from the transportation sector are the most significant factors affecting oil imports in the future *(25)*.

A recent study by the U.S. Congress Office of Technology Assessment (OTA) conducted in the aftermath of the Persian Gulf War paints a very serious picture of U.S. import vulnerability *(26)*. This study revisited a 1984 assessment of U.S. capability to replace imported oil in the face of a total shut-off of oil supplies from the Middle East. Congress requested this revised study to see how much things had changed between 1984 and 1991. The report's conclusions are disturbing. In 1984 OTA assumed a scenario of an immediate loss of oil of 3 million barrels per day, beginning in mid 1985 and continuing for at least five years. This scenario was equivalent to a loss of 70% of imported oil supplies and a loss of 20 percent of total U.S. oil supplies. Their analysis concluded that the United States had the capability to replace 3.6 million barrels per day over the projected period, exceeding the shortfall by a comfortable margin.

By 1990, the situation had changed a great deal. U.S. consumption had risen from 15 to 17 million barrels per day, while domestic production of oil dropped from 10.3 to 9.2 million barrels per day. Between 1984 and 1990, the share of U.S. oil consumption supplied by imports rose from 33% to 40%. Using the same oil import shut-off scenario in 1990 (i.e., a loss of all 15 million barrels of oil per day from the Persian Gulf), the study estimated that the shortfall in U.S. oil supplies could be as high as 5 million barrels per day. Such a curtailment would be devastating.

A shortfall of five million barrels per day in 1991 would still represent 70% of total imports; however, it would represent 30% of total supplies to the United States. Moreover, the study estimates that currently technology available to replace imports could only displace 2.9 million barrels per day within five years. Accounting for drops in domestic production, the replacement capability would be 1.7 to 2.8 million barrels per day. Thus, the U.S. today is now far less capable of responding to an oil crisis than it was in 1985.

Another way to look at the question of national security is how such increased dependence on oil translates to cost for the nation. The cost of oil imports involves more than just miliary expenses, though even these are considerable. The 1990-1991 Persian Gulf War was an inestimable cost in human life and actual costs. Yet, even this highly visible event was only a small fraction of the total U.S. security costs. Actual costs to defend oil supplies are buried in the total routine spending for defense. Experts disagree on how to apportion these costs to energy security. An average cost estimate from four recent analyses is $35 billion dollars per year. This is a conservative estimate because it is based strictly on the cost of defending the Persian Gulf during peacetime conditions *(27)*.

To these military costs can be added the cost of creating and maintaining the Strategic Petroleum Reserve. Between 1976 and 1990, the U.S. spent $40 billion on this reserve. Current spending is about $300 million per year *(27)*.

There are still other elements of the cost of U.S. imports which go beyond explicit spending on energy security. When the United States pays for imported oil, it transfers large amounts of its wealth to oil-producing countries which maintain monopolistic control over prices. This transfer of wealth is defined as the difference between prevailing oil prices and prices which might exist in a competitive market. Oak Ridge National Laboratory estimated in a 1993 study that the U.S. transferred $1.2-$1.9 trillion of its wealth to oil producers over the period from 1972 to 1991, averaging $60-$95 billion per year.

Both the U.S. and worldwide energy markets will, for the foreseeable future, be dominated by fossil fuels. This is an inevitable consequence of the huge fossil fuel resources available. Coal will dominate in the long term in both the United States and the world. The demand for crude oil in the United States will place the nation in a highly vulnerable position because of its increasing dependence on foreign oil and its shrinking domestic supplies. The costs to the United States in this situation in terms of national security and actual cost to our economy are tremendous.

Public Opinion on Energy, the Environment, and Renewable Fuels. Quantifying trends in public opinion is very difficult because there is more art than science involved in obtaining such information. Nevertheless, understanding public perception on issues related to renewable fuels technology is critical to determining the public's receptiveness toward and interest in the development and commercialization of bioconversion for fuels. It is the public, after all, who are the ultimate customers for the products and benefits of this technology.

Farhar *(28)* provides a comprehensive review of public opinion polls on energy and the environment to try to gauge trends in public opinion and to understand factors that have influenced opinion in these areas over the past 20 years. Here are some conclusions from this study:

- Concern for the environment is clearly on the rise. Relative percentages of people who feel environmental quality is declining versus those who feel it is the same or better have essentially reversed themselves since 1987, with a majority perceiving a deterioration of environmental quality in 1990. Although it is still not one of the top three national concerns, protection of the environment is climbing in relative importance.

- Most of the public are aware of waste disposal problems and favor an approach such as recycling to mitigate the problem.

- Waste-to-energy technology, while acceptable in principle, runs into the NIMBY (Not In My Backyard) syndrome when questions of local siting for such a facility are raised.

- Most people view global warming as a very serious threat, but understanding of causes and effects is not strong and misperceptions are common. Global warming seems to have superseded acid rain as the pressing environmental topic.

- The public supports paying more for energy and automobiles to protect the environment. An underlying lack of credibility among energy institutions concerning environmental issues is a barrier to overcome in the public's perception. The public overwhelmingly rejects a gasoline tax when it is listed among a host of options for federal deficit reduction, unless it can be tied to an environmental benefit such as a tax on carbon dioxide. In another survey, a majority expressed a willingness to pay two cents more per gallon for gasoline in order to reduce air pollution. When the price is bumped up to 20 cents per gallon, people are evenly split on their willingness to pay.

- The Persian Gulf War had a distinctive impact on public perceptions about energy. In February, 1991, just prior to the war, the public saw research on energy alternatives and conservation as a means of avoiding war. After the war, people saw the justification for military action to protect our foreign oil interests, but did not necessarily see the connection between this issue and domestic energy policies.

- The relative importance of energy and its relation to national security is declining in the public's mind. The current public perception is that near-term energy supplies are adequate, but concern over long-term energy security (in the 20-year to 50-year horizon) has been growing steadily since 1979.

- The public is basically uninformed about transportation energy issues and their relation to national energy security. Most people are unaware of the increasing U.S. dependence on foreign oil and the fact that it is driven by transportation demand. When it comes to alternative fuels such as methanol and ethanol, most people are unfamiliar with them as options for replacing petroleum. There is very little data available on people's views on policies that encourage the use of alternative fuels. One survey done in 1990 indicated that virtually everyone favored providing financial incentives for using or developing alternative fuels such as those derived from grain.

What does all this mean? The data compiled by Farhar is a patchwork of surveys addressing issues of energy and the environment in many different ways. Because these questions are framed in their own differing contexts from survey to survey, it is difficult to construct a coherent picture of public perception. Nevertheless, some general trends do emerge.

Quality of the environment and protection of our natural resources as a value for establishing priorities is rising in importance. In fact, there seems to be less fluctuation in the public's view about the environment than about other issues such as energy. Take, as an example, people's response to the question of whether or not environmental quality is rising or falling (see Figure 3). From 1983 to 1987, perceptions about environmental quality are remarkably unchanged. A consistent plurality saw environmental quality as declining. Since 1987, the fraction of the population who believe that environmental quality is getting worse has been on a steady rise.

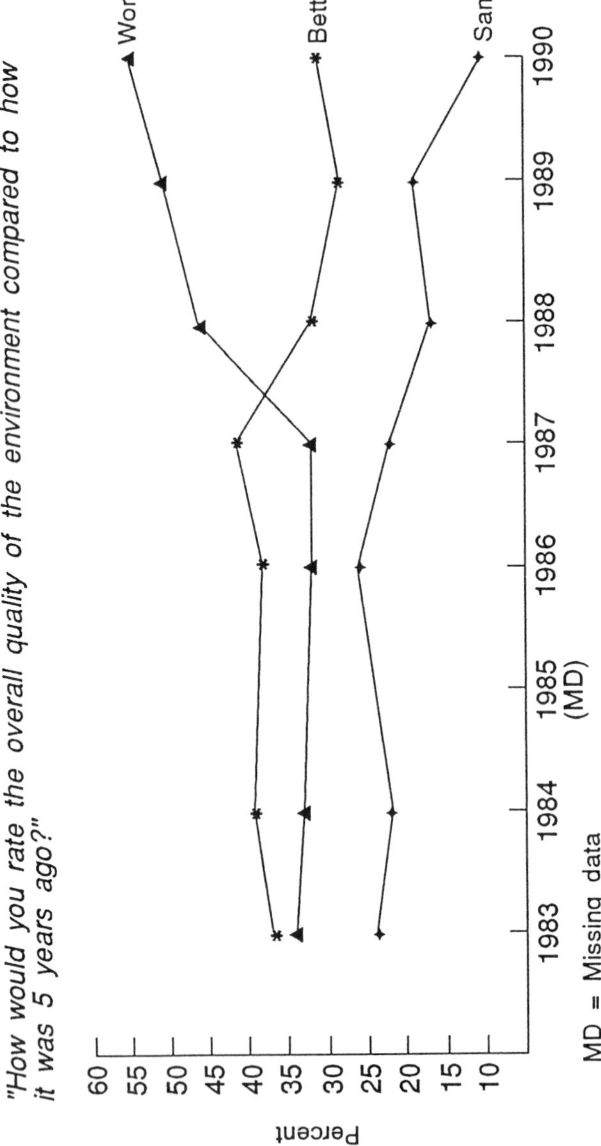

Figure 3: Public attitudes about the environment.

Compare opinion on environmental questions with perceptions about the potential for a future energy crisis (Figure 4). Responses are far more erratic, bouncing by as much as 20 and 30 percentage points from year to year during the 1970s. After a lull in the early to mid 1980's, a strong majority (almost 80%) was recognizing the seriousness of our energy situation by 1990, sparked, no doubt, by the Persian Gulf War. The differences in overall trends, to me, are striking. The rapidly changing attitudes toward energy reflect the "crisis mentality" context within which energy issues have always been portrayed -- oil embargoes, gasoline shortages, and finally, the Gulf War. It would be interesting to see if concerns about energy security have held at their 1990 high point since then. Despite attempts by many interest groups to frame environmental issues as "crises," the public seems able to recognize the complexities and longer term consequences of environmental problems. The result is a more consistently held view on the environment which I believe is a strong foundation for building long-term policy.

Public Policy Debate. I have tried thus far to separate trends in policy debate and legislation from specific trends related to energy and the environment. They are two different beasts which are easily confused, and I doubt that I have been entirely successful in making this distinction. I have attempted to draw a line between the scientific debate and the public policy debate on the status of energy supplies and the environment. Now, I turn my attention to the public policy debates that can impact bioconversion technology.

Let us begin with a general discussion of the role that science now plays in policy-making. In the 20th century, science continues to play a larger role in political and social issues because of the increasingly technological nature of our society. The Manhattan Project put scientists and engineers directly onto the center stage of politics. Scientists found themselves embroiled in moral and social conflicts. Their role in such controversies has only grown since then.

There is a symbiotic relationship between government and science, particularly in broad areas of policy such as the environment. Politicians and scientists tacitly work to encourage predictive research rather than to force unpleasant decisions today The benefits to each are obvious. Representative George Brown (D-CA) (Chairman of the House Committee on Science, Space and Technology) has spoken quite candidly on this point. From his perspective, politicians will do anything to put off hard decisions and scientists are naturally interested in maintaining their funding sources *(29)*. The call for more research is beneficial to each. My intent here is not to lay blame at anyone's feet, but rather to recognize a natural tendency in the interactions between scientists and politicians. Scientists with a sincere desire to do research for the public good and to solve social problems will admit this point *(8)*.

In policy debate, scientists are brought in to provide "expertise" to resolve a political conflict. This approach assumes that (1) experts can transfer "objective" information, (2) they can supply objective scientific tools to resolve complex issues, and (3) that scientific input is needed to resolve complex policy questions. Representative Brown negates these assumptions outright when he says, "In political debate, objective data do not exist. Successful politics, not good science, resolves conflicting issues"*(29)*. This no doubt sounds blasphemous, but his comments make some sense. In any debate, there is never a shortage of experts for *both* sides of an

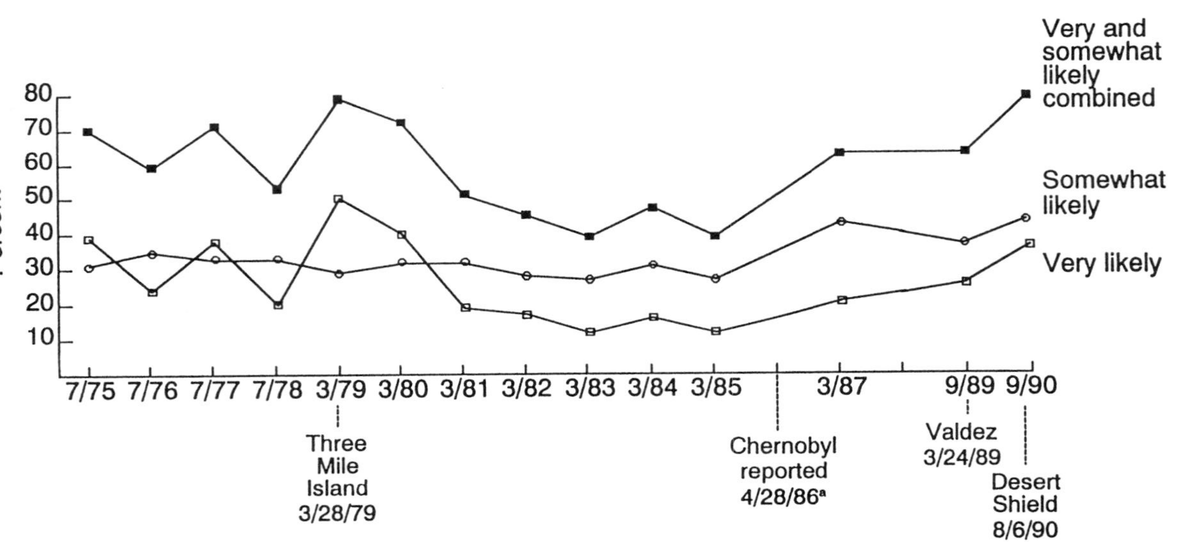

Figure 4: Public attitudes about near-term energy security.

issue. Attempts to provide objective information often make it "user-unfriendly, impractical, irrelevant, untimely, and incomprehensible" *(29)*. Brown recognizes that the predictive tools of science do more to highlight uncertainty than to reduce it. My review of complex issues like global climate change certainly bears this out. Brown's concern about the danger of continuing science's strictly "expert" role in policy is that it will lead to unrealistic expectations about what science can deliver to society, and ultimately create a strong "anti-science" backlash.

Are there signs of this "anti-science" sentiment in the debate over energy and environmental issues? I believe that there are. Take, for example, what *Science* recently referred to as the "Ozone Backlash" *(30)*. In 1992, scientists went to the public with dire predictions of an immediate threat of ozone depletion that might occur over populated areas, leading to serious health problems. This led to a lot of activity in the press and in the political arena. When a dramatic depletion did not occur, accusations about the motives of these scientists began to fly. At the Rio Summit, one of the leaders in ozone research arrived to present a paper only to find his work on chlorofluorocarbons (CFCs) being denounced as a "scam" on the basis that chlorine sources from volcanoes and other sources would swamp out the effect of CFCs on ozone. On a simplistic level, these arguments sound reasonable, but the questions are much more complex. The same argument was presented in Dixie Lee Ray's book *Trashing the Planet*. The mainstream press (*The Wall Street Journal, National Review, Omni Magazine, The Washington Post)* began picking up this theme as well. Enter the likes of Rush Limbaugh, who now regularly argues that ozone depletion by CFCs is a hoax perpetrated by "dunderhead alarmists and prophets of doom" *(30)* whose motives are strictly political.

This is the danger of advocating absolute positions on issues as complex as atmospheric ozone. I believe that similar controversies lie ahead for global climate change. Like the ozone depletion issue, global climate change is fraught with uncertainties that are well recognized by the scientific community. This did not stop researchers from writing comments like this in 1989:

> "The world is warming. Climatic zones are shifting. Glaciers are melting. Sea level is rising....these changes and others are already taking place and we expect them to accelerate over the next years as the amounts of carbon dioxide, methane and other trace gases accumulating in the atmosphere through human activities increase" *(9)*.

The authors go on to list dire consequences for agriculture, destruction of forests, flooding of coastal areas, and uncertainty of water supplies. Caveats and qualifications which may follow such definitive warnings are not likely to be heard. Almost five years later, these warnings take on a hollow ring. Addressing complex issues like global climate with a crisis mentality is a formula for disaster. This is a lesson that can be learned from our experience with the energy crisis in the 1970s, and with the issue of ozone depletion discussed above. Already, the debate on global climate change is heating up. Patrick Michaels, a meteorologist at the University of Virginia, has argued that the popular vision of global warming espoused by many scientists is highly flawed (12). Michaels predicts that 100 years from now it will be clear that data already existed to prove that the catastrophic vision of climate change was a failure. He goes on to say that the "exaggeration of environmental threats in the name of good intentions will have the unintended effect of harming science" *(12)*. As data come in ostensibly contradicting the general predictions that have been made by climatologists *(10-17)*, there is a serious risk that the debate will lose sight of the complexities of the issue of global climate change and revert to simplistic platitudes and predictions. The risks involved in global climate change are too great to allow the debate to take that route.

Legislative and Policy Trends. There are three major pieces of legislation that currently impact the development of bioconversion technology for fuels:

- The Clean Air Act Amendments of 1990 (1990 CAAA)
- The Energy Policy Act of 1992 (1992 EPACT)
- The Alternative Motor Fuels Act of 1988 (1988 AMFA)

The 1990 CAAA is perhaps the most significant of these. This law mandates that regions that do not meet National Ambient Air Quality Standards for ozone and for carbon monoxide use reformulated gasoline (for ozone) and oxygenated fuels (for carbon monoxide). Reformulated gasoline must contain at least 2% oxygen and it must meet the following performance standards:

- Aggregate emissions of volatile organic compounds (VOC's) and toxic emissions must be 15% below the emissions currently obtained with gasoline.
- Levels of NO_x cannot exceed 1990 levels for gasoline.

Toxic emissions include aromatics, polycyclic organics, 1,3-butadiene, acetaldehyde, and formaldehyde. These standards take effect in 1995. In the year 2000, these standards become more stringent. Aggregate emissions of VOCs and toxics must be reduced by 25%. Oxygenated fuel is required in CO nonattainment areas during the winter months and must contain at least 2.7% oxygen by weight. Ethanol and ethers of ethanol and/or methanol can be used as blending agents to meet this oxygen standard.

New standards for diesel fuel are also mandated under the 1990 CAAA. In 1998, 30% of fleet vehicles that refuel in a central location must meet clean vehicle standards. By 2000, this percentage goes up to 70%. In 1994, new standards for diesel fuel went into effect. Diesel must meet a 0.05% maximum sulfur content. This represents a ten-fold reduction in sulfur content. Diesel fuel is now also required to have a minimum cetane rating of 40. Cetane rating is roughly equivalent to an octane number for gasoline and is believed to be inversely related to the content of aromatics. By setting a minimum cetane rating, EPA intends to reduce aromatic content of the fuel. Finally, in 1995, EPA will require a 25% reduction in particulate emissions from urban buses.

In 1988, Congress passed the Alternative Motor Fuels Act, which was intended to encourage the use of alternative fuels. The law sets goals for purchase of alternative fueled vehicles, beginning with federal fleets. In 1993, the goal was that 10% of all federal fleet vehicles purchased would be able to run on an alternative fuel. This corresponds to roughly 5,000 vehicles. In 1998, 50% of fleet purchases must be alternative fueled vehicles, corresponding to around 25,000 vehicles. Currently, federal fleets include about 6,000 alternative fueled vehicles *(31)*.

The 1992 Energy Policy Act includes a requirement for the U.S. Department of Energy to establish state and local incentive programs for using alternative fuels. Also, the law directs DOE to establish low interest loan programs for small businesses to purchase alternative fueled vehicles. By 1996, businesses involved in alternative fuels handling, electricity, and petroleum importing must meet a goal of 30% alternatively fueled vehicles for its vehicle purchases. This percentage will continue to rise, and must be 90% by 1999. The 1992 EPACT also requires that 50% of all federal fleet vehicle purchases be alternative fueled vehicles that use a domestic fuel. President Clinton recently signed an executive order which increased the federal fleet goals by 50%. Finally, the 1992 EPACT mandates a reduction in petroleum dependency. By 2000, the United States aims to replace 10% of projected fuel consumption with alternative domestic sources, with a goal of 30% by the year 2010.

In addition to these three recent legislative actions, there are tax incentives available at the federal level for fuels with a renewable alcohol content of at least 10%. These fuels are exempted from part of the federal excise tax on gasoline. Blenders and sellers of renewable alcohols are eligible for income tax credits. All of these incentives are scheduled to expire in the year 2000.

These pieces of legislation contain attempts to translate environmental protection and energy security into a cost of doing business. The inevitability of higher pollution costs for industry makes incentives built into the 1990 CAAA to minimize cost a landmark development in economically efficient environmental policy *(22)*. The concept of correcting the marketplace for its inability to recognize the cost or value of energy and environmental resources is receiving a great deal of attention now (5,6,22,32-35). Vice President Gore has characterized our current economic system as "partially blind" *(5)*. It sees some things but not others. There are thousands of examples of hidden costs that the marketplace conveniently ignores: the potential cost of global warming, health risks of air pollution or ozone depletion, the depletion of fresh water supplies and other environmental resources. More conservative and business-oriented thinkers like management guru Peter F. Drucker now recognize that the marketplace has no incentive to "not pollute"; and, in fact, that it provides competitive advantage to those who do pollute and do not pay. In Drucker's words, "to treat environmental impacts as 'externalities' can ... no longer be justified theoretically" *(34)*.

Perhaps some of these costs are not quite as well hidden as they used to be. The insurance industry, for one, is grappling with the cost of unusually high rates of flooding and droughts in the United States over the past ten years. Is this simply an aberration in natural disaster activity or is it a result of global warming? The answer is not clear. But global climate change clearly has the attention of the insurance industry, which is accustomed to assigning probabilities of risk and cost to such situations. They are not about to ignore measurably higher risks of this nature. Franklin Nutter, President of the American Reinsurance Association, has remarked that "global warming could bankrupt the industry" *(36)*. This organization is considering playing a key advocacy role in the debate on global warming.

Another element of environmental and energy policy development that continues to receive more attention is the concept of "no-regrets" policies *(5,8,22,37)*. In this approach, a policy aimed at dealing with a future potential problem like global warming is designed to have other types of benefits. Many companies have discovered this approach in dealing with pollution problems that, in the end, involved solutions that saved them money (5). Approaches that combine benefits such as energy security, environmental protection and economic benefits such as job creation or even federal deficit reduction end up being far more palatable.

The "no-regrets" approach has now seen the light of day in the political arena. In the early days of the Clinton Administration, federal deficit reduction was, in fact, tied to energy security and environmental problems. Three options for deficit reduction were floated including a gasoline tax, a foreign import fee, and broad-based energy taxes. The last item included a so-called "carbon" tax on fuel which would have put a price tag on carbon dioxide emissions, a fuel sales tax and a general Btu tax on the energy value of fuel. Opposition by coal-producing states and utilities made the carbon tax and other broad based energy taxes impossible to sell *(38)*. In the end, linking federal deficits and energy successfully, the Administration managed to include a gasoline tax as part of a federal deficit reduction package that just barely squeaked through Congress. This demonstrates that such arguments can be used successfully, even when used to support something as generally unpopular as a gasoline tax.

Risk analysis and management must also be a component of any policy strategy *(39)*. This approach allows for situations that do not involve absolute, definable consequences. It is a matter of weighing both the probability and the seriousness of consequences associated with something like global warming against the cost of corrective actions to deal with it. Given sufficient information, most people are capable of making a sound and balanced decision about risk. Some have argued that this more common sense approach is what is lacking in policy decisions regarding the spending of ever more limited federal funds on environmental problems. The spending of billions of dollars on cleanup of contaminated federal sites should, for instance, be decided on the basis of the probability and seriousness of environmental or health risks and the numbers of people potentially harmed versus the cost of the cleanup *(40)*.

Situation Audit for Bioconversion of Fuels: Internal Factors

"Internal factors," for the purposes of this article, are technology related. I focus here on bioconversion technologies for two transportation fuels:

Ethanol
Biodiesel

A common theme for bioconversion technologies like the ones discussed here is that they all have their roots in antiquity *(37)*. In many ways, this is a comforting thought, in that it suggests that basic bioconversion processes are far from new or novel concepts. They are, to the contrary, quite tried and true. This does not mean that these technologies require no effort to commercialize, for it is the pursuit of these technologies for replacement of relatively cheap (but not renewable) petroleum sources that introduces the challenge. Furthermore the recent advances of modern biotechnology provides a solid reason to expect dramatic improvement in these bioconversion routes. This frames the question differently. These technologies require improvements that will bolster their position in the fuel marketplace, and the tools for achieving such substantial changes are promising.

For each of the fuel products, I will try to address the following questions:

- What is the commercial status of the product?
- What are the fuel characteristics of the product (and, perhaps more to the point, what are the advantages and disadvantages of the product vis-a-vis its petroleum counterpart)?
- What is the prognosis for technology improvement?

Bioconversion for Production of Ethanol--Commercial Status. Even in the United States, applications of biomass-derived products for fuels and chemicals have a surprisingly long history. In the case of ethanol, industrial production in the United States pre-dates the Civil War. In the late 1850's production was almost 340 million liters per year (90 million gallons per year.) Ethanol was used as an illuminant and as an industrial chemical.

Until the late 1970s, government tax policy in the U.S. was often a hindrance to development of the industry. At the start of the Civil War, a hefty liquor tax was levied by the U.S. government to support the war effort which caused the industry to virtually disappear. The timing was particularly bad because the first oil wells in Pennsylvania had just been struck and the first petroleum product, kerosene, had just hit the market. At the turn of the century, this tax was repealed and the industry began to re-establish itself. By the end of World War I, use of ethanol in munitions manufacture and as a fuel supported a 190 million liters per year (50 million gallons per year) industry *(42)*. When the first Ford Model T was introduced, it was actually designed to operate on ethanol as well as gasoline. The onset of prohibition in the 1920s put a damper on ethanol production, but it managed to come back again after the repeal of prohibition, rising to a level of 2.3 billion liters (600 million gallons) per year during World War II. After the war, cheap foreign oil imports killed this industry.

In the late 1970s, the Carter administration introduced tax credits and exemptions that encouraged the use of gasohol, a 10% blend of ethanol in gasoline. The Administration's efforts were aimed at reducing U.S. reliance on Middle East oil supplies. Since then, the fuel ethanol industry has grown from literally nothing to a capacity of 3.8 billion liters per year (more than 1 billion gallons per year). Today almost all of the fuel ethanol produced in the United States is derived from grain (mainly corn). It is used as a gasoline extender, as an octane enhancer and as an oxygenate. In all three cases, the ethanol is used in 10% blends with gasoline because of the tax incentive *(43)*.

The largest market for fuel ethanol today is in Brazil, where production is based exclusively on sugarcane. The history of fuel ethanol production in Brazil dates back to the turn of the century. Unlike the situation in the United States, Brazil used government policy early on to promote the development of a fuel ethanol industry. During World War I, use of ethanol as a fuel was compulsory in many areas of the country. By 1923, production reached 150 million liters per year (40 million gallons per year). In 1931, the government mandated the use of 5% ethanol blends. By 1941, capacity in Brazil had risen to 650 million liters per year (170 million gallons per year.) The government policies that encouraged the development of an ethanol industry in that country stemmed from a long-held goal to reduce Brazilian dependence on foreign oil.

In 1975, Brazil established a program to make ethanol its primary transportation fuel for light duty vehicles (the *Pro-Alcool* Program). By 1989, this effort boosted Brazil's fuel ethanol industry to a production level of 12 billion liters per year (3.2 billion gallons per year). In 1979, purely gasoline-operated cars were no longer available. Two options for use of ethanol now exist. It can be used in a 22% blend of ethanol in gasoline (which the Brazilians also call "gasohol") or as a "neat" (pure) hydrous ethanol fuel. In 1985, the sale of neat ethanol cars outpaced the sale of gasohol cars, a trend which continued until 1990 when a shortage of ethanol occurred *(44)*.

It is important to recognize that the success of ethanol in Brazil is due to what we would view in the United States as draconian measures on the part of the government. Artificially suppressed ethanol prices and policies within the government-owned oil industry requiring minimum purchases of ethanol were used to assure success of the industry. It is, in some ways, an extreme example of applying a "no regrets" energy policy because the government identified multiple domestic goals as part of its programs: energy security, development of domestic markets for its largest agricultural product (sugarcane), generation of jobs, and reduced trade deficits.

Ethanol as a fuel or as a fuel component is a commercial reality in the United States and Brazil. Ethanol has had a rocky history in the United States. Ethanol's commercial future remains tenuous because its financial success is tied to specific tax incentives, which may well disappear after the year 2000. Finally, ethanol in the United States has yet to break the barrier into the marketplace as a primary fuel. By contrast, ethanol as a fuel in Brazil is much more secure. The strong efforts of the government have produced a transportation infrastructure that now relies on ethanol.

Bioconversion for Ethanol Production--Fuel Characteristics. Important fuel properties include:

- Resource base
- Engine performance and air emissions characteristics

The resource base for ethanol depends on the technology used. Sugar prices in the United States are controlled at a level that currently makes sugarcane unattractive as a feedstock for ethanol production. Today's industry uses processes which can convert the starch fraction in corn to ethanol. The current production level of one billion gallons of ethanol per year corresponds to 400 million bushels of corn per year. The United States Department of Agriculture estimates that ethanol capacity could be increased to approximately 15 to 23 billion liters (4 to 6 billion gallons) per year without adversely affecting grain prices *(45)*. This means that, at best, the existing corn ethanol industry has the potential to supply roughly half of the current gasoline market with enough anhydrous ethanol for use in 10% blends *(46)*. As a neat fuel, this would represent less than 5% of the gasoline market.

To be a significant player in the transportation fuel market, ethanol needs a resource base much broader than the current industry has at its disposal. Technology for utilizing cellulosic biomass as a feedstock would expand the resource base to accommodate most of the gasoline market needs in the United States. Table I shows estimates of the fraction of the gasoline market that could be met with various sources of biomass. Excess cropland and available low-cost (waste) feedstocks are probably adequate to meet the entire needs of the gasoline market. Low-cost feedstocks, which provide near-term opportunities for economic ethanol production, could supply almost 30% of the gasoline needs. Thus, from a resource standpoint, ethanol from biomass is well positioned to compete with gasoline as a reliable fuel source (in terms of capacity). It has the added benefit of being a renewable source.

Fuel properties for ethanol, ethyl *tert*-butyl ether (ETBE) and gasoline are shown in Table II *(46)*. ETBE is an ether derivative of ethanol that is gaining increased attention as a fuel additive for gasoline because of its ability to significantly reduce evaporative CO emissions, and to improve octane. It is analogous to methyl-*tert* butyl ether (MTBE) which already has wide acceptance in the marketplace today. A quick survey of the numbers reveals a few salient comparisons. Ethanol has a lower volumetric energy density, about two-thirds that of gasoline, which reduces travel miles per tank of fuel. Its higher octane number indicates improved engine performance over gasoline. Vapor pressure of ethanol is much lower than that of gasoline, but in blends of 10%, the overall vapor pressure of the fuel is actually higher than that of unblended gasoline.

Ethanol is used today in 10% blends without any engine modifications. The use of direct blends allows engines to run leaner and reduce carbon monoxide by 10 to 30%. Direct blends do, as indicated above, increase evaporative emissions, as measured by Reid vapor pressure (Rvp). The net effect on smog generation from the use of these fuels is unclear. For instance, some modelling studies predict a net reduction in smog in urban areas when 10% ethanol blends are used instead of gasoline *(48)*.

To use a neat hydrous ethanol fuel would require changes in design and in some materials of construction for the engine, though these modifications are not especially difficult or expensive *(49)*. Because ethanol engine designs are not widely available in the United States, this represents a disadvantage for ethanol relative to gasoline. Nevertheless, experience with ethanol shows that it will improve engine performance (5-10% increase in engine specific power) and that it will allow 10-20% better thermal efficiency than gasoline. This improved efficiency reduces the problem of lower energy density. In Brazil, on average, 1.19 liters of ethanol are equivalent to one liter of gasoline. This ratio may be 1.2 to 1 in the U.S. *(47)*. Ethanol has many of the same highly desirable characteristics as methanol from an emissions viewpoint. Methanol's biggest benefit is that it reduces volatile organic compounds (VOCs) by 30% compared to gasoline. The combustion of ethanol produces aldehydes along with the unburned alcohol in the exhaust emissions *(49)*. Aldehydes, which may have potential health effects, are also produced photochemically from hydrocarbon emissions associated with gasoline *(19)*. Control of aldehyde emissions from ethanol combustion may be achieved through the use of catalytic converters *(50)*. Theoretically, NO_x emissions will be lowered in ethanol engines because of the lower combustion temperature *(51)*, but data on U.S. engines are required to assess this.

Table I: Potential For Ethanol Production from Cellulosic Biomass Based on net energy available from each source after energy for biomass production and yields from conversion process are taken into account *(47)* and on 2010 projections for light duty vehicle energy demand *(23)*.

Source	Percentage (%) of Gasoline Market
Wastes	27%
Agricultural	12
Forestry	10
Municipal Solid Waste	5
Cropland	**65 - 146**
Estimated Excess in 2010	41 - 81
Potential	24 - 65
Forest Land	**16**
TOTAL	**108 - 189**

Table II. Properties of Ethanol, ETBE and Gasoline

Property	Ethanol	ETBE	Gasoline
Formula	C_2H_5OH	$(CH_3)_3COC_2H_5$	C_4 - C_{12}
Molecular Weight	46	102	----
Specific Gravity	0.79	0.75	0.72-0.78
Heat of Combustion (kJ/kg)	26,860	36,280	41,800-44,000
Octane Number	98	110	88
Reid Vapor Pressure (kPa)	16 (pure)	30 (pure)	55-103
	83-186 (10% blend with gas)	21-34 (12.7% blend with gas)	

Furthermore, the reduced NO_x achieved by lower combustion temperature may well be negated in ethanol engines designed with higher compression ratios for greater thermal efficiency. Emissions data from hydrous ethanol and 20% gasohol cars in Brazil are shown in Table III *(40)*. The data confirm the ability to get significant reductions in CO and hydrocarbon. NO_x emissions, however, actually increase. In addition, the expected increase in aldehyde emissions is not seen. Because Brazilian cars are not equipped with electronic fuel injection or with catalytic converters, it is not clear that these results would be reproduced in U.S. automobiles.

Getting a true picture of the environmental and energy-related benefits of biomass-derived ethanol requires looking at the "full life cycle" of the fuel, which considers production and distribution impacts in addition to automobile tailpipe emissions. This is an approach that, from a policy perspective, is recognized as the only realistic way to compare the environmental and resource benefits of any product or process *(52)*. Biomass-derived ethanol offers an opportunity to provide a truly renewable energy source for transportation. This has several implications in full life-cycle analysis, one of which is a reduction in the level of carbon dioxide accumulation as a result of combustion. This attribute is common to all renewable fuels, but to varying degrees. Life-cycle analyses on existing corn ethanol technology have focused on the net energy balance of ethanol as a fuel. This question has been debated since the early days of gasohol in the United States *(53)*. And the debate continues. Pimentel *(54)* has most recently argued the case that corn ethanol is not a source of renewable fuel because it uses more energy than it produces in the form of fuel ethanol. He estimates that 72% more energy goes into producing the fuel than is available in the final fuel product. His arguments have been countered by Ahmed and Morris *(55)* who argue that, in the early days of gasohol, inefficiencies in the process were far more rampant than they are today, both in the corn field and in the conversion process. They estimate that there is a net energy gain of 15% associated with corn ethanol. At the same time, they recognize the real promise of "sustainability" in fuel production lies with the development of bioconversion technology for cellulosic biomass.

For conversion of cellulosic biomass to ethanol, the U.S. Department of Energy has conducted a fairly rigorous life-cycle analysis which considered air, water and solid waste pollutants, as well as overall energy and chemical inputs *(56,57)*. To make the analysis more meaningful, a full fuel-cycle analysis of E95 (an ethanol fuel containing 5% gasoline) was done in parallel with a fuel-cycle analysis for production of reformulated gasoline. The basic elements of the fuel cycle for each included the following:

- Feedstock Production
    - Crop production and harvesting for ethanol
    - Oil drilling for reformulated gasoline
- Feedstock Transport
    - Shipping of domestic crop to bioconversion facility for ethanol
    - Shipping of crude oil (foreign and domestic) to refineries
- Fuel Production
    - Bioconversion for ethanol
    - Oil refining for gasoline

- Fuel Distribution (from production facility to pump)
- Fuel End Use (combustion in light-duty vehicles)

The most striking differences between reformulated gasoline and E95 are (not surprisingly) in net carbon dioxide emissions and fossil fuel consumption. The production and use of ethanol made from biomass generates only one-tenth the amount of carbon dioxide that is generated by reformulated gasoline. This reflects the fact that the carbon dioxide generated by combustion in vehicles and in the production and distribution steps for the fuel are offset tremendously by the amount of carbon dioxide that is consumed by the energy crops during cultivation. Fossil-fuel inputs for the E95 fuel are one-quarter of those required for reformulated gasoline. Approximately 80% of the total energy inputs to E95 are renewable. Though this latter result may seem trivial or obvious, it is an important quantitative input required in making policy decisions regarding sustainable or renewable energy production.

Some important air pollutants also show significant differences between the two fuels. Sulfur oxide (SO_x) emissions for ethanol are 60% to 80% lower than that of reformulated gasoline. Volatile organics for ethanol are 13%-15% less than those for reformulated gasoline. For carbon monoxide and NO_x, the differences between the two fuels are marginal.

The overall benefits of ethanol as a fuel include the potential for sustainable production sufficient to meet current and projected demands now met with gasoline, as well as environmental advantages such as dramatically reduced greenhouse-gas emissions, and substantially lower emissions of volatile organics and SO_x. In urban areas, tailpipe reduction of CO may also have regional environmental benefits. Disadvantages of neat ethanol as a fuel are that it will require 25% greater refueling frequency and that it may have higher aldehyde emissions at the tailpipe. In the near term, direct blending of ethanol poses a potential environmental problem because, at the concentrations currently used, the resulting vapor pressure of the blend is higher than the vapor pressure of blended fuel. This is seen as a potential problem since evaporative emissions from gasoline are known to contribute to smog generation in urban areas.

Bioconversion for Ethanol Production--Prognosis for Technology Improvement. The technology for production of ethanol from biomass can be generalized as indicated in Figure 5. In essence, all processes involve the recovery of sugars from biomass followed by fermentation of the sugars to ethanol and recovery of the ethanol. The sugars can be in the form of biopolymers (as in the case of wood or grains) or as simple sugars (as in the case of sugarcane). Improvements in the process must involve:

- Utilization of higher-volume and lower-cost sources of biomass
- Improvements in yields and rates of sugar recovery from biomass
- Improvements in yields and rates of ethanol production from sugars
- Improvements in the efficiency of ethanol purification and recovery

Table III: Ethanol in Brazil
Pollutants in Grams per Kilometer

Vehicle Type:	CO	Hydro-carbons	NO_x	Aldehydes
Gasoline Before 1980	54	4.7	1.3	0.05
20% Gasohol	22	2.0	1.9	0.02
Ethanol	16	1.6	1.8	0.06

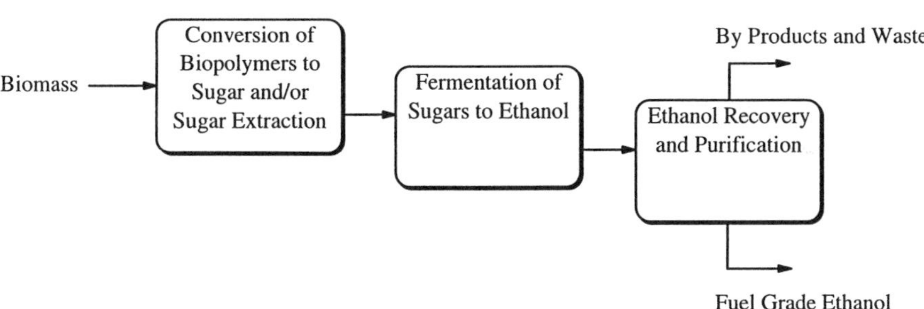

Figure 5: Generalized process scheme for ethanol-from-biomass.

Of course, each of these four improvements can have an impact on the others. For example, finding a cheaper, higher-volume feedstock is not a guarantee of improvement. If the yields and rates of sugar and ethanol recovery are significantly worse for this new feedstock, one may be no better off than one was with one's original feedstock. Any prognosis for improvement of ethanol production technology must therefore consider all four of these areas.

Current technology for ethanol production in the United States is based on conversion of starch to ethanol. Starch, a widespread polymeric storage form of sugar in plants, is readily available because of the intensive cultivation of starch-producing grains like corn for food production. The starch is present in specialized tissue in the form of granules that are readily extracted from plants. There are two classes of processes used: wet milling and drying milling operations. Wet milling represents roughly two-thirds of the industry. Its advantage is that it views ethanol more as a byproduct, relative to other higher value products that are generated as a result of wet milling. These include corn oil, corn gluten meal, and corn gluten feed. For both wet and dry milling, the process steps for ethanol production are based on the same fundamental technologies. The starch is treated with starch-degrading enzymes to produce simple sugars, which can then be converted to ethanol in a straightforward and well-established yeast fermentation *(41,58)*. Traditional distillation can be used to remove most of the water from the fermentation step. A final dehydration step, typically done using molecular sieves, removes the last few percent of water to produce anhydrous ethanol *(59)*.

Developing the technology so that it can accommodate more than starch and sugar-containing biomass is critical. This will not only open up the resource base to include a broader range of waste and non-food related crops, but will also allow for production of ethanol from additional (and presently inaccessible) forms of sugar in current feedstocks such as corn and sugarcane (59).

What are these other forms of sugar? The major components of terrestrial plants are two families of sugar polymers--cellulose and hemicellulose. Table IV shows typical composition data for various forms of biomass. There is remarkable similarity in the overall composition among these different forms.

Cellulose fibers comprise 4%-50% of the total dry weight of stems, roots and leaves. These fibers are embedded in a matrix of hemicellulose and phenolic polymers *(58)*. Cellulose is a polymer composed of six-carbon sugars, mostly glucose. Hemicellulose is a polymer of sugars, but the types of sugar vary with the source of biomass. With the exception of softwoods, xylose is the predominant component in hemicellulose. The phenolic polymers, known as lignin, cannot be used for ethanol production.

So, our prognosis boils down to this: What is the potential for utilizing cellulosic and hemicellulosic sugars for ethanol production? The good news is that work on utilizing the sugars available from cellulose and hemicellulose is not new. Attempts to produce ethanol from cellulosic biomass were first done in the late 1890s. At that time, Simonsen in Germany reported yields of 7.6 liters of pure ethanol per 100 kg of dry wood (or almost 20 gallons per ton of wood) *(62)*. Throughout World War I, two facilities in the southern U.S. were operated on the basis of this early experimental work, using waste sawdust as a feedstock. These facilities were able to ferment the pentose sugars at yields of around 7.6 to 10 liters per 100 kg of dry wood *(63)*.

The bad news is that efforts to commercialize this technology in the U.S. have since been through an economic and political roller coaster ride which has never allowed enough continuous and focussed effort to ever see the development of the technology through. After World War I, a curtailment of milling operations in the South, combined with dropping prices for molasses and the introduction of petroleum-derived fuel products, forced the two existing operations to close down *(63)*. The advent of World War II brought back shortages of fuel and food supplies, resulting in a renewed effort to develop wood sugar to ethanol technology sponsored by the War Production Board in Washington and the U.S. Department of Agriculture. This research ran the gamut from bench to pilot-scale studies. It included studies of woods and agricultural residues *(64-70)*. After World War II, all of the commercial and government- sponsored research came to a halt. The oil embargoes of the 1970's reawakened our concern over energy supplies, and re-kindled research efforts on cellulosic biomass to ethanol, an effort which lost some of its momentum by the 1980s. Today, that research and development effort is slowly gaining momentum once again.

Where does the development of this technology stand today? Research during the 1970s and 1980s has taken advantage of advances in enzymology, biotechnology and process technology to greatly improve the potential yield and productivity of ethanol production from cellulosic biomass *(71,72)*. In the 1990s, genetic engineering advances in the field of biotechnology are now being brought to bear on various aspects of the process to further advance the technology *(43,73)*.

A general schematic of the conversion process options being considered today is shown in Figure 6. A physical and/or chemical pretreatment of some type is required to make the cellulose in biomass more amenable to hydrolytic attack. A great deal of work on biomass-to-ethanol processes has focused on the use of acid hydrolysis to produce fermentable sugars from cellulose, including much of the early commercialization efforts *(43,62-72)*. In this case, pretreatment is typically some form of size reduction to facilitate acid uptake in the biomass. Soluble xylose and other sugars from hemicellulose can also be fermented to ethanol. Lignin can be burned as fuel to power the rest of the process, converted to octane boosters *(74)*, or used as feedstocks for the production of other chemicals *(75)*.

Acid hydrolysis processes include both dilute and concentrated acid variations. The former requires high temperatures (200°-240° centigrade). Under these severe conditions, sugars also degrade to unwanted by-products. There are some limited markets for these by-products, though none of these markets are sufficient to support a large-scale fuel ethanol industry *(43)*. Dilute acid processes are near term commercial options because they are reasonably well developed *(59,71,72)*. Concentrated acid processes can be carried out at much milder conditions and with little sugar degradation. However, the high volume of acid catalyst used makes the economics of these processes problematic unless a cost-effective means of recovering and recycling the acids can be developed.

One of the discoveries that had a profound effect on biomass-to-ethanol technology was the discovery of cellulose-degrading organisms by researchers at the U.S. Army Natick Laboratories, who were concerned with finding a solution to the

Table IV: Typical Composition of Cellulosic Biomass

Biomass Type	Cellulose	Hemicellulose	Lignin	Other
Agricultural Residue *(60)*	38%	32%	17%	13%
Hardwood Energy Crops *(60)*	50%	23%	22%	5%
Herbaceous Energy Crops *(60)*	45%	30%	15%	10%
Waste Paper *(61)*	47-53%	10-13%	18-25%	1-10%

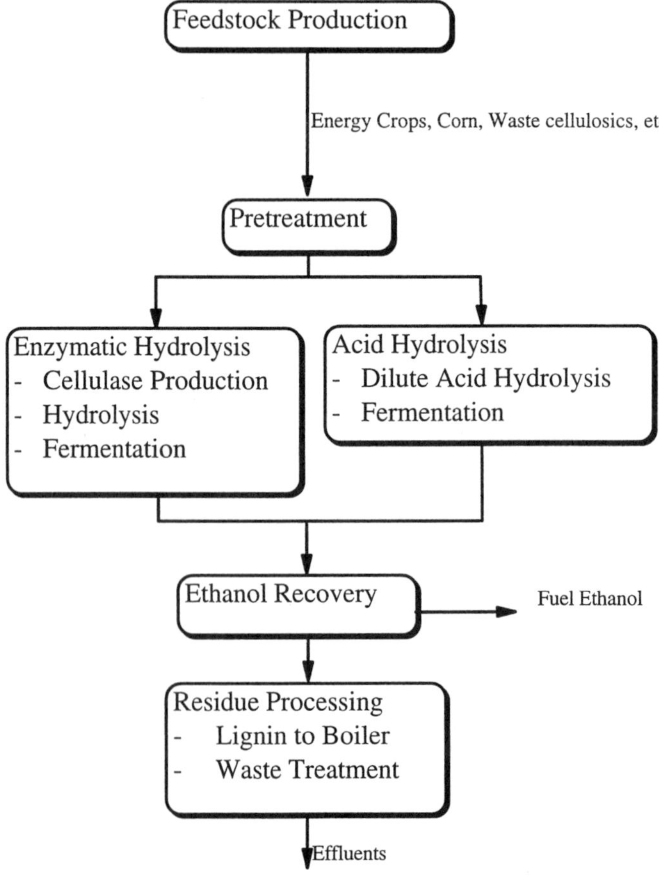

Figure 6: Ethanol-from-biomass process options.

problems of severe biological degradation of canvas tents and supplies experienced by army troops encamped in jungle locales *(72)*. This led to the isolation and development of fungi with enhanced levels of expression of cellulases, a family of enzymes that work in concert to break down cellulose to its simple sugar components. These enzymes allow degradation of cellulose to occur under mild conditions, in contrast to the severe conditions of temperature, pressure, and pH required in dilute acid hydrolysis of cellulose. In addition, these enzymes catalyze highly specific reactions, producing only one product and are required in much smaller quantities compared to concentrated-acid processes. Research and development for enzyme-based biomass-to-ethanol processes has tremendous potential for improving the efficiency and economics of this technology.

Biomass is naturally resistant to enzymatic attack (a fact that makes sense from the point of view of survival and evolutionary development). Therefore, enzyme-based processes require pretreatment processes which, in addition to simple size reduction of the feedstock, open up the hemicellulose-lignin matrix that surrounds the cellulose. A variety of techniques have now been developed to achieve this, including steam-explosion *(72)*, dilute acid pretreatment *(76)*, ammonia fiber explosion *(77)*, and organosolv *(66)*. The dilute acid pretreatment, which thus far appears to be the most economic alternative *(78)*, operates under milder conditions than those used to achieve cellulose hydrolysis. The major benefit of this process seems to be the removal of hemicellulose and its conversion to soluble sugars.

Production of cellulases for hydrolysis of cellulose is a costly step in the enzymatic process. The best available sources of commercial cellulases today are genetically altered strains of *Trichoderma reesei* (derived from the original U.S. Army Natick Lab isolates) *(79)*. In nature, fungi such as *T. reesei* tend to produce more cellulases than other organisms. Recently, bacteria have been discovered that produce more potent forms of cellulases, and which have other desirable characteristics, such as enhanced temperature tolerance and stability *(80,81)*. The goal of current research is to use genetic engineering techniques to improve the productivity and yield of cellulases and to produce more potent cellulase preparations that work more effectively at lower enzyme loadings.

So far, I have spoken of the enzymatic hydrolysis of cellulose and the fermentation of sugar to ethanol as two separate steps in the process. One of the more promising avenues of research in conversion of biomass to ethanol involves finding ways to combine these two biochemical steps in a single reactor. The simultaneous saccharification and fermentation (SSF) process is an example of this. This concept, originally developed by Gulf Oil and the University of Arkansas *(82,83)*, eliminates the problem of product inhibition encountered with cellulase systems. In simple terms, cellulase systems have mechanisms that slow down their activity when sugar (the end product of hydrolysis) begins to accumulate. By combining the enzymes with yeast that ferment the sugars, it is possible to prevent a buildup of sugar, and maintain faster rates of hydrolysis. A longer-term alternative is the development of so-called "direct microbial conversion" processes, which utilize organisms capable of carrying out both production and fermentation of sugar to ethanol *(84)*. Ultimately, genetic engineering may play an important role in developing organisms with this type of combined enzyme production.

Conversion of the five-carbon sugars from hemicellulose is another area of research with the potential to dramatically improve the efficiency of biomass-to-ethanol conversion. Until recently, xylose and other five-carbon sugars could not be effectively utilized, creating a serious loss in potential yield and a cost to the process in waste processing of these sugars. Yeast strains such as *Candida shehatae, Pichia stipitis, and Pachysolen tannophilus* have recently been identified which are capable of directly fermenting xylose to ethanol under carefully controlled microaerophilic conditions *(85-87)*. Other fungi and thermophilic bacteria have also been identified that are capable of anaerobically fermenting xylose to ethanol. Finally, genetic engineering techniques have been used to express genes from the glucose-fermenting bacterium *Zymomonas mobilis* in *Escherichia coli*, resulting in an organism that can directly ferment xylose to ethanol *(88)*. This is a very promising approach to fermentation of xylose.

Other process improvements which need attention include finding new uses for lignin (beyond its use as a fuel in ethanol conversion facilities). There is also room for improving the overall energy efficiency of the process *(89)*. This latter point is critical since reductions in the energy requirements of the process have a direct impact on the sustainability of ethanol-from-biomass technology, a key to understanding ethanol's role in the long term policy questions about energy.

The prognosis for technology improvement in ethanol production looks very good. To put it in perspective, Figure 7 shows the improvements in ethanol yield from wood over the last 100 years of research on this technology. Despite the erratic level of activity on ethanol-from-biomass research, yields have risen from the level of 8 liters of pure ethanol per 100 kg of dry wood from the first experiments performed at the end of the last century to almost 40 liters per 100 kg based on laboratory results available in 1992 *(90)*.

Yield is not all of the story. The ultimate measure of technological advance is in the cost of the technology. Even by this measure, great improvements have been made, and will likely continue to be made. In 1980, the cost of producing ethanol from wood was estimated to be $0.95 per liter ($3.60 per gallon.) Estimated cost in 1992 is $0.32 per liter ($1.22 per gallon), a 66% reduction in cost. Sensitivity studies on the SSF process suggest that additional cost reductions are possible which could bring the cost of ethanol down to $0.18 per liter ($0.67 per gallon). This would make ethanol competitive with projected prices for gasoline in the year 2000 *(60)*.

The cost reductions that have been documented for the SSF process are based on a disparate collection of laboratory results for individual aspects of the ethanol conversion process. What is needed now is larger-scale demonstration of the fully integrated process of pretreatment of biomass to final ethanol recovery. Such an effort has yet to be made. The first demonstration of this technology will likely be done using low-cost feedstocks such as municipal solid waste or waste paper. Likewise, these low-cost feedstocks will be the first commercial entree for this bioconversion technology.

In the scheme of things, even with the enhancements in the process already developed in the laboratory, the long term picture for ethanol is quite promising. This technology has barely scratched the surface as far as the opportunities for improvement and enhancement that exist as a result of the advances that have been made and continue to be made in the field of biotechnology and genetic engineering.

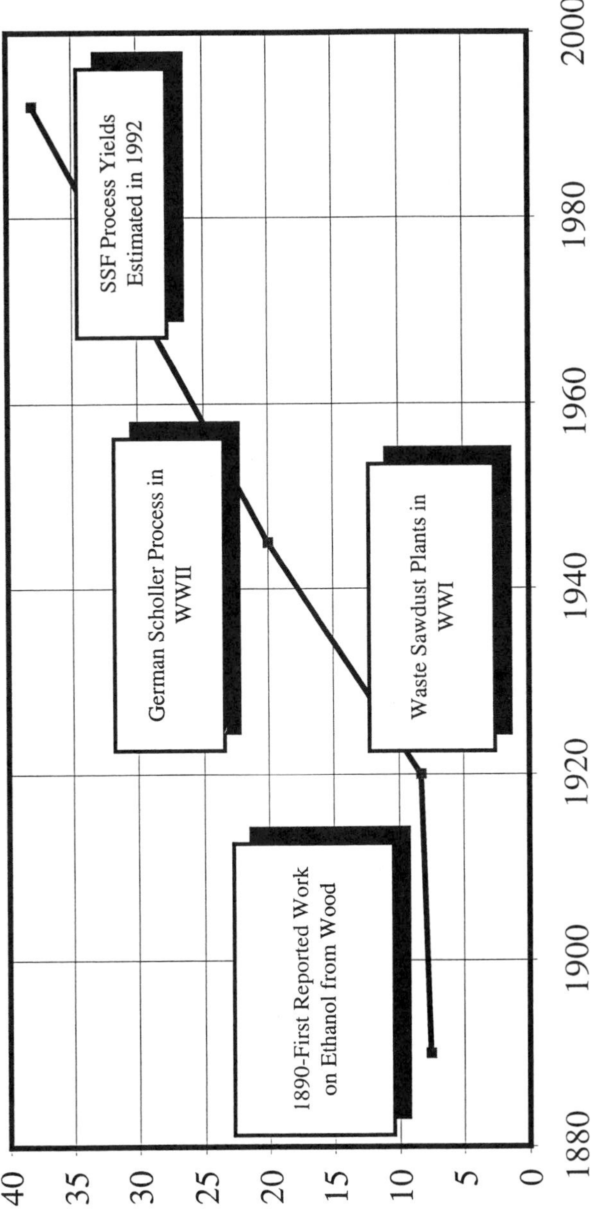

Figure 7: Yields of ethanol from wood (Liters/100 kg): 100 years of progress.

Bioconversion for Production of Biodiesel--Commercial Status. The first order of business is to identify what is meant by "biodiesel." There is not really one single recognized definition. In its most general sense, biodiesel refers to any fuel substitute for diesel engines that is "renewable" or biologically derived. According to the most popular definition currently being applied, biodiesel is an esterified lipid (natural oil) derived from vegetable oils and animal fats *(91-93)*.

At the risk of stating the obvious, I will make the point that the market for diesel fuel is vastly different from that of gasoline. The total diesel fuel market is less than half the size of the gasoline market. It is expected to grow proportionately faster than the gasoline market by 2010, reaching a total size of 220 billion liters per year (58 billion gallons per year) *(23)*. Approximately half of this demand is in the transportation sector *(94)*.

To fully understand the differences between these two fuels, it is important to understand the differences in engine design for gasoline versus diesel fuel. The conventional gasoline engine is based on the Otto cycle. It consists of an intake stroke at essentially constant pressure, during which the fuel-air mixture flows into the cylinder. In the second stroke, this mixture is compressed adiabatically. The fuel is then spark-ignited. Combustion occurs so rapidly (relative to the piston's movement) that combustion occurs essentially at constant volume. The work from the engine is produced as the combustion gases expand. Diesel engines differ from the Otto engine primarily because they are not spark-ignited. Instead, compression of air is used to reach a temperature sufficient to cause the fuel to self-ignite. Fuel is added at the end of the compression process, and is added at a rate such that work is produced by expansion at constant pressure.

Differences in fuel characteristics have a major impact on efficiency of these two engines. The diesel engine uses a fuel that will not "pre-ignite" at high compression ratios. The higher the compression ratio, the greater the work that can be derived from the fuel. Diesel engines can operate at much higher compression ratios than gasoline engines before problems with pre-ignition occur. Thus, diesel engines are generally more efficient than gasoline engines. The best Otto-cycle engines operate at thermodynamic efficiencies of 20-25%, while the diesel engine can operate at efficiencies of greater than 40% *(95)*.

Biodiesel is a relatively new product in the transportation fuel market. Although it, too, has some surprisingly old roots, its history is not nearly as long and complex as ethanol. Still there are some parallels. As in the case of ethanol and its early history with the Model T, vegetable oil-based fuels played a role in the early days of the diesel engine. Rudolph Diesel first tested his new engine in 1893. He is reported to have demonstrated his engine at the Paris Exhibition in 1900 using peanut oil. With the introduction of cheap petroleum-derived fuels, this approach was quickly abandoned, and the development of the engine went hand-in-hand with the development of petroleum-refining technology. As in the case of ethanol, vegetable oils did find increased use as fuels during the 1940s, when World War II put a premium on food and fuel supplies in many parts of the world. The Japanese are said to have been reduced to extracting natural oils from pine needles in order to supply airplane fuel for their beleaguered air force by the end of the war *(96)*. Once the "emergency" conditions no longer existed, plentiful petroleum supplies displaced natural oils in the fuel market. The use of vegetable oils either as pure fuels or in blends with diesel fuel has been demonstrated in numerous countries around the world *(97)*.

The newest market for biodiesel (in the form of esterified vegetable oils) is in Europe, where government policies have positioned this fuel to compete head-to-head with diesel fuel. Today, rapeseed-derived biodiesel is available "at the pump" in many European countries. Total capacity in Europe today is around 440 million liters per year *(98)*.

In the United States, the market for biodiesel is just beginning to develop and efforts to commercialize a soybean-oil-derived fuel are well underway *(99)*. Interest in biodiesel among urban bus-transit operators is growing rapidly because these fuels can be used to meet new stricter standards on urban bus emissions due to go into effect in 1995 *(100-104)*. Soybean producers are targeting the urban transit bus market as an initial entree for the use of soy-derived biodiesel in 20% blends with diesel fuel. Blending reduces the net cost of the fuel, while still taking advantage of the environmental benefits of biodiesel *(93)*. Other niche opportunities for biodiesel include the use of biodiesel in environmentally sensitive areas and in marine applications *(104)*. The concept of environmentally friendly biodiesel has even captured the imagination of the U.S.'s newest billionaire, cellular phone mogul Craig McCaw, who recently ruminated over dreams of building a company based on soybean oil as a source of nonpolluting fuel for diesel engines *(105)*.

Bioconversion for Production of Biodiesel--Fuel Characteristics. As in the case of ethanol, it is important to understand both the available resource base for biodiesel and the engine performance and emissions characteristics of this fuel.

Brown *(94)* has estimated that the total U.S. supply of vegetable oil and tallow is equivalent to 11 billion liters (3 billion gallons) per year of biodiesel. Setting aside concerns about other existing demands for these materials, it is clear that these sources, while adequate for meeting near-term market demands such as the urban transit market, are not adequate to meet the demands of the total diesel fuel market in the transportation sector.

Where is the resource base for biodiesel to come from? Traditional sources of natural oils could be supplemented by the use of microalgae, which are capable of producing oil photosynthetically from carbon dioxide. Microalgae could be used to remove carbon dioxide released from fossil-fuel-fired power plants. In this scenario, mitigation of release of the greenhouse gas carbon dioxide from power plants could be achieved while producing a diesel fuel substitute *(106)*. This approach takes advantage of the availability of a relatively concentrated source of carbon dioxide *(107)* which may at some point in the relatively near future become a financial liability to electric utilities. Feinberg and Karpuk (108) have estimated that with the use of semi-arid lands and associated saline water supplies in New Mexico and Arizona in conjunction with carbon dioxide generated by power plants in these states, microalgae farming could produce two quads (10^{15} Btu's) per year of energy (2.1 Exajoules/year). Such farms would need only 0.25% of the land area in these states. Their survey of ten states with reasonable conditions of sunlight and temperature for algae farming indicates enough carbon dioxide from electric utilities to support biodiesel production in excess of the current and projected U.S. diesel fuel market for transportation *(109)*.

Table V summarizes the basic fuel properties of soybean and rapeseed-oil derived ester fuels, as well as #2 diesel. Given the chemical similarity of lipids from microalgae and lipids from vegetable sources, little difference in fuel properties is anticipated between algal and vegetable-derived lipids, though ultimately this will have to be established.

Unmodified vegetable oils have been used for short periods of time, but testing of these oils in the 1940s showed that they had deleterious effects on the long-term performance of diesel engines (*97*). The data in Table V demonstrate the problem. The unmodified oils have viscosities well above that of regular diesel fuel. The increased viscosities lead to greater carbon buildup and plugging of injectors. Esterified oils do not have this problem.

The esterified oils have much higher cetane numbers, an indication that they perform much better as fuels than their unmodified oil counterparts. In fact, the cetane number of biodiesel is generally higher than that of #2 diesel. Density and energy content of the ester fuels are very similar to diesel fuel. Flash points are lower, implying that this fuel, like ethanol, is safer than its petroleum counterpart.

The higher pour point and cloud point suggest that these fuels may have a problem with gelling in extremely cold weather conditions. Sulfur content of the esterified oils is well within the new specifications for low sulfur diesel in the United States and much lower than pre-1993 #2 diesel. Engine tests conducted on biodiesel thus far confirm one of the biggest advantages of biodiesel--its ability to be used in existing diesel engines with only minor modifications and with no negative impacts on long-term engine performance and wear *(110,112,113)*.

The data from Europe on emissions characteristics of biodiesel look very encouraging, though none of this data can be translated directly into U.S. EPA-certified measurements because of differences in analytical protocol *(113,114)*. Table VI lists data collected by the Institut Francais du Petrole de Paris *(113)* that are typical of results reported on rapeseed-derived biodiesel.

The biggest relative improvement is in particulates, which are roughly 50% less for biodiesel as compared to diesel fuel. This 50% reduction in particulates has been seen elsewhere in Europe (97,114). Likewise, biodiesel carbon monoxide levels are reduced to 30-40% of those from diesel fuel, probably because of the oxygenated nature of this fuel. Hydrocarbons are reduced by 25%, though this probably does not tell the whole story since aromatic hydrocarbons are virtually nonexistent in biodiesel. NO_x levels are always slightly higher for biodiesel than for diesel. Except for NO_x, emissions from diesel engines are dramatically improved by the use of biodiesel.

In the United States, emissions testing has focussed on how to achieve a 25% reduction in particulates using a 20% soybean-derived biodiesel blends, with no increase in NO_x emissions *(93)*.

As the data in Table VII suggest, this can be done with a combination of a catalytic converter and an engine timing change.

Few comprehensive life-cycle analyses have been conducted for biodiesel to determine its net environmental impact and energy balance. This is a high priority for research on vegetable-oil-derived biodiesel, because there is some controversy about the net energy benefits of this fuel (not unlike the questions raised about corn ethanol). For example, a recent study conducted in Germany concluded that the net greenhouse-gas emissions for rapeseed-based biodiesel were actually positive *(115)*.

Table V
Fuel Properties of Biodiesel vs. #2 Diesel *(98)*

Property	Soybean Oil *(110)*	Soy Ester	Rapeseed Ester	#2 Diesel
Cetane Rating	38	52 (48 min)*(111)*	54.5	47.8
Flash Point (°C)	219	171	83.9	80.0
Cloud Point (°C)	-4	1.1	-2.2	-12.2
Pour Point (°C)	N/A	-1.1	-9.4	-28.9
Viscosity(cSt,40° C)	37.0	4.1	6.0	1.6-60
Sulfur (wt%)	N/A	0.04 (0.02 min.) *(111)*	<0.005	0.29 (0.05max for low sulfur)
Density (kg/liter)	0.923	0.86-0.9[94]	0.87	0.85
Heat of Combustion (kJ/kg gross)	N/A	38,460	35,376	38,536

Table VI
Pollutants from Biodiesel (grams/kWh)

Pollutant	Biodiesel	Diesel
Carbon Monoxide	2.7	7.9
Hydrocarbons	1.3	1.6
NO_x	5.3	4.6
Particulates	0.36	0.67

Table VII
Pollutant Reductions Achieved with 20% Blends of Biodiesel

Pollutant (g/bhp-hr)	Baseline Diesel No Engine Modifications	20% Blend with 3 degree retardation in injector pump timing	20% Blend with timing change and catalytic converter
Particulates	0.261	0.216 (-17.2%)	0.191 (-26.8%)
CO	1.67	1.50 (-9.8%)	0.45 (-72.8%)
Hydrocarbons	0.45	0.38 (-14.2%)	0.12 (-73.2%)
NO_x	4.46	4.25 (-4.6%)	4.32 (-3.1%)

The reason for this is the high estimate of carbon dioxide generated as a result of European agricultural practices for rapeseed. It is likely that the net greenhouse-gas emissions for soybean in the United States would be more favorable because U.S. farmers generally use less fertilizer and tend toward "no till" or "low till" practices. Furthermore, for soybeans, no nitrogen fertilizers are needed *(116)*. This supposition is supported by a recent study which concludes that the net energy balance for soy-derived biodiesel in the United States is positive *(117)*. This study, commissioned by the National SoyDiesel Development Board, found that the net ratio of energy produced as fuel versus energy put into the cultivation, harvesting, and production of fuel was 1.44:1 on average. When credit for meal and glycerine byproducts are accounted for, the energy ratio is 2.51:1. Potential improvements in industry practice for both soybean and fuel production could bring this ratio up to 4.1:1.

Theoretical considerations for microalgae-based biodiesel production suggest that the net carbon dioxide balance for this fuel should be highly favorable *(93)*. From one vantage point, this technology effectively cuts in half the carbon dioxide emissions associated with coal-powered electric utilities. Carbon dioxide trapped at the utility is ultimately released through combustion of biodiesel. The net effect is that the carbon dioxide associated with the coal has been recycled and used a second time for energy production. The integration of microalgae mass culture with coal-fired power plants reduces CO_2 from 86-97 kg/gigajoule to 39-49 kg/gigajoule. This would move coal from being the most carbon-intensive fossil fuel to being the least *(106)*. Given the size of existing coal reserves in the U.S. and in the rest of the world, this is an important potential contribution to the reduction of future carbon dioxide emissions. Microalgae processes require more scrutiny to quantify this benefit.

The overall benefits of biodiesel as a sustainable fuel (i.e., its ability to reduce greenhouse-gas emissions and provide a net positive source of energy) look promising, but have not been adequately quantified for either terrestrial crops or microalgae. Environmental benefits of this fuel at the tailpipe are very good. Biodiesel may prove to be one of the most effective ways to reduce particulate emissions from diesel engines. Particulates are major contributors to the levels of visible smog and to health-related aspects of urban air pollution. At the same time, use of biodiesel would reduce air toxic emissions, overall hydrocarbons, and carbon monoxide. NO_x reductions are not inherently better with this fuel and would likely need to be addressed in other ways. Finally, biodiesel is a fuel that the transportation diesel fuel users can use today. The fuel production technology is available and engine designs will require no major changes. Although near term fuel supplies (vegetable oils and animal fats) are limited to a few percent of the total market demand, future supplies of fuel can be increased through the use of bioconversion technology based on microalgae. Microalgae technology simultaneously solves two major problems: (1) development of domestic renewable sources of diesel fuel, and (2) mitigation of greenhouse gases from electric utilities.

Bioconversion for Production of Biodiesel--Prognosis for Technology Improvements. Figure 8 shows the overall approach to production of biodiesel. The scope of this technology, as it is presented here, is different from that of ethanol, in that it includes photosynthetic conversion of carbon dioxide to oils, followed by subsequent conversion to a suitable biodiesel fuel. Today, oil production is based on

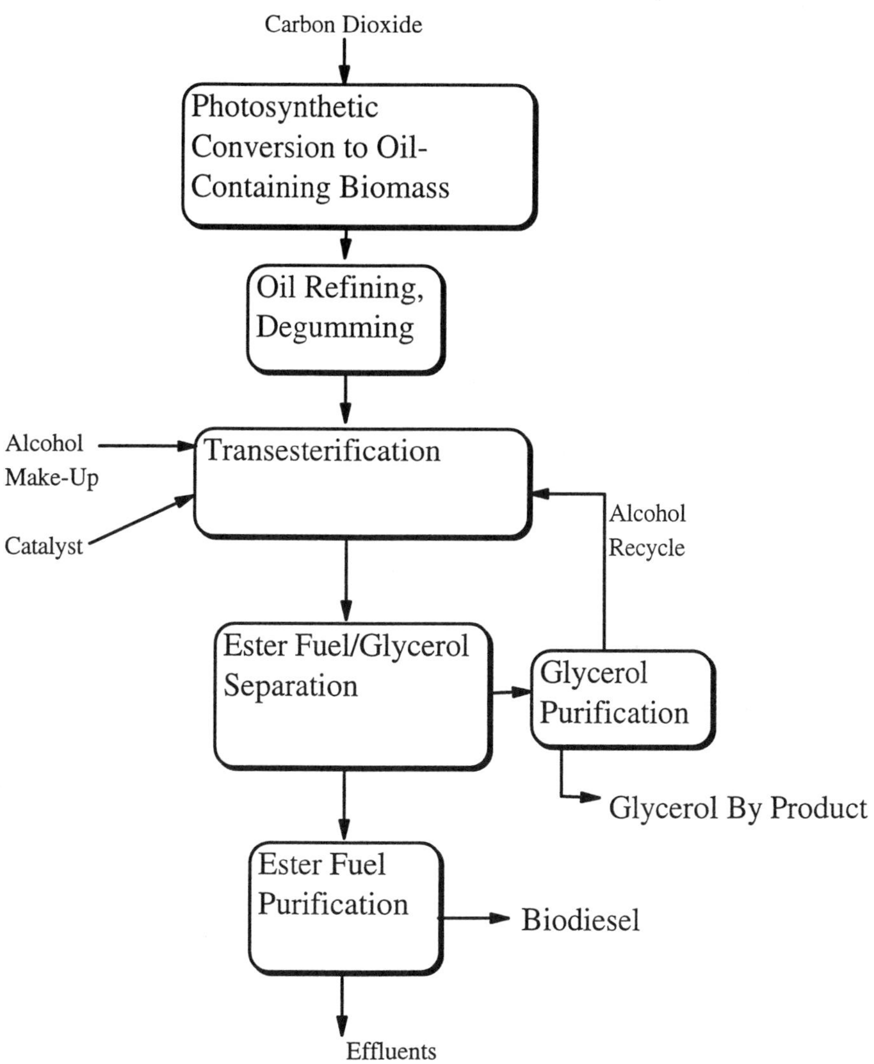

Figure 8: Biodiesel production.

existing vegetable oil cultivation and processing. The primary vegetable oil crop in the United States is soybean. Oil in soybeans comprises about 18% of their total dry weight. Waste sources of natural oils (from restaurants) and animal fats are lower-cost options which can be used to bring down the cost of fuel. Oil extraction processes for soybeans and other vegetables are well established. Hexane extraction is most commonly used. Conversion of the oils to a diesel fuel is also based on well-established chemistry currently practiced on a large scale.

Oil production, if not done using terrestrial plants, can be achieved using microalgae. Algae, like many terrestrial plants, have the capability to take carbon dioxide and convert it photosynthetically into lipids. Development of mass cultivation techniques for algae did not really begin until after World War II *(118,119)*. The use of microalgae as a fuel source was first proposed almost 40 years ago for the production of methane gas *(120)*. Since then, microalgae have been studied for production of protein supplements, enzymes, pigments, polysaccharides, hydrocarbons, fatty acids, and glycerol (118-121). In 1979, DOE began to pursue a research program on the use of algae for fuel production *(120)*. As issues of global warming increased in importance, the use of microalgae for capture of carbon dioxide in power plants also received a great deal of attention, both here and in Japan, where a multiyear, multimillion dollar program has been put in place to develop microalgae technology for CO_2 capture *(122-126)*. Thus, interest in microalgae technology is only now reaching a high level. Because of the inherently unexplored nature of this research field, the potential for improvement is great, especially in the area of microalgal oil production.

Microalgae are the most primitive form of plants. Almost 20,000 species have been identified. These fall into three main groups *(120)*:

- The blue-green algae (Cyanophyceae). Much closer to bacteria in structure and organization, these algae play an important role in fixing nitrogen from the atmosphere.

- The green algae (Chlorophyceae). These are the most common algae, especially in freshwater, and can occur as single cells or as colonies. Green algae typically form starches as their major storage product

- The golden algae (Chrysophyceae). These algae have more complex pigment systems, and can appear yellow, brown, or orange in color. These algae can form lipids as their major storage product. The oils in these cells can constitute 60% or more of their dry weight. Diatoms, a common group of algae, normally form silica shells around their cell mass.

While the mechanism of photosynthesis in microalgae is similar to that of higher plants, they are generally much more efficient converters of solar energy because of their simple cellular structure. In addition, because the cells grow in aqueous suspension, they have more-efficient access to water, carbon dioxide, and other nutrients. For these reasons, microalgae are capable of producing 30 times the amount of oil per hectare of land that can be obtained using traditional terrestrial crops *(127)*.

Another advantage of microalgae is that they can achieve extremely high productivities in saline water, at salt levels twice as high as that of seawater. This opens up the possibility of utilizing low quality land and water resources, such as those found in the arid and semi-arid regions of the U.S. Southwest, which would otherwise have no other economic value. Microalgae could improve the productivity of desert land 70-fold. With specially designed photobioreactor technology, this improvement could be as much as 160-fold (106).

The most cost-effective approach to mass culture of microalgae involves the use of outdoor raceway ponds *(120,122,128-130)*. These ponds are typically shallow, baffled raceways using slowly rotating paddle wheels to maintain circulation and to keep algae in suspension. Carbon dioxide from the flue gas of a power plant would be directly injected into the ponds as a carbon source, while other nutrients would be added to optimize growth. Tests in the lab have demonstrated that, even without genetic altering, strains that normally produce 5%-20% of their body weight as oil can be coaxed into producing 60% of their body weight as oil. In outdoor culture, oil contents of up to 40% have been achieved. Progress on improving productivities in these ponds has also been significant. Early results were on the order of 1-5 g/m^2/day. Today, yields of 15-25 g/m^2/day are not unusual.

A major focus of research on the use of microalgae for biodiesel production is improving our understanding of what controls lipid productions in these cells. Environmental factors have a major impact on lipid production, and algal metabolisms exhibit a remarkable ability to respond very rapidly to these factors *(131)*. For example, many strains show increased lipid production as a result of nitrogen depletion. When cells experience a shift from a surfeit of nitrogen supply to a depletion of nitrogen, oil content will increase anywhere from 50 to 400%. In diatoms, a depletion of silica in the growth medium will also trigger oil production. Temperature can cause a shift in the type of lipids produced, with more unsaturated fatty acids produced at lower temperatures. In general, lipids seem to be synthesized in response to conditions in which the rate of energy input in a system exceeds the cellular capacity to utilize this energy. Lipids are a highly efficient storage form for this excess energy.

A long term approach to improving product yields is artificial combination of desirable traits in a single species via genetic engineering. These techniques are new in phycology *(131)*. Unlike microbial and even animal cell systems, genetic engineering techniques and transformation systems for algae need to be developed, making progress more difficult, but certainly not insurmountable. A better understanding of lipid metabolism in algae is also needed to better target genes for genetic transformation. Some progress has been made in demonstrating the ability to transfer genes in algae *(132)*. In order to establish stable expression of foreign genes in microalgae, marker systems need to be developed. While efforts to develop these markers are still proceeding, researchers have recently succeeded in isolating and cloning a gene expressing a key lipid synthesis enzyme (acetyl-coenzyme A carboxylase) in the alga *Cyclotella cryptica (133)*. Future work is aimed at establishing stable expression of this gene to determine if over-production of this enzyme leads to increased oil synthesis.

Production of biodiesel from lipids is based on a chemical reaction known as "transesterification" *(93,97,110-114,117,133,134)*. The chemistry of this reaction is illustrated in Figure 9. Lipids from plants, animal fats and microalgae are classified as triglycerides, which consist of three long chain fatty acids tied to a glycerol backbone. In the transesterifcation reaction, triglycerides are reacted with an alcohol, producing three fatty acid-alcohol esters and a glycerol molecule. The reaction is catalyzed with a soluble acid or base catalyst. This reaction is often referred to as "bucket chemistry" because of its simplicity. It is practiced today at the large scale in the production of cosmetics. The biggest challenges facing this process are finding a use for the glycerol by-product and finding ways to recover the catalyst used. Because of the quantities involved, uses for the glycerol by-product are probably the dominant concern in the economics of the process.

Future enhancements of the process may include the development of biological catalysts for the reaction. The potential of research on lipases as biological catalysts stems as much from the novelty and unexplored nature of this area as from an understanding of the specific advantages known about these enzymes. Only recently has it been discovered that lipase enzymes can catalyze esterification reactions when placed in an organic solvent with minimal water present *(135)*. The advantages identified for enzymatic transesterification include the fact that it can be carried out at mild conditions, that it simplifies glycerol purification and that it makes catalyst recovery possible *(136)*.

The economics of microalgae-based biodiesel production have not been adequately assessed since 1986 *(120)*. At that time, the cost of biodiesel from microalgae was estimated to be $1.33 per liter ($5.00 per gallon) considerably above diesel fuel prices. If research goals identified at that time were met, it was estimated that the cost of the fuel would drop to $0.43 per liter ($1.62 per gallon). Most of these goals focussed on biological parameters for the algae, which are certainly central to the process. Other issues such as finding improved extraction and conversion steps were not included. The potential for improvement in all these areas is great. Another factor which comes into play is the possibility that, in the not-too-distant future, carbon dioxide capture may become a financial issue for electric utilities. Looked at from the vantage point of carbon dioxide removal, microalgae may well be a promising alternative to other technologies proposed *(107)*.

Strategic Issues for Bioconversion Technology

One way to sum up the "situation audit" is to develop a "WOTS UP" analysis *(1)*. In this analysis, information from the situation audit is organized in terms of:

- Weaknesses (Internal)
- Opportunities (External)
- Threats (External)
- Strengths (Internal)

For bioconversion technology (as limited to biodiesel and ethanol technology for this discussion), this analysis is presented in Table VIII.

As I pointed out at the outset, an analysis like this is fraught with potential pitfalls. The merits of the items which I have summarized in Table VIII could, each and everyone, be debated. Still, it is a starting point for discussion. This list invites

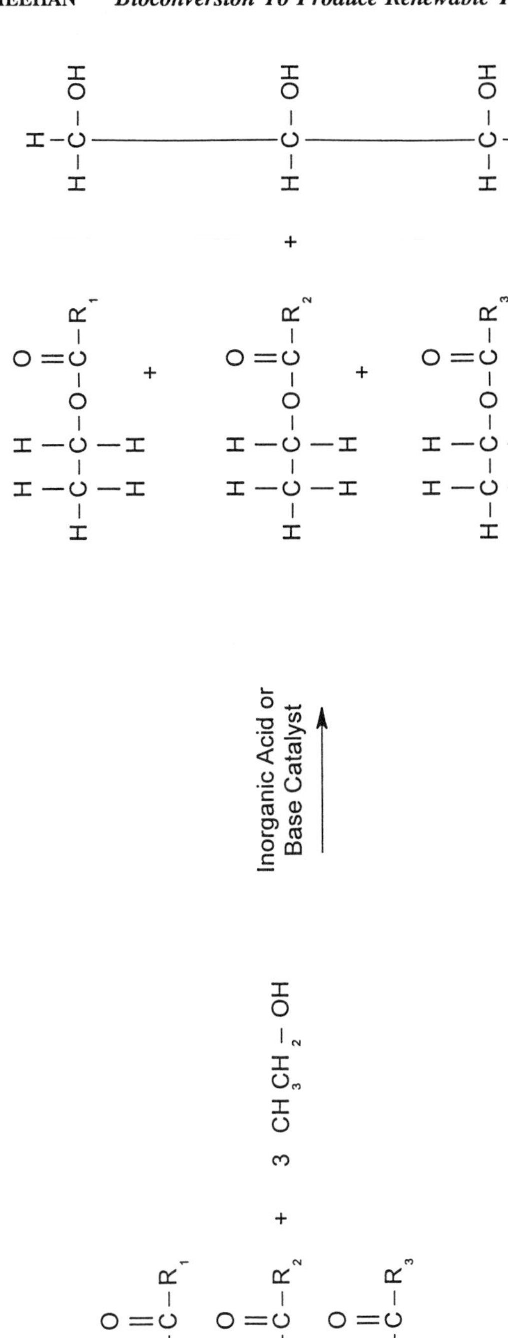

Figure 9: The chemistry of biodiesel production from natural oils.

Table VIII
WOTS-UP Analysis for Bioconversion Technology for Fuels

Strategic Issue	WOTS (Weaknesses, Opportunities, Threats, Strengths)
Environmental Quality	Opportunities - Historic levels of CO_2 buildup in last 100 years - Deterioration of urban air quality Threats - Inherent uncertainty in predicting global climate change - Recent slowing of greenhouse gas buildup - Uncertainty in assessing impact of air pollutants Strengths - Potential CO_2 reduction of bioconversion technology - Reduction in air pollutants associated with biofuels Weaknesses - Uncertainty in CO_2 impacts of bioconversion technology - Inability to deal with NO_x problem
Energy Trends	Opportunities - Diminishing world supply of petroleum - Even tighter U.S. domestic petroleum supply - Demand for liquid fuels due to transportation - Unprecedented foreign import vulnerability Threats - Abundant worldwide fossil-fuel reserve (coal) - "Cheap" price of fossil fuels Strengths - Renewable/sustainable supply for biofuels - Domestic base for biofuels feedstocks Weaknesses - Energy crop industry not established

Continued on next page

Table VIII *Continued*
WOTS-UP Analysis for Bioconversion Technology for Fuels

Strategic Issue	WOTS (Weaknesses, Opportunities, Threats, Strengths)
Public Opinion	Opportunities - Consistent growth in concern over environment - Recognition of long term energy security issue - Desire to balance environmental and energy issues - Recognition of waste disposal problems Threats - Short term attitude about energy - Ignorance about transportation sector effect on energy - History of "crises" involving energy - Weak understanding of global climate issues Weaknesses - No public awareness of bioconversion technology Public Policy Debate and LegislationOpportunities - Tougher Clean Air Act provisions for transportation - Energy Policy Act calls for sustainable domestic fuels - Alternative Motor Fuels Act - Tax Incentives for renewable fuels like ethanol - Support of policies to set a cost for pollution Threats - Loss of tax incentives for renewable ethanol in 2000 - Distrust of science in public policy - Continued role of the scientist as "expert"
Technology Development	Opportunities - Rapid pace of genetic engineering developments - Growth of biotechnology industry Strengths - Existing corn ethanol industry base - Promising new biodiesel market in U.S. - Growing European biodiesel industry - History of improved efficiency for ethanol techn, - Symbiotic role for biodiesel from algae and coal use - Compatibility of biodiesel in existing diesel engines - Excellent prognosis for future improvements Weaknesses - Prospects for biodiesel from algae long term only - Long, inconsistent development effort for ethanol - No "neat fuel" infrastructure for ethanol/biodiesel - No vehicles for ethanol available in U.S.

a whole level of discussion which I have not even ventured into: the question of developing strategies to take advantage of the strengths and opportunities and to deal with threats and weaknesses. Then, there is the question of priorities. Which of the items listed above represent the most pressing concerns to deal with? For instance, I view the weak foundation for public discussion of global climate change to be of immense importance. Somehow, the discussion of global warming and its potential solutions (such as bioconversion technologies for fuel production) needs to be put on the table in the form of a balanced risk assessment in which all the potential outcomes and the uncertainties are laid out. The same can be said for energy security issues. We need to dispel the platitudes and warnings of imminent crisis, and foster an admittedly subjective and *political* discussion in which the risks of global warming and energy vulnerability come to the surface. At the same time, the public needs to understand the promise and the pitfalls of renewable fuels technology. Today, it is not even an item for discussion in the public arena.

I hope that I have at least provided a sense for the types of issues that face bioconversion technology development today. The reader is free to criticize my observations and conclusions, hopefully with the intent of improving our understanding of our situation. If I have spurred more debate, then I have done well. In the words of the educator and philosopher Mortimer J. Adler:

> *Let us engage in the serious business of conducting our discussion rationally and logically, to discover the truth about points on which we differ (137).*

Literature Cited:

1. Steiner, G. *Strategic Planning: What Every Manager Must Know*; The Free Press: New York, NY, 1979.
2. Bryson, J. *Strategic Planning for Public and NonProfit Organizations: A Guide to Strengthening and Sustaining Organizational Achievement*; Jossey-Bass: San Francisco, CA, 1988.
3. Finney, G.; Owens, P. "Strategic Scheduling for Social Initiatives"; In *Smooth Sailing with Project Management: 24th Annual Seminar/Symposium 1993 Project Management Institute Proceedings*; Project Management Institute: Drexel Hill, PA, 1993.
4. Hein, P. "The Road to Wisdom," *Life,* October 14, 1966.
5. Gore, A. *Earth in the Balance: Ecology and the Human Spirit*; Houghton Mifflin: New York, NY, 1992.
6. Schneider, S. *Science* **1989,** *243,* 771-781.
7. Revelle, R. *Scientific American* **1982**, *247*(2), 35-43.
8. Sarmiento, J. *Chemical and Engineering News* **1993,** *71*(22), 30-43.
9. Houghton, R.; Woodwell, G. *Scientific American* **1989**, *260*(4), 36-44.
10. Kerr, R. *Science* **1993**, *260,* 890-893.
11. Oreskes, N.; Shrader-Frechette, K.; Belitz, K. *Science* **1994**, *263,* 641-646.
12. Michaels, P. *National Geographic Research and Exploration* **1993,** *9*(2), 222-233.
13. Schneider, S. *Science* **1994**, *263,* 341-347.

14. Schneider, S. *Nature* **1992**, *356*, 11-12.
15. Powell, C. *Scientific American* **1994**, *270*(3), 22-28.
16. Kerr, R. *Science* **1993**, *260*, 1232.
17. Kerr, R. *Science* **1994**, *263*, 751.
18. Revelle, R.; Suess, H. *Tellus* **1957**, *9/1*, 18-21.
19. Seinfeld, J. *Science* **1989**, *243*, 745-752.
20. Lents, J.; Kelly, W. *Scientific American* **1993**, *269*(4), 32-29.
21. Anderson, T. Personal Communication, 1993.
22. Darmstadter, J.; Fri, R. *Annual Review of Energy and the Environment* **1992**, *17*, 45-76.
23. *Annual Energy Outlook: 1993;* U.S. Department of Energy, Energy Information Administration: Wadhington, DC.
24. *Annual Energy Review: 1991*; U.S. Department of Energy, Energy Information Administration.
25 Lynd, L; Cushman, J.; Nichols, R.; Wyman, C. *Science* **1991**, *253*, 1318-1323.
26. *U.S. Oil Import Vulnerability: The Technical Replacement Capability;* Congress of the United States Office of Technology Assessment: Washington, DC, 1992.
27. Geller, H.; DeCicco, J.; Laitner, S.; Dyson, C. *Twenty Years After the Oil Embargo; The Cost of U.S. Oil Import Dependence and How They Can Be Reduced*; Public Citizen, Inc. and American Council for an Energy-Efficient Economy: Washington, DC, October 1993.
28. Farhar, B. "Trends in Public Perceptions and Preferences on Energy and Environmental Policy"; NREL/T7P-461-485; National Renewable Energy Laboratory, Golden, CO, 1993.
29. Brown, G., *Chemical and Engineering News* **1993**, *70*(22), 9-11.
30. Taubes, G., *Science* **1993,** *260*, 1580-1583.
31. Goodman, B.; Putsche, V.; Sheehan, J. "Renewable Fuels from Biomass"; *Chemistry in Britain;* In Press.
32 Harvey, H. *Solar Today*, Nov/Dec **1993**, 23-25.
33. Holmberg, W., *Energy--Time to Change the Counting System: A Call for the Quantification of Risks and Benefits of Energy Production and Use;* American Biofuels Association: Arlington, VA, 1993.
34. Drucker, P., The New Realities: In Government and Politics/In Economics and Business/In Society and World View; Harper and Brothers: New York, NY, 1989.
35. Stavins, R. *Annual Review of Energy and the Environment* **1992**, *17*, 187-210.
36. Linden, E., *Time,* March 14, **1994.**
37. Chartier, P.; Saranne, D. *Biofutur* **1992** *113*, 16-25.
38. Goodgame, D., *Time,* February 15, **1993**, 25-27.
39. Morgan, M., *Scientific American* **1993**, *269*(1), 32-41.
40. Abelson, P. *Science* **1994**, *263*, 591.
41. Rhodes, A.; Fletcher, D. Principles of Industrial Microbiology; Pergamon Press: New York, NY, 1966.
42. Morris, D. Presented at the Governors' Ethanol Coalition Meeting, Peoria, IL, December 1993.
43. Wyman, C.; Goodman, B. In *Opportunities for Innovation: Biotechnology;* Busche, R. Eds. U. S. Department of Commerce: Gaithersburg, MD, 1993.

44. Goldemberg, J.; Monaco, L.; Macedo, I. In *Renewable Energy: Sources for Fuels and Electricity;* Johansson, T.; Kelly, H.; Reddy, A.; Williams, R., Eds. Island Press: Washington, DC, 1993.
45. *Ethanol's Role in Clean Air,* USDA Backgrounders Series 1102-89, United States Department of Agriculture, Washington, DC 1989.
46. Wyman, C.; Hinman, N. *Applied Biochemistry and Biotechnology* **1990**, *24-25*, 735-753.
47. Lynd, L. *Science* **1991**, *253*, 1318-1323.
48. Wyman, C.; Bain, R.; Hinman, N.; Stevens, D. In *Renewable Energy: Sources for Fuels and Electricity;* Johansson, T.; Kelly, H.; Reddy, A.; Williams, R. Eds; Island Press: Washington DC, 1993.
49. *The Motor Gasoline Industry: Past, Present and Future.* Energy Information Administration; U.S. Department of Energy, Washington DC, DOE/EIA-0539 January 1991.
50. Goodman, B., Presented at the National Meeting of the American Chemical Society, Denver, CO, March 28-April 2, 1993.
51. *Encyclopedia of Chemical Technology: Alcohol Fuels to Toxicology* (supplement), Grayson, M., Ed., Wiley: New York, NY, 1984.
52. Hunt, R.; Sellers, J.; Franklin, W. In *Environmental Impact Assessment Review,* Elsevier Science Publishing Co.: New York, NY, **1992**.
53. Energy Resource Advisory Board, *Gasohol,* U.S. Department of Energy, Washington DC, 1980.
54. Pimentel, D. *Focus* **1992**, *2/3,* 36, 38-43.
55. Ahmed, I.; Morris, D. *Focus* **1992**, *2/3,* 37, 44-49.
56. Bull, S. Presented at the National Meeting of the American Chemical Society, Denver, CO, March 28-April 2, 1993.
57. Tyson, K.; *Fuel Cycle Evaluations of Biomass - Ethanol and Reformulated Gasoline,* Vol 1. NREL/TP-463-4950. National Renewable Energy Laboratory, Golden, CO, 1993.
58. Grohman, K.; Himmel, M. In *Enzymes in Bioconversion:* Leatham, G.; Himmel, M. Eds. ACS Symposium Series No. 460. American Chemical Society: Washington DC, 1991.
59. Wyman, C. *Applications of Cellulose Conversion Technology to Ethanol Production from Corn;* National Renewable Energy Laboratory, Golden, CO, 1992.
60. *Ethanol Annual Report: FY 1990*, Texeira, R.H.; Goodman, B.J. Eds, Solar Energy Research Institute, SERI/TP-231-3996, Golden, CO, 1990.
61. Bergeron, P.; Riley, C. *Wastepaper as a Feedstock for Ethanol Production,* National Renewable Energy Laboratory, NREL/TP-232-4237, Golden, CO, 1991.
62. Simonsen, *Z. Angew. Chem.* **1898**, 195,962,1007.
63. Peterson, W.; Frazier, W. *Industrial and Engineering Chemistry*, **1945**, *37*(4), 30-35.
64. Sherrard, E.; Kressman, F. *Industrial and Engineering Chemistry*, **1945**, *37*(4), 5-8.
65. Faith, W. *Industrial Engineering Chemistry*, **1945**, *37*(4), 9-11.
66. Harris, E.; Beglinger, E.; Hajny, G.; Sherrard, E. *Industrial Engineering Chemistry*, **1945**, *37*(4), 12-23.

67. Dunning, J.; Lathrop, E. *Industrial and Engineering Chemistry*, **1945**, *37*(4), 24-29.
68. Plow, R.; Saeman, J.; Turner, H.; Sherrard, E. *Industrial and Engineering Chemistry* **1945**, 36-43.
69. Saeman, J. *Industrial and Engineering Chemistry* **1945**, *37*(4), 43-51.
70. Hasche, R. *Industrial and Engineering Chemistry* **1945**, *37*(4), 51-54.
71. Wright, J. *Energy Progress* **1988**, *8/2*, 71-78.
72. Wright, J. *Chemical Engineering Progress* August, 1988.
73. *NREL Research Brief: Mini-Manhattan Project for Cellulases*; NREL/MK-336-5676; National Renewable Energy Laboratory, Golden, CO, 1993.
74.. Bergeron, P.; Hinman, N. *Applied Biochem. Biotech.* **1990**, *24/25*, 15-29.
75. "New Uses for Lignin Must Compete with Entrenched Applications." *Industrial Bioprocessing*, Jan. 1994, *16*(1), Technical Insights, Inc: Englewood, NJ. , 1994.
76. Torget, R.; Werdene, P.; Grohmann, K. *Applied Biochem. Biotech.* **1990**, *24/25*, 115-126.
77. Holtzapple, M.; Jun, J.; Patibandla, S.; Dale, B. *Applied Biochemistry and Biotechnology* **1991**, 28/29, 58-74.
78. Schell, D.; Torget, R.; Power, A.; Walter, P.; Grohmann, K.; Hinman, N. *Applied Biochemistry and Biotechnology* **1991,** 28/29, 87-97.
79. *NREL Research Briefs: Mini Manhattan Project* for Cellulases; NREL/MK-336-5676. National Renewable Energy Laboratory, Golden, CO, 1993.
80. Shiang, M.; Linden, J.; Mohagheghi, A.; Rivard, C.; Grohmann, K.; Himmel, M. *Applied Biochemistry and Biotechnology* **1990**, 24/25, 223-235.
81. Himmel, M. Presented at the 1993 DOE Automotive Contractors' Meeting, Dearborn, MI, October, 1993.
82. Gauss, W.F.; Suzuki, S.; Takagi, M. *Manufacture of Alcohol from Cellulosic Materials Using Plural Elements,* U.S. Patent #3,990,944, Nov. 9, 1976.
83. Takagi, M.; Abe, S.; Suzuki, S.; Emert, G.; Yata, N. *A Method for Production of Alcohol Directly from Cellulose Using Cellulase and Yeast,* Proceedings, Bioconversion Symposium, IIT, Delhi, India, **1977,** pp 551-571.
84. Hogsett, D.; Ahn, H.; Bernardez, T.; South, C.; Lynd, L. *Applied Biochemistry and Biotechnology* **1992**, *34/35*, 527-541.
85. Skoog, K.; Hahn-Hag, B. *Enzyme Microb. Technol.* **1988,** *10*, 66.
86. Prior, B.; Kilian, S.; du Preez, J.; *Process Biochemistry* **1988**, 21-32.
87. Jeffries, T., In *Yeast*: *Biotechology and Biocatalysis;* Verachtert, H.; DeMot, R. Eds. Marcel Dekker: New York, NY, 1990.
88. Ingram, L.; Sewell, G.; Conway, T.; Clark, D.; Preston, J.; *Appl. Environ. Microbiol.* **1987**, *53*(110), 2420.
89. Stone, K.; Lynd, L. In *Proceedings of the DOE Automotive Technology Contractors' Meeting*, in Press.
90. Hinman, N. *Applied Biochemistry and Biotechnology* **1992**, *34/35*, 639-649.
91. Clements, D. In *Proceedings of Technology for Expanding the Biofuels Industry* Chicago, IL, April 21-22, 1992, sponsored by the U.S. Department of Energy, the U.S. Department of Agriculture and the Renewable Fuels Association.

92. Holmberg, W.; Gavett, E.; Morrill, P.; Peeples, J. In *Proceedings of the First Biomass Conference of the Americas: Energy, Environment, Agriculture, and Industry*. National Renewable Energy Laboratory, Golden, CO, 1993, NREL/CP-200-5768, pp 876-890.
93. *Biodiesel: A Technology, Performance, and Regulatory Overview.* National Soy Diesel Development Board, Jefferson City, MO, 1994.
94. Brown, L., In *Proceedings of the First Biomass Conference of the Americas: Energy, Environment, Agriculture, and Industry*. National Renewable Energy Laboratory, Golden, CO, 1993, NREL/CP-200-5768, pp 902-908.
95. "Internal Combustion Engines"; In *Funk and Wagnalls New Encyclopedia*, Funk and Wagnalls: New York, NY, 1979.
96. *History of the Petroleum Industry*, PBS Television Documentary, 1993.
97. Shay, E. *Biomass and Bioenergy* **1993**, *4*(4), 227-242.
98. Chowdhury, J., Fouhy, K. *Chemical Engineering* **1993,** 35-39.
99. Johannes, K., Speech at the 1993 World Conference on Refinery Processing and Reformulated Gasolines. San Antonio, TX, March 1993.
100. Caruana, C. *Chemical Engineering Progress* **1992,** *88/27*, 14-18.
101. "Sioux Falls Converts Entire Bus Fleet to Blend of Biodiesel Fuel," *New Fuels Report*, **July 12, 1993.**
102. "Illinois Farmers Begin One-Year Trial of Biodiesel," *St. Louis Post-Dispatch*, August 18, 1993.
103. "California to Operate 250 Mass Transit Buses on 20 Percent Biodiesel," *New Fuels Report*, **September 20, 1993.**
104. *The United States SoyDiesel Market Analysis.* Information Resources: Washington DC, 1993.
105. Miller, A.; Crockett, R.; Olson, E. *Newsweek*, August 30, **1993**, 47.
106. Chelf, P.; Brown, L.; Wyman, C. *Biomass and Bioenergy* **1993**, *4*, 175-183.
107. Herzog, H.; Drake, E.; Tesler, J. *Carbon Dioxide Mitigation: Priorities for Research and Development.* Draft MIT Energy Laboratory: Cambridge, MA, **1993**.
108. Feinberg, D.; Karpuk *CO_2 Sources for Microalgae-Based Liquid Fuel Production,* National Renewable Energy Laboratory, Golden, CO, **1990**, SERI/TP-232-3820.
109. Sheehan, J. Calculations based on Feinberg and Karpuk data and yield assumptions from Herzog, reference *107*.
110. Reed, T., In *Proceedings of the First Biomass Conference of the Americas: Energy, Environment, Agriculture, and Industry,* National Renewable Energy Laboratory, Golden, CO, 1993, NREL/CP-200-5768, pp 797-814.
111. "MidWest Biofuels, Inc. Biodiesel Quality Specifications"; In *Biodiesel Information Kit*, National SoyDiesel Development Board: Jefferson City, MO. 1993.
112. Peterson, C. *Transactions of the American Society of Agricultural Engineers* **1986**, *29*(5), 1413-1422.
113. *Biodiesel Commercialization.* Interchem Environmental: Overland Park, KS, 1992.

114. Van Dyne, D., "Biodiesel Production Potential from Industrial Rapeseed in the Southeastern United States, Tennessee Valley Authority, Muscle Shoals, AL 1992, TV-86444V.
115. "German Report Expected to be Negative on Biodiesel from Rape Seed Oil", *New Fuels Report,* September 7, **1992**, 13/36.
116. Fletchner, M.; Gushee, D. *Biodiesel Fuel: What Is It? Can It Compete?* CRS Report to Congress. Congressional Research Service Library of Congress, Washington DC, 1993.
117. Ahmed, I. *How Much Energy Does It Take to Make a Gallon of Soy Diesel?* Institute for Local Self-Reliance: Washington DC, 1994.
118. Abeliovich, A. In *Polysaccharides from Microalgae: A New Agroindustry*; Ramus, J.; Jones, M. Eds. Duke University: Beaufort, NC, 1988, pp 142-143.
119. Badour, S.; Gedavari, H.; Waygood, E. *A Review of Mass Cultivation of Microalgae,* Department of Botany, University of Manitoba; Winnipeg, Manitoba, Canada, 1972.
120. Neenan, B.; Feinberg, D.; Hill, A.; McIntosh, R.; Terry, K. "Fuels from Microalgae: Technology Status, Potential, and Research Requirements"; National Renewable Energy Laboratory, Golden, CO, 1986, SERI/SP-231-2550.
121. Gudin, C.; Thepenier, C.; Chaumont, D.; Berson, X. In *Polysaccharides from Microalgae: A New Agroindustry;* Ramus, J.; Jones, M., Eds. Duke University; Beaufort, NC, 1988, pp 133-141.
122. Brown, L.; Jarvis, E. In *Industrial Application of Single Cell Oils*; Kyle, D., Ratledge, C., Eds. American Oil Chemists' Society: Champaign, IL, 1992.
123. Negoro, M.; Shioji, N.; Ikuta, Y.; Makita, T.; Uchiumi, M. *Applied Biochemistry and Biotechnology* **1992,** *34/35*, 681-692.
124. Nishikawa, N.; Hon-Nami, K.; Hirano, A.; Ikuta, Y.; Hukuda, Y.; Negoro, M.; Kaneko, M.; Hada, M. *Energy Conservation and Management* **1992**, *23*(5-8,) 553-560.
125. Takano, H.; Takeyama, H.; Nakamura, N.; Sode, K.; Burgess, J.; Manabe, E.; Hirano, M.; Matsunaga, T. *Applied Biochemistry and Biotechnology* **1992**, *34/35*, 449-458.
126. Watanabe, Y.; Ohmura, N.; Saiki, H. *Energy Conservation and Management* **1992**, *23*(5-8,) 545-552.
127. Brown, L., In *Aquatic Species Project Report: FY* 1989-90, Brown, L; Sprague, S., Eds.; National Renewable Energy Laboratory, Golden, CO. 1992, NREL/MP-232-4174.
128. Weissman, J., In *Polysaccharides from Microalgae: A New Agroindustry*; Ramus, J.; Jones, M., Eds. Duke University: Beaufort, NC, 1988, pp 123-132.
129. Weissman, J., In *Aquatic Species Project Report: FY 1989-90*, Brown, L; Sprague, S., Eds.; National Renewable Energy Laboratory, Golden, CO. 1992, NREL/MP-232-4174.
130. Laws, E.; Taguchi, S.; Hirata, J.; Pang, L. *Biomass* **1988**, *16*, 19-32.
131. Roessler, P., *J. Phycology* **1990**, *26,* 393-399.
132. Dunahay, T.; Jarvis, E.; Zeiler, K.; Roessler, P.; Brown, L. *Applied Biochemistry and Biotechnology* **1992**, *34/35*, 331-339.
133. Roessler, P.; Ohlrogge, J. *The Journal of Biological Chemistry* **1993**, *268*, 19254-19259.

134. Howell, S. *Technical and Economic Review of Community Based Biodiesel Plants and Their Assumptions*; National Soy Diesel Development Board: Jefferson City, MO, 1993.
135. Zaks, A.; Klibanov, A. *Proceedings of the National Academy of Science* **1985**, *82*, 3192-3196.
136. Mittlenach, M. *J. Amer. Oil Chemists' Society* **1990**, *67*(3), 168-170.
137. Adler, M., *Reforming Education: The Opening of the American Mind*; MacMillan Publishing: New York, NY, 1988.

RECEIVED May 23, 1994

Cellulase Biochemistry and Cloning

Chapter 2

Cellulase and Xylanase Systems of *Thermotoga neapolitana*

Jin Duck Bok[1], Steven K. Goers[2], and Douglas E. Eveleigh[1]

[1]Biochemistry and Microbiology, Cook College, Rutgers University, New Brunswick, NJ, 08903–0231
[2]Kraft-General Foods, Inc., 250 North Street, White Plains, NY 10625

Thermotoga neapolitana and *T. maritima*, hyperthermophiles that grow optimally at 80°C, have very similar cellulase systems based on SDS-PAGE/zymogram analysis. *T. neapolitana* has at least three types of cellulase components, including an aryl ß-glucosidase (BglA), a ß-glucosidase with cellobiohydrolase activity (BglB) and endoglucanase (EndoA and B). The Sephadex adsorption properties of EndoB allowed a facile purification from both *T. neapolitana* and *T. maritima*. All enzymes exhibited very high thermostability, e.g. EndoB $t_{1/2}$= 8 h at 100°C. The specific activities (CMC U/mg) of both endoglucanases are very high: 1,200 for EndoA and 1,500 for EndoB. These properties make *Thermotoga* endoglucanases potentially useful for industrial application. To facilitate production, these endoglucanase genes and those of BglA and B have been cloned in *E. coli*. Furthermore, an endo-acting ß-1,4-xylanase from *T. neapolitana* was discovered and found to have remarkable thermal stability ($t_{1/2}$=130 h at 82°C), and pH and temperature optima respectively of 5.5 and 85°C. This xylanase has little endoglucanase activity and has potential application for biological bleaching in the pulp and paper industry.

Lignocellulose represents an important renewable energy source. Despite the current surplus of oil and the corresponding stability in energy-related prices, the conversion of wood, municipal cellulosic wastes or agricultural residues to fermentable sugars continues to be studied intensely, as ethanol can be derived from them and is a practical liquid transportation fuel (*1*). In the variety of processes which are targeted to produce ethanol and other chemicals based upon the enzymatic hydrolysis of lignocellulosic materials, an expensive step is the production of cellulases (*1-2*).

Cellulases are about 100 times less efficient (specific activity) compared to amylases, a major difference being that starches can be effectively treated by heating to enhance the accessibility of the amorphous substrate to the enzymes while more drastic measures (e.g. steam explosion) are required to increase the accessibility of the recalcitrant crystalline lignocellulosic materials (*2-3*). As the abundant cellulosics remain a major potential energy resource, studies for gaining more efficient cellulases continueunabated (*4-6*). In a different vein, cellulases are also employed in laundry detergents (*7*).

0097–6156/94/0566–0054$08.00/0

Cellulase

Cellulase systems (endoglucanase, cellobiohydrolase, ß-glucosidase) that degrade cellulose, the most abundant biopolymer on Earth, have been more extensively studied in fungi (*8-10*), especially *Trichoderma reesei* (*11-13*), with more recent studies of those from bacteria (*14-18*). Initially it was considered that fungi are more hydrolytic than the most active cellulolytic bacteria (*19*), but it is now realized that this "higher activity" was mostly due to the greater amount of extra cellular protein secreted by fungi (*20*). In fact, cellulolytic fungi (e.g. *T. reesei* mutant strains) are probably among the greatest producers of extracellular protein identified to date (20-35g/l extracellular protein predominantly as cellulases) (*2, 21*). However, a serious limitation in the large-scale application of mesophilic fungal cellulases in biomass conversion is their relatively low temperature stability under usage and recycling conditions, and their relatively low specific activity (*22-23*). To date the search for the optimal industrial cellulase system with such desirable characteristics as broad pH optimum, high specific activity and high stability under harsh application conditions (e.g. high temperature) has been only partially realized. Some cellulases have high specific activity while others are very stable to extreme conditions, yet no cellulase has been found with both characteristics.

Xylanase: biobleaching of wood pulp

ß-1,4-Xylanases with broad specificity can be usefully employed especially when used in conjunction with other hydrolytic enzymes for: (i) the bioconversion of forestry and agricultural wastes for fermentation feedstocks, (ii) the hydrolysis of pulp waste liquors and wood extractives to monomericsugars for use in the production of single cell protein, and (iii) in the food and beverage industry for the clarification/liquefaction of fruit and vegetable juices, modification of bread doughs, and to improve the separation of flour starch granules from gluten (*25-28*). Xylanases have also been proposed to improve the digestibility of animal feeds by breaking down the lignocellulose complex and increasing the accessibility of cellulose and xylan for degradation by ruminant bacteria. The pulp and paper industry has extended this latter concept in biobleaching of wood pulps. Crude paper pulps are susceptible to discoloration (yellowing) due to the oxidation of lignin complexes. Xylanases (1,4-ß-D-xylan xylanohydrolase, EC 3.2.1.8) have been proposed to circumvent this problem through "biobleaching". Thus, xylanase by cleaving the hemicellulose that cross-links lignins to cellulose in the crude paper pulps, releases xylan-lignin complexes leading to brighter pulps. Concomitantly this allows reduction in the use of chlorine bleach of up to 25-40%, and additionally lowers the production of pollutants (*29-33*). Indeed xylanase catalyzed pre-bleaching has been identified as one of the biotechnological processes most likely to be applied to the paper industry (*34*). In addition, xylanases are being exploited in the production of xylan-free dissolving pulps for acetate, cellophane, and rayon materials (*34-35*). For such applications, xylanases free of cellulase activity are essential because contaminating cellulase will damage the cellulose fibers and consequently the quality of the paper product.

The search for the "perfect" cellulase and xylanase continues, and has recently been directed to hypothermophiles from Bacterial and Archaebacterial Domains, that grow at 80°C or higher. *Thermotoga neapolitana* is an anaerobic hyperthermophilic, Eubacterium, capable of growth up to 90°C and appears to be a useful source of thermostable enzymes (*36-39*). It ferments a variety of carbon sources including simple sugars (glucose, xylose, cellobiose) and also homo- (cellulose, laminaran, starch) and hetero-polysaccharides (lichenan, oat glucan, xylan) (*40-42*). The xylanase from *T. neapolitana* is specific for xylan and does not attack cellulose. We now present the initial characterization of the cellulase and xylanase systems of *T. neapolitana*.

Results and Discussion

Comparison of the Cellulase Systems of *T. neapolitana* and *T. maritima*. *T. neapolitana* NS-E and *T. maritima* MSB-8 were isolated by Drs. Huber, Jannasch and Stetter, and cultured according to their basic protocols (39; *41*). Culture broths, and cell associated and intracellular fractions were examined for cellulase activity. The sodium-dodecylsulfate-polyacrylamide gel electro-phoresis (SDS-PAGE)/zymogram technique (for review see *43*) was used to separate and detect the individual cellulase components from the crude cell extract of *T. neapolitana* and *T. maritima* (Figure 1; Table I). After washing the gel to remove SDS and to allow renaturation of the proteins, the gel was incubated with the fluorogenic substrate, methylumbelliferyl-cellobiopyranoside (MUC) for 5-10 min. Four activity bands (58.3, 30.3, 29.3 and 23.6 kD bands) were clearly detected with similar intensity from batch cultures of *T. neapolitana.* Subsequent to MUC zymogram analysis, the same gel was incubated with MU-glucopyranoside (MUG) to detect ß-glucosidase. A 93.3 kD band was detected and the 58.3 kD band was enhanced (Figure 1). A carboxymethyl-cellulose (CMC)/agar overlay for 30 min at 75°C followed by Congo Red staining (*44-45*) showed overlapping activity with the MUC 30.3, 29.3 and 23.6 kD bands. This preliminary data suggested at least three major types of cellulase were present in *T. neapolitana* including a ß-glucosidase (BglA, 93.3), a ß-glucosidase with "cellobiohydrolase" activity (BglB, 58.3 kD) and endoglucanases (EndoA 30.3, 29,3, and EndoB 23.6 kD). The *T. maritima* crude cell extracts showed a very similar activity pattern in the SDS-PAGE/zymograms. In this species two bands exhibited the same mobility/activities to BglA and BglB (93.3 and 58.3 kD) of *T. neapolitana.* In *T. maritima* the larger EndoA bands (29.3 and 30.3 kD) present in *T. neapolitana*, could not be detected. However, two EndoB components (23.6 and 24.7 kD) were found in *T. maritima* (Figure 1), apparently equivalent to EndoB of *T. neapolitana.* EndoB was purified from *T. maritima* crude cell extract in a single step by selective adsorption chromatography on Sephadex G-50, and slow release by continued washing with buffer. The two endoglucanases were eluted slowly as almost pure proteins with nearly a thousand-fold purification (data not shown). This Sephadex adsorption/ elution also was used to purify EndoB from *T. neapolitana*, but EndoA (two bands) did not bind to Sephadex. Sephadex binding has also been used to purify *Cellulomonas fimi* endoglucanase C (CenC) (*46*). Purified EndoB from both species showed high specific activity on CMC substrate; over 1,200 for *T. maritima* and 1,500 for *T. neapolitana.*

Overall these data suggest that *T. neapolitana* and *T. maritima* share a very similar cellulase system. *T. neapolitana* has one extra endoglucanase (EndoA) (Figure 1; Table I). Considering the rather low (24-32%) overall DNA homology between *T. maritima* MSB 8 and *T. neapolitana* NS-E (*39,42*), this sharing of cellulase components is rather striking. *T. neapolitana* and *T. maritima* share 3 out of 4 cellulase components, but as *T. neapolitana* has an extra endoglucanase we decided to study its cellulase system.

Thermotoga neapolitana **ß-glucosidases (BglA and BglB).** Two ß-glucosidases, BglA and BglB were partially purified from the crude cell extract of cellobiose-grown *T. neapolitana*, using Sepharose Q Fast Flow-FPLC (FFQ-FPLC), gel filtration on BioGel P-60, chromatofocusing using PBE-94 and Polybuffer 74 (Pharmacia, Piscataway, NJ) and preparative PAGE (5%). After this four step purification, BglA and BglB were clearly separated from each other and also the other cellulase components. These partially purified enzymes were used to characterize their substrate specificity (Table II) and thermostability (Figure 2).

Both enzymes are mostly (80%) found free in cytoplasm and were not secreted to the culture broth. The 93 kD BglA had 14 times higher specific activity on pNP-ß-D-glucopyranoside (pNP-glu) than to cellobiose, suggesting that it is an aryl-ß-glucosidase or perhaps an exoglucosidase (*47*).

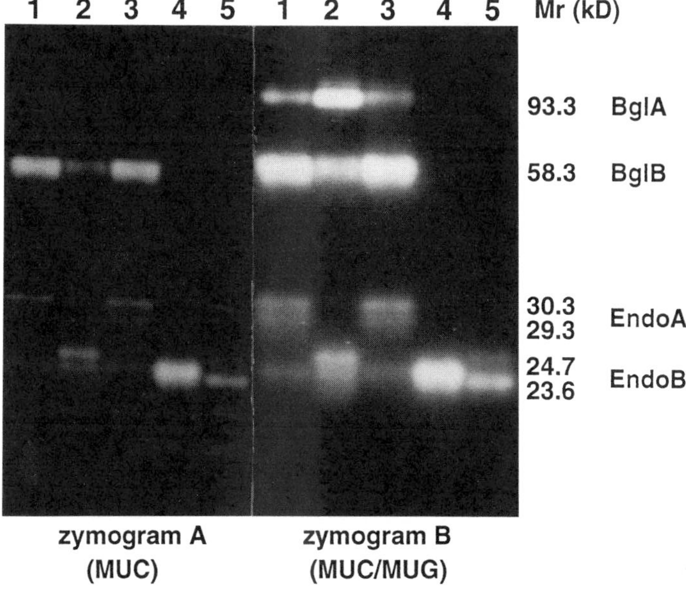

Figure 1. Analysis of cellulase components from *Thermotoga neapolitana* and *T. maritima* by SDS-polyacrylamide gel electrophoresis (SDS-PAGE)/zymograms. Zymogram A with MUC developed for 7 min at 80°C. Zymogram B: zymogram A followed by development with MUG for 5 min at 80°C. Lanes: 1, *T. neapolitana* crude cell extract I (7 µg); 2, *T. maritima* crude cell extract (7 µg) ; 3, *T. neapolitana* crude cell extract II (different batch culture from I)(7 µg); 4, EndoB from *T. neapolitana* (0.2 µg); 5, EndoB from *E. coli* clone # 083 (0.2 µg).

Table I.
Cellulase Components of *Thermotoga neapolitana* and *T. maritima* Revealed by SDS-PAGE / Zymogram Analysis

Enzyme[a]	*T. neapolitana* (kD)	*T. maritima* (kD)
BglA	93.3	93.3
BglB	58.3	58.3
EndoA	30.3	ND[b]
	29.3	
EndoB	23.6	24.7
		23.6

a. Bgl, β-glucosidase; Endo, carboxymethyl cellulase.
b. ND, not detectable

Table II.
Substrate Specificity of Two ß-glucosidases of *Thermotoga neapolitana*

Substrate[a]	BglA (U/mg)[b]	BglB (U/mg)[c]
pNP-glu	69.6	132
pNP-CB	5.6	33
pNP-lac	0.114	38
oNP-glu	ND[d]	158
oNP-gal	ND[d]	138
cellobiose	4.87	48
lactose	ND[d]	51
CMC	0.31	0.3

[a] pNP-glu, pNP-β-D glucopyranoside; -CB, -cellobiopyranoside; -lac, -lactopyranoside; -gal, -galactopyranoside; CMC, carboxymethylcellulose.

[b] assay conditions; 50 mM sodium acetate buffer, pH 4.8, containing 1 mM pNP substrate, 40 mM cellobiose or lactose, or 0.6% CMC at 85°C for 15 min in 1 ml assay volume.

[c] assay conditions, 50 mM sodium citrate buffer, pH 5.2 containing 5 mM pNP- or oNP-monosaccharide, 2.5 mM pNP-disaccharide, 80 mM cellobiose or lactose, or 0.6% CMC at 84°C for 15 min in 1 ml assay volume.

[d] ND, not determined.

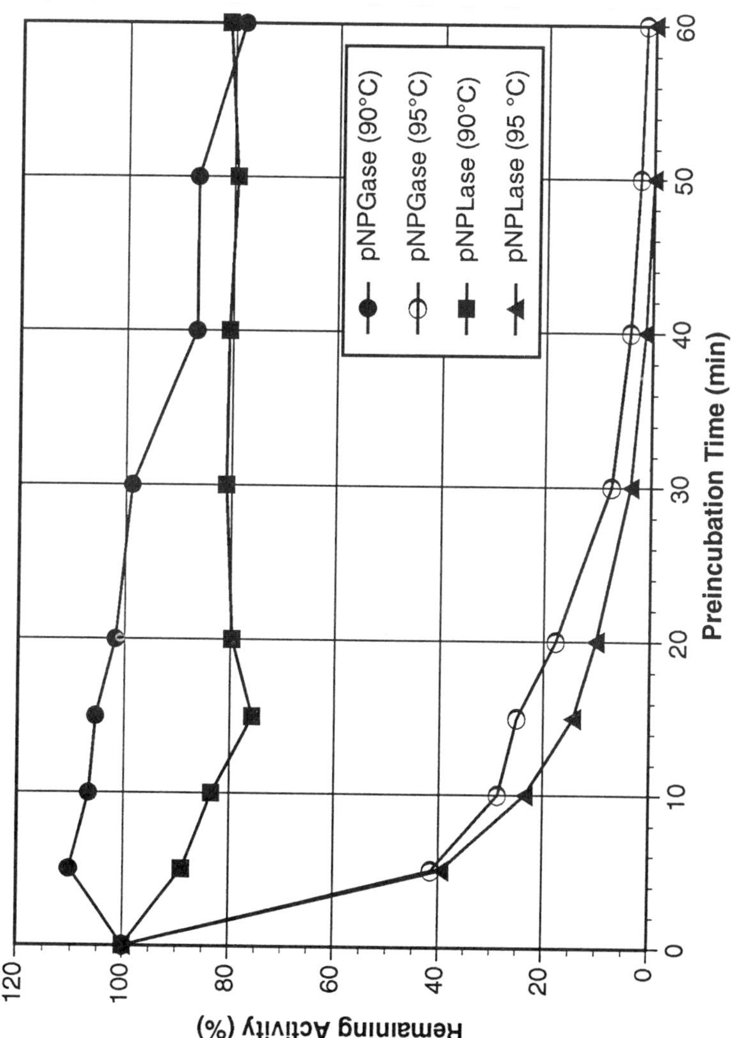

Figure 2. Thermostability of BglA and BglB from *Thermotoga neapolitana*. Enzymes were preincubated at 90 or 95°C in 50 mM sodium acetate buffer, pH 4.8 and residual activity towards pNP-glu (BglA) or pNPL (BglB) was assayed at 85°C in the same buffer.

The 58 kD BglB is active towards pNP-ß-D-cellobiopyranoside (pNPC "cellobiohydrolase") as well as towards pNP-glu and cellobiose. This can be simplistically interpreted as cellobiohydrolase (pNPC) and also ß-glucosidase (pNP-glu and cellobiose) activity. However, attack of pNPC could be by ß-glucosidase via sequential cleavage of glucose from pNPC at its non-reducing end with final release of *p*-nitrophenol. Thus ß-glucosidase could show low activity towards pNPC. However, the BglB showed high activity on pNP-lactoside (pNPL), in which the galactosyl moiety at the non-reducing end classically blocks the attack by ß-glucosidase thus eliminating the possibility of sequential cleavage (*48*). To elucidate the mechanism BglB attack towards pNPC as well as pNPL, BglB was tested on various substrates (Table II). BglB showed similar specific activity on oNP-glu and oNP-gal. Furthermore, BglB hydrolyzed lactose at a very slightly higher rate compared to cellobiose, while it showed a similar rate of attack towards pNPC and pNPL. These combined data indicate that BglB has a broad substrate specificity and that it attacks pNPC and pNPL via sequential cleavage of glucose or galactose from their non-reducing end with final release of *p*-nitrophenol.

BglA has an optimal pH around 5.0 to 5.2 and BglB around 6.2 to 6.7 (in citrate-phosphate buffer; 49). Both BglA and BglB have temperature optima around 85-90°C in 50 mM sodium acetate buffer, pH 4.8.

Thermostability was determined by preincubating the enzymes at preset temperatures in 50 mM sodium acetate buffer, pH 4.8 and measuring the residual activity at 85°C (Figure 2). Both enzymes showed very high thermostability even at 90°C ($t_{1/2}$ @ 90°C = 133 min for BglA). Using PNPL as a substrate, the BglB lost about 20% activity in first 10 min but the activity stabilized at 80% after an initial one hour incubation, suggesting that two species may be present, one of them being more thermally stable. However, at 95°C, both enzymes lost 50% of activity in less than 5 min. ß-Glucosidase is often the least stable member of the cellulolytic enzyme system (*50*). Even so the thermostability of BglA and BglB are among the highest in this class. A thermostable ß-glucosidase was isolated from *Caldocellum saccharolyticum* which was cloned and expressed in *E. coli* showed Topt of 85°C and $t_{1/2}$ @ 80°C=14 h (*51*). Another thermostable ß-glucosidase has been isolated from *Thermotoga* sp. strain FjSS3-B-1 (*50*). This enzyme was a cell-associated 75 kD protein. On purification, this enzyme became much less thermostable: $t_{1/2}$ @ 80°C=47 min compared to $t_{1/2}$ @ 90°C=8 h of the partially purified enzyme. However its thermostability was dramatically increased to $t_{1/2}$ @ 90°C=2.5 h by the presence of bovine serum albumin. The authors suggested that the added protein may prevent the adsorption of the enzyme on the microfuge tube wall, noting that Triton X-100 (0.01%) also restored the enzyme "stability". This increased thermostability could also be via increased protein-protein hydrophobic interaction. Thermostability was further increased by including 14% (w/v) betaine or trehalose to $t_{1/2}$ @ 98°C=15 h. Based on the molecular weight in SDS-PAGE, substrate specificity and enzyme localization (cell-associated vs. cytoplasmic), the ß-glucosidase from *Thermotoga* sp. strain FjSS3-B.1 appears distinct from both BglA and BglB of *T. neapolitana*. The most thermophilic and thermally stable ß-glucosidase is apparently from *Pyrococcus furiosus* (*52*), having a temperature optimum of 102-105°C at optimal pH 5.0 although it is relatively thermally unstable at this pH/buffer concentration ($t_{1/2}$ @ 110°C=30 min in 150 mM sodium citrate, pH 5.0). In contrast in 150 mM Tris buffer pH 8.5, its $t_{1/2}$ increased to 85 h @ 100°C and 13 h @ 110°C. It is unclear what is responsible for the enhanced stability in Tris buffer. Generally much less protein is thermally precipitated at alkaline than in acidic pH. Direct comparisons of thermostability by half life can be hampered by the different conditions used in each case (pH, type of buffer, additives, etc.). We have yet to address the precise roles of pH and buffer on the thermostability of the *Thermotoga* enzymes.

Thermotoga neapolitana **Endoglucanases (EndoA and B).** Two types of endoglucanase, EndoA and EndoB, were purified from the cell extract of cellobiose-grown *T. neapolitana* cells. About 60% of the CMCase was found extracellularly or cell associated (including periplasmic location) and the remainder was intracellular. Separation of these fractions was by the osmotic shock method (*53*). Quantitation in relation to location of the enzymes is complicated by the relative ease of cell lysis (unpublished observation), and furthermore about 10% of the apparently cell associated enzymes can be eluted from the cells simply by washing with buffer or distilled water. Purification of EndoA was by FFQ-FPLC, gel filtration on BioGel P-60 and preparative PAGE, while for EndoB FFQ-FPLC and affinity adsorption on Sephadex G-50 were employed (detailed methods and results will be published elsewhere). Purified EndoB showed a remarkable temperature optimum (110°C) and thermostability ($t_{1/2}$ @ 100°C, >8 h; @ 106°C, 2.2 h in 20 mM MOPS buffer, pH 7.0). These are the highest temperature optimum and thermostability reported for endoglucanases (Table III).

Table III. Thermostability of Endoglucanases

Organism	T_{opt}, °C	$t_{1/2}$[a] hr (°C)	Ref
Trichoderma reesei	60	INAC[b] (70)	22
Microbispora bispora	NR[c]	0.3 (80), 0.2 (100)	54
Thermoanaerobacter cellulolyticus	80	>5 (70), 1.5 (80)	55
Acidothermus cellulolyticus	83	1 (85), 0.2 (90)	56
NA 10	80	>3 (80)	57
Thermotoga neapolitana EndoB	110	>8 (100), 2.2 (106)	this study

[a]$t_{1/2}$ = half life, hours at the preincubation temperature.

[b]INAC, rapid inactivation of the enzyme.

[c]NR, not reported.

Both EndoA and B showed high specific activity towards CM-cellulose: around 1,200-1,500 U/mg at 95°C. The majority of endoglucanases have specific activities of 20-100 U/mg, though most are assayed only at 50°C. Thus specificity activity 1,200 is over 60-fold higher than that *T. reesei* endoglucanases (Table IV).

Ruttersmith and Daniel (*63*) reported a highly thermostable 35 kD "cellobiohydrolase" from *Thermotoga* sp. strain FjSS3-B.1. This enzyme showed higher activity towards CMC than pNPC or amorphous cellulose. Furthermore, it degraded CMC at least 6%, thus also suggesting it might be endoglucanase rather than an exo-splitting cellulase. In contrast, to date we have been unable to detect an enzyme in *T. neapolitana* NS-E or *T. maritima* MSB-8 with such substrate specificity. Whatever the name of the FjSS3-B.1 enzyme, it exhibited very similar Topt and thermostability to that of EndoB, having Topt at 105°C and $t_{1/2}$ @ 108°C, 70 min in 0.1 M MOPS buffer, pH 7.0. It showed lower specific activity towards CMC (38 U/mg) and also to pNPC (2.3 U/mg).

Table IV. Activities of Endoglucanases[a]

Organism	Enzyme	Sp Act[b] (IU/mg)	Assay T. (°C)	Ref
T. reesei QM9414	ENDO I	13.1	30	58
T. reesei QM9414	ENDO II	20.1	30	58
Cellulomonas fimi	CENC	368	37	46
C. thermocellum	EGD	428	60	59
Streptomyces lividans 66	ENDO	539	50	60
Thermomonospora fusca	E1	1,100	55	61
P. fluorescens v. cellulosa	CELA	1,150	30	62
P. fluorescens v. cellulosa	CELC	1,860	30	62
T. neapolitana	EndoA	1,200	95	this study
T. neapolitana	EndoB	1,500	95	this study

[a] Comparative selection of enzymes was based on the apparent high activity.*Trichoderma* endoglucanases as the reference standard.
[b] Specific activity, IU/mg protein towards CMC.

Xylanase. Xylanase activity is induced when *T. neapolitana* is grown on xylose (11-fold) or xylan (6.6-fold) as a carbon source compared to glucose grown cells. When grown on xylose, *T. neapolitana* expresses one ß-1,4-xylanase and at least one ß-1,4-xylosidase. The xylanase was purified 28-fold from cell extracts of a 50 liter fermentation of *T. neapolitana.* The ß-1,4-xylanase is an acidic (pI 4.9), monomeric protein (37 kD) with strict specificity for ß-1,4-linked xylose polymers. The pH and temperature optima are 5.5 and 85°C, respectively. The enzyme exhibits remarkable thermal stability with $t_{1/2}$=130 h at 82°C. The mode of action is endo-acting as determined by HPLC analysis of xylan degradation products (X_7-X_2) with no hydrolysis of xylobiose.

Total yields of xylanase are extremely low (0.024 U/ml) though with relatively poor growth yields (2 x 10^8 cells/ml). Generally, microbial yields of xylanase are in the range of 0.1 - 50 U/ml. Exceptionally productive strains and also some mutants cultured under optimal conditions include: *Aureobasidium pullulans* Y-2311 (380 U/ml) [*64*]; *Aspergillus awamori* mutant AANTG43 (820 U/ml) (*65*), *A. fumigatus* VTT-D-82195 (223 U/ml) (*66*) and *Trichoderma reesei* mutants (600 U/ml) (*66*). However the stability of enzymes from hyperthermophiles is dramatically increased in comparison to such fungal xylanases, e.g. for *Thermotoga* sp. strain FjSS3-B.1, $t_{1/2}$=90 min @ 95°C (*67*), and for the current *T. neapolitana,* $t_{1/2}$=130 h @ 82°C. Greater production of the *T. neapolitana* xylanase enzyme is being pursued through cloning of the xylanase gene in *E. coli.* Initial clones give greater xylanase yields (2 U/ml). The cloned enzyme possesses essentially the same remarkable temperature stability at 80°C as the native enzyme. The enzyme is specific for xylan and does not attack cellulose, and hence will not directly produce adverse effects towards paper pulp if used for biopulping. Further evaluation of the cloned *T. neapolitana* xylanase under the rigorous conditions occurring in a paper mill are planned.

Acknowledgments:

This study was supported by the National Renewable Energy Laboratory, Golden, CO., and the New Jersey Agricultural Experiment Station, publication no. D-01111-02-93.

Literature Cited.

1. Lynd, L.R.; Cushman, J.H.; Nichols, R.J.; Wyman, C.E. *Science* **1991,** *251*, 1318-1323.
2. Saddler, J.N. *Microbiol. Sci.* **1986,** *3,* 84-87.
3. Sinitsyn, A.P.; Gusakov, A.V.; Yu Vlasenko, E. *Appl. Biochem. Biotechnol.* **1991,** *30*, 43-59.
4. Béguin, P. *Ann. Rev. Microbiol.* **1990,** *44,* 219-248.
5. *Biosynthesis and Biodegradation of Cellulose*; Haigler, C.H.; Weimer, P.J. Eds., Marcel Dekker: New York, NY, 1990.
6. Walker, L.P.; Wilson, D.B. *Bioresource Technol.* **1991**, *36,* 3-14;
7. Shikata, S.; Saeki, K.; Okoshi, H.; Yoshimatsu, T.; Ozaki, K.; Kawai, S.; Ito, S. *Agric. Biol. Chem.* **1990,** *54,* 91-96.
8. Coughlan, M.P. In *Microbial Enzymes and Biotechnology*; 2nd edition; Fogarty, W.M.; Kelly, C.T., Eds., Elsevier Applied Science: London, UK , 1990; pp. 1-36.
9. Goyal, A.; Ghosh, B.; Eveleigh, D.E. *Bioresource Technol.* **1991**, *36*, 37-50.
10. Kloysov, A.A. *Biochemistry* **1990,** *29*, 10577-10585.
11. Kubicek, C.P. *Adv. Biochem. Eng./Biotechnol.* **1992,** *45,* 1-27.
12. Nevalainen, K.M.H.; Penttillä, M.; Harkki, A.; Teeri, T.T.; Knowles, J. In *Molecular Industrial Mycology*; Leong, S.A.; Berka, R.M., Eds.; Marcel Dekker: New York, NY, 1991; pp 129-148.
13. *Cellulases: Biochemistry, Genetics, Physiology and Application*; Kubicek, C. P.; Esterbauer, H; Eveleigh, D.E.; Steiner, W.; Kubicek-Pranz, E.M., Eds.; Royal Chemical Society: London, UK, 1990.
14. Coughlan, M.P.; Mayer,.F. In *The Prokaryotes*; 2nd edition; Balows, A.; Trüper, H.G.; Dworkin, M.; Harder, W.; Schleifer, K.-H., Eds.; Springer Verlag: New York, NY, 1991, Vol. I; pp 460-516.
15. Gilbert, H.J.; Hazlewood, G.P. *J. Gen. Microbiol.*, **1993**, *139*, 187-194.
16. Gilkes, H.J.; Kilburn, D.G.; Miller, R.C., Jr.; Warren, R.A.J. *Bioresource Technol.* **1991,** *36*, 21-36.
17. Robson, L.M.; Chambliss, G.H. *Enzyme Microb. Technol.* **1989,** *11*, 626-644.
18. Stutzenberger, F. In *Microbial Enzymes and Biotechnology*; 2nd edition; . Fogarty, W.M.; Kelly, C.T., Eds., Elsevier: New York, NY, 1990; pp 37-70.
19. Bolobova, A.V.; Klesov, A.A. *Prikladnaya Biokimiya Microbiologiya*, 1990, *26*, 253-258 (English translation).
20. Lamed, R.; Kenig, R.; Morag, E.; Calzada, J.F.; DeMicheo, F; Bayer, E.A. *Appl. Biochem. Biotechnol.* **1991**, *27*, 173-183.
21. Pourquié, J.; Desmarquest, J.-P. In *Enzyme Systems for Lignocellulose Degradation*; Coughlan, M.P., Ed.; Elsevier Applied Science: New York, NY, 1989; pp. 283-292.
22. Durand, H.; Soucaille, P.; Tiraby, G. *Enzyme Microb. Technol.* **1984,** *6*, 175-180.
23. Margaritis, A.; Merchant, R.F.J. *CRC Crit. Rev. Biotechnol.* **1986,** *4*, 327-367.
24. Ball, A.S.; McCarthy, A.J. *J. Appl. Bacteriol.* **1989**, *66,* 439-444.
25. Biely, P. In *Enzymes in Biomass Conversion*; Leatham, G.F.; Himmel M.E., Eds.; American Chemical Society: Washington, D.C., 1991, Vol. 460; pp 408-417.
26. Dziezak, J.D. *Food Tech.* **1991,** *45*, 77-82.
27. Senior, D.J.; Mayers, P.R.; Saddler, J.N. In *Plant Cell Wall Polymers: Biogenesis and Biodegradation*; Lewis, N.G.; Paice, M.G., Eds.; American Chemical Society: Washington, D.C., 1989, Vol. 399; pp 630-640.

28. Zeikus, J.G.; Lee, C.; Lee, Y.-E.; Saha, B.C. In *Enzymes in Biomass Conversion*; Leatham, G.F.; Himmel; M.E., Eds.; American Chemical Society: Washington, D.C., 1991, Vol 460; pp 36-51.
29. Bajpal, P.; Bajpal, P.K. *Process Biochem.* **1992**, *27*, 319-325.
30. Jeffries, T.W. In *Emerging Materials and Chemicals from Biomass*; Rowell, R.M.; Narayan, R.; Schults, T., Eds.; American Chemical Society: Washington, D.C., 1991, Vol 476; pp 313-329.
31. Paice, M.G.; Bernier, R., Jr.; Jurasek, L. *Biotech. Bioeng.* **1988,** *32*, 235-239.
32. Senior, D.J.; Mayers, P.R.; Saddler, J.N. *Biotech. Bioeng.* **1991**, *37,* 274-279.
33. Viikari, L.; Kantelinen, A.; Poutanen, K.; Ranua, M. In *Biotechnology in Pulp and Paper Manufacture*; Kirk, T.K.; Chang, H.-M., Eds.; Butterworths-Heinemann: Boston, MA, 1990; pp 145-143.
34. Wong, K.K.Y.; Saddler, J.N. *Crit. Rev. Biotech.* **1992,** *12*, 413-435.
35. Zamost, B.L.; Nielsen, H.K.; Starnes, R.L. *J. Ind. Microbiol.* **1991,** *8,* 71-82.
36. Belkin, S.; Wirsen, C.O.; Jannasch, H.W. *Microbiol.* **1986**, 51, 1180-1185.
37. Coolbear, T.; Daniel, R.M.; Morgan, H.W. *Adv. Biochem. Eng./ Biotechnol.* **1992,** *45*, 57-98.
38. Huber, R.; Stetter, K.O. In *Thermophilic Bacteria;* Kristjansson, J.K., Ed.; CRC Press, Inc.: Boca Raton, FL, 1992; pp 185-194.
39. Jannasch, H.W.; Huber, R; Belkin, S.; Stetter, K.O. *Arch. Microbiol.* **1988**, *150*, 103-104.
40. Goers, S.K. *Ph.D. Thesis*; Rutgers University: New Brunswick, NJ, 1993.
41. Huber, R.; Langworthy, T.A.; Konig, H.; Thomm, M.; Woese, C.R.; Sleytr, U.B.; Stetter, K.O. *Arch. Microbiol.* **1986,** *144*, 324-333.
42. Windberger, E.; Huber, R.; Trincone, A.; Fricke, H.; Stetter, K.O. *Arch. Microbiol.* **1989,** *151,* 506-512.
43. Coughlan, M.P. In *Methods in Enzymology*; Wood, W.A.; Kellog, S.T., Eds.; Academic Press: San Diego, CA, 1988, Vol. 160; pp 135-144.
44. Béguin, P. *Anal. Biochem.* **1983**, *131*, 333-336.
45. Wood, P.J.; Erfle, J.D.; Teather, R.M. In *Methods in Enzymology*; Wood, W.A.; Kellogg, S.T., Eds.; Academic Press: San Diego, CA., 1988, Vol. 160; pp 59-74.
46. Moser, B.; Gilkes, N.R.; Kilburn, D.G.; Warren, R.A.J.; Miller, R.C., Jr. *Appl. Environ. Microbiol.* **1989**, *55*, 2480-2487.
47. Goyal, A.K.; Wright, R.M.; Hu, P.; Eveleigh, D.E. *Post-conference Proc. Ann. Automotive Technology Development Contractors Coordination Meeting*; Society of Automotive Engineers: Warrendale, PA, 1991; pp. 761-769.
48. Deshpande, M.V.; Pettersson; L.G.; Eriksson, K.-E. In *Methods in Enzymology;* Wood, W.A.; Kellog, S.T., Eds.; Academic Press: San Diego, CA, 1988, Vol. 160; pp 126-130.
49. Perrin, D.D.; Dempsey, B. *Buffers for pH and Metal Ion Control*; Chapman and Hall: London, UK, 1974.
50. Ruttersmith, L.D.; Daniel, R.M. *Biochim. Biophys. Acta* **1993**, *1156,* 167-172.
51. Love, D.R.; Streiff, M.B. *Bio/Technol.* **1987,** *5*, 384-387.
52. Kengen, S.W.M.; Leusink, E.J.; Stams, A.J.M.; Zehnder, A.J.B. *Eur. J. Biochem.* **1993**, *213,* 305-312.
53. Neu, H.C.; Heppel, L.A. *J. Biol. Chem.* **1965**, *240*, 3685-3692.
54. Waldron, C.R.; Becker-Vallone, C.A.; Eveleigh, D.E. *Appl. Microbiol. Biotechnol.* **1986**, 24, 477-486.

55. Honda, H.; Naito, H.; Taya, M.; Iijima, S.; Kobayashi, T. *Appl. Microbiol. Biotechnol.* **1987**, 25, 480-483.
56. Tucker, M.P.; Mohagheghi, A.; Grohmann, K.; Himmel, M.E. *Bio/Technol.* **1989**, 7, 815-820.
57. Honda, H.; Saito, T.; Iijima, S.; Kobayashi, T. *Appl. Microbiol. Biotechnol.* **1988,** 29, 264-268.
58. Voragen, A.G.T.; Beldman, G.; Rombouts, F.M. *Methods in Enzymology*; Wood, W.A.; Kellog, S.T., Eds.; Academic Press: San Diego, CA, 1988, Vol. 160; pp. 243-251.
59. Joliff, G.; Beguin, P.; Juy, M.; Ryter, A.; Poljak, R,; Aubert, J.-P. *Bio/Technology*. **1986,** *4*, 896-900.
60. Theberge, M.; Lacaze, P.; Shareck, F.; Morosoli, R.; Kluepfel, D. *Enzyme Microb. Technol.* **1989,** 11, 626-644.
61. Wilson, D.B. In *Methods in Enzymology*; Wood, W.A.; Kellog, S.T.; Eds; Academic Press: San Diego, CA, 1988, Vol. 160; pp 314-323.
62. Yamane, K.; Suzuki, H. In *Methods in Enzymology*; Wood, W.A.; Kellog, S.T., Eds.; Academic Press: San Diego, CA, 1988, Vol. 160; pp 200-210.
63. Ruttersmith, L.D.; Daniel, R.M. *Biochem.J.* **1991**, *277*, 887-890.
64. Leathers, T.D.; Kurtzman, C.P.; Detroy, R.W. In *Biotechnol. Bioeng. Symp. No. 14*; Scott, C.D., Ed.; John Wiley & Sons: New York, NY, 1984; pp. 225-240.
65. Smith, D.C.; Wood, T.M. *Biotech. Bioeng.* **1991,** *38,* 883-890.
66. Bailey, M.J.; Poutanen, K. *Appl. Microbiol. Biotechnol.* **1989**, *30*, 5-10.
67. Simpson, H.D.; Haufler, J.R.; Daniel, R.M. *Biochem. J.* **1991**, *277*, 413-417.

RECEIVED February 25, 1994

Chapter 3

Structure–Function Studies of Endo-1,4-β-D-glucanase E2 from *Thermomonospora fusca*

Mike Spezio, P. Andrew Karplus, Diana Irwin, and David B. Wilson

Section of Biochemistry, Molecular and Cell Biology, Cornell University, Ithaca, NY 14853

Six different cellulases have been purified to homogeneity from the culture supernatant of a protease negative mutant of the soil bacterium, *Thermomonospora fusca* (*1,2*). Three of these enzymes are endoglucanases, two are exocellulases and one is an exocellulase with some endoglucanase activity (*3*). All six enzymes contain a cellulose binding domain joined by a flexible linker to a catalytic domain. When the distributions of reducing sugars between the solution and filter paper in filter paper assays of E2 and E2cd were determined, there were two soluble sugars produced for each insoluble sugar with E2 while with E2cd there was only one soluble sugar produced for each insoluble sugar (*3*). This result seems reasonable since the cellulose binding domain should attach the enzyme to a specific site on the substrate for a longer time than would the catalytic domain alone.

The six cellulases isolated from *T. fusca* culture broth comprise a well-studied bacterial cellulase system. E2 is the smallest *T. fusca* cellulase and has a molecular weight of 42,000. It contains an N-terminal catalytic domain of 31,000 joined by a 26 amino acid linker to a C-terminal cellulose binding domain (*4*). E2 has specific activities of 370, 170, and 0.85 μmoles/min/μmole on carboxymethylcellulose (CMC), swollen cellulose and filter paper respectively (*3*) as assayed by the procedure of Ghose (*5*). A proteolytic derivative of E2, the E2 catalytic domain (E2cd), which is missing the cellulose binding domain and linker region has specific activities of 344, 113 and 0.50 μmoles/min/μmole on the same set of substrates but on filter paper it cannot give 5% digestion so the specific activity was calculated from the activity given by 0.6 nmoles of enzyme. E2 has been shown to invert the configuration of the anomeric carbon during hydrolysis (*6*). Selected properties of E2 and other *T. fusca* cellulase enzymes are given in Table I.

0097–6156/94/0566–0066$08.00/0

Table I. Properties of Individual Cellulases*

Protein	MW (kD) *a*	Extinction coefficients (molar) *b*	Stereo chemistry *c*	Activity CMC	Specific Activity (μmol cellobiose/min-μmol) swollen cellulose	filter paper	p-NP-cellobiose
E1	101.2	208,000	inversion	5410.0	362.0	0.182 *d*	40.4
E2	43.0	80,000	inversion	369.0	168.0	0.846	*g*
E2cd	30.0	57,600	inversion	344.0	113.0	0.501 *d*	*g*
E3	65.0	145,000	*f*	1.3 *d*	1.6	0.303 *d*	*g*
E4	90.2	214,000	*f*	122.0	34.9	0.565 *d*	*g*
E5	46.3	97,000	retention	2840.0	90.4	0.832	14.8
E5cd	34.4	70,300	retention	2480.0	85.3	0.573 *d*	14.4
E6	106.0	385,000	*f*	64.2 *e*	83.5	0.863 *e*	0.04

[a]MWs for E3 and E6 were estimated from SDS polyacrylamide gels.
[b]Extinction coefficients were calculated from the number of trp and tyr residues in proteins. E3 and E6 extinction coefficients were estimated from amino acid analysis.
[c]ref. *3.*
[d]Target % digestion could not be achieved. In this case, filter paper specific activities were calculated using digestion achieved by 0.6 nmoles of enzyme in 16 hours.
[e]Contaminating CMC activity as determined by CMC overlays of native page gels.
[f]Not determined.
[g]Activity below detectable limits (i.e., <0.02).
*Some data adapted from reference *3.*

The E2 gene has been cloned (*7*) and sequenced (*4*) and the sequence is shown in Figure 1. It can be expressed at a high level in *Streptomyces lividans* which completely secretes E2 into the culture supernatant (*7*). *T. fusca* E2 is a glycoprotein containing about 2% carbohydrate while E2 produced by *S. lividans* is not glycosylated. However, we have not detected any difference in the activities of the two forms of the enzyme. The amino acid sequence of E2cd clearly belongs in cellulase family B (*8*); and it has 26% amino acid identity to another family member, *Trichoderma reseei* CBHII. The amino acid sequence of the E2 cellulose binding domain is similar to that of many bacterial cellulose binding domains, including all three from *T. fusca* (*3*), two from *Cellulomonas fimi* (*9*), and one from *Pseudomonas cellulosa* (*10*). When the DNA sequence encoding the E2 cellulose binding domain and linker region was joined to the DNA encoding a *Prevotella ruminicola* CMCase catalytic domain, the resulting hybrid cellulase had nearly ten times more activity on ball milled and amorphous cellulose than the original enzyme (*11*). An unusual property of E2 is that it is present as a homodimer in the pH range from 5.5 to 9 at temperatures below 65°C (*12*). E2 gives strong synergism in the hydrolysis of filter paper with a number of exocellulases including E3 and E4 from *T. fusca* and CBHI and CBHII from *Trichoderma reesei* (*3*). E2 has the highest specific activity in Avicel fragmentation of any enzyme that has been tested (*13*).

All attempts to crystallize E2 were unsuccessful and at this time no two domain cellulase has been crystallized. However, we were able to crystallize E2cd. Excellent crystals that diffract to 1.35Å were obtained by the hanging drop procedure from ammonium sulfate at pH 4.0 using microseeding. Three rounds of microseeding were required to obtain good crystals initially, each time using the improved crystals as seeds. Using the good crystals, a single round of microseeding yields good crystals after a period of three weeks. A typical crystal size is 0.15 x 0.40 x 1.8 mm^3. The crystals contain only 36% solvent and their space group is $P2_1$, with a=43.35Å, b=65.94Å, c=43.41Å and β=107.69°. There is one molecule per asymmetric unit which is consistent with our solution studies, as E2 is a monomer at pH 4.0.

The crystals appear to catalyze the hydrolysis of cellotetraose to cellobiose so that the enzyme retains catalytic activity in the crystal. A 2.8Å resolution electron density map was determined from phases derived by isomorphous replacement using uranium, mercury, gold and platinum derivatives. The uranyl derivative gave the most signal, and a nearly complete set of anomalous scattering data was collected for it. The structure of the CBHII catalytic domain had already been determined (*14*) and we used this structure to help us fit the amino acid sequence of E2cd into our electron density map. The initial model did not refine well so we calculated a new density map using solvent flattening and including data to 2.6Å resolution (*15*). When the model was rebuilt into this map the coordinates refined well, and we have refined the structure using data to 1.8Å resolution. The R-value for all reflections between 10 and 1.8Å is 18.4% and the model contains 73 bound water molecules and one bound sulfate. The overall root-mean-square deviations from ideality for the E2cd model are 0.014Å for the bond lengths, 2.8° for the bond angles and 1.1° for fixed dihedral angles. All of the 286 residues in the model fall into acceptable regions of a Ramachandran plot except for Asp^2, His^{83} and Trp^{162}. Asp^2 and His^{83} are in flexible regions judging by their higher temperature factors. Trp^{162} is in the active site and is well ordered. The main chain torsion angles for Trp^{162} (ϕ = -146°, Ψ=105), although

1 *GTACCCGGGGATCCTCTAGAGT*CGATATGGATGATCTGACGTCTGAATCCCCTTGTCACCCTAGACATTCACCCATTTTGTCGCTTTTAC 90

91 GGCTTTCTTTGGGAGTTCTCCGTTTCACCAAGGAACAAAACCGCAACGGAGAGTAGGCGCGGTCTTTACAGCTCCCTTGCCAATGGTTAT 180

181 CGTCCGAACGGAAAACGATCTGGGAGCGCTCCCAGCCATGCGCTCCTCTTCGTGCCCCTCACTTCTTTTGAGCCTTGTGCTCGTTAGGAG 270

271 CCCCGAATGTCCCCCAGACCTCTTCGCGCTCTTCTGGGCGCCGCGGCGGCGGCCTTGGTCAGCGCGGCTGCTCTGGCCTTCCCGTCGCAA 360
M S P R P L R A L L G A A A A A L V S A A A L A F P S Q

361 GCGGCGGCCAATGATTCTCCGTTCTACGTCAACCCCAACATGTCCTCCGCCGAATGGGTGCGGAACAACCCCAACGACCCGCGTACCCCG 450
A A A N D S P F Y V N P N M S S A E W V R N N P N D P R T P

451 GTAATCCGCGACCGGATCGCCAGCGTGCCGCAGGGCACCTGGTTCGCCCACCACAACCCCGGGCAGATCACCGGCCAGGTCGACGCGCTC 540
V I R D R I A S V P Q G T W F A H H N P G Q I T G Q V D A L

541 ATGAGCGCCGCCCAGGCCGCCGGCAAGATCCCGATCCTGGTCGTGTACAACGCCCCGGGCCGCGACTGCGGCAACCACAGCAGCGGCGGC 630
M S A A Q A A G K I P I L V V Y N A P G R D C G N H S S G G

631 GCCCCCAGTCACAGCGCCTACCGGTCCTGGATCGACGAATTCGCTGCCGGACTGAAGAACCGTCCCGCCTACATCATCGTCGAACCGGAC 720
A P S H S A Y R S W I D E F A A G L K N R P A Y I I V E P D

721 CTGATCTCGCTGATGTCGAGCTGCATGCAGCACGTCCAGCAGGAAGTCCTGGAGACGATGGCGTACGCGGGCAAGGCCCTCAAGGCCGGG 810
L I S L M S S C M Q H V Q Q E V L E T M A Y A G K A L K A G

811 TCCTCGCAGGCGCGGATCTACTTCGACGCCGGCCACTCCGCGTGGCACTCGCCCGCACAGATGGCTTCCTGGCTCCAGCAGGCCGACATC 900
S S Q A R I Y F D A G H S A W H S P A Q M A S W L Q Q A D I

901 TCCAACAGCGCGCACG[illegible]TATCGCCACCAACACCTCCAACTACCGGTGGACCGCTGACGAGGTCGCCTACGCCAAGGCGGTGCTCTCGGCC 990
S N S A H G I A T N T S N Y R W T A D E V A Y A K A V L S A

991 ATCGGCAACCCGTCCCTGCGCGCGGTCATCGACACCAGCCGCAACGGCAACGGCCCCGCCGGTAACGAGTGGTGCGACCCCAGCGGACGC 1080
I G N P S L R A V I D T S R N G N G P A G N E W C D P S G R

1081 GCCATCGGCACGCCCAGCACCACCAACACCGGCGACCCGATGATCGACGCCTTCCTGTGGATCAAGCTGCCGGGTGAGGCCGACGGCTGC 1170
A I G T P S T T N T G D P M I D A F L W I K L P G E A D G C

1171 ATCGCCGGCGCCGGCCAGTTCGTCCCGCAGGCGGCCTACGAGATGGCGATCGCCGCGGGCGGCACCAACCCCAACCCGAACCCCAACCCG 1260
I A G A G Q F V P Q A A Y E M A I A A G G T N P N P N P N P

1261 ACGCCCACCCCCACTCCGACCCCCACGCCGCCTCCCGGCTCCTCGGGGGCGTGCACGGCGACGTACACGATCGCCAACGAGTGGAACGAC 1350
T P T P T P T P T P P P G S S G A C T A T Y T I A N E W N D

1351 GGCTTCCAGGCGACCGTGACGGTCACCGCGAACCAGAACATCACCGGCTGGACCGTGACGTGGACCTTCACCGACGGCCAGACCATCACC 1440
G F Q A T V T V T A N Q N I T G W T V T W T F T D G Q T I T

1441 AACGCCTGGAACGCCGACGTGTCCACCAGCGGCTCCTCGGTGACCGCGCGGAACGTCGGCCACAACGGAACGCTCTCCCAGGGAGCCTCC 1530
N A W N A D V S T S G S S V T A R N V G H N G T L S Q G A S

1531 ACAGAGTTCGGCTTCGTCGGCTCTAAGGGCAACTCCAACTCTGTTCCGACCCTTACCTGCGCCGCCAG*GGGGATCCTCTAGAGTCGACCT* 1620
T E F G F V G S K G N S N S V P T L T C A A

1621 *GCAGGCATGC* 1630

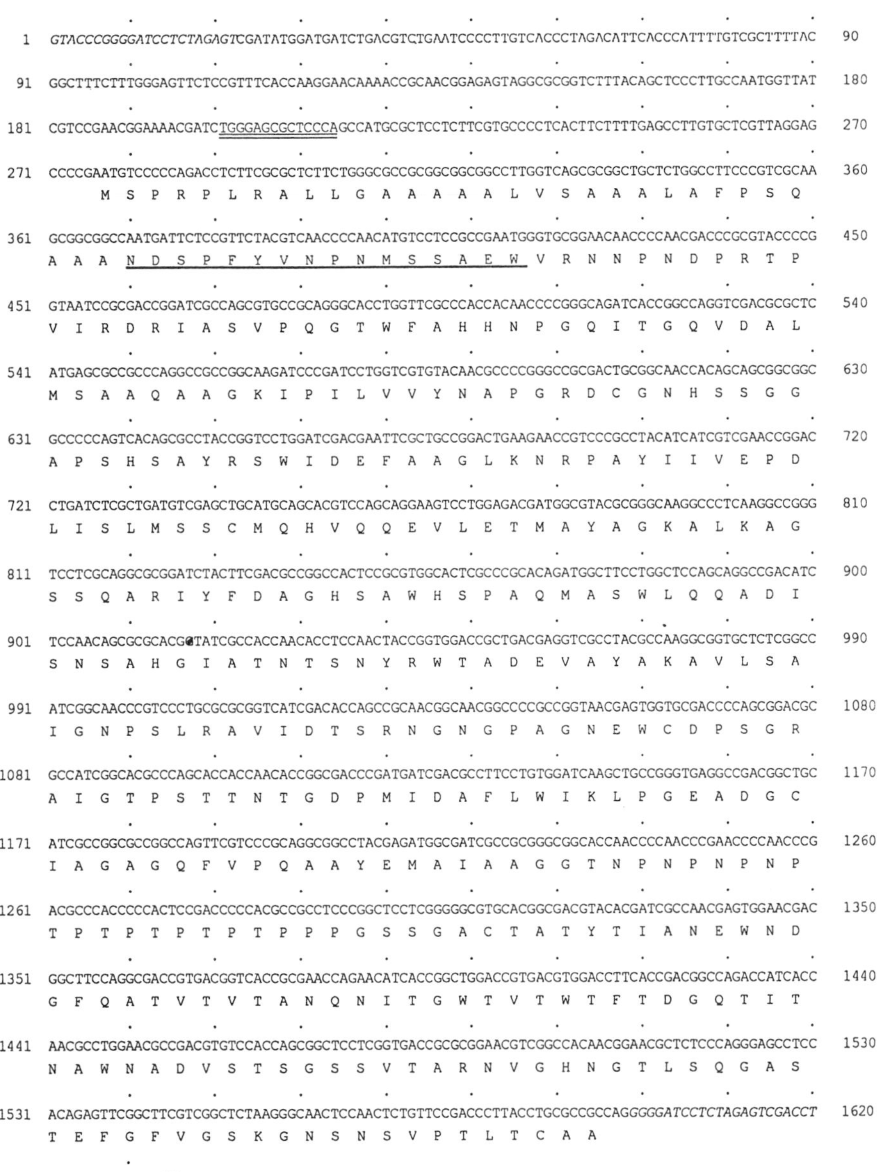

Figure 1. E2 DNA and protein sequence. A 14-base sequence, present in the regulatory binding site of the E2 gene of *T. fusca*, is double underlined. The N terminal sequence of the mature protein is underlined. The vector sequence is italicized. (Reproduced with permission from reference 4. Copyright 1991 American Society for Microbiology.)

unusual, have been seen for active site residues in other well-refined protein structures (*16*).

The molecular weight of the 286 residues is 30,410, which was very close to the 30k daltons measured by SDS gel electrophoresis. Figure 2 shows the E2cd model. E2cd is globular and slightly ellipsoidal with dimensions of 53 by 38 by 38Å. There is a pronounced cleft which extends across the molecule and is 10Å wide and 11Å deep at its deepest point (*17*). We know that the active site is in this cleft both because it is in the same position as the active site of CBHIIcd and because cellobiose binds into the cleft. This cleft contains a number of aromatic side chains which function in binding glucosyl units into the active site and several carboxyl side chains some of which appear to function in catalysis. Asp^{117} in E2 is in the identical location to Asp^{221} of CBHIIcd which was shown by site directed mutagenesis to be essential for catalysis (*14*). The other three Asp residues present in the active site of CBHIIcd near the proposed cleavage site of bound cellulose are all conserved in the E2 sequence. In E2cd, Asp^{156} and Asp^{265} are in the same positions as Asp^{263} and Asp^{401} in CBHIIcd while Asp^{79} is not in the same position as Asp^{175} in CBHIIcd. Since Asp^{156} is completely buried in E2cd it is not likely to function directly in catalysis. Asp^{79} is displaced from the catalytic center and is involved in a crystal contact so it is unlikely to move in the crystal or function directly in catalysis. This leaves Asp^{265} as the most likely residue to function as a general base in catalysis.

The E2 cleft is much deeper than the groove present in the active site of *Clostridium thermocellum* Cel D which is the only other endocellulase whose structure has been determined (*17*). Another difference between these two enzymes is that Cel D has binding sites for six glucose residues while E2 has sites for only four. E2cd is an atypical α/β barrel containing eight β-strands and eight major and two minor helices. Like CBHIIcd but unlike most α/β barrel proteins (*19,20*) the barrel closure has only one hydrogen bond which is between the amide of Gly^{39} and the carbonyl oxygen of Trp^{257}. Strand VIII is antiparallel to strand VII and forms a β bridge with two hydrogen bonds between Thr^{244} and Phe^{255}. Ser^{243} in strand VIII forms hydrogen bonds with Pro^{37} and Trp^{257} which also help to close the barrel. The active site of E2cd is formed by four loops present at the C-terminal end of the β sheets. A topology diagram of the chain fold is shown in Figure 3, and this structure is very similar to that of CBHIIcd especially in the β-strand regions, where there is an average rms deviation of 0.7Å between the two structures.

The major difference between CBHIIcd and E2 is that in CBHIIcd the active site is present in a tunnel rather than a cleft, providing a clear structural basis for the functional difference between the endocellulolytic activity of E2, which can cleave the cellulose chain at any site, and the exocellulolytic activity of CBHIIcd, which cleaves off cellobiose units from the nonreducing end of a cellulose chain. While much of the difference between the two active sites is due to a 20 amino acid loop that is present in CBHIIcd but not in E2cd, the rest of the difference is due to the fact that a loop which is present in both enzymes is folded back from the cleft in E2cd, while it occludes part of the cleft in CBHIIcd. This can be seen in Figure 4 which shows the peptide backbone of CBHIIcd in blue and that of E2cd in yellow. The E2cd loop which blocks the cleft in CBHIIcd is also the binding site for the bound sulfate and it is possible that in solution this loop is more flexible and could partially close the

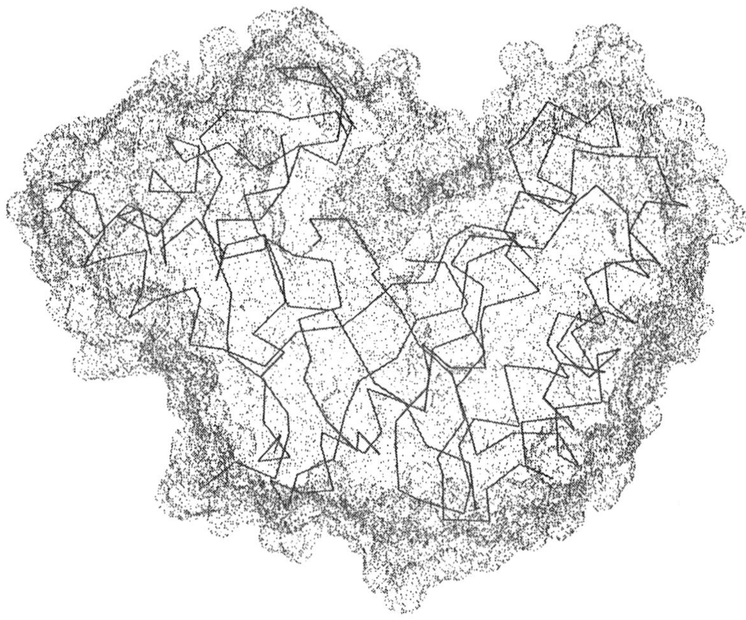

Figure 2. Model of E2cd showing the α-carbon backbone and the accessible surface area.

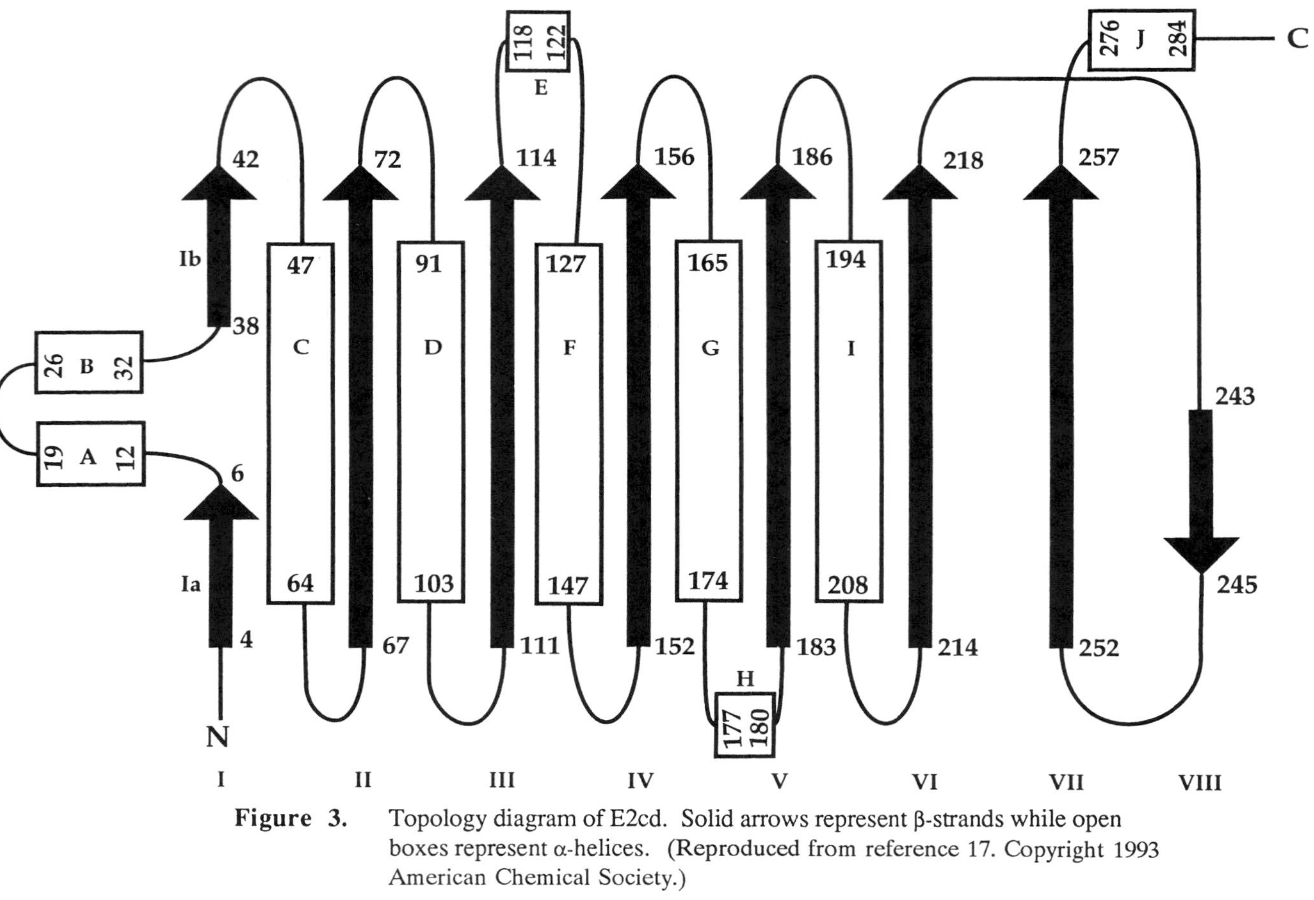

Figure 3. Topology diagram of E2cd. Solid arrows represent β-strands while open boxes represent α-helices. (Reproduced from reference 17. Copyright 1993 American Chemical Society.)

active site after substrate binding. At this time we have no direct evidence for such motion.

The structure of E2cd crystallized in the presence of 5mM cellobiose was determined and cellobiose was bound into the active site cleft in subsites A and B identified in the CBHIIcd structure (*14*). There was only ca. 50% occupancy by cellobiose in the co-crystal. E2cd did not bind glucose. The fact that E2 activity is not inhibited by glucose while it is competitively inhibited by cellobiose is consistent with these results. In contrast, CBHIIcd binds glucose into site A and cellobiose into sites C and D.

As has been seen with many sugar binding proteins (*21*) a Trp residue, Trp^{41}, is packed against the non-reducing glucosyl unit of cellobiose. Trp^{162} is also available to bind to a sugar residue in site D. Lys^{259} and Glu^{263} are hydrogen bonded to hydroxyl groups on the non-reducing end of cellobiose in site A. We have isolated a mutant of E2 in which Glu^{263} is changed to Gly (*12*) and find that the mutant enzyme has the same V_{max} and K_m CMC and swollen cellulose activity as wild type E2. The reducing end of the bound cellobiose contacts Ser^{189} through a bound water molecule and is also hydrogen bounded to Asp^{265} and Lys^{259}.

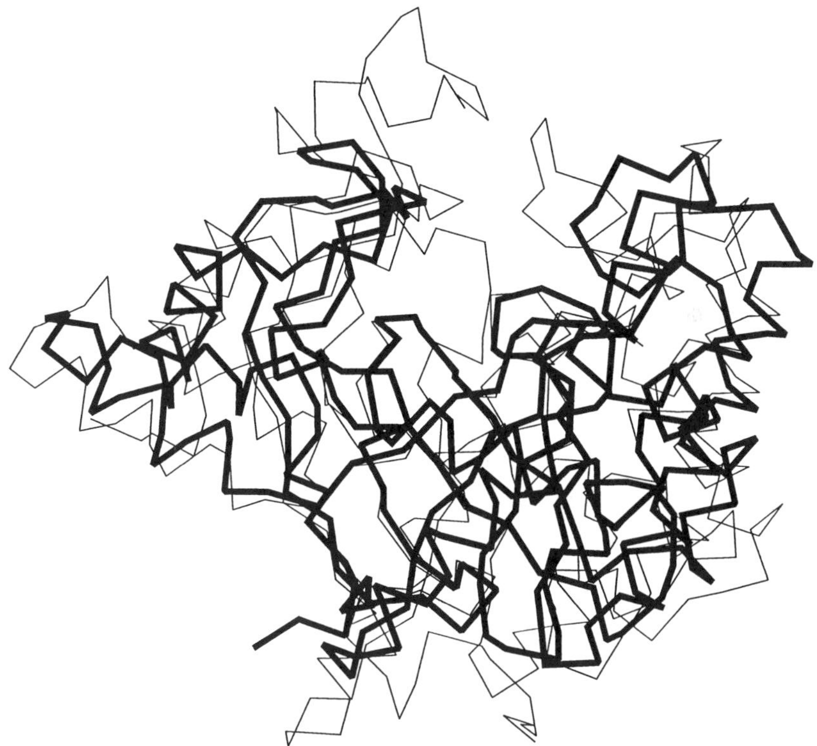

Figure 4. Overlay of the α-carbon backbones of E2cd and CBHII. E2cd is the heavy line, CBHII is the thin line. Note that the loop in CBHII at the top left of the cleft is not present in E2cd, while the loop on the right side of the cleft is present yet adopts a conformation which allows for a more open active site.

Literature Cited

1. Calza, R.E.; Irwin, D.C.; Wilson, D.B. *Biochemistry* **1985,** *24,* 7797-7804.
2. Wilson, D.B. *Crit. Rev. Biotechnol.* **1992,** *12,* 45-63.
3. Irwin, D.; Spezio, M.L.; Walker, L.; Wilson, D.B. **1993,** submitted.
4. Lao, G.; Ghangas, G.S.; Jung, E.D.; Wilson, D.B. *J. Bacteriol.* **1991,** *173,* 3397-3407.
5. Ghose, T.K. *Pure Appl. Chem.* **1987,** *59,* 257-268.
6. Gebler, J.; Gilkes, N.R.; Claeyssens, M.; Wilson, D.B.; Beguin, P.; Wakarchuk, W.W.; Kilburn, D.G.; Miller, R.C., Jr.; Warren, R.A.J.; Withers, S.G. *J. Biol. Chem.* **1992,** *267,* 12559-12561.
7. Ghangas, G.S.; Wilson, D.B. *Appl. Env. Microbiol.* **1988,** *54,* 2521-2526.
8. Henrissat, B.; Claeyssens, M.; Tomme, P.; Lemesle, L.; Mornon, J.-P. *Gene* **1989,** *81,* 83-95.
9. Wong, W.K.R.; Gerhard, B.; Guo, Z.M.; Kilburn, D.G.; Warren, R.A.J.; Miller, Jr., R.C. *Gene* **1986,** *44,* 315-324.
10. Hall, J.; Gilbert, H.J. *Mol. Gen. Genet.* **1988,** *213,* 112-117.
11. Maglione, G.; Matsushita, O.; Russell, J.B.; Wilson, D.B. *Appl. Env. Microbiol.* **1992,** *58,* 3593-3597.
12. McGinnis, K.; Kroupis, K.; Wilson, D.B. **1993**, submitted.
13. Walker, L.P.; Wilson, D.B.; Irwin, D.C.; McQuire, C.; Price M. *Biotech. and Bioenerg.* **1992,** *40,* 1019-1026.
14. Rouvinen, J.; Bergfors, T.; Teeri, T.; Knowles, J.K.C.; Jones, T.A. *Science* **1990,** *249,* 380-385.
15. Wang, B.-C. *Methods Enzymol.* **1985**, *115*, 90-117.
16. Karplus, P.A.; Schultz, G.E. *J. Mol. Biol.* **1987**, *195*, 701-725.
17. Spezio, M.; Wilson, D.B.; Karplus, P.A. *Biochem.* **1993,** In press.
18. Juy, M.; Amit, A.G.; Alzari, P.M.; Poljak, R.J.; Claeyssens, M.; Beguin, P.; Aubert, J.-P. *Nature* **1992,** *357,* 89-91.
19. Branden, C. *Current Opin. Struct. Biol.* **1991**, *1*, 978.
20. Farber, G.K.; Petsko, G.A. *Trends Biochem. Sci.* **1990**, *15*, 228.
21. Vyas, N.D. *Current Opin. Struct. Biol.* **1991**, *1*, 732-740.

RECEIVED May 9, 1994

Chapter 4

Role of Cellulose-Binding Domain of Cellobiohydrolase I in Cellulose Hydrolysis

T. R. Donner[1], B. R. Evans[2], K. A. Affholter[2], and J. Woodward[2,3]

[1]Department of Chemistry, University of Tulsa, Tulsa, OK 74104
[2]Chemical Technology Division, Oak Ridge National Laboratory, Oak Ridge, TN 37831–6194

The kinetics of the cellulose-binding domain's ability to adsorb onto microcrystalline cellulose (Avicel) and its effects on the surface structure of the cellulose fibers were studied. The catalytic domain of *Trichoderma reesei* cellobiohydrolase I was rendered inactive by modification with a water-soluble carbodiimide. After modification, the cellobiohydrolase I possessed no ability to hydrolyze the model substrate p-nitrophenyl-β-D-cellobioside (PNPC). However, the modified cellobiohydrolase I was still capable of adsorbing onto microcrystalline cellulose. Scanning electron microscopy showed that whereas native CBH I smoothed the surface of cotton fibers, catalytically inactivated CBH I was without effect.

The cellulase enzyme complex secreted by the fungus *Trichoderma reesei* catalyzes the hydrolysis of insoluble crystalline substrates to glucose (*1*). These complexes can be separated by various chromatographic procedures into three major types of catalytic activity: cellobiohydrolase, endoglucanase, and ß-glucosidase. Approximately 80% of this cellulase complex is composed of two cellobiohydrolases, termed CBH I and II, of which CBH I comprises 60% (*2*). Small-angle X-ray scattering studies have shown that both cellobiohydrolase I and II are shaped like a tadpole (*3,4*) consisting of a cellulose-binding domain, or "tail" region, and a catalytic domain, or "head" region. The exact mechanism by which these domains hydrolyze insoluble cellulose fibers, including their effect on the microscopic structure of the insoluble cellulose, is unknown. However, it has been shown recently that the cellulose-binding domain of a bacterial endoglucanase plays a direct role in the disruption of the Ramie cellulose fiber structure during hydrolysis (*5*). This apparent disruption of Ramie cellulose fibers by the cellulose-binding domain supports the conclusion that the cellulose-binding domain's purpose in the degradation of insoluble cellulose may be two-fold. Not only does the cellulose-binding domain secure the enzyme to the

[3]Corresponding author

0097–6156/94/0566–0075$08.00/0

surface of cellulose fibers, but its apparent ability to disrupt the surface fibers may render crystalline celluloses more susceptible to hydrolysis. The "tail" region of cellobiohydrolase I possibly possesses a "wedge-like structure" (*6*), which aids it in physically prying apart the surface fibers of crystalline cellulose.

It was postulated over 40 years ago by Elwyn Reese (*7*) that the C_1 factor of cellulases renders crystalline celluloses susceptible to hydrolysis. These recent data support the conclusion that the cellulose-binding domain could be the C_1 factor of cellulases. This study was performed to determine the physiological function of the cellulose-binding domain of cellobiohydrolase I from *T. reesei.*

Materials and Methods

Purification and Assay of Cellobiohydrolase I. The crude *T. reesei* cellulase preparation was obtained as a gift from Novo Nordisk Bioindustrials. CBH I was purified from this preparation by chromatofocusing as described previously (*8*). Fractions containing CBH I were further purified using a PD-10 column with Sephadex equilibrated with 50 mM ammonium carbonate pH 5.0. The CBH I was then lyophilized. It was assayed as described previously using p-nitrophenyl-β-D-cellobioside (PNPC) as the substrate (*9*).

Modification of Cellobiohydrolase I. Approximately 5.0 mg of CBH I was dissolved in 3.0 mL of nanopure water, pH 5.0. One-ethyl-3-(3-dimethylaminopropyl)-carbodiimide (EDC) was added to the solution to a final concentration of 0.1 M. The solution was then incubated at room temperature (25°C) for 48 h, and the pH was maintained at 5.0 by titrating with 0.5 M HCl. At intervals of time, the catalytic activity of the modified enzyme was tested using the model substrate PNPC. Quenching of the reaction occurred during the assay, which was conducted in 50 mM sodium acetate buffer, pH 5.0. A 25-μL of the reaction mixture containing approximately 40 μg protein was used in the assay mixture (0.5 mL total volume) for the determination of residual CBH I activity.

Scanning Electron Microscopy (SEM). In separate test tubes, two batches of cotton fibers (1.0 mg) were incubated at 40°C in either 0.25 mL modified CBH I (2.38 mg/mL) or 0.06 mL native CBH I (10.56 mg/mL) for 72 h. Fifty mM sodium acetate, pH 5.0, was added to each reaction to a total reaction volume of 1.0 mL. Flat pieces of both native and modified enzyme-treated cotton fibers were mounted on separate aluminum stubs and coated at a 15°-angle with Pt-C. The samples were examined with an ETAC autoscan scanning electron microscope operated at 20 kV. Samples were observed at magnifications of 600 and 1200.

Analytical Techniques. The protein concentration was determined using the Coomassie Blue reagent (Bio-Rad Laboratory) according to the method of Bradford using almond β-glucosidase as the standard. Catalytic activity of CBH I after incubation with EDC was determined at 55°C and pH 5.0 by measuring the hydrolysis of 20 mM PNPC at pH 5.0. The fluorescence spectra of both the native and modified enzyme were recorded using a Perkin-Elmer LS-5 spectrofluorimeter. The samples were excited at 280 nm, and the fluorescence emitted was measured between 310 nm and 480 nm. Both the modified and native enzyme were tested for their ability to adsorb onto microcrystalline cellulose (Avicel PH-105, 10 mg/mL). For details, see reference 10. The absorbance at 280 nm was recorded for each sample before the addition of Avicel. After the incubation of CBH I with Avicel, the reactions were spun down in a centrifuge, and the supernatant was filtered through a 0.2-µm Acrodisc. Binding of the enzyme to Avicel was determined by measuring the absorbance of the supernatant at 280 nm and extrapolating the bound protein from that of the remaining free protein.

Results and Discussion

Inactivation of CBH I_{native}. The purified native CBH I was reacted with EDC for approximately 48 h. Throughout the reaction, the pH was adjusted with 0.5 M HCl to keep the optimal pH for the reaction at 5.0. Complete inactivation of the enzyme was observed after 48 h, although >95% of activity was lost after 3 h (Figure 1). It should be noted that the modified enzyme possessed no catalytic activity at both pH 5.0 and 7.5. After modification, CBH I also lacked the ability to degrade barley β-glucan as determined by viscosity measurement (*10*), as well as PNPC. The reason for the inactivation of CBH I by EDC is apparently due to the modification of an essential carboxyl group within the catalytic site (*11*). The number of carboxyl groups in CBH I modified by reaction with EDC was not determined.

Characterization of CBH $I_{modified}$ and CBH I_{native}. Isoelectric focusing of the modified CBH I displayed a dramatic increase in its isoelectric point when compared to accepted values for the pI of the native enzyme (data not shown). This apparent alkalinity may be attributed to the conversion of the carboxyl groups of the native enzyme to the N-acylurea derivatives (*12*). Under mild, rather than denaturing, conditions only the most reactive carboxyl group sites will be modified. The measured isoelectric point of the modified CBH I was found to be between pH 8.5 and 9.3, while that of the native CBH I remained consistent, with accepted values between pH 3.6 and 4.0.

The fluorescence spectra of both the modified and native CBH I, excited at 280 nm, gave peaks of maximum emission at 340 nm. However, the peak at 340 nm of the modified CBH I was dramatically lower than that of native CBH I (Figure 2). This change in fluorescence may be due to a conformational change of the enzyme during the modification with EDC.

SDS-PAGE analysis was used to determine the molecular weights of both modified and native CBH I. It was found that each enzyme had a molecular weight of approximately 67,000 g/mol, which was in basic agreement with the accepted value for *T. reesei* CBH I.

Adsorption Isotherms. The binding of native and modified CBH I to Avicel as a function of protein concentration is shown in Figure 3. It is obvious that although the modified enzyme does bind to Avicel, it does so to a considerably lesser extent than native CBH I. The reason for this is unknown but may be related to the difference

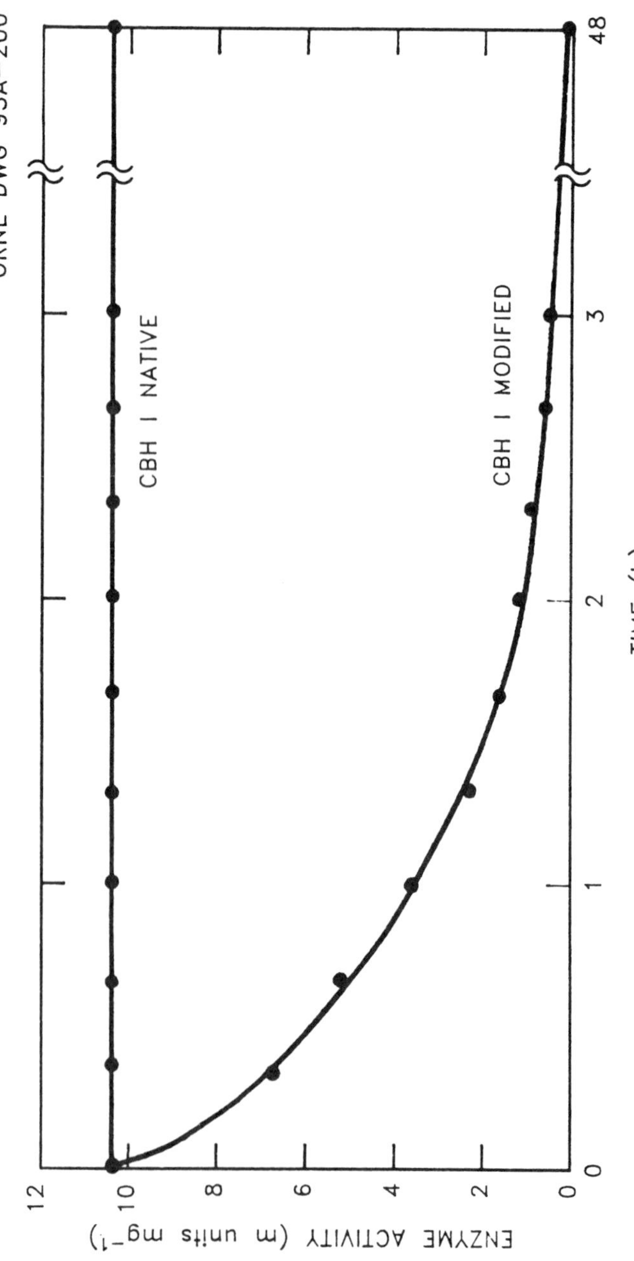

Figure 1. The effect of chemical modification of *T. reesei* CBH I with EDC on its ability to hydrolyze PNPC.

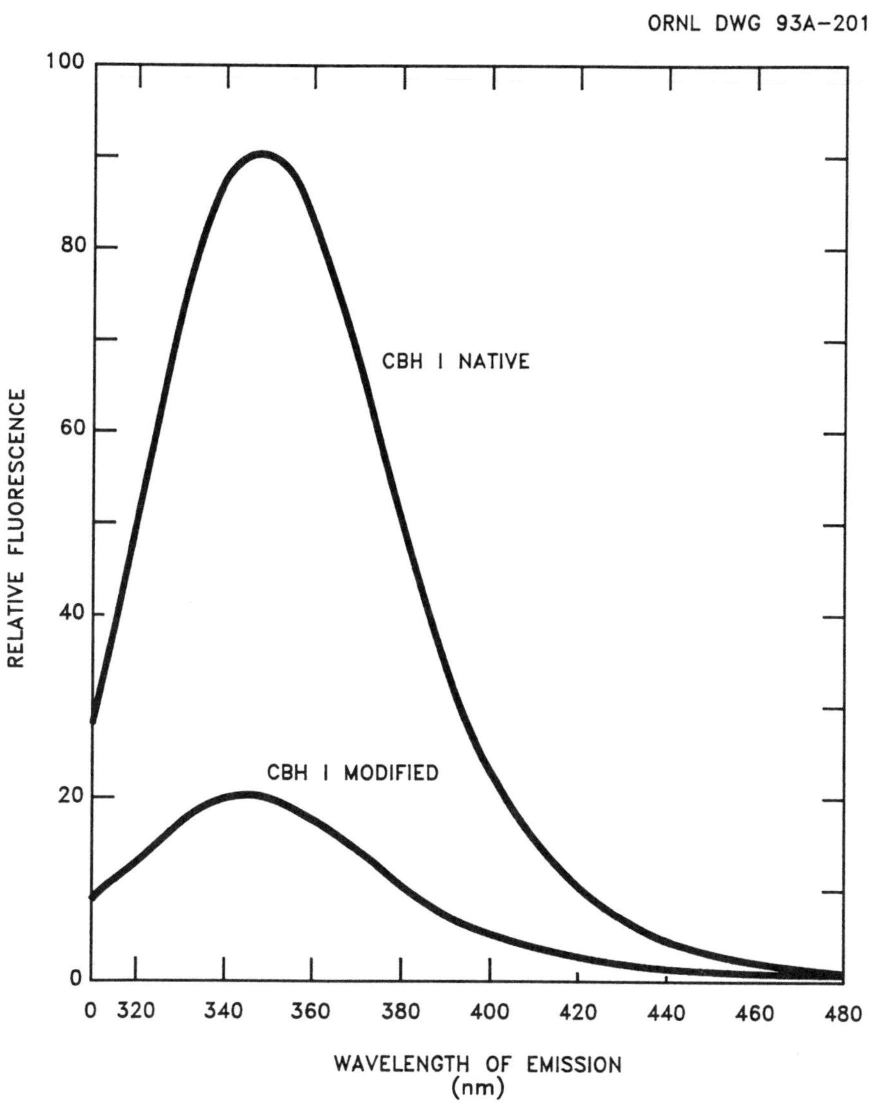

Figure 2. Fluorescence spectra of native and chemically modified CBH I. Protein concentration 100 µg/mL.

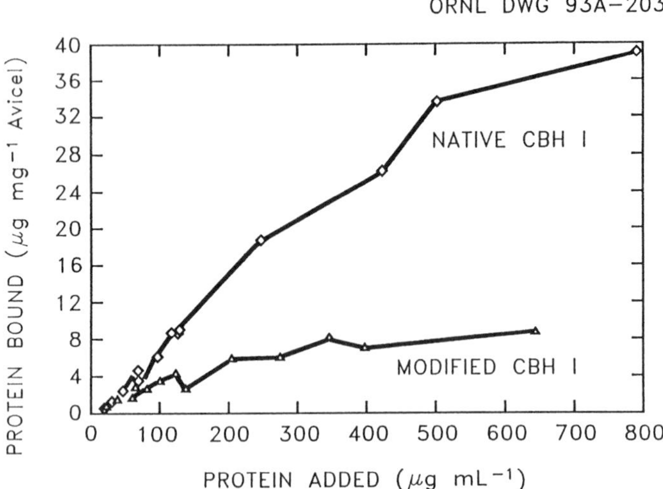

Figure 3. The adsorption of native and modified CBH I to Avicel as a function of protein concentration.

in conformation that exists between the native and modified enzymes. Another explanation could be the possible reaction between EDC and tyrosine residues in the cellulose-binding domain of CBH I (*12*). Tyrosines are important for binding of CBH I to cellulose (13). The concentration of adsorption sites on Avicel for CBH I was reported to be 70 μg/mg Avicel (14). The data shown here for native CBH I indicate that saturation of Avicel had not occurred.

SEM Observations. Scanning electron micrographs of cotton linters that were untreated and treated with native or modified CBH I are shown in Figures 4 and 5. The interpretation of these observations is that untreated cotton linters possess macrofibrils protruding from the surface of the fiber which are removed by the catalytic action of native CBH I. This results in the smoothing appearance of the fiber surface and is similar to the observations made by Din *et al.* (*5*) when the Ramie cellulose fibers were treated with the catalytic domain of an endoglucanase from *Cellulomonas fimi*. No such action takes place resulting from the incubation of modified CBH I on the cotton linters. There is also no evidence for any disruption of the fiber surface by either the native or modified enzyme.

Conclusions

The chemical modification of *T. reesei* CBH I by EDC results in a loss in its catalytic activity. The enzyme thus modified had no obvious effect on the appearance of cotton linters. The conclusion that the cellulose-binding domain of CBH I has no physical effect on the surface of cotton cellulosic fibers is tempered by the finding that modification also resulted in a change in the conformation of the enzyme and also in the reduced binding of the enzyme to cellulose.

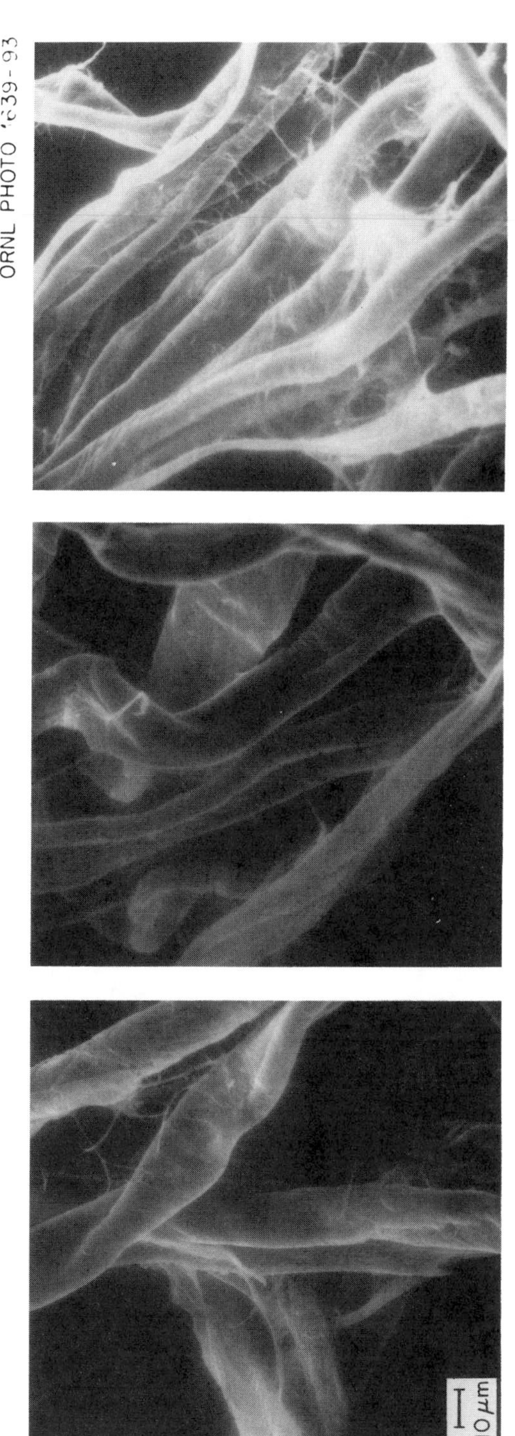

Figure 4. Scanning electron micrographs of cotton linters untreated and treated with native and modified CBH I. Magnification x600.

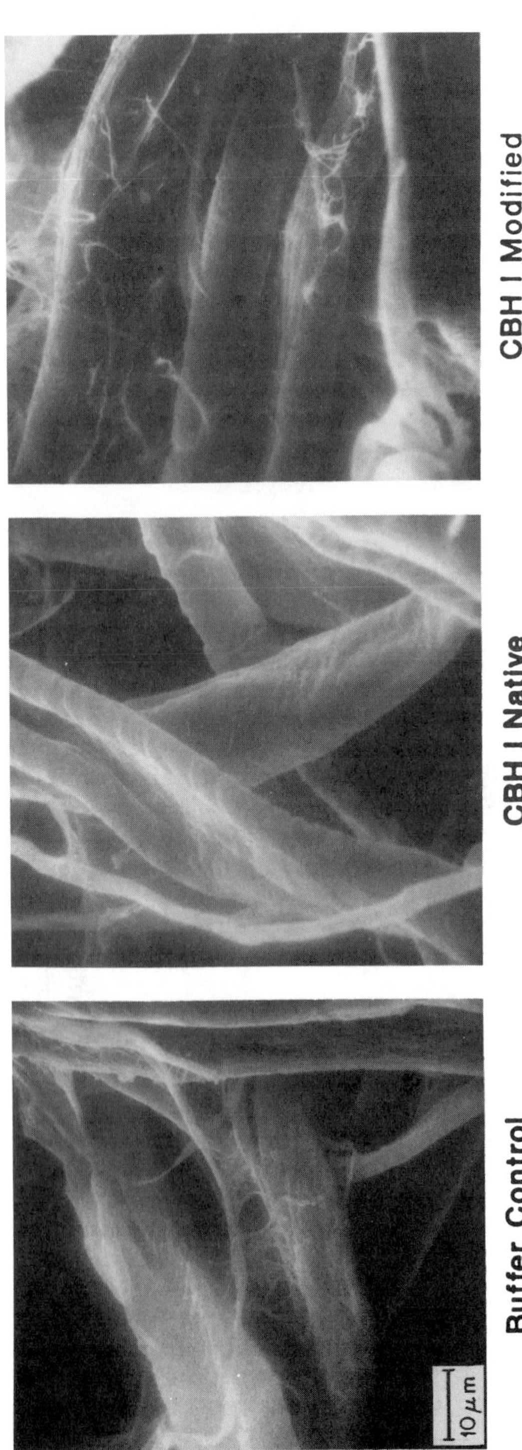

Figure 5. As for Figure 4 except magnification x1200.

Acknowledgments

The authors thank Donna Reichle for editorial assistance and Debbie Weaver for secretarial assistance. This work was supported by the Chemical Sciences Division, Office of Basic Energy Sciences, U.S. Department of Energy, under contract DE-AC05-84OR21400 with Martin Marietta Energy Systems, Inc., and the National Renewable Energy Laboratory.

Literature Cited

1. Wood, T.M. *Biochem. Soc. Trans.* **1992**, *20*, 46-53.
2. Goyal, A.; Ghosh, B.; Eveleigh, D. *Bioresource Technol.* **1991**, *36*, 37-50.
3. Schmuck, M.; Pilz, I.; Hayn, M.; Esterbauer, H. *Biotechnol. Lett.* **1986**, *8*, 397-402.
4. Abuja, P.M.; Pilz, I.; Claeyssens, M.; Tomme, P. *Biochem. Biophys. Res. Comm.* **1988**, *156*, 180-185.
5. Din, N.; Gilkes, N.R.; Tekant, B.; Miller, R.C. Jr.; Warren, A.J.; Kilburn, D.G. *BioTechnology* **1991**, *9*, 1096-1099.
6. Kraulis, P.J.; Clore, G.M.; Nilges, M.; Jones, T.A.; Peterson, G.; Knowles, J.; Gronenborn, A.M. *Biochemistry* **1989**, *28*, 7241-7257.
7. Reese, E.T.; Siu, R.G.H.; Levinson, H.S. *J. Bacteriol.* **1950**, *59*, 485-497.
8. Offord, D.A.; Lee, N.E.; Woodward, J. *Appl. Biochem. Biotechnol.* **1991**, *28/29*, 377-386.
9. Woodward, J.; Lee, N.E.; Carmichael, J.S.; McNair, S.L.; Wichert, J.M. *Biochim. Biophys. Acta* **1990**, *1037*, 81-85.
10. Woodward, J.; Affholter, K.A.; Noles, K.K.; Troy, N.T.; Gaslightwala, S.F. *Enzyme Microb. Technol.* **1992**, *14*, 625-630.
11. Tomme, P.; Claeyssens, M. *FEBS Lett.* **1989**, *243*, 239-243.
12. Means, G.E.; Feeney, R.E. In *Chemical Modification of Proteins;* Holden-Day: San Francisco, CA, 1971.
13. Teeri, T.T.; Reinikainen, T.; Ruohonen, L.; Jones, T.A.; Knowles, J.K.C. *J. Biotechnol.* **1992**, *24*, 169-176.
14. Tomme, P.; Heriban, V.; Claeyssens, M. *Biotechnol. Lett.* **1990**, *12*, 525-530.

RECEIVED March 1, 1994

Chapter 5

CelS: A Major Exoglucanase Component of *Clostridium thermocellum* Cellulosome

Kristiina Kruus, William K. Wang, Pei-Ching Chiu, Joting Ching, Tzuu-Yi Wang, and J. H. David Wu

Department of Chemical Engineering, University of Rochester, Rochester, NY 14627–0166

The CelS protein (M_r = 82,000) is the most abundant protein species secreted by *Clostridium thermocellum.* It has been identified as a key catalytic component of the *C. thermocellum* cellulosome, an extremely complicated and large protein aggregate responsible for the degradation of crystalline cellulose. We have proposed that CelS acts synergistically with CelL (M_r = 250,000), another major cellulosome component, to degrade crystalline cellulose through a novel enzyme-anchor mechanism distinguishable from that of the fungal cellulase system. However, the function and the properties of this key catalytic component remain to be elucidated. We have recently cloned the *celS* gene. Its DNA sequence cannot be classified into any of the known cellulase families. In fact, the *celS* sequence has led to the discovery of a new cellulase family. Two previously un-characterized gene fragments from other bacteria have been found to contain sequences highly homologous to the *celS* sequence. Furthermore, the N-terminal amino acid sequence of a novel exo-β-glucanase (Avicelase II) purified from *C. stercorarium* matches that of CelS. The members of this new cellulase family are so far found only in strict anaerobic bacteria. CelS may therefore be one of the key components distinguishing the anaerobic cellulase system from its fungal counterpart. Expression of the *celS* gene in *Escherichia coli* resulted in the formation of inclusion bodies. The solubilized and refolded gene product displayed activity typical of an exo-β-glucanase. Thermostability was enhanced by Ca^{++}, as for the crude enzyme. The *celS* gene therefore encodes a novel exoglucanase.

Clostridium thermocellum, an anaerobic, thermophilic, cellulolytic, and ethanogenic bacterium, produces an extremely complicated extracellular cellulase complex called a cellulosome (*1*). The cellulosome is formed by more than fourteen cellulase

0097–6156/94/0566–0084$08.00/0

components with a total molecular weight of millions (*2-4*), and accounts for most of the Avicelase activity (activity against Avicel, a microcrystalline cellulose preparation) in the culture filtrate of this bacterium (*5*). Many enzymatic properties displayed by the cellulosome are significantly different from those of the better-studied fungal cellulase system; In particular, the clostridial cellulase appears to have a higher specific activity against crystalline cellulose (*6*).

Aggregation of cellulase components has been found in other cellulolytic bacteria (*7,8*). With their unique multi-component structure and properties, these bacterial cellulase systems likely belong to a different and less well-studied class of cellulase systems. Difficulty in purifying individual cellulosome components has posed serious technical challenges and impeded progress in elucidating the mechanism and structure of the cellulosome. Molecular cloning approaches have led to the discovery of many endoglucanase genes, but have given no clue to the mechanism of the cellulosome.

Initial insight into this complicated enzyme complex has been obtained by the identification of the two major cellulosome components, CelL (S_L or CipA) and CelS (or S_S), which degrade crystalline cellulose synergistically (*5*). An "anchor-enzyme model" has been proposed to explain the synergism between these two subunits (*9*), in which CelL functions as an anchor on the cellulose surface for CelS, the catalytic subunit. This mechanism is clearly different from the well-accepted, exo-endo synergism model based on the fungal cellulase system. Recent progress in cloning and analyzing the DNA sequences of both *celL* (*10*) and *celS* (*11,12*) genes has confirmed the enzyme-anchor model and expanded it into a more sophisticated model (*13*). This expanded enzyme-anchor model involves specific receptor-ligand interactions as the mechanism of the complex formation between the catalytic and the anchorage subunits. The enzyme-anchor model therefore provides a structural basis for the unique properties of the enzyme. Further characterization of CelS and CelL, which could now be made available in the recombinant form, will eventually unveil the detailed and sophisticated mechanism by which the cellulosome degrades cellulose.

In this article, recent progress in cloning and expressing the *celS* gene in *E. coli* and the initial characterization of the gene product will be briefly presented.

CelS Belongs to a New Cellulase Family

The Uniqueness of the *celS* DNA Sequence. Due to the complexity and the stability of the *C. thermocellum* cellulosome, purification of individual components has proven to be extremely difficult. Molecular cloning and expression of individual genes in a heterologous host appears to be the only feasible approach for elucidating the functions of cellulosome components and the possible synergism between them. Using the carboxymethylcellulose (CMC) and Congo-Red plate screening method (*14*), many cellulase genes have been cloned. It is not surprising that all the genes obtained are endo-enzymes since the screening method is specific for endoglucanase activity. However, it is somewhat puzzling that the bacterium harbors so many *cel* genes with seemingly similar function.

Many of the DNA sequences of the cloned *C. thermocellum cel* genes have been determined. These genes fall into known cellulase families on the basis of primary structure homology (Table 1; *15*). Most of the genes (*celB*, *celC*, *celE*, *celG*, and *celH*) belong to Family A. There are two genes (*celD* and *celF*) in Family E and one each in Family D (*celA*) and Family F (*xynZ*), respectively. They therefore share homology with themselves and with cellulases of other microorganisms including *Trichoderma reesei*, *Bacillus sp.*, *Schizophyllum commune*, and *Erwinia chrysanthemi*, (Family A); *Cellulomonas uda* (Family D); *Pseudomonas fluorescens* (Family E); and *Cellulomonas fimi* and *Cryptococcus* (Family F).

In contrast to the endoglucanase genes described above, the *celS* gene was cloned by hybridization to a degenerate oligonucleotide probe derived from its N-terminal amino acid sequence. The DNA sequence of *celS* (*11*) does not have significant homology with other *cel* genes cloned from *C. thermocellum* or other microorganisms except the conserved, duplicated sequence at its C-terminus (discussed later). It therefore does not belong to any known cellulase family (Families A - F; *15*) reported so far. It is clear that the probe hybridization method has yielded a novel cellulase gene (Table 1).

Table 1. Cloned *C. thermocellum* Endoglucanase and Xylanase Genes

Gene Product	Size (b.p.)	Predicted M.W. (Da)	Observed M.W. (Da)	Cellulase[1] Family	Reference
CelA	1,344	52,503	56,000	D	(*16,17*)
CelB	1,689	63,857	66,000	A	(*18,19*)
CelC	1,032	40,439	38,000	A	(*20,21*)
CelD	1,947	72,344	65,000	E	(*22,23*)
CelE	2,442	90,211	-	A	(*24*)
CelF	2,219	87,409	-	E	(*25*)
CelG	1,698	63,128	66,000	A	(*26*)
CelH	2,702	102,301	-	A	(*27*)
XynZ	2,511	92,159	90,000	F	(*28*)
CelS	2,241	80,670	82,000	NEW	(*11,12*)

[1]Henrissat et al., (*15*).

A New Cellulase Family. Since CelS has been shown to play a critical role in the degradation of crystalline cellulose, it would be surprising if the *celS* gene is only found in *C. thermocellum* and not in other cellulolytic microorganisms. The *celS* and its related genes may have escaped cloning because the CMCase activity screening method is commonly used in cloning cellulase genes. Indeed, sequences highly homologous to the *celS* sequence have been identified from at least two other bacterial genes (Figure 1), the ORF1 of *C. cellulolyticum* (*29*) and the *celA* of *Caldocellum saccharolyticum* (*30*). These two gene fragments are each located next to an endoglucanase gene and were therefore cloned during the cloning of the endoglucanase gene. The complete open reading frames of both homologous

```
CelS 386 EFYQWLQSAEGGIAGGATNSWNGRYEKYPAGTSTFYGMAYVPHPVYADPG 435
A      1 EFYQWLQSAEGAIAGGATNSWNGRYEAVPSGTSTFYGMGYVENPVYADPG  50
B      1                                                 G   1

CelS 436 SNQWFGFQAWSMQRVMEYYLETGDSSVKNLIKKWVDWVMSEIKLYDDGTF 485
A     51 SNTWFGMQVWSMQRVAELYYKTGDARAKKLLDKWAKWINGEIKFNADGTF 100
B      2 SNTWFGFQAWSMQRVAEYYYVTGDKDAGALLEKWVSWIKSVVKLNSDGTF  51
C     50                                          IKFNGDGTF  58

CelS 486 AIPSDL-EWSGQPDTW--TGTYTGNPNLHVRVTSYGTDLGVAGSLANALA 532
A    101 QIPSTI-DWEGQPDTWNPTQGYTGNANLHVKVVNYGTDLGCASSLANTLT 149
B     52 AIPSTL-DWSGQPDTW--NGTYTGNPNLHVKVVDYGTDLGITASLANALL  99
C     59 GNYMTGFDDFGQINVW--NIS                               77

CelS 532 TYAAATERWEGKLDTKARDMAAELVNRAWYNFYCSEGKGVVTEEARADYK 582
A    150 YYAAKSG------DETSRQNAQKLLDAMWNNY--SDSKGISTVEQRGDYH 191
B    100 YYSAGTKKY-GVFDEEAKNLAKELLDRMWK--LYRDEKGLSAPEKRADYK 147

CelS 583 RFFEQEVYYVPAGWSGTMPNGDKIQPGIKFIDIRTKYRQDPYYDIVYQAYL 632
A    192 RFLDQEVFVPAGWTGKMPNGDVIKSGVKFIDIRSKYKQDPEWQTMVAALQ 241
B    148 RFFEQEVYIPAGWTGKMPNGDVIKSGVKFIDIRSKYKQDPDWPKLEAAYK 197

CelS 633 RGEAPVLNYHRFWHEVDLAVAMGVLATYFPDMTYKVPGTPSTKLYGDVND 682
A    242 AGQVPTQRLHRFWAQSEFAVANGVYAILFPDQG-------PEKLLGDVNG 284
B    198 SGQVPEFRYHRFWAQCDIAIVNATYEILFGNQ COOH              229

CelS 683 DGKVNSTDAVALKRYVLRSGISINTDNADLNEDGRVNSTDLGILKRYILK 732
A    285 DETVDAIDLAILKKYLLNSSTTINTANADMNSDNAIDAIDYALLKKALLS 334

CelS 733 EIDTLPYKN COOH                                      741
A    335 IQ COOH                                             336
```

Figure 1. Amino acid sequence comparison between CelS, the partial ORF (ORF1) preceding the *celCCC* gene of *C. cellulolyticum* (A), the partial ORF (also ORF1 or *celA*) preceding the *manA* gene of *Caldocellum saccharolyticum* (B), and the 28 amino acid cartridge in an endoglucanase of *Bacteroides ruminicola* (C). Boxed amino acids are identical or have similar chemical properties. Numbers indicate the position, within the sequence of each protein, of the first or last amino acid shown on a line. For (A) and (B), the numbers start with the first amino acid residues reported. For (C), the numbering refers to the second encoding ORF. Similar residues were: V,L,I,M,F; R,K; D,E; N,Q; Y,F,W; S,T; G,A. (Taken from reference *11*)

sequences have not been published. There were no clues to the functions of these two homologous open reading frames when they were originally cloned and sequenced. However, based on the degree of homology with *celS*, these two open-reading-frames most likely code for cellulolytic enzymes. The structure of the *celS* gene has thus led to the discovery of a new cellulase family.

It is interesting to note that, although *C. thermocellum* is a thermophilic bacterium with the optimum growth temperature of 60°C, *C. cellulolyticum* is mesophilic. On the other hand, *C. saccharolyticum* has an optimum growth temperature of 68 - 70°C. The *celS* gene family therefore appears to have evolved to function in at least three different temperature ranges. Furthermore, the *celA* gene of *C. saccharolyticum* appears to codes for a "hybrid protein" with its C-terminal half resembling that of CelS and its N-terminal half homologous to an endoglucanase gene (*cenB* of *C. fimi*; *31*). The *celA* is therefore likely a product of "domain shuffling" involving part of the CelS and an endoglucanase. Additionally, unlike CelS and ORF1 of *C. cellulolyticum*, the *celA* gene of *C. saccharolyticum* does not encode the C-terminal conserved, duplicated sequence. As will be discussed below, this conserved, duplicated sequence is the binding ligand to CelL. This indicates that CelS family may function either alone (as in the case of CelA) or by forming a multi-subunit complex with other cellulase components (as in the cases of CelS and ORF1). Finally, all three bacteria are strict anaerobes, suggesting that CelS may be unique to the anaerobic cellulase system.

The Conserved Binding Ligand Sequence. The genetic evidence that CelS forms a protein complex with CelL, which is essential if the enzyme-anchor model is to be considered valid, is that CelS contains a conserved, duplicated sequence at its C-terminus. It has been shown (*32*) that this conserved, duplicated sequence serves as a binding ligand to the receptor sites of the CelL protein as depicted in the modified enzyme-anchor model (*13*). The list of the genes containing this ligand sequence continues to expand. A total of eighteen genes are compiled in Figure 2. Twelve of them are from *C. thermocellum*, five from *C. cellulolyticum*, and one from *C. cellulovorans*. This suggests that the cellulolytic enzymes of these clostridia are organized as described by the modified enzyme-anchor model. These cellulase systems therefore likely represent a special class of cellulase adopting the enzyme-anchor mechanism which clearly distinguishes them from the fungal system. A CelL-like protein (CbpA) has been discovered in *C. cellulovorans* (*33*). A CelS-like protein (ORF 1) has been found in *C. cellulolyticum* (*29*). It will not be surprising if both the CelL and CelS equivalents are discovered later in both bacteria.

Cloning and Expression of *celS* in *E. coli*

PCR Cloning and Expression of *celS*. To obtain the intact *celS* gene, as opposed to the four overlapping clones used earlier to determine the sequence (11), we used the Polymerase Chain Reaction (PCR) technique in which two sequences flanking the *celS* gene were used as the PCR primers and the *C. thermocellum* genomic DNA as the PCR template. The PCR-amplified gene product was cloned into the *E. coli* expression vector pRSET-B for expression (Figure 3). The resulting DNA construct was sequenced to ensure that fusion of the *celS* structural gene to the expression

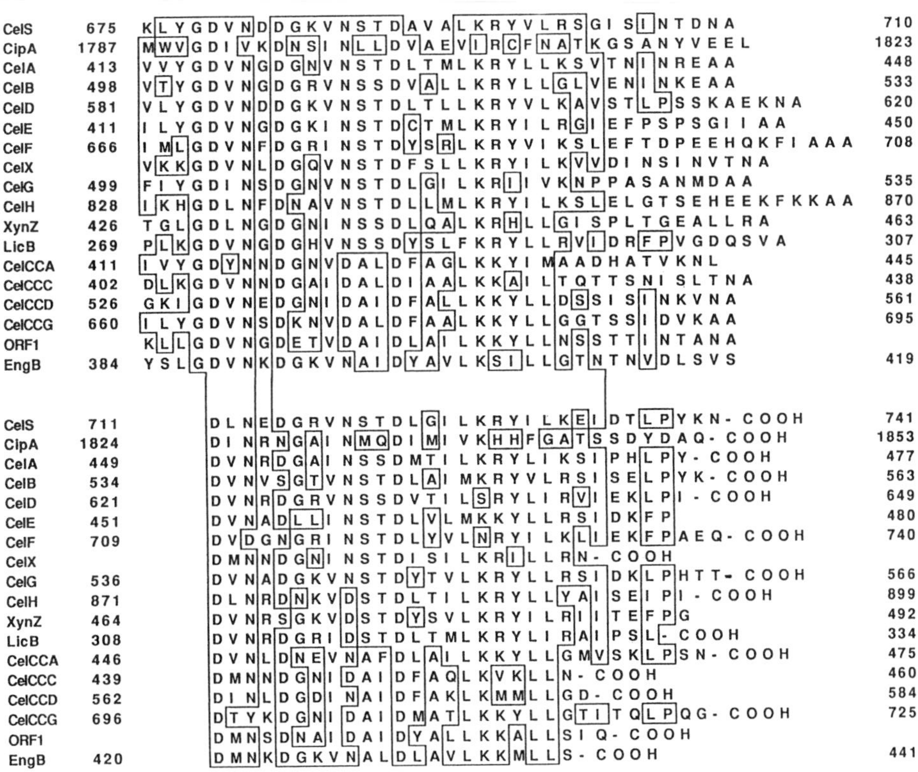

Figure 2. The alignment of the conserved duplicated sequence between CelS, CipA, CelA, CelB, CelD, CelE, CelF, CelG, CelX, CelH, XynZ, and LicB (*38*) of *C. thermocellum*, CelCCA, CelCCC, CelCCD, CelCCG, and ORF1 of *C. cellulolyticum* (*29, 39, 40*), and EngB of *C. cellulovorans* (*41*). Boxed amino acids are identical or have similar chemical properties. Numbers indicate the position, within the sequence of each protein, of the first or the last amino acid shown on a line. Similar residues are: V, L, I, M, F; R, K; D, E; N, Q; Y, F, W; S, T. CipA is also named as CelL or S_L.

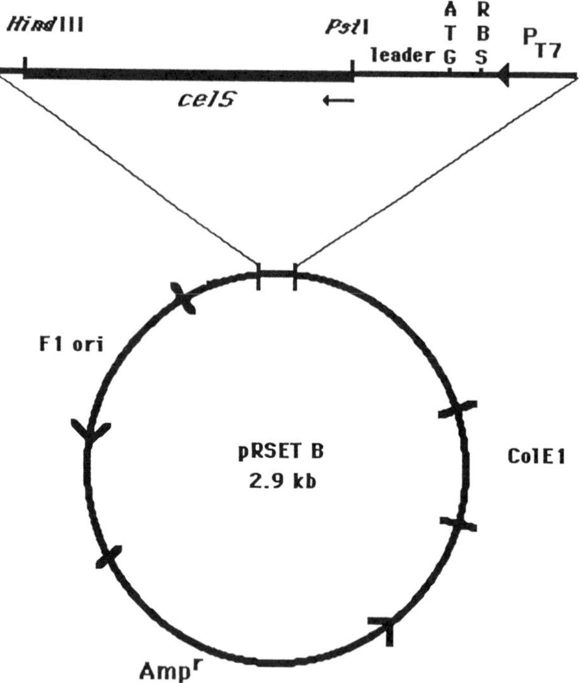

Figure 3. Map of the *celS*-pRSET B clone. The 2.1 kb *Pst*I-*Hind*III *celS* PCR fragment was cloned into the 2.9 kb pRSET B expression vector. The cloned gene was inserted in-frame with the initiation ATG codon. Expression was controlled with the promotor recognized by T7 RNA polymerase. The resulting gene product contained a 4 kDa N-terminal fusion encoded by the leader sequence. The plasmid was selected by growth on media containing the antibiotic ampicillin. The figure is not drawn to scale.

vector is in frame (data not shown). An additional protein species (M_r = 86,000) was produced in the *E. coli* strain harboring the insert gene (Lanes 3 and 4, Figure 4). The expression product was slightly larger than native CelS (Lane 1, Figure 4) due to the fused sequence from the expression vector. The identity of the gene product was verified by Western blotting, using a polyclonal antibody preparation specific to CelS. The antibody specifically recognized the rCelS (data not shown), indicating its authenticity.

Phase-contrast microscopy revealed that the rCelS protein was produced in the form of inclusion bodies in the host cells. Under an electron microscope, the inclusion body appeared as an amorphous structure inside the *E. coli* cell (Figure 5). The inclusion bodies existed even when the conserved, duplicated sequence at the C-terminus was deleted (data not shown).

The Cellulase Activity of rCelS. To obtain an active protein, the inclusion bodies were isolated by centrifugation and solubilized in 5M urea, which was later removed by dialysis. This resulted in a protein preparation containing predominantly rCelS (Lane 2, Figure 4). The preparation was subsequently used for characterizing the enzyme activity of CelS.

Activity on Various Cellulose Substrates. The rCelS produced little reducing sugar when incubated with carboxymethylcellulose (CMC; Table 2). However, when Avicel was used as the substrate, almost twice as much reducing sugar was released (Table 2). A more dramatic increase in reducing sugar production was observed on amorphous cellulose (phosphoric acid-swollen Avicel). These results are typical of an exoglucanase, indicating that CelS is not an endoglucanase (commonly referred to as CMCase) as are many other *cel* gene products identified from *C. thermocellum* (Table 1). Its lack of a significant activity on CMC explains why the *celS* gene could not have been cloned by an CMC-Congo red screening method.

Table 2. Activity of rCelS on Various Cellulosic Substrates

	Reducing Sugar Released[1] Glucose Equivalent (µg/ml)		
Incubation Time (hours)	0	15	24
CMC	0	6.2	6.8
Avicel	0	9.5	13
Amorphous Cellulose	0	39	48

[1]Incubation was carried out at 60°C in a succinate buffer (pH 5.7) which contained 12 mM Ca^{++} and 0.5% (w/v) substrate.

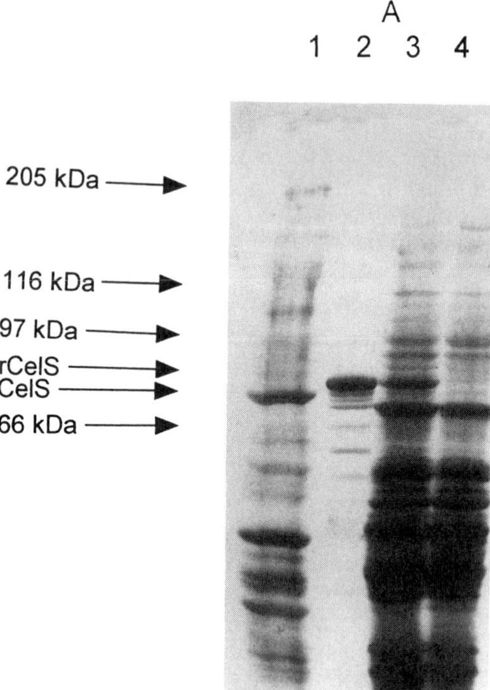

Figure 4. SDS-PAGE analysis of the *celS* expression in *E. coli*. Lane 1, the crude *C. thermocellum* culture filtrate; Lane 2, the partially purified rCelS; Lane 3, the lysate of *E. coli* strain harboring the recombinant plasmid containing the *celS* gene; Lane 4, the lysate of *E. coli* strain harboring the control plasmid without the *celS* gene. The proteins were electroblotted to a PVDF membrane before staining with Coomassie blue.

Effect of Ca^{++}. Ca^{++} has been found to stimulate the Avicelase activity of the crude *C. thermocellum* cellulase (*6*). The effect of Ca^{++} on the rCelS activity was therefore examined using amorphous cellulose as the substrate. As shown in Figure 6, Ca^{++} (12 mM) did not have a significant effect at 60°C under the assay conditions. However, at higher temperature settings, the rCelS activity was stimulated by the Ca^{++} (Figure 6), indicating that Ca^{++} enhanced the thermostability of the enzyme. The optimum temperature was 70°C in the presence of Ca^{++} and was lower in the absence of Ca^{++}. This observation is consistent with the effect of Ca^{++} on the crude enzyme (*6*) and the S8-tr protein (*34*) discussed below.

A New Family of Exo-β-glucanase. While CelS is produced by the *C. thermocellum* ATCC 27405, the S8 protein is the major component of the cellulosome of the YS strain. A truncated form of S8 (S8-tr, M_r = 68,000) has been purified by partial proteolysis of the cellulosome and the subsequent purification of the dissociated proteolysis product (*34*). The purified S8-tr displays typical cellobiohydrolase activities. Its thermostability is enhanced by Ca^{++}. Although the reported apparent molecular weight of S8 appears to be different from that of CelS (75,000 vs. 82,000), both protein species migrate at the same rate when subjected to the SDS-gel electrophoresis on the same gel (Figure 7). Furthermore, S8 has an N-terminal amino acid sequence identical to that of CelS (Figure 8; *35*). Although the cloning and DNA sequence of the S8 gene have not been reported, these two proteins appear to be identical. As indicated in Figure 7, S8 and CelS are both the most abundant protein species of the YS and ATCC strains, respectively. The next major protein species are CelL (of the ATCC 27405 strain) or S1 (of the YS strain). It is interesting to note that the S1 protein is slightly but noticeably smaller than CelL (Figure 7). The high level production of both protein species (CelS/CelL or S8/S1) in two different strains suggest that they play important roles in the cellulolytic activity as described in the enzyme-anchor model.

Besides the S8 protein, the *C. stercorarium* Avicelase II (M_r = 87,000; *36*) also has an N-terminal amino acid sequence highly homologous to that of CelS (Figure 8). The purified Avicelase II shows exoglucanase activities. However, unlike the conventional cellobiohydrolase which preferentially releases cellobiose from the non-reducing ends of cellulose chains, this enzyme releases cellobiose, cellotriose and cellotetraose from the non-reducing ends depending on the type of substrate. Avicelase II therefore appears to represent a novel type of exoglucanase. A name of "cellodextrinohydrolase" has been proposed to describe this new exoglucanase. The novel action of Avicelase II is consistent with the novelty of the CelS sequence. In addition, as in the case of CelS, the thermostability of Avicelase II is enhanced by Ca^{++}. However, the gene sequence of this enzyme has not been reported, preventing sequence comparison beyond the N-terminal peptide. Furthermore, Avicelase II is slightly larger than CelS (87,000 vs. 82,000). Finally, there is a significant difference between the two proteins: CelS is part of a cellulase aggregate and Avicelase II appears to be a free enzyme. In fact, no protein aggregation has been reported in *C. stercorarium*.

It appears that at least four microorganisms produce CelS-like protein (*C. thermocellum* [CelS or S8], *C. cellulolyticum* [ORF-1], *C. stercorarium* [Avicelase II], and *C. saccharolyticum* [CelA]), although CelS so far is the only member with

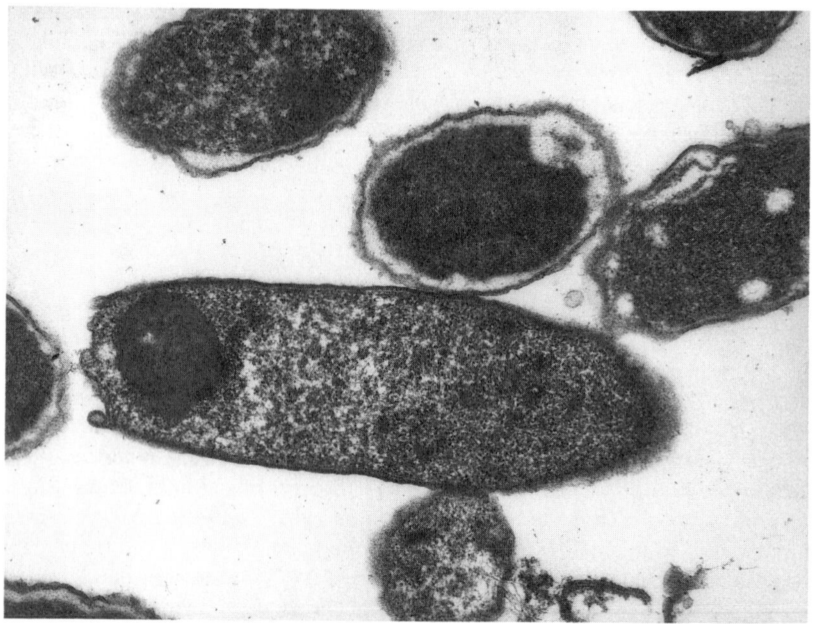

Figure 5. The transmission electron micrograph showing the inclusion body formation in a *E. coli* cell expressing the *celS* gene.

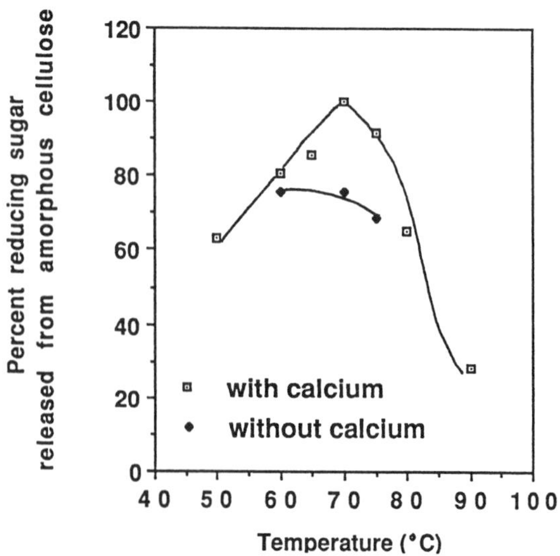

Figure 6. The effects of temperature and Ca^{++} on the activity of rCelS on the amorphous cellulose. The assay tubes contained 0.5% (w/v) phosphoric acid-swollen Avicel. The incubation time was 15 hours. The Ca^{++} concentration was 12 mM.

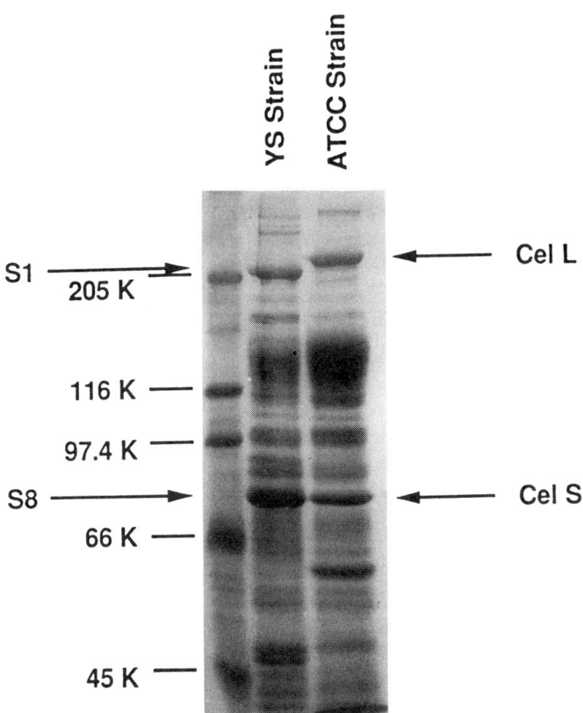

Figure 7. SDS-PAGE pattern of the total extracellular proteins of *C. thermocellum* ATCC 27405 and YS strains, indicating that CelS (S_S) and S8 have the same apparent molecular weight and that CelL (S_L or CipA) is noticeably larger than S1.

```
CelS          28  G P T K A P T K D G T S Y K D L F L E L Y G K I K D P K N G Y F S P D E G I P Y H S I  70
S8             1  G P T K A P T K D G T S Y K D L F X E                                                  19
Avicelase II   1              X S D D P Y K Q R F L E L W E E L H D P S N G Y F S - X H G I P Y H A V  35
```

Figure 8. The alignment of the N-termini of CelS and S8 from the ATCC 27405 and YS strains of *C. thermocellum*, respectively, and Avicelase II from *C. stercorarium*. Boxed amino acids are identical or have similar chemical properties. Numbers indicate the position, within the sequence of each protein, of the first or last amino acid shown on a line. Similar residues were: V,L,I,M,F; R,K; D,E; N,Q; Y,F,W; S,T; G,A. The first 27 amino acid residues of CelS presumably form the signal peptide for secretion. The N-terminus of the native CelS starts at the residue 28.

complete DNA and amino acid sequences reported. These proteins form a new family of exo-β-glucanases that may have a novel mode of action different from that of conventional cellobiohydrolases. All four bacteria are anaerobic, indicating that CelS may be a major component contributing to the unique properties of the anaerobic bacterial cellulase system.

Conclusion

Compared to the well-studied *T. reesei* cellulase system, the *C. thermocellum* cellulase exhibits a higher specific activity toward crystalline cellulose. It also displays other unique properties such as requiring Ca^{++} and reducing agent for maximum activity, and inhibition by non-polar chelating agents (*6,37*). As we have previously proposed, those unusual properties are rooted in the unique structure of the cellulosome, best described by the enzyme-anchor model involving predominantly two proteins, CelS and CelL.

It is clear now that CelL, a major component of the cellulosome, serves as a scaffolding protein of the cellulosome and an anchor for the catalytic subunits on the cellulose surface. Inclusion of this scaffolding/anchorage subunit in the *C. thermocellum* cellulase system clearly distinguishes it from the fungal cellulases.

CelS, the most abundant protein species secreted into the culture supernatant, has been identified to be the essential catalytic subunit for crystalline cellulose degradation. Cloning of the *celS* gene has led to the discovery of a new cellulase family. The members of this family so far have only been found in anaerobic bacteria. It codes for an exoglucanase clearly distinguishable from its fungal counterpart. The properties of the *C. thermocellum* cellulase system may therefore be further defined by the presence of CelS in the complex.

It appears that an exciting, novel mechanism of enzymatic cellulose degradation is in the process of being unveiled. The mechanism involves an anchor/scaffolding protein (CelL) and a novel exoglucanase (CelS). The availability of the *celS* gene and the rCelS protein will facilitate further characterization of its function and interactions with CelL, and provide further insights into this intriguing, sophisticated, and non-conventional cellulolytic process which has only begun to be explored recently.

Acknowledgments

The research project on cellulase in our laboratory is financially supported by NREL (XAC-3-13419-01) and USDA (Alcohol Fuels Program; 93-37308-8988). W.K.W. appreciates the support of the Link Foundation Energy Fellowship. K.K. appreciates the scholarship from the Finnish Academy.

Literature Cited

1. Lamed, R.; Setter, E.; Kenig, R.; Bayer, E.A. *Biotechnol. Bioeng. Symp.* **1983**, *13*, 163-181.
2. Lamed, R.; Setter, E.; Bayer, E.A. *J. Bacteriol.* **1983**, *156*, 828-836.
3. Hon-Nami, K.; Coughlan, M.P.; Hon-Nami, H.; Ljungdahl, L.G. *Arch. Microbiol.* **1986**, *145*, 13-19.

4. Coughlan, M.P.; Hon-Nami, K.; Hon-Nami, H.; Ljungdahl, L.G.; Paulin, J.J.; Rigsby, W.E. *Biochem. Biophys. Res. Commun.* **1985**, *130*, 904-909.
5. Wu, J.H.D.; Orme-Johnson, W.H.; Demain; A.L. *Biochemistry* **1988**, *27*, 1703-1709.
6. Johnson, E.A.; Sakajoh, M.; Halliwell, G.; Madia, A.; Demain, A.L. *Appl. Environ. Microbiol.* **1982**, *43*, 1125-1132.
7. Shoseyov, O.; Doi, R.H. *Proc. Natl. Acad. Sci. U.S.A.* **1990**, *87*, 2192-2195.
8. Ljungdahl, L.G.; Eriksson, K.-E. *Adv. Microb. Ecol.* **1985**. *8*, 237-299.
9. Wu, J.H.D.; Demain; A.L. In *Biochemistry and Genetics of Cellulose Degradation*; Aubert; J.-P.; Beguin, P.; Millet, J., Eds.; Academic Press: London, 1988; pp 117-131.
10. Gerngross U.T.; Romaniec, M.P.M.; Kobayashi, T.; Huskisson, N.S.; Demain, A.L. *Mol. Microbiol.* **1993**, *8*, 325-334.
11. Wang, W.K.; Kruus, K.; Wu, J.H.D. *J. Bacteriol.* **1993**, *175*, 1293-1302.
12. Wang, W.K.; Wu, J.H.D. *Appl. Biochem. Biotechnol.* **1993**, *39/40*, 149-158.
13. Wu, J.H.D. In *Biocatalyst Design for Stability and Specificity*; Himmel, M.E. and Georgiou, G., Eds.; American Chemical Society: Washington, DC, 1993, Vol. 516; pp 251-264.
14. Cornet, P.; Millet, J.; Beguin, P.; Aubert, J.-P. *Bio/Technology* **1983**, *1*, 589-594.
15. Henrissat, B.; Claeyssens, M.; Tomme, P.; Lemesle, L.; Mornon, J.-P. *Gene* **1989**, *81*, 83-95.
16. Beguin, P.; Cornet, P.; Aubert, J-P. *J. Bacteriol.* **1985**, *162*, 102-105.
17. Schwarz, W.; Grabnitz, F.; Staudenbauer, W.L. *Appl. Env. Microbiol.* **1986**, *51*, 1293-1299.
18. Grepinet, O.; Beguin, P. *Nucleic Acids Res.* **1986**, *14*, 1791-1799.
19. Beguin, P.; Cornet, P.; Millet, *J. Biochimie* **1983**, *65*, 495-500.
20. Schwarz, W.H.; Schimming, S.; Rucknagel, K.P.; Burgschwaiger, S.; Kreil, G.; Staudenbauer, W. *Gene* **1988**, *63*, 23-30.
21. Petre, D.; Millet, J.; Longin, R.; Beguin, P.; Girard, H.; Aubert, J-P. *Biochimie* **1986**, *68*, 687-695.
22. Joliff, G.; Beguin, P.; Aubert, J.-P. *Nucleic Acids Res.* **1986**, *14*, 8605-8613.
23. Joliff, G.; Beguin, P.; Juy, M.; Millet, J.; Ryter, A.; Poljak, R.; Aubert, J.-P. *Bio/Technology* **1986**, *4*, 896-900.
24. Hall, J.; Hazlewood, G.P.; Barker, P.J.; Gilbert, H.J. *Gene* **1988**, *69*, 29-38.
25. Navarro, A.; Chebrou, M.-C.; Beguin, P.; Aubert, J.-P. *Res. Microbiol.* **1991**, *142*, 927-936.
26. Lemaire, M.; Beguin, P. *J. Bacteriol.* **1993**, *175*, 3353-3360.
27. Yague, E.; Beguin, P.; Aubert, J.-P. *Gene* **1990**, *89*, 61-67.
28. Grepinet, O.; Chebrou, M.-C.; Beguin, P. *J. Bacteriol.* **1988**, *170*, 4582-4588.
29. Bagnara-Tardiff, C.; Gaudin, C.; Belaich, A.; Hoest, P.; Citard, T.; Belaich, J.-P. *Gene* **1992**, *119*, 17-28.
30. Luthi, E.; Jasmat, N.B.; Grayling, R.A.; Love, D.R.; Bergquist, P.L. *Appl. Environ. Microbiol.* **1991**, *57*, 694-700.
31. Gibbs, M.D.; Saul, D.J.; Luthi, E.; Bergquist, P.L. *Appl. Environ. Microbiol.* **1992**, *58*, 3864-3867.
32. Fujino, T.; Beguin, P.; Aubert, J.P. *FEMS Microbiol. Lett.* **1992**, *94*, 165-170.

33. Shoseyov, O.; Takagi, M.; Goldstein, M.A.; Doi, R.H. *Proc. Natl. Acad. Sci. USA* **1992**, *89*, 3483-3487.
34. Morag, E.; Halevy, I.; Bayer, E.A.; Lamed, R. *J. Bacteriol.* **1991**, *173*, 4155-4162.
35. Morag, E.; Bayer, E.A.; Hazlewood, G.P.; Gilbert, H.J.; Lamed, R. *Appl. Biochem. Biotechnol.* **1993**, *43*, 147-151.
36. Bronnenmeier, K.; Rucknagel, K.P.; Staudenbauer, W.L. *Eur. J. Biochem.* **1991**, *200*, 379-385.
37. Johnson, E.A.; Demain, A.L. *Arch. Microbiol.* **1984**, *137*, 135-138.
38. Schimming, S.; Schwarz, W.H.; Staudenbauer, W.L. *Eur. J. Biochem.* **1992**, *204*, 13-19.
39. Faure, E.; Belaich, A.; Bagnara, C.; Gaudin, C.; Belaich, J.-P. *Gene* **1989**, *84*, 39-46.
40. Shima, S.; Igararashi, Y.; Kodama, T. *Gene* **1991**, *104*, 33-38.
41. Foong, F.; Hamamoto, T.; Shoseyov, O.; Doi, R. *J. Gen. Microbiol.* **1991**, *137*, 1729-1736.

RECEIVED February 17, 1994

Chapter 6

Plant Endo-1,4-β-D-glucanases

Structure, Properties, and Physiological Function

David A. Brummell, Coralie C. Lashbrook, and Alan B. Bennett

Mann Laboratory, Department of Vegetable Crops, University of California, Davis, CA 95616

Multiple genes encoding distinct endo-1,4-β-D-glucanases have been characterized in plants. The enzymes are of approximately 50 to 70 kDa, and possess discrete amino acid domains that are conserved with microbial cellulases of the E_2 subgroup. Low levels of extractable endo-1,4-β-D-glucanase activity have been found in virtually all plant organs, with higher levels being associated with certain developmental processes which involve modifications in cell wall structure, such as cell elongation, leaf abscission and fruit ripening. In many cases, the developmentally regulated changes in endo-1,4-β-D-glucanase activity have been shown to result from hormone-mediated changes in gene expression. Activities of an auxin-induced, ethylene-antagonized endo-1,4-β-D-glucanase have been correlated with the cell expansion process, and activities of an ethylene-induced, auxin-antagonized type have been correlated with organ abscission and fruit ripening. The enzymes that have been cloned possess predicted signal sequences, and several have been immunolocalized to the cell wall where they presumably act against cell wall polysaccharides. Their precise *in vivo* substrates have not yet been established, although they do not appear capable of acting alone on native crystalline cellulose. It is possible that the multiplicity of genes and their diverse pattern of developmental regulation contributes to the disassembly of distinct components of the plant cell wall.

Plants have long been known to possess enzyme activities capable of the *in vitro* hydrolysis of soluble cellulose derivatives such as carboxymethylcellulose (CMC) (*1,2*). This review will describe the endo-1,4-β-D-glucanases (1,4-β-D-glucan-4-glucanohydrolase, EC 3.2.1.4) of higher plants, but will not cover the mixed linkage endo-1,3/1,4-β-D-glucanases (EC 3.2.1.73) described in monocotyledonous plants (*3,4*). Even though the latter enzymes may cleave 1,3/1,4-β-glucans at 1,4-β linkages

0097–6156/94/0566–0100$09.80/0

(*3*), 1,3/1,4-β-glucanases share no sequence homology with endo-1,4-β-glucanases, but rather share homology with the 1,3-β-glucanases (*5,6*) which are involved in plant defense against fungal attack. Where appropriate, mention will be made of plant xyloglucan endotransglycosylase (XET) (*7-9*), the activity of which can resemble that of endo-1,4-β-glucanase in some *in vitro* assays. The role of endo-1,4-β-glucanases in plant development was last reviewed in 1991 (*10*).

Cellulases have been classified into two groups (*11*), C_1 which can degrade native crystalline cellulose, and C_x which hydrolyse only soluble substituted cellulose derivatives. True cellulases capable of the complete hydrolysis of crystalline cellulose to glucose, as found in bacterial and fungal cells, are usually multi-enzyme complexes composed of endo-1,4-β-glucanases, cellobiohydrolases and cellobiases possessing synergistic activity (*12*). Plant endo-1,4-β-glucanases are of the C_x type, since there is no compelling evidence that they can degrade crystalline cellulose. Plant endo-1,4-β-glucanases cloned so far do not possess the cellulose binding domain which is attached to the catalytic core of many microbial cellulases (*13,14*), and which confers upon them the ability to bind tightly to and attack crystalline cellulose (*12*). C_x-cellulases adsorb weakly to and are essentially inactive against crystalline cellulose, with hydrolysis not usually exceeding 2-5% and even this low level of activity probably being due to attack on amorphous regions of the substrate (*12*). Thus the term cellulase when applied to plant endo-1,4-β-glucanases is imprecise and ambiguous, but has become well established. In this review, the term endo-1,4-β-glucanase will be used when referring to the enzymes and genes found in plants, and the term carboxymethylcellulase (CMCase) or C_x-cellulase used when referring to an extractable activity determined *in vitro* versus a soluble substituted cellulosic substrate.

Relationships of Plant and Microbial Endo-1,4-β-glucanases

Primary sequence comparisons and hydrophobic cluster analysis have revealed that the catalytic cores of bacterial, fungal and plant endo-1,4-β-glucanases can be grouped into families that share catalytic properties (*15*). Most endo-1,4-β-glucanases can be placed into one of six major families (*13,15*), although additional families have been suggested in order to accommodate the few exceptions (*14*).

At present, four plant endo-1,4-β-glucanase cDNA clones have been sequenced in entirety, one from ripening avocado fruit (avoCel1) (*16*), one from bean abscission zones (BAC 1) (*17*) and two from ripening tomato fruit (tomCel1 and tomCel2) (*18*). In addition to these four clones, partial length clones have been obtained from a tomato fruit cDNA library (tomCel3 and tomCel4), from a soybean leaf abscission zone cDNA library (SAC1), and from a gynoecium cDNA library of the orchid *Doritaenopsis*, a monocotyledonous plant (orcCel1) (Table I).

The plant endo-1,4-β-glucanases cloned to date have been assigned to the E family which is comprised of several bacterial cellulases and one from a unicellular eukaryotic slime mold (Table I); as yet no fungal cellulases have been assigned to this family. Family E has been further subdivided into subgroups E_1 and E_2 (*13,15*), with *P. fluorescens* subsp. *cellulosa* CelA, *B. fibrisolvens* Ced1, *C. fimi* CenC and *C. thermocellum* CelD in subgroup E_1, and the remaining bacterial endo-1,4-β-glucanases, the spore germination-specific protein of *Dictyostelium discoideum*, and

Table I. E Family Endo-1,4-β-glucanases. This family consists of two subgroups, with subgroup E_2 possessing both microbial and plant members. Although shown separately in this table for emphasis, no structural or functional separation between microbial and plant E_2 endo-1,4-β-glucanases is intended to be implied.

Species	Enzyme	Abbreviation[a]	Reference
Subgroup E_1			
Bacteria			
Butyrivibrio fibrisolvens H17c	Ced1	*Bfi*Ced1	*19*
Cellulomonas fimi	CenC	*Cfi*CenC	*20*
Clostridium thermocellum	CelD	*Cth*CelD	*21*
Pseudomonas fluorecens subsp. *cellulosa*	CelA	*Pfl*CelA	*22*
Subgroup E_2			
Bacteria			
Cellulomonas fimi	CenB	*Cfi*CenB	*23*
Clostridium thermocellum	CelF	*Cth*CelF	*24*
Clostridium stercorarium	CelZ	*Cst*CelZ	*25*
Fibrobacter succinogenes AR1	$endA_{FS}$	*Fsu*EndA	*26*
Thermonospora fusca	E_4[b]	*Tfu*E4	*27*
Protista			
Dictyostelium discoideum	SGSP270-6	*Ddi*270-6	*28*
Plant			
Doritaenopsis sp.	Cel1[b]	orcCel1	[c]
Glycine max	SAC1[b]	SAC1	*29*
Lycopersicon esculentum	Cel1	tomCel1	*18*
	Cel2	tomCel2	*18*
	Cel3[b]	tomCel3	[d]
	Cel4[b]	tomCel4	[e]
Persea americana	Cel1	avoCel1	*16*
	Cel2	avoCel2	*30*
Phaseolus vulgaris	BAC1	BAC1	*17*

[a]Abbreviations as used in Figures 2, 3, Table II

[b]Partial sequence

[c]Bui, A.Q. and O'Neill, S.D., personal communication

[d]Brummell, D.A., Catala, C., Lashbrook, C.C. and Bennett, A.B., unpublished

[e]W. Schuch, personal communication

the plant endo-1,4-β-glucanases in subgroup E_2 (*31*). One of the characteristics of E_1-type endo-1,4-β-glucanases is the presence of an N-terminal extension, as shown in Figure 1, which E_2-type endo-1,4-β-glucanases lack (*24*). Plant endo-1,4-β-glucanases consist essentially of only the catalytic core, and lack the cellulose binding domains, extension regions and anchor motifs which are often observed in bacterial cellulases (Figure 1). Consequently plant endo-1,4-β-glucanases generally have a maximum size of around 70 kDa (excluding possible glycosylation), whereas microbial cellulases can be considerably larger and, in bacteria, may be further organised via the anchor motifs into cellulosomes comprised of up to 15-20 individual polypeptides (*12-14*).

Despite great differences between bacterial and plant endo-1,4-β-glucanase polypeptides, analysis of their catalytic cores shows that several amino acid domains are conserved. Two highly conserved domains are depicted in Figure 1, and two less conserved domains in Figure 2. It is likely that these domains contain amino acid residues important for catalytic activity, substrate binding, or protein tertiary structure.

The crystal structure of CelD of *C. thermocellum*, a member of the E_1 family, has been determined (*32*). CelD was found to be a globular molecule, with the catalytic domain composed of an α-barrel of 12 helices in which the substrate binding site appeared as a groove on one side. Site-directed mutagenesis and chemical modification techniques identified several amino acid residues as being potentially important in catalysis. His-516 (the conserved histidine in Figure 2, left-hand domain) appeared to be in the active center of CelD, although it was not absolutely essential for catalysis (*33*). Mutation of Glu-555 (the conserved glutamic acid in Figure 2, right-hand domain) to Ala reduced the k_{cat} of CelD by at least 4000-fold, suggesting participation in the catalytic mechanism (*34*). Strong inactivation was also observed when either Asp-198 or Asp-201 (the two aspartic acids of the conserved DAGD motif seen in Figure 1, 5'-domain) was changed to Ala, and it was suggested that one of these two residues may be involved in stabilizing the positively-charged transition state of the substrate (*34*). The side chains of all four of these amino acid residues (Asp-198, Asp-201, His-516 and Glu-555) face the substrate, as determined from the crystal structure (*32*), and all are conserved in the four plant endo-1,4-β-glucanase sequences (Figures 1 and 2).

Phylogenetic analysis of the amino acid sequences of fifteen E family endo-1,4-β-glucanase catalytic domains resulted in the unrooted phylogenetic tree depicted in Figure 3. The tree indicated early divergence into four main branches. Two contained the members of the E_1 subgroup (*C. fimi* CenC, *P. fluorescens* CelA, *C. thermocellum* CelD and *B. fibrisolvens* Ced1), another branch contained the bacterial and protist members of the E_2 subgroup, while the fourth branch contained the plant E_2 subgroup endo-1,4-β-glucanases. The branch containing plant endo-1,4-β-glucanases diverged into three main sublineages, one containing tomato tomCel1 and bean BAC1, another containing avocado avoCel1 and tomato tomCel2 and tomCel4, and a third, more divergent, sublineage contained tomato tomCel3 and orchid orcCel1. Other sublineages within the plant endo-1,4-β-glucanases may remain to be discovered. Based on this tree, it appears that tomato (a dicotyledonous plant) tomCel3 is more closely related to orchid (a monocotyledonous plant) orcCel1 than it is to other tomato endo-1,4-β-glucanases, suggesting that the polymorphism of plant endo-1,4-β-glucanase genes arose prior to the divergence of monocotyledonous and dicotyledonous plants.

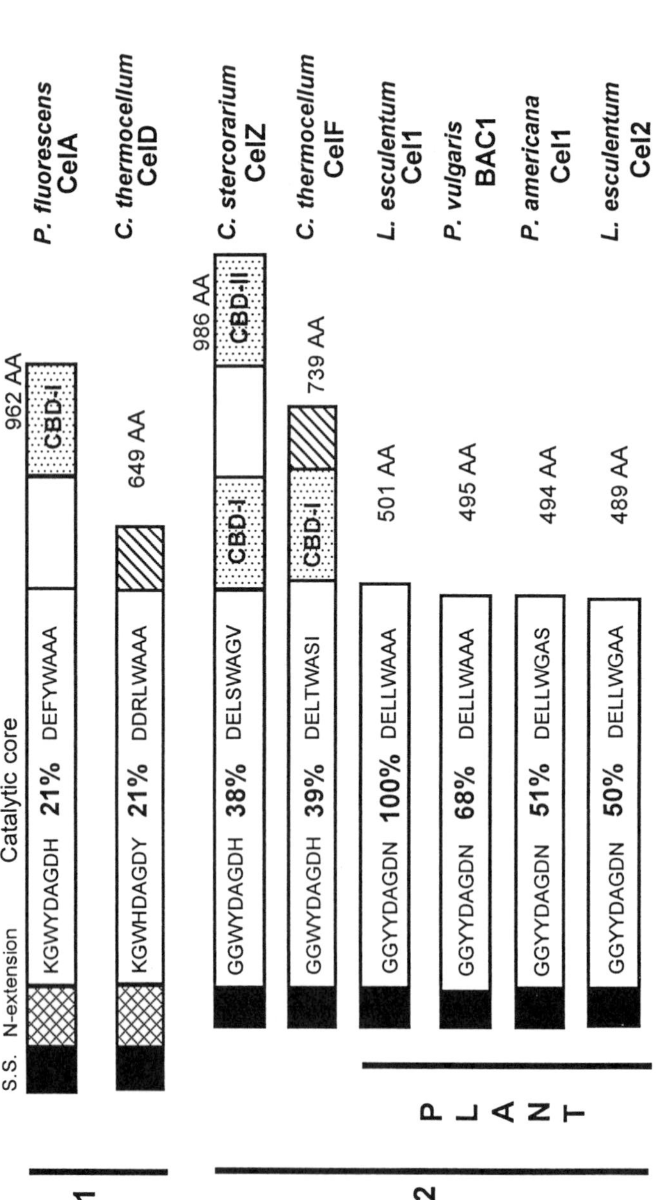

Figure 1. Schematic representation of selected endo-1,4-β-glucanases of the E family. Two highly conserved amino acid domains are indicated. The percentage figures show the amino acid identity of the catalytic cores relative to tomato (*L. esculentum*) Cel1. For full species names see Table I. CBD, cellulose-binding domain; S.S., signal sequence. Data based partly on schemes in Navarro *et al.* (*24*) and Coutinho *et al.* (*20*).

```
CfiCenC     SYVTGWGEVASHQQHSR    ELTVNWNSALSWVAS
PflCelA     SFITGLGTNTVAQPHHR    EITINWNAPLAWVLG
CthCelD     SYVTGLGINPPMNPHDR    EIAINWNAALIYALA
BfiCed1     SYVTGNGEKAFKNPHLR    EITIYWNSPLVFALS
Ddi270-6    SFVVGMGPNYPINPHHR    EVATDYNAGFVGALA
CfiCenB     SYVVGFGANPPTAPHHR    EVATDYNAGFTSALA
CstCelZ     SYVVGFGANPPTAPHHR    EVACDYNAGFVGALA
CthCelF     SFVVGFGKNPPRNPHHR    EVANDYNAGFVGALA
tomCel1     SYMVGFGNKYPTKLHHR    EPTTYMNAAFIGSVA
BAC1        SYMVGFGSKYPKQLHHR    EPTTYINAAFVASIS
avoCel1     SYMVGFGERYPQHVHHR    EPATYINAPLVGALA
tomCel2     SYMVGYGPHFPQRIHHR    EPTTYVNAPLVGLLA
            *.. * *       * *    * .   *. .    .
```

Figure 2. Two amino acid domains conserved between twelve E family endo-1,4-β-glucanases. For abbreviations of organisms and enzymes see Table I. Positions showing the retention of identical residues are indicated by asterisks, and those with conserved amino acid substitutions by dots.

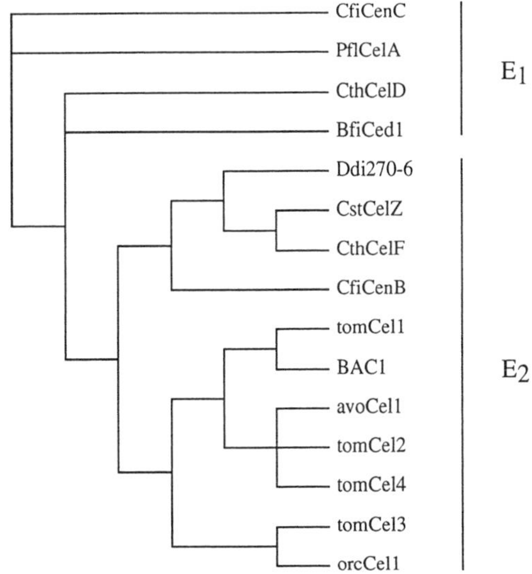

Figure 3. Phylogeny of truncated catalytic domains of E family endo-1,4-β-glucanases. Sequences were aligned using ClustalV (*35*) then truncated at the amino-terminal end of the DHXCWERPED domain found in plant endo-1,4-β-glucanases, or the conserved acidic amino acid equivalent (E-285 of *C. thermocellum* CelD) in the microbial members. Truncation at the carboxyl-terminal end was close to the end of the catalytic cores, at the absolutely conserved Asn shown in Figure 2, right-hand domain (N-561 of *C. thermocellum* CelD or its equivalent in other genes). The truncated segments represented approximately 75% of the catalytic domains of the plant endo-1,4-β-glucanases. Phylogeny was determined using Phylogeny Analysis Using Parsimony (*36*) employing a branch-and-bound search, the three resulting trees being reduced to a strict consensus phylogram. For abbreviations of organisms and enzymes see Table I.

Structure and Processing of Plant Endo-1,4-β-glucanases

The size and organization of plant endo-1,4-β-glucanases and the presence in the catalytic cores of amino acid domains conserved with related microbial endo-1,4-β-glucanases has been described above. However, the eukaryotic plant endo-1,4-β-glucanases may be subjected to various forms of post-transcriptional and post-translational regulation not present in prokaryotes.

Some of the properties of four plant endo-1,4-β-glucanases as deduced from their cDNA sequences are shown in Table II. These endo-1,4-β-glucanases are of similar size but have very different pI values, and only three of the four possess consensus N-X-S/T sites for potential N-glycosylation. It should be noted that pI values deduced from primary sequence information do not take into account the tertiary structure of the folded protein. The actual pI value of the native avocado protein is 4.7 or 5.95 (*37,38*) and that of the native bean abscission zone protein is 9.5 (*39*), as compared to predicted pIs of 5.3 and 7.8, respectively (Table II).

As assessed using the rules of von Heijne (*40*), avoCel1, BAC1, tomCel1 and tomCel2 all possessed predicted signal sequences, of 25, 23, 24 and 23 amino acids respectively. Immunological analysis of various purified forms and *in vitro* translation products suggested that the avocado endo-1,4-β-glucanase protein was synthesised as a 54 kDa pre-protein, which was processed by removal of the signal sequence to yield a mature polypeptide of 52.8 kDa (*41*). This polypeptide was then heavily glycosylated to a membrane-associated secretory form of 56.5 kDa, which by carbohydrate trimming yielded the 54.2 kDa mature protein. Partial endoglycosidase H digestion indicated that the mature glycoprotein possessed two high-mannose oligosaccharide side chains, a finding in agreement with the number of consensus N-glycosylation sites subsequently revealed by sequencing its cDNA (*16*). Binding to Concanavalin A-Sepharose has been used to purify avocado endo-1,4-β-glucanase (*42*), further showing that the mature protein is N-glycosylated.

Multiple forms of mature endo-1,4-β-glucanase protein have been noted in ripening avocado fruit (*43,44*). However, so far only a single cDNA (*16,45*) and a single gene (*30*) have been identified as contributing to the massive accumulation of endo-1,4-β-glucanase in ripe avocado fruit. Native isoelectric focusing followed by activity staining or immunostaining showed that at least 11 forms of endo-1,4-β-glucanase could be separated (*38*). All 11 forms were immunologically related, and possessed molecular masses of 54 kDa. However, their pI values ranged from 5.1 to 6.8, with predominant forms at pIs of 5.6, 5.8, 5.95 and 6.2. The pI 5.95 form was the major species, and the first to appear during ripening (*38,44*). Whether the existence of multiple forms is due to the activity of multiple genes, or to post-transcriptional processing of a single mRNA, or to post-translational modification of the protein, is not known.

Genomic DNA blot analysis suggested that single genes likely encoded each of BAC1 (*17*), tomCel1 and tomCel2 (*18*). In contrast, avoCel1 appeared to be encoded by a small family of genes (*16*). Screening of an avocado genomic library identified a gene termed *cel1* as coding for the avoCel1 cDNA (*30*). This gene consisted of eight exons and seven introns, and possessed a promoter displaying elements responsive to ethylene, a plant hormone. A second closely related gene, *cel2*, was divergent from *cel1* only at the 5' end, although no mRNA derived from *cel2* was

Table II. Physical Characteristics of Cloned Plant Endo-1,4-β-glucanases.

Endo-1,4-β-glucanase	Size of mRNA (kb)	Deduced amino acids[a] ss	mp	Predicted M_r (kDa)[b]	pI of deduced sequence[b]	Potential N-glycosylation sites[b]	Ref.
avoCel1	2.2	25	469	51.4	5.3	2	*16*
BAC1	2.0	23	472	51.8	7.8	0	*17*
tomCel1	1.9	24	477	52.7	7.3	4	*18*
tomCel2	2.1	23	466	51.5	8.2	1	*18*

[a]ss, predicted signal sequence; mp, mature protein.

[b]After removal of signal sequence (i.e., mature protein).

detected in ripe fruit. Whether *cel2* is expressed at different times or in other tissues, or represents an unexpressed pseudogene, has not been determined. A genomic fragment corresponding to a possible third gene has not been cloned.

Structure of Plant Cell Walls

Unlike microbial endo-1,4-β-glucanases, whose function is to attack exogenous substrates, plant endo-1,4-β-glucanases are presumed to act *in vivo* by selectively disassembling portions of their own cell wall during development (see section on Regulation and Physiological Role of Endo-1,4-β-glucanases). A basic model of the plant cell wall molecular structure is therefore needed in order to interpret the possible physiological functions of endo-1,4-β-glucanases.

The primary (growing) cell walls of plants are rigid yet dynamic structures composed of roughly equal quantities (around 30% for each) of cellulose, hemicellulosic polysaccharides and pectic polysaccharides, plus about 10% glycoprotein (*46*). In dicotyledonous plants and in the non-graminaceous monocotyledonous plants the predominant hemicellulose is xyloglucan, a polysaccharide consisting of a 1,4-β-glucan backbone substituted at regular intervals with single xylosyl residues, some Xyl side chains being further extended by Gal or Fuc-Gal (*46,47*). In contrast, the graminaceous monocots (family Poaceae) possess only small amounts of xyloglucan and instead the major hemicellulosic polysaccharide is glucuronoarabinoxylan, together with quantities of an unbranched, mixed-linkage 1,3/1,4-β-glucan present only during growth (*48*). The other main neutral (non-pectic) polysaccharides in primary walls are small amounts of essentially pure xylans, arabinans, galactans and arabinogalactans (*46*). The main pectic polysaccharides are polygalacturonic acid and rhamnogalacturonanans, the latter possessing side chains of arabinogalactan (*48*). Recent models of the expanding primary cell wall of dicots and non-graminaceous monocots agree that cellulose microfibrils are held together, and apart, by xyloglucan molecules which bind tightly to the surface of the microfibrils and stretch between them, locking together the microfibrils in a cellulose-xyloglucan network (*48-50*). This framework is itself embedded in a pectin matrix, and a protein network may be laid down as growth ceases (*48*). In the graminaceous monocots, glucuronoarabinoxylan appears to play a similar structural role to the xyloglucan of dicots and other monocots (*48*). Plant secondary (non-growing) cell walls contain a high proportion of cellulose, have heteroxylans and glucomannans possessing a low degree of substitution as the major matrix components, and become extensively lignified (*51*).

Perhaps the most likely substrate for endo-1,4-β-glucanases in the dicot and non-graminaceous monocot cell wall is xyloglucan, due to its structure, abundance and function in the wall. Xyloglucan is strongly bound to cellulose, requiring 24% KOH for its disassociation (*52*). Isolated xyloglucans exhibited pH-dependent binding to cellulose *in vitro*, suggesting association by hydrogen bonding (*53*). When xyloglucan was added to cellulose-forming cultures of *Acetobacter xylinum*, the observed partial disruption of the cellulose ribbon structure indicated that xyloglucan formed an association with cellulose at the time of synthesis (*53*). It is possible that some xyloglucan may be intercalated into the outer layers of cellulose microfibrils during synthesis, or bind to amorphous regions, since the 24% KOH used to release xyloglucan from cell walls would also cause some swelling of the microfibrils (*49*). Cells of pea epicotyl which had been extracted with various solvents to remove cytoplasmic contents and cell wall pectins consisted of a macromolecular complex of only xyloglucan and cellulose, yet retained mechanical structure and the shape of the cell (*52*). Visualization of xyloglucan in cell walls suggested that xyloglucan was distributed throughout the thickness of the wall and both coated and linked cellulose microfibrils at all stages of growth (*49,52-54*). The ratio of xyloglucan to cellulose was highest in the most rapidly-growing tissue, and declined in older regions of the stem (*52*). Xyloglucan thus appears to play a pivotal role in dicot primary cell wall architecture, and its metabolism may be essential for cell wall loosening in growth and cell wall disassembly in abscission and fruit ripening.

Biochemical Properties of Endo-1,4-β-glucanases

Although many studies have examined parameters such as pH optimum and substrate specificity in crude endo-1,4-β-glucanase preparations from various plant species and organs, it is likely that these activities were due to a combination of different enzymes, each with its own characteristics. Very few plant endo-1,4-β-glucanases have been purified sufficiently, and in adequate quantities, to allow accurate determinations of K_m and V_{max} versus different substrates, and none have so far been expressed in a heterologous system.

With the exception of the 20 kDa endo-1,4-β-glucanase of pea epicotyls, plant endo-1,4-β-glucanases that have been purified to homogeneity possessed apparent molecular weights of around 50 to 70 kDa (Table III). The native proteins exhibited a wide variation in isoelectric point, and were generally most active between pH 5 and 7, which presumably reflects the pH of the cell wall space where their substrates are located (*62*).

The relative substrate specificities of purified endo-1,4-β-glucanases have been determined in only a few studies, using endo-1,4-β-glucanases purified from avocado fruit, tobacco callus, and two endo-1,4-β-glucanases purified from pea epicotyl, a 20 kDa buffer-soluble form (BS endo-1,4-β-glucanase) and a 70 kDa buffer-insoluble, salt-soluble form (BI endo-1,4-β-glucanase) (Table IV). Although the available data is limited, in general the purified enzymes showed high activity *in vitro* versus CMC and mixed-linkage 1,3/1,4-β-glucan but exhibited differing affinities for xyloglucan, a potential *in vivo* substrate.

Purified endo-1,4-β-glucanases from pea and avocado rapidly reduced the viscosity of CMC solutions, whereas the increase in reducing groups was linear and relatively slow, with reducing power increasing even after the loss in viscosity was essentially complete (*63,65*). Such an action pattern is characteristic of the hydrolysis of internal linkages, showing that the enzymes are endohydrolases. Avocado fruit endo-1,4-β-glucanase was more than twice as active versus CMC with a degree of substitution of 0.4 (i.e., an average of 4 out of 10 Glc residues were substituted) than it was versus CMC with a degree of substitution of 0.7 (Table IV). This suggests that the enzyme preferred either longer unsubstituted regions of the molecule or regions of lower substitution. Hydroxyethylcellulose, which has bulkier side chains than CMC, was hydrolysed much less effectively by both avocado fruit and tobacco callus endo-1,4-β-glucanase.

Table III. Physical Properties of Native Endo-1,4-β-glucanases.

Plant	Tissue	M_r (kDa)	pI	pH optimum	Ref.
Pea	epicotyl	20	5.2	5.5-6.0	*55*
		70	6.9	5.5-6.0	*55*
Tobacco	callus	50, 52	8.2	5.5-6.5	*56*
Poplar	cultured cells	50	5.5	6.0	*57*
Lathyrus	anthers	49	8.0	-	*58*
		51	7.8	-	*58*
Bean	cotyledon	70	4.5	5.7-6.2	*59*
		70	4.8	4.8-5.6	*59*
	abscission zone	51	9.5	6.0-8.5	*60*
Sambucus	abscission zone	54	-	7.0	*61*
Avocado	fruit	49	4.7	-	*37*
		54	5.1-6.8	-	38

Dashes denote that values were not determined.

Table IV. Relative Substrate Specificities of Four Purified Endo-1,4-β-glucanases, Assayed By Increase In Reducing Sugar (RS[a]) Equivalents Using Glc As Standard. Data compiled from Wong *et al.* (*63*), Hayashi *et al.* (*64*), Hatfield and Nevins (*65*), Truelsen and Wyndaele (*56*).

Substrate	Pea stem BS	Pea stem BI	Avocado fruit	Tobacco Callus
	V_{max}[b]		µgRS/µgTS/µg prot	µgRS/h/ml
CMC (d.s. 0.7)	93	201	33.4	5.0
CMC (d.s. 0.4)	-	-	72.8	-
HEC	-	-	4.6	1.8
Insoluble β-glucan	13	80	-	0.7
Acid-swollen cellulose	22	87	0	-
Dicot cell wall xyloglucan	3.3	22.4	2.8	7.6
Amyloid seed xyloglucan	2.0	25.6	-	0
Poaceae 1,3/1,4-β-glucan	93	177	68.8	3.2
Lichenan	22	141	28.4	-
Xylan	0	0	0	0

[a]RS, reducing sugar; TS, total sugar; BS, buffer-soluble endo-1,4-β-glucanase; BI, buffer-insoluble, salt-soluble endo-1,4-β-glucanase; CMC, carboxymethylcellulose; HEC, hydroxyethylcellulose; d.s., degree of substitution; -, not determined.

[b]V_{max}, µmol Glc equiv./µmol enzyme/min at 35°C and optimal pH with saturating concentrations of substrate.

Insoluble β-glucans and acid-swollen cellulose were hydrolysed by both endo-1,4-β-glucanases from pea, as measured by an increase in reducing power, but without release of Glc, cellobiose or soluble cellodextrins, thus showing that the internal linkages of long chains were preferentially hydrolysed (*63*). No increase in reducing sugar in the supernatant was detected upon treatment of acid-swollen cellulose by endo-1,4-β-glucanase from avocado, although cleavage of some internal linkages in the insoluble substrate cannot be ruled out (*65*).

Dicot cell wall xyloglucan, which is substituted in an ordered manner on 3 out of 4 Glc residues (*47*), was hydrolysed by all four purified endo-1,4-β-glucanases though to varying extents. The BS and BI endo-1,4-β-glucanases of pea epicotyl possessed similar apparent K_m values versus dicot cell wall xyloglucan (3.2 and 2.5 $mg.ml^{-1}$, respectively), although the V_{max} value of the latter enzyme was much greater (3.3 and 22.4, respectively), suggesting that their turnover numbers were different (*64*). The two pea endo-1,4-β-glucanases both hydrolysed pea cell wall xyloglucan (though at a 10 to 30-fold lower rate than they did CMC) to yield equal quantities of xyloglucan oligosaccharides with nine ($Glc_4.Xyl_3$.Gal.Fuc) and seven ($Glc_4.Xyl_3$) sugar units (*64*). Small amounts of the dimer of 16 sugars were observed with partial digestion, as was found with digestion of pea xyloglucan with *Streptomyces* cellulase (*52*). The reducing ends of both xyloglucan oligosaccharides were a single unsubstituted Glc unit, cleavage thus being at internal linkages adjacent to unsubstituted Glc units in a manner identical to the cleavage of xyloglucan by fungal cellulases (*52*).

The apparent K_m values for the two pea endo-1,4-β-glucanases versus cellohexaose were similar to those versus CMC or xyloglucan (2.5 to 4.0 $mg.ml^{-1}$) (*64,66*). However, both enzymes had substantially (at least 10-fold) higher V_{max} values against cellohexaose, and it was shown that cellodextrins competitively inhibited the hydrolysis of CMC by the pea endo-1,4-β-glucanases. The degree of inhibition increased exponentially with degree of polymerization (D.P.) indicating that the enzymes bound preferentially to longer-chain cellodextrins (*63*). The relative activities of both enzymes versus a cellodextrin series showed that the enzymes were inactive versus cellobiose, with activity approximately doubling for each substrate from cellotriose to cellohexaose (cellohexaose is the largest cellodextrin which is soluble to an extent sufficient to allow measurement of kinetic properties). This indicated that the endo-1,4-β-glucanases possessed a binding site that recognized at least six consecutive 1,4-β-linked Glc units (*66*). Limited random substitution of the substrate at C-6 as in CMC evidently did not interfere with substrate binding or hydrolysis, provided that the degree of substitution was not too high or the side chains too bulky. Cellodextrins were cleaved preferentially at internal sites to give final products of mainly cellobiose and Glc, but the linkage adjacent to the reducing end of the cellodextrins was not cleaved, showing that these enzymes acted as true endohydrolases (*63*).

Endo-1,4-β-glucanase purified from tobacco callus was highly active against tobacco cell wall xyloglucan, at a rate exceeding that versus CMC, yet showed no activity against amyloid seed xyloglucan, a storage xyloglucan obtained from seeds of nasturtium and tamarind or their relatives. Amyloid xyloglucan is substituted on average at 3.2 out of 4 Glc residues although its side chains lack fucose (*67*), whereas tobacco cell wall xyloglucan has been reported to contain long stretches of unsubstituted Glc residues (*68*). The pI 9.5 endo-1,4-β-glucanase purified from bean abscission zones was also inactive versus amyloid seed xyloglucan (*60*). In contrast, the two pea endo-1,4-β-glucanases hydrolyzed amyloid seed and dicot cell wall xyloglucan equally effectively. This suggests that the tobacco callus and the bean abscission zone endo-1,4-β-glucanase require a less substituted substrate than do the pea enzymes, and may act only upon a substrate with more than one non-xylosylated Glc at the site of hydrolysis. An endo-1,4-β-glucanase purified from poplar suspension-culture cells also showed limited activity against amyloid seed xyloglucan, at less than 2% of its activity against CMC (*57*). Purified avocado fruit endo-1,4-β-glucanase was only weakly active against soybean hypocotyl cell wall xyloglucan (*65*), and no degradation of avocado fruit cell wall xyloglucan was detected *in vitro* (*42*).

The mixed-linkage 1,3/1,4-β-glucan lichenan and the 1,3/1,4-β-glucan from monocots of the family Poacaea (i.e., the grasses, generally oat or barley) were effective substrates for all the purified endo-1,4-β-glucanases (Table IV), since they contain consecutive 1,4-β-linkages. Graminaceous monocot 1,3/1,4-β-glucan consists of randomly-arranged cellotriose and cellotetraose oligosaccharides separated by single 1,3-β-linkages (*65*). The two pea endo-1,4-β-glucanases degraded barley 1,3/1,4-β-glucan by endohydrolysis, as followed by viscosity loss (*63*). Exhaustive digestion of oat 1,3/1,4-β-glucan by purified avocado fruit endo-1,4-β-glucanase produced a range of oligosaccharides of D.P. 4 or greater, with only 1,4-β-linked Glc residues at the reducing termini (*65*). The hydrolytic action pattern suggested that avocado endo-1,4-β-glucanase hydrolysed an internal 1,4-β-linkage in consecutively-linked 1,4-β-glucosyl residues of D.P. 4 or greater. This contrasts with the cellulase from *Streptomyces,* which also required a minimum of a tetrasaccharide but cleaved at a 1,4-β-linkage on the non-reducing terminal side of two adjacent 1,4-β-Glc residues, and thus produced oligosaccharides with either 1,4-β- or 1,3-β-Glc at the reducing end (*65*).

No hydrolytic activity of purified endo-1,4-β-glucanases was detected against 1,4-β-xylan, or against the 1,3-β-glucans laminarin and pachyman (*63,65*).

In addition to endo-1,4-β-glucanase activity, xyloglucan endotransglycosylase (XET) activity has been purified from plants. XET from nasturtium seed was a 31 kDa protein bearing no sequence homology with endo-1,4-β-glucanases (*69*). Purified nasturtium XET acted *in vitro* as a xyloglucanase by catalyzing endohydrolysis of xyloglucan and consequent viscosity loss. Its activity in this assay was strongly promoted by the addition of xyloglucan oligosaccharides which acted as alternative acceptors and rapidly reduced the average M_r of the xyloglucan substrate (*9,70*). At higher substrate concentrations the enzyme was capable of transglycosylating xyloglucan oligosaccharide dimers to monomers and higher M_r forms, whereas at low substrate concentrations only hydrolysis was observed (*70*). Unlike XET from nasturtium, XET purified from bean (*Vigna*) epicotyl did not exhibit hydrolase activity versus xyloglucan in the absence of added xyloglucan oligosaccharides, and could not use xyloglucans of smaller than 10 kDa (equivalent to a D.P. of the Glc backbone of about 32) as donor substrate (*8*). In crude extracts from pea epicotyl (*71*), bean leaf (*72*) and tomato fruit (*73*) xyloglucan endohydrolase activity was dramatically stimulated by added xyloglucan oligosaccharides, suggesting that the observed activity was attributable to XET.

Distribution of Endo-1,4-β-glucanases

Endo-1,4-β-glucanase activities have been detected in extracts from many different plant organs and tissues, and in numerous species (Table V). However, direct comparisons of absolute endo-1,4-β-glucanase activities between different studies are complicated by the use of different methods of extraction, different assays (reduction in viscosity or increase in reducing power of substituted glucan chains), different substrates (usually various types of CMC or xyloglucan), and to the use of different or inadequate definitions of units of activity.

In growing plant tissues such as hypocotyls, epicotyls and stems, low levels of extractable CMCase activity have been detected. In pea and soybean the maximal levels of CMCase activity were found in the regions of maximal cell expansion (*2,75,102*). High levels of CMCase activity were identified in some differentiating tissues where cell wall breakdown occurred, particularly in phloem (*79*), in articulated laticifers (*80*), and in waterlogged sunflower stem during the development of aerenchyma (*82*).

Numerous studies have shown that CMCase activity in leaf, flower or fruit abscission zones remains low until the abscission process begins, when high activity levels accumulate (*85,103*). The level of endo-1,4-β-glucanase in immature fruits has generally been reported as low, but increases as ripening progresses. Ripening avocado fruit accumulate exceptionally high levels of CMCase, more than 700 times higher than those in ripe tomato (*37*).

In germinating seeds of *Datura*, increased CMCase activity preceded endosperm softening and radicle protrusion (*83*), suggesting a possible role in the weakening of the endosperm layers covering the radicle. In seeds which possess large reserves of storage xyloglucan, such as nasturtium and tamarind, an apparent xyloglucan-specific endo-1,4-β-glucanase has been described (*104*), but subsequent work by the same group has shown this enzyme to be XET (*69,70*).

Regulation and Physiological Role of Endo-1,4-β-glucanases

With the exception of a few unconfirmed reports, evidence suggests that plant endo-1,4-β-glucanases do not act *in muro* to significantly degrade crystalline cellulose. However, extractable endo-1,4-β-glucanase activity has been detected in association with several plant developmental processes where changes in other cell wall components occur, although the conclusion that endo-1,4-β-glucanases play a critical, or even an important, role in these processes should not be drawn in the absence of more compelling evidence. Extractable CMCase activities described in the early literature probably represented the combined action of several differently-regulated endo-1,4-β-glucanases, unknown levels of XET activity and even the possible presence of microbial contamination in assays performed over several days. Emphasis should thus be given to studies where particular endo-1,4-β-glucanases were purified, or cloned, or levels assessed using specific antibodies. The hormonal regulation and potential involvement of endo-1,4-β-glucanases in plant development is discussed below.

Table V. Occurrence of Endo-1,4-β-glucanase Activities in Higher Plants.

Stage of Development	Organ	Species	Reference
Growing tissues	Epicotyl	Pea	*2, 74*
	Hypocotyl	Soybean	*75*
	Stem	Bean	*76*
	Apical bud	Soybean	*29*
	Leaf	Tobacco	*77*
		Rice	*78*
	Petiole	Bean	*76*
	Root	Tobacco	*77*
Differentiating tissues	Xylem, phloem	Sycamore	*79*
	Articulated laticifers	Rubber	*80*
		Poppy	*81*
	Aerenchyma	Sunflower	*82*
Ageing tissues	Seeds	Bean	*59*
		Datura	*83*
	Flower parts	*Lathyrus*	*58*
		Bean	*84*
	Leaf abscission zones	Bean	*85*
		Orange	*86*
		Cotton	*87*
		Coleus	*87*
		Peach	*88*
		Sambucus	*61*
		Soybean	*29*
	Flower abscission zones	Tomato	*89*
		Soybean	*29*
	Fruit abscission zones	Orange	*90, 91*
		Peach	*88*
	Ripening fruit	Tomato	*1, 92*
		Date	*93*
		Peach	*94*
		Strawberry	*95*
		Avocado	*96*
		Japanese pear	*97*
		Papaya	*98*
		Muskmelon	*99*
		Blackberry	*100*
		Mango	*101*

Cell Expansion. In order for plant cells to irreversibly expand the cell wall must in some way be loosened, even while rapid net synthesis is occurring. It is not known whether endo-1,4-β-glucanase plays a role in cell wall loosening, but in etiolated pea epicotyls extractable endo-1,4-β-glucanase activity was highest in the region of maximal cell expansion and declined in regions of the epicotyl where growth rate was low (*2,102*). In etiolated soybean hypocotyl, levels of total extractable endo-1,4-β-glucanase activity (versus CMC) followed a similar pattern, but levels of extractable xyloglucanase activity (versus soybean cell wall xyloglucan) showed an inverse correlation with growth rate (*75*). Whether the observed xyloglucanase activity was due to endo-1,4-β-glucanase or XET was not determined.

Dramatic stimulation of endo-1,4-β-glucanase mRNA and activity levels in growing tissues by the plant hormone auxin has been observed. Early work by Maclachlan and others showed that exogenous application of high concentrations of auxin to etiolated epicotyls of pea seedlings dramatically increased both lateral growth and the amount of extractable CMCase activity (*74,102,105,106*). Lower concentrations of auxin resulted in much more modest increases in endo-1,4-β-glucanase activity that were proportional to total growth (*107*). Two forms of auxin-induced endo-1,4-β-glucanase were separated on the basis of extractability with salt, a buffer-soluble form (BS endo-1,4-β-glucanase) with a pI of 5.2 and an apparent M_r of 20 kDa, and a buffer-insoluble, salt-soluble form (BI endo-1,4-β-glucanase) with a pI of 6.9 and an apparent M_r of 70 kDa (*55*). The mRNA for BS endo-1,4-β-glucanase increased 10-fold between 24 and 48 h of auxin treatment (*108*), and after 5 d the activities of both endo-1,4-β-glucanases had increased 100-fold (*55*). Virtually no pre-existing translatable mRNA for BS endo-1,4-β-glucanase was detected in zero-time control tissue (*108*). Since endo-1,4-β-glucanase mRNA levels began to rise without delay after auxin treatment and increased several fold before an increase in enzyme activity was detected at 24 h, it was suggested that endo-1,4-β-glucanase synthesis may be regulated at both the transcriptional and translational levels *in vivo* (*108*). Auxin also increased extractable buffer-soluble CMCase activity, though to a much lesser extent, in differentiating callus of rice, a monocotyledonous plant (*109*).

During a 48 h auxin treatment of pea epicotyls, lateral swelling was promoted, levels of extractable endo-1,4-β-glucanase activity increased 85-fold, and the M_r of extractable xyloglucan was reduced from a D.P. of 760 to 170, whereas cellulose D.P. showed little change (*64*). Induction of endo-1,4-β-glucanase activity was elicited by concentrations of auxin high enough to promote radial expansion rather than elongation of the epicotyls, but did not appear to be mediated through increased ethylene production since treatment with ethylene alone caused a similar lateral swelling response but was without effect on basal CMCase levels (*106,110*). Treatment of pea epicotyls with ethylene in addition to auxin strongly reduced the increase in CMCase activity evoked by auxin (*106*). In contrast, a 3-d treatment of sunflower stem with ethylene substantially increased the total extractable CMCase activity (*82*). Application of auxin and ethylene to bean stem and petiole tissue suggested that auxin enhanced the formation of acidic pI endo-1,4-β-glucanases, whereas ethylene stimulated the formation of a basic pI endo-1,4-β-glucanase (*76*). Two acidic, auxin-stimulated endo-1,4-β-glucanases of apparent M_r of 70 kDa have been identified in bean, one with a pI of 4.8 found in young rapidly-growing tissue, and the other with a pI of 4.5 found associated with plasma membranes (*60*).

Gibberellic acid caused a small promotion of CMCase specific activity in growing rice leaves (*78*), and a small increase in total activity but a decrease in specific activity in growing dwarf pea internodes (*111*).

Antibodies raised against pea endo-1,4-β-glucanases showed that the BI endo-1,4-β-glucanase was localized at the inner surface of the cell wall, which is thought to be the load-bearing region of the wall (*112*), while the BS endo-1,4-β-glucanase was found in ER and ER vesicles (*113*). Cell wall xyloglucan has been considered to be the most likely substrate for pea epicotyl endo-1,4-β-glucanase activity *in vivo* (*66*), and correlations have been made between auxin-stimulated growth, xyloglucan metabolism and endo-1,4-β-glucanase activity. Despite the tight binding of xyloglucan to cellulose, cellulose-bound xyloglucan was still capable of hydrolysis by endo-1,4-β-glucanase (*64*). Indeed, the auxin-induced BI endo-1,4-β-glucanase of pea had a greater affinity for pea xyloglucan (apparent K_m 2.5 $mg.ml^{-1}$) than it did for cellohexaose (apparent K_m 3.8 $mg.ml^{-1}$), which together with the preferential hydrolysis of xyloglucan from xyloglucan-cellulose macromolecular complexes suggested that xyloglucan may be the more accessible and preferred substrate *in vivo* (*64*). The regions of xyloglucan molecules spanning different microfibrils, and potentially forming load-bearing bonds, may be the most accessible to endo-1,4-β-glucanase.

In pea epicotyl segments, pulse-chase labeling studies indicated that a labeled xylose- and glucose-containing polymer declined in pectinase-soluble fractions and increased in water-soluble fractions after a 7-h treatment with auxin (*114*). The increase in putative soluble xyloglucan was detected within 15 min of auxin treatment (*115*), about the same lag time as that between auxin presentation and the increase in growth rate (*62*). The change in xyloglucan metabolism increased with increasing auxin dose and growth, and was due to a direct effect of auxin itself and not to cell expansion since osmotically blocking growth with mannitol did not prevent the increase in soluble xyloglucan (*115*). The same effect was seen in apoplastic fluids centrifuged from pea epicotyl segments treated with auxin for 30 min, where the xylose and glucose was shown to be in an ethanol-insoluble polymer which possessed the cellulose-binding characteristics of xyloglucan (*116*). In azuki bean epicotyl segments incubated in auxin for 3-4 h, a reduction in peak M_r of 14-26% relative to controls was noted in a 24% KOH-soluble xyloglucan fraction (*117,118*). A similar finding was made in oat coleoptiles (*119*), despite the low levels of xyloglucan in graminaceous monocots.

Treatment of abraded excised segments with various lectins inhibited auxin-stimulated growth, indicating the possible involvment in cell wall loosening of cell wall polymers (or glycoproteins) containing Man (or Glc) and Fuc in azuki bean epicotyl, and Man (or Glc), GlcNAc, GalNAc, and GlcUA in oat coleoptile (*120*). In azuki bean epicotyl segments, a Fuc-binding lectin suppressed the effects of auxin on growth, cell wall loosening (decrease in the minimum stress-relaxation time of the walls) and the reduction in the peak M_r of 24% KOH-soluble xyloglucan (*118*). Polyclonal antibodies raised against xyloglucan oligosaccharides also inhibited auxin effects on growth, cell wall loosening and xyloglucan peak M_r in azuki bean (*121*). In the coleoptiles of graminaceous monocots, the antibodies prevented the auxin-induced reduction in xyloglucan peak M_r but did not suppress growth or cell wall loosening, suggesting that xyloglucan metabolism is associated with cell wall loosening and growth in dicots but not in the Poaceae.

In pea epicotyl segments auxin caused a 2- to 3-fold decrease in the peak M_r of xyloglucan within 30 min, relative to controls without auxin (*122*). Increases in xyloglucan peak M_r were also observed, with newly-synthesized xyloglucan (determined by [^{3}H]-Fuc labeling) increasing in size from the equivalent of 9 kDa dextrans shortly after deposition to the equivalent of 30 kDa dextrans by 30 min, perhaps representing its integration into the wall. Auxin evoked a large reduction (to as small as 5 kDa) in the peak M_r of this newly-synthesized xyloglucan, beginning after 60 min. Auxin treatment also prevented the effect of a step increase in turgor, which in control tissue resulted in a 10-fold increase in M_r of previously-deposited xyloglucan within 30 min, followed by a decline back to its initial value by 90 min. The increases and decreases in xyloglucan chain length and the lack of a simple relationship between xyloglucan M_r and growth rate indicate that wall loosening must also involve factors other than endo-1,4-β-glucanase activity, two of which may be the incorporation into the wall of newly-synthesized low-M_r xyloglucan and xyloglucan transglycosylation.

While endo-1,4-β-glucanase is clearly capable of degrading cell wall xyloglucan, and while xyloglucan turnover may be associated with cell wall loosening, evidence that xyloglucan cleavage mediated by endo-1,4-β-glucanase *in vivo* is responsible for growth is lacking. The increase in extractable endo-1,4-β-glucanase activity by auxin takes several days, whereas growth and changes in xyloglucan M_r are stimulated in minutes or hours. Furthermore, the auxin-induction of endo-1,4-β-glucanase requires high concentrations of auxin, much higher than the 10 μM usually used in the studies showing promotion of growth and reduction in peak xyloglucan M_r. Auxin may promote the activation or secretion into the wall of low levels of pre-existing constitutive endo-1,4-β-glucanase, but the complex modifications of xyloglucan M_r described by Talbott and Ray (*122*) suggest that more than mere backbone cleavage is involved in xyloglucan metabolism. Whether growth is dependent mainly upon the action of endo-1,4-β-glucanases, other cell wall hydrolases, XET (*7*), enzyme-mediated loosening of hydrogen bonds between cellulose and xyloglucan (*123*), the incorporation of new cell wall polymers into the wall (*124*), other factors, or more likely all of these, has not been established.

Production of Oligosaccharins. Endo-1,4-β-glucanases can cleave dicot stem xyloglucan at every unsubstituted, i.e., at about every fourth, Glc residue to produce a mixture of hepta-and nonasaccharides (*64,125*). The fucosylated nonasaccharide ($Glc_4.Xyl_3$.Gal.Fuc) has been found to evoke small inhibitions of auxin-stimulated growth at low (nM) concentrations (*126,127*), and to cause small promotions of growth in the absence of added auxin at higher (μM) concentrations (*72*). Such biologically-active oligosaccharides have been termed oligosaccharins. The metabolism and structural requirements for oligosaccharin activity have recently been reviewed by Aldington and Fry (*128*).

Xyloglucan nonasaccharide has been detected *in vivo*, accumulating extracellularly to a steady-state concentration of 430 nM in the medium of suspension-cultured spinach cells (*129*). However, effects of the xyloglucan nonasaccharide on auxin-stimulated growth were observed only with the synthetic auxin 2,4-D, and not in experiments using the natural auxin IAA (*130*).

Addition of various xyloglucan oligosaccharides to crude pea or bean extracts rapidly stimulated viscosity loss versus xyloglucan but not versus CMC (*71,72*). While it is possible that this is due to some form of oligosaccharide-stimulated endo-1,4-β-glucanase, the incorporation of some of the oligosaccharides into high M_r xyloglucan suggests that their use as substrate by XET is more likely (*7*). How xyloglucan oligosaccharides affect growth *in vivo* is not known, but the promotion of growth at μM concentrations could conceivably be due to XET-mediated wall loosening by the transglycosylation of load-bearing xyloglucan molecules in the wall to added oligosaccharides. Endo-1,4-β-glucanases may thus play a role in the production of biologically-active xyloglucan oligosaccharides which could have growth-regulatory properties.

Abscission. During the shedding of certain plant parts, abscission zones develop in which localized cell wall breakdown occurs prior to separation of the subtending organ (*131*). The accumulation of endo-1,4-β-glucanase and the abscission process itself are accelerated by exogenous ethylene and retarded by exogenous auxin in leaf (*85*) and fruit (*86,91*) abscission zones. Inhibitors of ethylene synthesis or action delayed abscission (*132,133*) as well as suppressed the accumulation of endo-1,4-β-glucanase (*133*). Ethylene treatment promoted levels of extractable CMCase activity from bean leaf abscission zones in 6 h (*87*) and from fruit abscission zones of orange within 24 h (*91*), and experiments with protein synthesis inhibitors and labeled tracers suggested that the endo-1,4-β-glucanase activity was formed *de novo* during abscission (*86,87,91,134*). In bean leaf abscission zones at least two forms of endo-1,4-β-glucanase were detected, one with a pI of 4.5 and one with a pI of 9.5 (*39*). The pI 9.5 endo-1,4-β-glucanase had an apparent M_r of 51 kDa (*60*) and did not appear to be antigenically related to the 70 kDa pI 4.5 form (*135*). Low levels of CMCase activity present in non-abscising tissues were due to the pI 4.5 endo-1,4-β-glucanase (*39, 136*). Immunodetection of the pI 9.5 endo-1,4-β-glucanase indicated that this enzyme was virtually undetectable prior to the induction of abscission, but increased dramatically between 18 and 40 h of ethylene treatment (*136*). The pI 9.5 endo-1,4-β-glucanase correlated with the abscission process in appearance, specific localization and activity (*103,136,137*). In contrast, the amount of pI 4.5 endo-1,4-β-glucanase declined during abscission and was not correlated with weakening of the abscission zone (*136*). In leaf and fruit abscission zones of peach, endo-1,4-β-glucanases with pIs of 6.5 and 9.5 were detected (*88*).

The mRNA encoding the pI 9.5 endo-1,4-β-glucanase from bean leaf abscission zones showed regulation of abundance by ethylene (*138*). The mRNA and pI 9.5 endo-1,4-β-glucanase protein levels were detectable by 24 h of ethylene treatment, whereas a prior treatment with auxin or a subsequent treatment with the ethylene action inhibitor 2,5-norbornadiene reduced bean abscission zone endo-1,4-β-glucanase mRNA to undetectable levels (*138*). In tomato, very low levels of mRNA encoding two endo-1,4-β-glucanases, tomCel1 and tomCel2, were present in most tissues of the plant (*18*). However, relatively higher levels of both mRNAs were found in ageing abscission zones and fruit. In abscission zones maximal levels of tomCel1 mRNA were about 4-fold higher than those of tomCel2, and treatment with exogenous ethylene accelerated the time required for maximal accumulation of tomCel1 mRNA by about 2-fold. Treatment with 2,5-norbornadiene reduced both mRNAs to barely detectable levels, suggesting that transcription of both endo-1,4-β-glucanase genes was activated by ethylene.

In bean, immunological localization of pI 9.5 endo-1,4-β-glucanase protein using microscopy (*103*) or tissue prints (*137*) together with *in situ* hybridizations to localize BAC1 mRNA (*139*) showed that both protein and mRNA were found in the two or three cell layers adjacent to each side of the fracture plane and extended a few mm along the vascular traces on either side (Figure 4). How expression is induced and restricted so precisely to just these few cell layers is not known.

Endo-1,4-β-glucanase activity increased in abscission zones prior to or coincident with the weakening of the cell separation layer (*91,103,134*). Increases in endo-1,4-β-glucanase activity associated with abscission were detected only in a fraction requiring high salt for extraction; i.e., presumably associated with the cell wall (*134*). Ethylene treatment increased the release or secretion into extracellular fluids of CMCase activity (*140*), shown subsequently by isoelectric focusing to possess a pI of 9.5 (*39*). The decline in break strength of the zones correlated with levels of extracellular rather than intracellular endo-1,4-β-glucanase (*140*). The important role played in abscission by pI 9.5 endo-1,4-β-glucanase was shown by Sexton *et al.* (*103*), who injected antibodies raised against the purified protein into bean explant abscission zones, reducing extractable CMCase activity by 75% and increasing average break strength from 49 g in pre-immune serum controls to such an extent that 90% of zones did not break even if 200 g was applied.

In most abscission zones, cell wall breakdown appeared limited to the middle lamella along which the fracture usually occurs, leaving a high proportion of cells on the fracture faces intact (*141,142*). Abscission thus occurs between intact, living cells and is not due to a localized form of senescence or cell death. Structural studies suggested that cell separation occurred primarily as the result of dissolution of the middle lamella region of the walls (*142*), which is composed mainly of pectic polymers lacking the bonds normally attacked by endo-1,4-β-glucanases. However polygalacturonase activity in abscission zones was also promoted by ethylene, and increased prior to abscission (*91,143*). As break strength fell and polygalacturonase activity rose, the amount of water-extractable pectic sugars increased while there was a corresponding decline in the pectic fractions extractable with dilute acid (*144*), indicating breakdown of pectic chains. Endo-1,4-β-glucanase activity may thus play an important role in abscission, but may not be sufficient in itself to cause cell separation. Durbin *et al.* (*136*) have suggested that endo-1,4-β-glucanase action may make the wall more permeable to other enzymes, increasing the pore size of the wall and perhaps allowing polygalacturonase to reach its site of action. Most evidence of cell wall disassembly in abscission zones is from cytochemical observations, and data on *in situ* changes in M_r of potential endo-1,4-β-glucanase substrates, such as xyloglucan, is urgently needed.

Fruit Ripening. As fruits ripen, changes occur in color, aroma and flavor, and textural and structural alterations lead to a loss of firmness (*145*). Correlations have been made between increasing levels of extractable CMCase activity and fruit softening during ripening in date (*93*), tomato (*92*), peach (*94*), avocado (*96*), mango

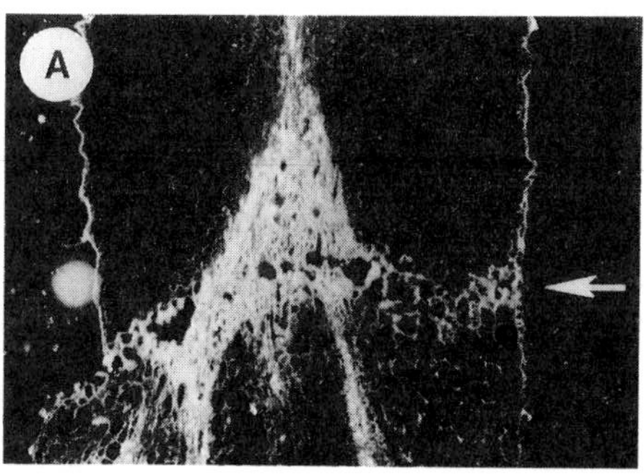

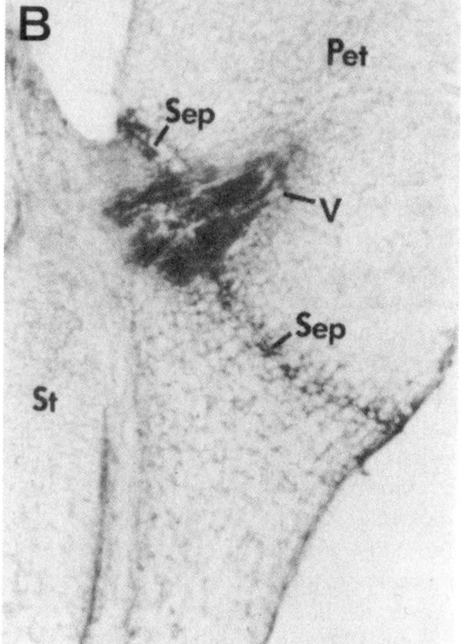

Figure 4. Bean abscission zones probed with a radioactive antisense cDNA probe to localize BAC1 endo-1,4-β-glucanase mRNA (panel A), and with a rabbit antibody followed by goat anti-rabbit IgG-peroxidase to localize pI 9.5 endo-1,4-β-glucanase protein (panel B). In panel A, the separation layer is marked with an arrow; in panel B, Pet, petiole; Sep, separation layer; St, stem; V, vascular trace. (Reproduced with permission from: panel A, Tucker *et al.* (*139*), Copyright 1991 Springer-Verlag; panel B, Sexton *et al.* (*103*), Copyright 1980 Macmillan Magazines Limited).

(*101*), and other fruits (*145-147*). The cell wall changes that lead to fruit softening are not known and may differ between species (*145*), although there are significant similarities between the two best-studied systems, tomato and avocado.

Levels of extractable CMCase activity rose in ripening fruit in a manner correlated with ethylene evolution (*96,98*), and treatment of tomato (*92*) and mature avocado (*96*) fruit with exogenous ethylene strongly increased extractable CMCase activity levels. In tomato, extractable CMCase activity increased in immature developing fruit, declined and then rose again during ripening (*92,148*). Treatment of tomato fruit with gibberellic acid or auxin delayed softening and the rise in endo-1,4-β-glucanase levels (*92*). The levels of two endo-1,4-β-glucanase mRNAs, tomCel1 and tomCel2, increased during tomato fruit ripening (*18*). The mRNA corresponding to tomCel1 showed a transient rise during the late phases of fruit expansion, with re-induction during early stages of fruit ripening. In contrast, tomCel2 mRNA did not appear until the initiation of fruit ripening, with levels increasing progressively as ripening proceeded. The maximal levels of tomCel2 mRNA were approximately 20-fold higher than those of tomCel1 mRNA, the opposite of their relative abundances in abscission zones. Levels of both tomCel1 and tomCel2 mRNA were strongly depressed by 2,5-norbornadiene, suggesting that both genes are activated by ethylene in ripening fruit .

In tomato, the rise in endo-1,4-β-glucanase during ripening was followed by a rise in polygalacturonase activity (*92*), and pectic cell wall components exhibited degradation during ripening (*149*). However, genetic down-regulation of polygalacturonase levels in wild-type fruit and its over-expression in the non-softening *rin* mutant of tomato (which lacks the enzyme) showed that polyuronide degradation by polygalacturonase was not sufficient for fruit softening, and contributed primarily to tissue deterioration very late in ripening (*147*). As ripening progressed, no change in M_r of a cell wall hemicellulose fraction extractable by 4 M KOH was detected but substantial depolymerization was seen in a 8 M KOH-extractable fraction, which was presumably more firmly attached to cellulose (*150*). Using an iodine staining method specific for xyloglucan, comparisons of cell wall fractions from green and red fruit showed that a high molecular weight xyloglucan fraction was degraded during ripening (*151*). Multiple endo-1,4-β-glucanase genes (Table I) and multiple activities have been described in tomato, with higher activity levels observed in locules than in pericarp (*73,152*). However, the non-softening *rin* mutant apparently possesses near wild-type levels of extractable CMCase activity (*153,154*), suggesting that endo-1,4-β-glucanase-mediated polysaccharide depolymerization is not solely responsible for fruit softening in tomato. XET activity has also been detected, which may participate in xyloglucan metabolism (*73*).

Avocado fruit ripen only after detachment from the tree, after which levels of endo-1,4-β-glucanase and polygalacturonase activity begin to rise (*155*). Extractable CMCase activity started to increase 3 d before polygalacturonase activity could be detected (*155*) and rose to very high levels, coincident with fruit softening (*96*). The levels of an endo-1,4-β-glucanase mRNA (avoCel1) increased more than 50-fold during fruit ripening (*45*), and accumulation of the corresponding protein was due to *de novo* synthesis (*43*). The major avocado endo-1,4-β-glucanase protein possessed a pI of 4.7 and apparent M_r of 49 kDa (*37*) or pI of 5.95 and apparent M_r of 54 kDa (*38*), depending on cultivar. Hybrid selection of mRNA with the avoCel1 cDNA clone followed by *in vitro* translation suggested that treatment of pre-climacteric fruit with ethylene induced the appearance of a single endo-1,4-β-glucanase mRNA within 8 h, but that several mRNAs were apparent by 30 h of ethylene treatment or in ripe, untreated fruit (*44*). Maintenance of pre-climacteric avocado fruit in low oxygen atmospheres, which interferes with the synthesis and action of ethylene and retards fruit ripening and softening, strongly reduced both the accumulation of endo-1,4-β-glucanase transcripts and the levels of immunologically-detectable endo-1,4-β-glucanase proteins (*156*). In excised disks of avocado fruit, the accumulation of endo-1,4-β-glucanase activity, immunologically-detectable endo-1,4-β-glucanase protein and softening were prevented by aminoethoxyvinylglycine (AVG, an inhibitor of ethylene synthesis), 2,5-norbornadiene (an inhibitor of ethylene action) and auxin (an inhibitor of ripening) (*157*). However, at the concentrations used, these ethylene antagonists did not suppress the accumulation of endo-1,4-β-glucanase mRNA, indicating that endo-1,4-β-glucanase gene expression is not controlled exclusively at the transcriptional level. It was suggested that ethylene at low concentrations controls transcription, with higher concentrations mediating post-transcriptional events including translation (*157*).

The avocado endo-1,4-β-glucanase has been immunolocalized to the cell wall (*158*). Observations by electron microscopy suggested that the first visible change in the wall during softening was a breakdown of the pectic middle lamella (*159*), followed by reductions in wall matrix material and disruption of the wall's fibrillar organization (*96,159*). Extensive depolymerization of avocado polyuronides was evident in a chelator-extractable cell wall fraction (*149*). When hemicelluloses, a substantial proportion of which was xyloglucan, were extracted from cell walls by 4 M or 8 M KOH they showed a much more limited general decline in average M_r during ripening, the peak M_r falling from 54 to only 43 kDa, expressed as dextran equivalents (*42*). Separation of CMCase activity from other cell wall hydrolases revealed that the purified avocado endo-1,4-β-glucanase caused no observable decrease in avocado xyloglucan M_r *in vitro*, although a small decrease in M_r of mid-sized, non-xyloglucan polysaccharides occurred. Furthermore, the CMCase-depleted fraction brought about the same small reductions in xyloglucan M_r as did the total protein extract. Similarly, Hatfield and Nevins (*65*) found that purified avocado endo-1,4-β-glucanase showed very limited activity against soybean hypocotyl cell wall xyloglucan, and when added to an avocado fruit cell wall preparation caused a release of polymers containing Ara and Gal but only a minor release of Glc. Although it cannot be ruled out that exogenous endo-1,4-β-glucanase did not penetrate to the *in vivo* sites of action, or that changes in non-released polymers were responsible for the release of the polymers observed, these results all suggest that the depolymerization of xyloglucan in softening avocado did not occur through the exclusive action of endo-1,4-β-glucanase. The purified enzyme did however have a high activity versus 1,4-β-Glc linkages (*65*), and it is possible that it may be able to attack avocado xyloglucan in combination with other enzymes, for example after glycosidases have reduced its degree of substitution.

Decreases in hemicellulose polymerization during softening have also been reported in strawberry (*160*), pepper (*161*) and muskmelon (*162*). However, in muskmelon levels of extractable CMCase activity declined during softening (*99*). The contribution of endo-1,4-β-glucanases to cell wall disassembly in ripening fruits remains obscure. Little information is available on fruit cell wall composition and structure, and the precise *in vivo* substrates of endo-1,4-β-glucanase and the cell wall changes that lead to wall softening have not been established.

Cell Wall Dissolution. In certain very specialized cells, areas of the cell wall can disappear completely during development. This happens during the development of the flower reproductive organs, where high levels of an endo-1,4-β-glucanase antigenically related to abscission zone endo-1,4-β-glucanase have been found (*58,84*). Similar breakdown occurs during perforation of the end walls of living phloem sieve tubes, and in the complete dissolution of end walls in the production of dead xylem vessels and articulated laticifers. High levels of CMCase have been detected in the sap of these vessels (*79,80*). In developing poppy articulated laticifers, CMCase was cytochemically localized only in patches of the wall where recently-formed perforations were visible (*81*). How endo-1,4-β-glucanase expression is localized so precisely and how other regions of the wall or neighboring cells escape degradation is not known. Presumably even cellulose microfibrils are hydrolysed in these circumstances; possibly endo-1,4-β-glucanase can act on the glucan chains of the microfibrils after other enzymes have reduced the degree of crystallinity.

Conclusions

Two categories of plant endo-1,4-β-glucanase have so far been established, based on their expression patterns and hormonal regulation. One group is found mainly in rapidly-growing tissues and is induced to high levels by auxin and antagonized by ethylene. The second group is found in aging tissues like abscission zones and fruit, and is induced to high levels by ethylene and antagonized by auxin. The processes that they are correlated with, cell expansion and abscission or ripening respectively, share the same antagonistic hormonal regulation, suggesting that the endo-1,4-β-glucanases contribute to these physiological processes. However, much has yet to be learned about the roles played by endo-1,4-β-glucanases in plant development, and some important questions remain unanswered, including:

i). How many endo-1,4-β-glucanase genes are present in plants? At present, four divergent cDNA clones have been isolated from tomato (Table I), and preliminary evidence suggests that at least several more exist.

ii). Which genes are involved in which developmental processes, and what factors are involved in their regulation? How is expression localized to such an extent that only the end walls of vascular vessels are degraded, or expression limited to just a few cell layers in abscission?

iii). What are the substrates for plant endo-1,4-β-glucanases *in vivo*? The endo-1,4-β-glucanases from pea epicotyl are active *in vitro* versus xyloglucan, but that from avocado fruit is not and its natural substrate remains undefined.

iv). Are the actions of endo-1,4-β-glucanases dependent upon the prior actions of other cell wall hydrolases, and precisely which changes are responsible for cell wall loosening and disassembly?

The answers to these and similar questions are required to fully assess the role of endo-1,4-β-glucanases in plant cell wall metabolism and cell development. The availability of cloned genes corresponding to many of the major plant endo-1,4-β-glucanases promises to provide the experimental basis for obtaining these answers.

Acknowledgments

This review was prepared with support by grants from Unilever/Van den Bergh Foods and USDA-NRICGP grant number 91-37304-6508. We thank Dr Ann Powell for the provision of computing facilities, and Drs. S.D. O'Neill and W. Schuch for supplying unpublished sequence information.

Literature Cited

1. Hall, C.B. *Nature* **1963**, *200*, 1010-1011.
2. Maclachlan, G.A.; Perrault, J. *Nature* **1964**, *204*, 81-82.
3. Woodward, J.R.; Fincher, G.B. *Carbohydr. Res.* **1982**, *106*, 111-122.
4. Hatfield, R.; Nevins, D.J. *Carbohydr. Res.* **1986**, *148*, 265-278.
5. Fincher, G.B.; Lock, P.A.; Morgan, M.M.; Lingelbach, K.; Wettenhall, R.E. H.; Mercer, J.F.B.; Brandt, A.; Thomsen, K.K. *Proc. Natl. Acad. Sci. USA* **1986**, *83*, 2081-2085.
6. Wolf, N. *Mol. Gen. Genet.* **1992**, *234*, 33-42.
7. Smith, R.C.; Fry, S.C. *Biochem. J.* **1991**, *279*, 529-535.
8. Nishitani, K.; Tominaga, R. *J. Biol. Chem.* **1992**, *267*, 21058-21064.
9. Farkas, V.; Sulova, Z.; Stratilova, E.; Hanna, R.; Maclachlan, G. *Arch. Biochem. Biophys.* **1992**, *298*, 365-370.
10. Maclachlan, G.; Carrington, S. In *Biosynthesis and Biodegradation of Cellulose*; Haigler, C.H.; Weimer, P.J., Eds.; Marcel Dekker: New York, NY, 1991; pp 599-621.
11. Reese, E.T.; Siu, R.G.H.; Levinson, H.S. *J. Bacteriol.* **1950**, *59*, 485-497.
12. Klyosov, A.A. *Biochemistry* **1990**, *29*, 10577-10585.
13. Beguin, P. *Annu. Rev. Microbiol.* **1990**, *44*, 219-248.
14. Gilkes, N.R.; Henrissat, B.; Kilburn, D.G.; Miller, R.C.; Warren, R.A. *J. Microbiol. Rev.* **1991**, *55*, 305-315.
15. Henrissat, B.; Claeyssens, M.; Tomme, P.; Lemesle, L.; Mornon, J.P. *Gene* **1989**, *81*, 83-95.
16. Tucker, M.L.; Durbin, M.L.; Clegg, M.T.; Lewis, L.N. *Plant Mol. Biol.* **1987**, *9*, 197-203.
17. Tucker, M.L.; Milligan, S.B. *Plant Physiol.* **1991**, *95*, 928-933.
18. Lashbrook, C.C.; Gonzalez-Bosch, C.; Bennett, A.B. *In preparation* **1994**.
19. Berger, E.; Jones, W.A.; Jones, D.T.; Woods, D.R. *Mol. Gen. Genet.* **1990**, *223*, 310-318.

20. Coutinho, J.B.; Moser, B.; Kilburn, D.G.; Warren, R.A.J.; Miller, R.C. *Mol. Microbiol.* **1991**, *5*, 1221-1233.
21. Joliff, G.; Beguin, P.; Aubert, J.P. *Nucl. Acids Res.* **1986**, *14*, 8605-8613.
22. Hall, J.; Gilbert, H.J. *Mol. Gen. Genet.* **1988**, *213*, 112-117.
23. Meinke, A.; Braun, C.; Gilkes, N.R.; Kilburn, D.G.; Miller, R.C.; Warren, R.A.J. *J. Bacteriol.* **1991**, *173*, 308-314.
24. Navarro, A.; Chebrou, M.C.; Beguin, P.; Aubert, J.P. *Res. Microbiol.* **1991**, *142*, 927-936.
25. Jauris, S.; Rucknagel, K.P.; Schwarz, W.H.; Kratzsch, P.; Bronnenmeier, K.; Staudenbauer, W.L. *Mol. Gen. Genet.* **1990**, *223*, 258-267.
26. Cavicchioli, R.; East, P.D.; Watson, K. *J. Bacteriol.* **1991**, *173*, 3265-368.
27. Lao, G.; Ghangas, G.S.; Jung, E.D.; Wilson, D.B. *J. Bacteriol.* **1991**, *173*, 3397-3407.
28. Giorda, R.; Ohmachi, T.; Shaw, D.R.; Ennis, H.L. *Biochemistry* **1990**, *29*, 7264-7269.
29. Kemmerer, E.C.; Tucker, M.L. *Plant Physiol.* **1994**, *104*, 557-562.
30. Cass, L.G.; Kirven, K.A.; Christoffersen, R.E. *Mol. Gen. Genet.* **1990**, *223*, 76-86.
31. Tomme, P.; van Beeumen, J.; Claeyssens, M. *Biochem. J.* **1992**, *285*, 319-324.
32. Juy, M.; Amit, A.G.; Alzari, P.M.; Poljak, R.J.; Claeyssens, M.; Beguin, P.; Aubert, J.P. *Nature* **1992**, *357*, 89-91.
33. Tomme, P.; Chauvaux, S.; Beguin, P.; Millet, J.; Aubert, J.P.; Claeyssens, M. *J. Biol. Chem.* **1991**, *266*, 10313-10318.
34. Chauvaux, S.; Beguin, P.; Aubert, J.P. *J. Biol. Chem.* **1992**, *267*, 4472-4478.
35. Higgins, D.G.; Bleasby, A.J.; Fuchs, R. *CABIOS* **1992**, *8*, 189-191.
36. Swofford, D.L. *PAUP: Phylogenetic Analysis Using Parsimony*, vers. 3.0s; Illinois Natural History Survey: Champaign, IL, 1991.
37. Awad, M.; Lewis, L.N. *J. Food Sci.* **1980**, *45*, 1625-1628.
38. Kanellis, A.K.; Kalaitzis, P. *Plant Physiol.* **1992**, *98*, 530-534.
39. Reid, P.D.; Strong, H.G.; Lew, F.; Lewis, L.N. *Plant Physiol.* **1974**, *53*, 732-737.
40. von Heijne, G. *Nucl. Acids Res.* **1986**, *14*, 4683-4690.
41. Bennett, A.B.; Christoffersen, R.E. *Plant Physiol.* **1986**, *81*, 830-835.
42. O'Donoghue, E.M.; Huber, D.J. *Physiol. Plant.* **1992**, *86*, 33-42.
43. Tucker, M.L.; Laties, G.G. *Plant Physiol.* **1984**, *74*, 307-315.
44. De Francesco, L.; Tucker, M.L.; Laties, G.G. *Plant Physiol. Biochem.* **1989**, *27*, 325-332.
45. Christoffersen, R.E.; Tucker, M.L.; Laties, G.G. *Plant Mol. Biol.* **1984**, *3*, 385-391.
46. McNeil, M.; Darvill, A.G.; Fry, S.C.; Albersheim, P. *Annu. Rev. Biochem.* **1984**, *53*, 625-663.
47. Hayashi, T. *Annu. Rev. Plant Physiol. Plant Mol. Biol.* **1989**, *40*, 139-168.
48. Carpita, N.C.; Gibeaut, D.M. *Plant J.* **1993**, *3*, 1-30.
49. McCann, M.C.; Wells, B.; Roberts, K. *J. Cell Sci.* **1990**, *96*, 323-334.
50. Talbott, L.D.; Ray, P.M. *Plant Physiol.* **1992**, *98*, 357-368.

51. Bacic, A.; Harris, P.J.; Stone, B.A. In *Carbohydrates*; Preiss, J., Ed.; Biochemistry of Plants: A Comprehensive Treatise, Vol. 14; Academic Press: San Diego, CA, 1988; pp 297-371.
52. Hayashi, T.; Maclachlan, G. *Plant Physiol.* **1984**, *75*, 596-604.
53. Hayashi, T.; Marsden, M.P.F.; Delmer, D.P. *Plant Physiol.* **1987**, *83*, 384-389.
54. Moore, P.J.; Staehelin, L.A. *Planta* **1988**, *174*, 433-445.
55. Byrne, H.; Christou, N.V.; Verma, D.P.S.; Maclachlan, G. *J. Biol. Chem.* **1975**, *250*, 1012-1018.
56. Truelsen, T.A.; Wyndaele, R. *J. Plant Physiol.* **1991**, *139*, 129-134.
57. Nakamura, S.; Hayashi, T. *Plant Cell Physiol.* **1993**, *34*, 1009-1013.
58. Sexton, R.; del Campillo, E.; Duncan, D.; Lewis, L.N. *Plant Sci.* **1990**, *67*, 169-176.
59. Lew, F.T.; Lewis, L.N. *Phytochemistry* **1974**, *13*, 1359-1366.
60. Durbin, M.L.; Lewis, L.N. *Methods Enzymol.* **1988**, *160*, 342-351.
61. Webb, S.T.J.; Taylor, J.E.; Coupe, S.A.; Ferrarese, L.; Roberts, J.A. *Plant Cell Environ.* **1993**, *16*, 329-333.
62. Brummell, D.A.; Hall, J.L. *Plant Cell Environ* **1987**, *10*, 523-543.
63. Wong, Y.S.; Fincher, G.B.; Maclachlan, G.A. *J. Biol. Chem.* **1977**, *252*, 1402-1407.
64. Hayashi, T.; Wong, Y.S.; Maclachlan, G. *Plant Physiol.* **1984**, *75*, 605-610.
65. Hatfield, R.; Nevins, D.J. *Plant Cell Physiol.* **1986**, *27*, 541-552.
66. Maclachlan, G. *Methods Enzymol.* **1988**, *160*, 382-391.
67. Gidley, M.J.; Lillford, P.J.; Rowlands, D.W.; Lang, P.; Dentini, M.; Crescenzi, V.; Edwards, M.; Fanutti, C.; Reid, J.S.G. *Carbohydr. Res.* **1991**, *214*, 299-314.
68. Mori, M.; Eda, S.; Kato, K. *Carbohydr. Res.* **1980**, *84*, 125-135.
69. de Silva, J.; Jarman, C.D.; Arrowsmith, D.A.; Stronach, M.S.; Chengappa, S.; Sidebottom, C.; Reid, J.S.G. *Plant J.* **1993**, *3*, 701-711.
70. Fanutti, C.; Gidley, M.J.; Reid, J.S.G. *Plant J.* **1993**, *3*, 691-700.
71. Farkas, V.; Maclachlan, G. *Carbohydr. Res.* **1988**, *184*, 213-219.
72. McDougall, G.J.; Fry, S.C. *Plant Physiol.* **1990**, *93*, 1042-1048.
73. Maclachlan, G.; Brady, C. *Aust. J. Plant Physiol.* **1992**, *19*, 137-146.
74. Fan, D.F.; Maclachlan, G.A. *Can. J. Bot.* **1966**, *44*, 1025-1034.
75. Koyama, T.; Hayashi, T.; Kato, Y.; Matsuda, K. *Plant Cell Physiol.* **1981**, *22*, 1191-1198.
76. Linkins, A.E.; Lewis, L.N.; Palmer, R.L. *Plant Physiol.* **1973**, *52*, 554-560.
77. Tracey, M.V. *Biochem. J.* **1950**, *47*, 431-433.
78. Prakash, L.; Prathapasenan, G. *Ann. Bot.* **1990**, *65*, 251-257.
79. Sheldrake, A.R. *Planta* **1970**, *95*, 167-178.
80. Sheldrake, A.R.; Moir, G.F.J. *Physiol. Plant.* **1970**, *23*, 267-277.
81. Nessler, C.L.; Mahlberg, P.G. *Amer. J. Bot.* **1981**, *68*, 730-732.
82. Kawase, M. *Amer. J. Bot.* **1979**, *66*, 183-190.
83. Sanchez, R.A.; de Miguel, L.; Mercuri, O. *J. Exp. Bot.* **1986**, *37*, 1574-1580.
84. del Campillo, E.; Lewis, L. N. *Plant Physiol.* **1992**, *99*, 1015-1020.
85. Horton, R.F.; Osborne, D.J. *Nature* **1967**, *214*, 1086-1088.
86. Ratner, A.; Goren, R.; Monselise, S.P. *Plant Physiol.* **1969**, *44*, 1717-1723.
87. Abeles, F.B. *Plant Physiol.* **1969**, *44*, 447-452.

88. Bonghi, C.; Rascio, N.; Ramina, A.; Casadoro, G. *Plant Mol. Biol.* **1992**, *20*, 839-848.
89. Tucker, G.A.; Schindler, C.B.; Roberts, J.A. *Planta* **1984**, *160*, 164-167.
90. Rasmussen, G.K. *Plant Physiol.* **1975**, *56*, 765-767.
91. Greenberg, J.; Goren, R.; Riov, J. *Physiol. Plant.* **1975**, *34*, 1-7.
92. Babbitt, J.K.; Powers, M.J.; Patterson, M.E. *J. Amer. Soc. Hort. Sci.* **1973**, *98*, 77-81.
93. Hasegawa, S.; Smolensky, D.C. *J. Food Sci.* **1971**, *36*, 966-967.
94. Hinton, D.M.; Pressey, R. *J. Food Sci.* **1974**, *39*, 783-785.
95. Barnes, M.F.; Patchett, B.J. *J. Food Sci.* **1976**, *41*, 1392-1395.
96. Pesis, E.; Fuchs, Y.; Zauberman, G. *Plant Physiol.* **1978**, *61*, 416-419.
97. Yamaki, S.; Kakiuchi, N. *Plant Cell Physiol.* **1979**, *20*, 301-309.
98. Paull, R.E.; Chen, N.J. *Plant Physiol.* **1983**, *72*, 382-385.
99. Lester, G.E.; Dunlap, J.R. *Sci. Hort.* **1985**, *26*, 323-331.
100. Abeles, F.B.; Takeda, F. *HortScience* **1989**, *24*, 851.
101. Abu-Sarra, A.F.; Abu-Goukh, A.A. *J. Hort. Sci.* **1992**, *67*, 561-568.
102. Fan, D.F.; Maclachlan, G.A. *Can. J. Bot.* **1967**, *45*, 1837-1844.
103. Sexton, R.; Durbin, M.L.; Lewis, L.N.; Thomson, W.W. *Nature* **1980**, *283*, 873-874.
104. Edwards, M.; Dea, I.C.M.; Bulpin, P.V.; Reid, J.S.G. *J. Biol. Chem.* **1986**, *261*, 9489-9494.
105. Fan, D.F.; Maclachlan, G.A. *Plant Physiol.* **1967**, *42*, 1114-1122.
106. Ridge, I.; Osborne, D.J. *Nature* **1969**, *223*, 318-319.
107. Davies, E.; Ozbay, O. *Z. Pflanzenphysiol.* **1980**, *99*, 461-469.
108. Verma, D.P.S.; Maclachlan, G.A.; Byrne, H.; Ewings, D. *J. Biol. Chem.* **1975**, *250*, 1019-1026.
109. Yoshida, K.; Komae, K. *Plant Cell Physiol.* **1993**, *34*, 507-514.
110. Hayashi, T.; Maclachlan, G. *Plant Physiol.* **1984**, *76*, 739-742.
111. Broughton, W.J.; McComb, A.J. *Ann. Bot.* **1971**, *35*, 213-228.
112. Taiz, L. *Annu. Rev. Plant Physiol.* **1984**, *35*, 585-657.
113. Bal, A.K.; Verma, D.P.S.; Byrne, H.; Maclachlan, G.A. *J. Cell Biol.* **1976**, *69*,97-105.
114. Labavitch, J.M.; Ray, P.M. *Plant Physiol.* **1974**, *53*, 669-673.
115. Labavitch, J.M.; Ray, P.M. *Plant Physiol.* **1974**, *54*, 499-502.
116. Terry, M.E.; Jones, R.L.; Bonner, B.A. *Plant Physiol.* **1981**, *68*, 531-537.
117. Nishitani, K.; Masuda, Y. *Plant Cell Physiol.* **1983**, *24*, 345-355.
118. Hoson, T.; Masuda, Y. *Physiol. Plant.* **1991**, *82*, 41-47.
119. Inouhe, M.; Yamamoto, R.; Masuda, Y. *Plant Cell Physiol.* **1984**, *25*, 1341-1351.
120. Hoson, T.; Masuda, Y. *Physiol. Plant.* **1987**, *71*, 1-8.
121. Hoson, T.; Masuda, Y.; Sone, Y.; Misaki, A. *Plant Physiol.* **1991**, *96*, 551-557.
122. Talbott, L.D.; Ray, P.M. *Plant Physiol.* **1992**, *98*, 369-379.
123. McQueen-Mason, S.; Durachko, D.M.; Cosgrove, D.J. *Plant Cell* **1992**, *4*, 1425-1433.
124. Brummell, D.A.; Hall, J.L. *Physiol. Plant.* **1985**, *63*, 406-412.
125. Bauer, W.D.; Talmadge, K.W.; Keegstra, K.; Albersheim, P. *Plant Physiol.* **1973**, *51*, 174-187.

126. York, W.S.; Darvill, A.G.; Albersheim, P. *Plant Physiol.* **1984**, *75*, 295-297.
127. McDougall, G.J.; Fry, S.C. *Planta* **1988**, *175*, 412-416.
128. Aldington, S.; Fry, S.C. *Adv. Bot. Res.* **1993**, *19*, 1-101.
129. Fry, S.C. *Planta* **1986**, *169*, 443-453.
130. Hoson, T.; Masuda, Y. *Plant Cell Physiol.* **1991**, *32*, 777-782.
131. Sexton, R.; Roberts, J.A. *Annu. Rev. Plant Physiol.* **1982**, *33*, 133-162.
132. Baird, L.M.; Reid, M.S.; Webster, B.D. *J. Plant Growth Regul.* **1984**, *3*, 217-225.
133. Sisler, E.C.; Goren, R.; Huberman, M. *Physiol. Plant.* **1985**, *63*, 114-120.
134. Lewis, L.N.; Varner, J.E. *Plant Physiol.* **1970**, *46*, 194-199.
135. del Campillo, E.; Durbin, M.; Lewis, L.N. *Plant Physiol.* **1988**, *88*, 904-909.
136. Durbin, M.L.; Sexton, R.; Lewis, L.N. *Plant Cell Environ.* **1981**, *4*, 67-73.
137. del Campillo, E.; Reid, P.D.; Sexton, R.; Lewis, L.N. *Plant Cell* **1990**, *2*, 245-254.
138. Tucker, M.L.; Sexton, R.; del Campillo, E.; Lewis, L.N. *Plant Physiol.* **1988**, *88*, 1257-1262.
139. Tucker, M.L.; Baird, S.L.; Sexton, R. *Planta* **1991**, *186*, 52-57.
140. Abeles, F.B.; Leather, G.R. *Planta* **1971**, *97*, 87-91.
141. Sexton, R. *Planta* **1976**, *128*, 49-58.
142. Sexton, R.; Hall, J.L. *Ann. Bot.* **1974**, *38*, 849-854.
143. Taylor, J.E.; Tucker, G.A.; Lasslett, Y.; Smith, C.J.S.; Arnold, C.M.; Watson, C.F.; Schuch, W.; Grierson, D.; Roberts, J.A. *Planta* **1990**, *183*, 133-138.
144. Morre, D.J. *Plant Physiol.* **1968**, *43*, 1545-1559.
145. Brady, C.J. *Annu. Rev. Plant Physiol.* **1987**, *38*, 155-178.
146. Huber, D.J. *Hort. Rev.* **1983**, *5*, 169-219.
147. Fischer, R.L.; Bennett, A.B. *Annu. Rev. Plant Physiol. Plant Mol. Biol.* **1991**, *42*, 675-703.
148. Hobson, G.E. *J. Food Sci.* **1968**, *33*, 588-592.
149. Huber, D.J.; O'Donoghue, E.M. *Plant Physiol.* **1993**, *102*, 473-480.
150. Tong, C.B.S.; Gross, K.C. *Physiol. Plant.* **1988**, *74*, 365-370.
151. Sakurai, N.; Nevins, D.J. *Physiol. Plant.* **1993**, *89*, 681-686.
152. Huber, D.J. *HortScience* **1985**, *20*, 442-443.
153. Buescher, R.W.; Tigchelaar, E.C. *HortScience* **1975**, *10*, 624-625.
154. Poovaiah, B.W.; Nukaya, A. *Plant Physiol.* **1979**, *64*, 534-537.
155. Awad, M.; Young, R.E. *Plant Physiol.* **1979**, *64*, 306-308.
156. Kanellis, A.K.; Solomos, T.; Mehta, A.M.; Mattoo, A.K. *Plant Cell Physiol.* **1989**, *30*, 817-823.
157. Buse, E.L.; Laties, G.G. *Plant Physiol.* **1993**, *102*, 417-423.
158. Dallman, T.F.; Thomson, W.W.; Eaks, I.L.; Nothnagel, E.A. *Protoplasma* **1989**, *151*, 33-46.
159. Platt-Aloia, K.A.; Thomson, W.W.; Young, R.E. *Bot. Gaz.* **1980**, *141*, 366-373.
160. Huber, D.J. *J. Food Sci.* **1984**, *49*, 1310-1315.
161. Gross, K.C.; Watada, A.E.; Kang, M.S.; Kim, S.D.; Kim, K.S.; Lee, S.W. *Physiol. Plant.* **1986**, *66*, 31-36.
162. McCollum, T.G.; Huber, D.J.; Cantliffe, D.J. *Physiol. Plant.* **1989**, *76*, 303-308.

RECEIVED April 18, 1994

Chapter 7

Approaches to Cellulase Purification

Christian Paech[1]

COGNIS, Inc., 2330 Circadian Way, Santa Rosa, CA 95407

Cellulases comprise a mixture of enzymes whose complexity is determined by gene structure and by post-translational events. They bring about the enzymatic degradation of cellulose the rate of which depends on the synergistic interaction of individual components involved. To study this phenomenon isolated, purified enzymes must be available. The merits of techniques employed in cellulase purification are discussed and a table is provided which lists the methods used in the preparation of a particular cellulase.

The need for purifying enzymes has been eloquently advocated by Arthur Kornberg (*1*) based on Ephraim Racker's dogma "don't waste clean thinking on dirty enzymes." The extent to which an enzyme can be purified is set by the sensitivity of detection methods although with rapid advances in analytical techniques this limit may be difficult to reach. The burden of enzyme preparation can be eased if only those impurities are eliminated that interfere with the analyte. Walking the fine line between removing and tolerating impurities is a practical approach to answering these questions: what degree of purity, stability, and authenticity is necessary, and what quantity of a protein is needed to unambiguously determine its performance and its molecular characteristics. These considerations should precede the bench work as they shape the separation strategy and determine the outcome of the attempted purification. Often analogy to an existing method is the best approach to purifying enzymes. In many cases, however, there is no precedence because the enzyme has never been described before or it is from a significantly different source material (e.g., a protease from an extracellular microbial broth *versus* a liquid laundry detergent). Then, general advice on how to practice the art and science of protein purification may be found in one of several excellent reviews on this subject (*2-8*).

[1]Current address: Genencor International, 180 Kimball Way, South San Francisco, CA 94080

0097–6156/94/0566–0130$13.22/0

Regardless of the approach, the first step in protein purification has the greatest impact on its outcome. Here, efforts should be made to remove non-proteinaceous contaminants - even at the expense of yield - as they might seriously interfere with subsequent steps. Equally important is to remember that the percentage of the desired protein lost in every step increases as purification proceeds, i.e. the total number of purification steps must be kept at a minimum. An overall yield of less than 1% after 5 to 6 steps is a sobering experience.

In any source material quite a few enzymes are unique in their physical-chemical properties and provide for easy purification. For example, alkaline proteases from the extracellular space of microbes often have an isoelectric point >10 and are recovered in pure form simply by cation exchange chromatography (*9*). Ribulose 1,5-bisphosphate carboxylase/oxygenase from higher plants is a soluble protein with a mass of 520,000 Da which lends itself to rapid purification by sucrose gradient centrifugation (*10,11*). There are no such advantages when the desired protein resides matrix-bound, is part of an isozyme mixture, or is subject to multiple post-translational modifications by proteases and de-glycosylating enzymes. A case in point are the enzymes involved in cellulose depolymerization. First, microorganisms accommodated the topographical diversity of the substrate, cellulose, by producing cellulases, that are structurally different but catalyze the same chemical reaction, the hydrolysis of a ß-1,4-glucosylic bond. Thus, one should *a priori* not expect a great diversity of protein structure among cellulases. Indeed, cellobiohydrolase II and endoglucanase I from *Trichoderma reesei* could only be separated by immunoadsorption (*12*). Second, some microorganisms coped with environmental adversity by forming tightly aggregated cellulolytic complexes which resist fragmentation into active components (*13-16*). Whether it is structural multiplicity or protein aggregation, nature clearly did not have the needs of a biochemist in mind when evolving these enzymes. Third, the ubiquitous protein recycling tools, proteases (and de-glycosylating enzymes in a supportive role), add another dimension to enzyme multiplicity. Successive removal of glyco residues and nicking of the protein structure rapidly degenerates the enzyme spectrum, a process particularly prevalent among cellulases. Clearly, fractionation of cellulases to determine genetic or process-related multiplicity presents a rather demanding separation problem.

The original observation by Reese and coworkers (*17,18*) on the accelerated cellulose degradation by *T. viride* enzymes has laid the foundation for the hypothesis that cellulases are acting synergistically on cellulose (*19-37*). This mutually supportive function of cellulases is difficult to measure without access to the components in pure form. However, the information is needed to improve on the commercial application of cellulases.

The Basic Cellulolytic System

The term 'cellulases' is a rather global expression for a consortium of enzymes which collectively brings about the degradation of cellulose to glucose. Cellulases have been identified in many organisms: fungi, bacteria, protozoa, nematodes, plants and many other organisms (*38*). While this list is constantly being amended or corrected (e.g. (*25*)) and though new ideas may come from understanding how various organisms have solved for themselves the problem of utilizing cellulose, it is clear that, for practical reasons, initially only microorganisms are being viewed as promising candidates for the production and the commercial application of their cellulases. This may change, however, as more genes for cellulases become available and the technology in metabolic engineering advances.

Most of the early knowledge on cellulases is derived from studies of fungal systems, of which *Trichoderma reesei* (formerly *T. viride*) has become the predominant industrial producer. The specific activity of fungal cellulases is generally low (<1 $U \cdot mg^{-1}$) but their secretion level easily exceeds 20 $mg \cdot ml^{-1}$ (*39,40*). This abundance of material has greatly aided the dissection of the cellulolytic complex and the development of models that describe action and interaction of cellulases (*17,22,33,41*). However, expanded studies have made it clear that the components of the cellulase complex are neither exclusively located in the extracellular space nor always present in constant proportions. Furthermore, fungi secrete non-aggregating cellulolytic enzymes while bacteria produce either tightly substrate-bound or highly aggregated cellulases. The latter have been termed cellulosomes. No consensus has been reached whether bacterial and fungal enzymes share a common mechanism of cellulose breakdown.

The basic enzymatic process for the depolymerization of cellulose requires three types of enzymes. Their properties (see below) are not free of exceptions but whether they reflect true characteristics of the respective enzyme is debatable on two grounds. One is that contaminating enzymes may be the source of the unexpectedly observed catalytic activity. Refined purification protocols could resolve this question. The other is that substrates used for the classification of these enzymes are often poorly defined and are rarely identical from one laboratory to another. To partially eliminate this problem chromogenic and fluorogenic substrates of well defined structure were recommended a decade ago (*42*). However, some of these substances are difficult to obtain and their cost exceeds by orders of magnitude that of traditional substrates (e.g., Avicel, CMC, filter paper, cotton fiber), reason enough for their slow introduction (*43*). For detailed discussions of assay procedures and their pitfalls the reader is referred to the literature (*44-62*).

Endo-1,4-ß-D-glucanase. The official name for this enzyme is cellulase (EC 3.2.1.4, systematically named 1,4-(1,3;1,4)-ß-D-glucan 4-glucanohydrolase), for short endoglucanase (EG). It should be noted that the Nomenclature Committee of the International Union of Biochemistry and Molecular Biology (IUBMB) in its latest edition of the Enzyme Nomenclature (*63*) recommends the name 'cellulase' for this enzyme, a somewhat unfortunate choice for a partial component of a system that is generally known as 'cellulases'. Therefore, the name endoglucanase will be retained in this chapter to avoid possible confusions.

EG randomly cleaves internal ß-1,4-glucosidic links of acid or alkali swollen cellulose, in carboxymethylcellulose (CMC), but not in crystalline or amorphous cellulose. EGs display a wide range of molecular masses (12 to 115 kDa) and isoelectric points (3-9), and a large variability in the degree of glycosylation, akin to ß-glucosidases, but an extreme load of carbohydrates has not been reported. Absence of glyco moieties has also been observed (*64-66*). EGs frequently are multiple domain proteins, although length and composition of the linker and size of the cellulose-binding domain varies (*67*). Many organisms produce more than one endoglucanase. However, only in one instance has the stereochemistry of two EGs from the same organism (*Thermomonospora fusca*) been investigated and found to be different (*68*). A recent survey (*69*) and subsequent data (*70,71*) imply that stereoselectivity of EGs (and other cellulases) is related to the structural determinants outlined by Henrissat (*72-75*). It remains to be seen whether the need for an efficient cellulose degradation has endowed all cellulolytic organisms with a set of EGs of different stereochemical mechanism.

Cellobiohydrolase. The official name for this enzyme is cellulose-1,4-ß-cellobiosidase (EC 3.2.1.91, systematically named 1,4-ß-D-glucan cellobiohydrolase, or for short cellobiohydrolase; occasionally also called exoglucanase). Cellobiohydrolase (CBH) removes cellobiose from the non-reducing end of cello-oligosaccharides and of crystalline, amorphous, and acid/alkali treated cellulose, but not from chemically modified cellulose (e.g., CMC).

CBHs of fungi are the most studied and best characterized members of the cellulase family. One enzyme from *Trichoderma reesei* yielded the first detailed crystal structure of any cellulase (*76,77*). *Trichoderma reesei* produces two CBHs which differ in primary structure, response to antibodies, and in mechanism. CHB I (the more acidic form) hydrolyses cellobiose with retention of configuration while CHB II inverts the configuration at the anomeric center of the scissile bond (*78*). Already in 1981 Wood hypothesized (*79*) on the basis of steric considerations that all cellulolytic systems should contain two mechanistically different cellobiohydrolases, because the unit cell of cellulose is cellobiose, not glucose. This imparts chirality on the ß-1,4-glucosylic bond of the cellobiose residue at the non-reducing end of cellulose and, thus, requires enzymes with different stereochemical mechanism. The discovery of two different CBHs in *Trichoderma reesei* neatly supports this theory but its general validity is still a matter of debate (*80*). For example, only one CBH (operating with retention of configuration (*81*)) has been identified among the secreted enzymes of *Cellulomonas fimi*. It would be of great interest to establish whether the *C. fimi* system really needs only one CBH or whether a second CBH eluded researchers because of the tenacity with which it binds to cellulose.

CBHs display a characteristic protein structure: the catalytic core domain is connected to a cellulose binding domain through a hinge (linker) region composed of proline and threonine residues. The linker sequence is thought to be the seat of most if not all glyco residues (*82*). Their absence renders the protein vulnerable to proteolytic attack by either specific proteases from the same organism (*83-85*), or by unrelated proteases, e.g. papain and trypsin (*86-92*). However, with or without glyco residues, proteolytic degradation is suppressed while the enzyme is substrate-bound. Nevertheless, extended fermentation times contribute to the multiplicity enzymes through de-glycosylation and limited proteolysis as recently shown for CBH I of *T. reesei* (*93*).

The molecular weight distribution of CBHs is more uniform than that of ß-glucosidases. Molecular masses range from 40 to 65 kDa, and the spectrum of isoelectric points centers closely around 5. Glycosylation of CBHs appears to be ubiquitous, although a non-glycosylated cellobiohydrolase is known (*94*).

The cellulose-binding domain of CBHs both complicates and facilitates the purification of CBHs. Recovery of these enzymes from enzyme-substrate complexes (e.g., CBH from *C. fimi*) and from ion exchange and size exclusion chromatography matrices can be dismal while separation of CBHs from other enzymes by hydrophobic interaction chromatography (HIC) can be superb.

ß-Glucosidase. The official name for this enzyme is ß-glucosidase (EC 3.2.1.21, systematically named ß-D-glucoside glucohydrolase, also known as cellobiase). ß-Glucosidase (BG) hydrolyses cellobiose to yield two molecules of glucose, which not only completes the depolymerization of cellulose, but also relieves cellobiohydrolase of product and endoglucanase of feedback inhibition.

The configuration at the anomeric center of glucose is retained. Fluorogenic and chromogenic glucose derivatives are the substrates of choice to distinguish BGs from any other member of the cellulolytic system. BGs occur intracellularly, cell wall bound, or in the extracellular space, they range in size from 12 to over 500 kDa and are typically monomeric polypeptides, although some of the larger enzymes are oligomeric in nature (*95-100*). The isoelectric point of ß-glucosidases centers around pI = 5 (±2). Glycosylation is not indispensable for these enzymes but can reach a staggering 90% (w/w) of the total enzyme mass (*101*). In spite of their vital role ß-glucosidases may comprise only as little as 1% of the total mass of cellulolytic enzymes. Many organisms produce more than one ß-glucosidase found either in the extracellular space or distributed over the intracellular and extracellular space, but whether they are genetically related has not been established.

Variation in the Mechanism of Cellulose Breakdown

An extended survey of cellulases shows that some organisms have developed alternate mechanisms for cellulose utilization. There are enzymes that act oxidatively on cellobiose (cellobiose oxidase, EC 1.1.3.25, cellobiose dehydrogenase (quinone), EC 1.1.5.1, cellobiose dehydrogenase (acceptor), EC 1.1.99.18), cleave cellobiose by phosphorylation (cellobiose phosphorylase, EC 2.4.1.20), remove glucose residues hydrolytically from cellulose chains (glucan-1,4-ß-D-glucosidase, EC 3.2.1.74), and degrade pretreated cellulose under alkaline conditions. All these enzymes play presently a subordinate role in the industrial saccharification of cellulose because either they are too slow (glucose removal from cellulose chains by glucan glucosidase) or their reaction products are not suitable for ethanol fermentation (glucose produced under alkaline conditions, cellobiono-1,5-lactone and glucose-1-phosphate from cellobiose oxidase and cellobiose phosphorylase, respectively). These enzymes may gain importance and, therefore, are included in Table I of this chapter.

The Separation Techniques

A standard protein purification involves one or several of 4 distinctly different procedures (homogenization, filtration, and centrifugation excluded): precipitation, ion exchange chromatography, affinity chromatography, and gel filtration (*102*). Reviews on cellulase purification confirm this conclusion (*103-107*) but point out that electrophoretic, chromatofocusing, and immunological techniques are in many instances indispensable for obtaining pure protein.

Whether purification of cellulases is attempted from commercial material, native strains, or from cloned organisms, the methods to be used must be tailored to the particulars of cellulases. Many technical aids such as dialysis tubing, gel filtration and ion exchange matrices, and ultrafiltration membranes taken for granted in a biochemist's laboratory are of limited use when purifying cellulases because they are mostly cellulose-based. Fortunately, suitable replacements are available for these products: agarose, acrylamide, dextran, and styrene based chromatography supports (Sepharose, Ultrogel AcA, BioGel A, Sephacryl, BioGel P, Trisacryl, Sephadex) and poly(ether)sulfone-based ultrafiltration membranes (e.g., PM-series by Amicon, Omega- and Nova-series by Filtron). Nitro-cellulose based collodion bags have been used for dialysis but they are not readily available.

The purpose of this review is to evaluate both traditional and less frequently used protein purification techniques as they pertain to cellulases. The survey will focus on the method of purification rather than the properties of the purified enzyme and the particulars of the source material. A table alphabetically arranged by organism will list the type of enzyme isolated and the methods of purification used. In view of the almost overwhelming literature on cellulases no claim is made as to the completeness of this table. Complementary reviews on cellulase (*25,67,104,105,108-123*) and their purification (*103-105,107,124-127*) are available.

Sample Preparation - from Commercial Materials. Many researchers have resorted to isolating enzyme from commercial samples. The obvious reasons are low cost and large quantities which eliminate the yield problem and, at least in principle, offer the opportunity to obtain superior preparations by selecting very stringent cuts at each step of the purification protocol. The disadvantage of this approach is that the enzyme profile in the starting material may be degenerated because commercial enzymes are produced to meet functional specifications, not to consist of homogeneous protein fractions. This multiplicity of cellulases arises from proteolytic (*86,106*) and glycolytic (*128*) attack by enzymes secreted by the production strain, and from chemical reactions during the processing of the culture filtrate.

Commercial products that are (or were) available for the isolation of cellulolytic enzymes include: Cellulosin AC-40 (*Aspergillus niger*, Ueda Kagakukogyo, Japan), Taka-cellulase (*Aspergillus niger*, Miles, USA), Cellulase A (*Aspergillus niger*, Amano Pharmaceutical, Japan), Driselase (*Irpex lacteus*, Kyowa Hakko, Japan), Sturge CP Cellulase (*Penicillium funiculosum*, Sturge Enzymes, United Kingdom), Cytolase CL (*Trichoderma longibrachiatum*, Genencor Int., USA), Laminex BG (ß-glucosidase from *Trichoderma longibrachiatum*, Genencor Int., USA), Celluclast (*Trichoderma reesei*, Novo-Nordisk, Denmark), Maxazyme (*Trichoderma reesei* QM9414, Gist-Brocades, Netherlands), cellulase extracts (*Trichoderma reesei*, Institute Français du Pétrole, France), cellulase extracts (*Trichoderma reesei* QM9414, VTT, Finland), Cellulase T (*Trichoderma viride*, Amano Pharmaceutical, Japan), Onozuka SS (*Trichoderma viride*, All Japan Biochemicals, Japan), Pancellase SS (*Trichoderma viride*, Yakult Biochemicals, Japan), and Meicelase P (*Trichoderma viride*, Meiji seika Kaisha, Japan). These products come as solids or in solution. In either case, a sample is diluted to 3 to 5 $mg \cdot ml^{-1}$ protein with a low molarity buffer (10 to 50 mM) or water at 4°C. Added to this suspension are 0.03% (w/v) sodium azide to prevent microbial growth (remember the known mutagenic effect of sodium azide and wear gloves at all times) and 2 mM Na_2-EDTA (to suppress metallo protease activity). The suspension is further supplemented with 1 mM phenylmethyl sulfonylfluoride (PMSF; to suppress serine proteases) added drop-wise from a 100 mM stock solution in isopropanol. Thiol proteases may be inhibited by leupeptine but its use is limited to special cases due to the high cost of the reagent. After stirring for 30 min the suspension is clarified by centrifugation. Depending on the purification protocol additional treatments may be necessary. In size exclusion chromatography (SEC), hydrophobic interaction chromatography (HIC), and affinity chromatography (AC) the crude cellulase solution is applied 'as is'. In ion exchange chromatography (IEC), adsorption chromatography on hydroxyapatite (HAC), and for electrophoretic methods the solution must be adjusted to a defined ionic strength. This process is occasionally called 'desalting' and can be accomplished by size exclusion chromatography on highly cross-linked supports, or by diafiltration (buffer exchange) in a tangential flow cell, a hollow fiber dialysis apparatus, or an ultrafiltration cell.

Solutions of commercial cellulases are often darkly colored. Fractionated precipitation (see below) with ammonium sulfate before chromatography may be beneficial but requires subsequent 'desalting' if IEC, HAC, or electrophoretic procedures were to follow as the next step. Precipitation with organic solvents or with polyethylene glycol avoids the 'desalting' step.

Sample Preparation - from Cultured Organisms. Depending on the location of the desired enzyme (or enzyme complex) the cytoplasmic fluid, the cell membrane, the periplasmic space, the extracellular broth, or the substrate particles (e.g., Avicel suspended in the fermentation broth) must be prepared for purification. Thus, the first step in processing of the culture broth is separation of biomass from culture fluid (or biomass and culture fluid from particulate cellulase-substrate complexes - see below). This is done by microfiltration but more commonly by centrifugation. It is not recommended to improve the separation of culture fluid from biomass and particulate substrate (e.g. wood chips) by surfactants as they may break open cells and release proteins that are difficult to purify away. However, ammonium sulfate at a concentration of 10% (w/v) has been used in the separation of culture fluid and biomass (*129,130*). Sometimes it is advantageous to filter the broth through nylon cloth before centrifugation. A continuous centrifuge of the Sharples type is the preferred instrument for separating large volumes of broth from biomass but many laboratories do not have access to this convenience. Before unloading centrifuge tubes it is a good idea to have a funnel with a glass wool pad or a nylon cloth handy in case lipophilic material floats on top of the supernatant or the pellet of the biomass is not well packed. The clarified broth often contains soluble compounds such as polyphenols (wood-cultures), nucleic acids (cell-free extracts), gums, slime and polysaccharides (extracellular fluids), that must be removed as they potentially interfere with subsequent purification steps. Polyphenols can be trapped with insoluble polyvinylpyrrolidone (1 to 3%, w/v), with up to 1% (w/v) polyethylene imine, or with 0.1 to 0.3% (v/v) of a beaded ion exchange material (Biocryl BPA, Supelco, Bellefonte, PA), and then removed by centrifugation. Slime, gums, and other polysaccharides are skimmed off or removed by centrifugation after the addition of 10 to 20% (v/v) ethanol to the culture fluid. Care must be taken that no (or only unwanted) proteins are precipitated (see below). Polysaccharides may also be removed by repeated freeze-thaw cycles which, however, requires prior knowledge on the enzyme's pH stability because buffers, particularly phosphate, undergo dramatic pH changes upon temperature change (*131*). Nucleic acids are precipitated with $MgCl_2$, protamine sulfate (0.1 to 1%, w/v) or streptomycin sulfate (1 to 4%, w/v), and polyethyleneimine (0.1 to 0.5%, w/v). Protamine is more efficient than streptomycin but precipitates also more protein.

If the biomass is of interest, the pellet is usually washed with buffer (to release cell-wall bound proteins) and then sonicated (cell wall bound cellulosomes and other protein aggregates, e.g. *Clostridium* sp., *Ruminoccus* sp.), treated with a cell-wall degrading enzyme preparation followed by osmotic shock (periplasmic space), or disintegrated by mechanical force (ball mill, French press) for intracellular proteins.

Concentration. In contrast to commercial samples (which allow for some flexibility in setting the protein concentration needed for the first step in purification) freshly prepared culture fluids are often too low in protein concentration for successful precipitation or chromatography. Ultrafiltration in pressured cells, in tangential flow cells, or by hollow fiber systems, and lyophilization are the methods of choice. Other techniques applicable to small sample sizes are dialysis in collodion bags against solid polyethylene glycol, and mixing of the clarified supernatant with highly cross-linked polyacrylate gel (*132*) or with temperature-sensitive hydrogels (*133*).

The main point to remember in ultrafiltration is to avoid cellulose-based membranes. Even for the purification of BGs or when working well outside the pH optimum (e.g. pH 9 (*134,135*)) it is advisable to choose poly(ether)sulfone membranes because absence of CBHs and EGs is never certain. Regardless of such precautions ultrafiltration results in loss of protein and activity mostly due to the leakiness of membranes (Gaussian distribution of pore sizes around the nominal value of the molecular-weight-cut-off) and the physical forces acting on proteins in the retentate. Losses larger than 5% should not occur but may be prevented by the inclusion glucose (*136*) and lactose (*137*) as stabilizers.

In lyophilization the presence of sodium azide (explosive) is a safety concern, and high concentrations of salts and cryoprotectants (polysaccharides, polyols) may even prevent the freeze-drying. Fungal enzymes seldom loose activity because culture fluids tend to have a stabilizing effect on proteins. Even in distilled water a crude mixture of cellulases from *Talaromyces emersonii* survives lyophilization (*135*). Culture fluids of *Clostridium thermocellum*, however, suffer large losses of activity upon freeze-drying. Removal of salts by dialysis and addition of 10 mM dithiothreitol greatly increases the yield (*138*). Flash evaporation under reduced pressure at 30 to 35°C has been successfully applied to extracts from *Pseudomonas fluorescens* var. *cellulosa* (*139*). Spray-drying is rarely available to bench-scale research but its merit in concentrating cellulase solutions has been evaluated (*140*).

Precipitation. Precipitation is probably one of the earliest separation techniques used in protein purification (for a review see (*141*)). Bulk precipitation conveniently facilitates buffer exchange and concentration of protein, and is the method of choice when the subsequent step promises a sharp resolution, for example, the separation of glycosylated and non-glycosylated proteins by affinity chromatography on a concanavalin A column (*142*). However, potentially large solubility differences of components in cell-free extracts or extracellular fluids invite fractionated precipitation by gradual addition of the precipitant. With a wide panel of precipitating reagents to choose from, and convenient methods to control precipitation parameters (stirring, weight, volume, pH and temperature), fractionated precipitation is indicated when the protein of interest assumes a unique place with respect to size, solubility, or isoelectric point among the unwanted proteins, but one has to be prepared for low yields (*143*).

There are five general methods for precipitating proteins: addition of salts, organic solvents, or nonionic polymers, change of pH, and heat treatment. Immunoprecipitation with antibodies is time-consuming and leads often to tight complexes that require chaotropic reagents for the resolution of the desired protein. Advances in monoclonal antibody technology, e.g. cloning (*144*), may soon provide 'designer antibodies' with properties tailored to the particular separation needs.

Salt Precipitation. The most popular reagent for precipitating proteins is ammonium sulfate. To a large extent empirical in nature this technique has yielded great results throughout the history of biochemistry (*145-149*). For example, the specific activity of alcohol dehydrogenase (molecular mass 145 kDa, pI 5.4-5.8) from extracts of baker's yeast was doubled just by shifting the pH from 6.5 to 5.9 prior to precipitation (*148*).

In agreement with a recent survey (*102*), ammonium sulfate precipitation has been the most frequently used method for recovering cellulases from cell free extracts and extracellular fluids. The salt was introduced as solid or added as saturated solution. The latter has the advantage that a particular pH can be established prior to precipitation while the addition of salt may require adjustment of pH during precipitation. One drawback of working with a saturated ammonium sulfate solution is dilution of protein (which reduces precipitation) and the disproportionate increase in total volume as the degree of saturation increases (which sets the upper practical limit of use to around 50% of saturation).

The ammonium sulfate precipitate of a protein is usually collected by centrifugation. This is possible, because in most cases the precipitant's specific gravity is significantly above that of even the saturated ammonium sulfate solution (1.235 $g \cdot cm^{-3}$). However, in situations where centrifugation does not produce a sediment, the precipitate may be collected by filtration over sintered glass, over glass fiber filters, or over a layer of celite on top of regular filter paper (*150*). This last technique is particularly useful because the coarseness of the diatomaceous earth guarantees reasonable filtration rates compared to glass fiber filters and sintered glass (which easily clog).

There is no general rule that indicate at which concentration a particular cellulase will precipitate but, in general, solubility of proteins is lowest at the isoelectric point, larger proteins tend to precipitate at lower salt concentration then smaller proteins, and glycosylation increases solubility. More than half of all publications concerned with cellulase purification (Table I) report protein precipitation by ammonium sulfate, but there is a large void in experimental detail. Often, the pH of the solution neither before nor after addition of ammonium sulfate is given. In no case, pH-dependent precipitation was attempted, although one would think that any starting material is conducive to a quick analytical survey in these days of pocket-sized pH-meters, minielectrodes, rapid electrophoretic and isoelectrophoretic techniques, and convenient

Table 1. Characteristics and Purification Protocols for Cellulases Reported in the Literature.

Organism	Enzyme	Location	M_r (kDa)	pI	Purification Methods	Reference
Alcaligenes faecalis	BG	i	120		p-strep,p-AS(40-60),SEC,IEC	*(292)*
A. faecalis	BG	e	50		p-strep,p-AS(35-55),IEC,SEC	*(293)*
Arxula adeninovorans	BG	e	570		p-acet,IEC,HAC,SEC	*(294)*
	BG	i	290			
Aspergillus aculeatus	all	e			p-AS(80),p-etha,IEC,SEC,IEF-col	*(168,295)*
	EG-I	e	109	4.7		*(296,297)*
	EG-II	e	112	4.0		
A. aculeatus	BG(gl)	e	133	4.7	p-etha,IEC,SEC,IEF	*(261)*
Aspergillus ficum	CBH-I	e	128		p-AS(0-60),p-acid,SEC,IEC,SEC	*(180)*
	CBH-II	e	50			
Aspergillus fumigatus	EG(no gl)	e	12	7.1	lyo,SEC,IEC,AC-conA	*(203)*
	CBH	e	66	4.02		
	BG	e	340	5		
A. fumigatus	EG(no gl)	e	13	7.1	lyo,SEC,AC-conA	*(298)*
A. fumigatus	BG(gl)	e	2x170		p-acet,IEC,SEC,HAC	*(95)*
A. fumigatus	BG	e	4x95	4.5	p-AS(80),SEC,IEC,HAC	*(299)*
Aspergillus japonicus	BG	e	240		p-etha,SEC,IEC,SEC	*(300)*
Aspergillus nidulans	BG-I	e	125	4.4	PAGE-prep,IEC	*(301)*
	BG-II	e	50	4.0		
A. nidulans	BG-I,-II	e	14,26		p-AS(40-60),SEC,IEC,PAGE-prep	*(158)*
	CBH-I,-II,-III	e	29,72,138			
	EG-I,-II	e	25,32			
Aspergillus niger	EG-I(gl)	e			Chrofo	*(258)*
	EG-II(gl)	e				
A. niger	EG(no gl)	e	31	3.67	p-AS(60-80),IEC,SEC,SEC,SEC	*(302,303)*
A. niger	BG(gl)	e	2x73		p-AS(40-60),SEC,SEC	*(304)*
A. niger	BG	e	230		SEC	*(305)*

Continued on next page

Table 1. *Continued*

Organism	Enzyme	Location	M_r (kDa)	pI	Purification Methods	Reference
A. niger	EG-I	e			AC-cell	*(185)*
A. niger	BG(gl)	e	96	4.6	SEC,chrofo	*(306)*
	BG-II(gl)	e	96	3.8		
A. niger	EG	e	31		p-AS(20-50),SEC	*(307)*
A. niger (Cellulosin)	BG	e	2x120	4.0	SEC,IEC,Chrofo,AC-cellobiamine	*(97)*
A. niger (taka-cellulase)	CBH	e	46	3.3	p-AS(30-80),SEC,IEC,IEC,HAC	*(283)*
A. niger (technical)	EG	e	26		p-AS(80),SEC,IEC,SEC,AC-cell,HAC,SEC	*(191)*
Aspergillus oryzae	BG	e	2x118	4.3	p-acet,IEC,HAC	*(98)*
Aspergillus terreus	EG-I(gl)	e	78		lyo,p-AS(20-80),SEC,IEC	*(308)*
	EG-II(gl)		16			
	CBH(gl)		70			
	BG(gl)		90			
Avocado	EG		49	4.7	AC-cell	*(309)*
Bacillus sp.	CMCase(gl)	e	30	4.8	p-AS(70),AC-conA	*(241)*
Bacillus sp. KSM-330	CMCase	e	42	>10	IEC	*(310,311)*
Bacillus sp. KSM-635	CMCase	e	130		IEC,SEC	*(312)*
	CMCase	e	103			
Bacillus sp. PKM-5430	CMCase	e	27	4.5	p-AS(90-100),SEC	*(313)*
Bacillus sp. 1139	CMCase	e	92	3.1	IEC,SEC,IEC	*(314)*
Bacillus lautus	EG-I	cloned	74		AC-cell,IEC,SEC	*(315)*
	EG-II	cloned	57			
Bacillus licheniformis	EG	e	25	8.5	p-AS(80),p-acid,IEC	*(316)*
Bacillus subtilis	EG	e	23		p-AS(30-40),AC-cell,IEC,IEC	*(219)*
Bacillus subtilis DLG	EG	e	55		p-AS(70-100),SEC,IEC-HPLC,SEC	*(317)*
Bacterium	EG	all			PAGE-prep	*(268)*
Bacterium (shipworm)	EG-I	e	58	3.83	lyo,IEC-HPLC,SEC-HPLC	*(318)*
Bacterium (thermophilic)	BG	i	43	4.55	son,SEC,IEC,HACSEC	*(319)*
Bacterium (thermophilic)	BG	cloned	52		p-heat(70/65),p-AS(75),IEC,SEC-HPLC	*(320)*
Bacterium (thermophilic)	EG-I	e	91	3.85	p-AS(40),IEC	*(321)*
Bacteroides ruminicola	CMCase	cloned			French,HAC,p-AS(35),SEC	*(195)*
	CMCase	e	88		HAC,SEC,IEC	*(194)*

Continued on next page

Table 1. *Continued*

Organism	Enzyme	Location	M_r (kDa)	pI	Purification Methods	Reference
Bacteroides succinogenes	EG-I	e	65	4.8	IEC,HAC,Chrofo,SEC,HIC	*(187,188)*
	EG-II	e	118,94	9.4		
B. succinogenes	CMCase	e	45		IEC,IEC,HIC	*(322)*
B. succinogenes	CD	e	40	4.9	IEC,HAC,Chrofo	*(323,324)*
	CBH	e	75	6.7	IEC,HAC,SEC	*(263)*
Bifidobacterium breve	BG	i	2x48		son,p-AS(30-70),IEC,SEC,HAC	*(325)*
	BG	i	450			
Botryodiplodia theobromae	EG	e	46		lyo,SEC,IEC,PAGE-prep	*(326)*
	CBH/CMC	e	5,10,15,30			
	BG	e	350			
	BG	e	45			
	BG	e	8x12		SEC,IEC	*(96)*
Butyrivibrio fibrisolvens	complex	e	700		HAC	*(327)*
B. fibrisolvens	EG-A	cloned	49		son,IEC,IEC	*(328)*
Calf liver	BG	cytosol	52		p-acid,p-heat(53/30),IEC,AC-glucit	*(234)*
Candida albicans	CBH(no gl)	e	38		IEC,SEC	*(329)*
Candida guilliermondii	BG	i	48		p-PEI,p-AS(75),AC-salicine	*(231)*
Candida molischiana	BG(gl)	e	2x64	3.7	p-acet,IEC,SEC	*(171)*
Candida wickerhamii	BG	e	97		p-AS(0-80),IEC,SEC	*(330)*
C. wickerhamii	BG(gl)	bound	2x94	3.89	p-acet,IEC-HPLC,SEC	*(166)*
Cellulomonas sp.	EG,CBH	e	44-140		IEC,HAC	*(163)*
Cellulomonas sp. 21399	EG-B	e	67	4.3	AC-cell,SEC,IEC,AC-conA	*(331)*
	EG-C	e	25			
	EG-A(gl)	e	78,63,56		AC-cell,AC-conA,IEC	*(217)*
Cellulomonas fermentans	EG-I	e	40		IEC,IEC-HPLC,SEC-HPLC	*(202)*
	EG-II	e	58			
Cellulomonas fimi	EG-C	cloned	120		osmotic,AC-immuno,IEC	*(275)*
C. fimi	EG+/-gl	e			GH-extr,AC-conA,AC-immuno	*(248)*
C. fimi	CBH(gl)	cloned	55	4.8	AC-cell,GH-extr	*(271)*
C. fimi	EG-B	e	110	4.6	AC-cell,IEC	*(332)*
	EG-D	e	75	4.8	AC-cell,IEC	*(273)*

Continued on next page

Table 1. *Continued*

Organism	Enzyme	Location	M_r (kDa)	pI	Purification Methods	Reference
	EG-C	e	115	4.2	AC-seph,IEC,SEC	*(333)*
C. fimi	EG-B	cloned	110	4.3	AC-cell,GH-extr	*(220)*
Cellulomonas flavigena	CMCase	e	20		p-acet,SEC,IEC,PAGE-prep	*(269)*
C. flavigena	EG-I	e			AC-seph,SEC	*(204)*
	EG-II(gl)	e			GH-rel,IEC,SEC	
	EG-III(gl)	e			GH-rel,SEC,SEC	
Cellulomonas uda	EG	e	66	4.4	p-AS(10-50),SEC,IEC	*(334,335)*
Cellvibrio gilvus	CB-phos	e	2x85	6.0	p-prot,p-AS(40),HIC,p-AS(55),IEC	*(336)*
Chaetomium cellulolyticum	EG-I(gl)	e			p-AS(10-90),AC-conA,PAGE-prep,AC-dye	*(242)*
Chaetomium thermophile	BG-I	bound			IEC,IEC,SEC	*(337)*
	BG-II	bound				
	BG-III	e				
C. thermophile	EG-I(gl)	e	36		p-AS(70),SEC,IEC,IEC,SEC,SEC	*(154)*
	CBH-I(gl)	e	60		p-AS(80),SEC,IEC,SEC,SEC,IEC	*(153)*
	CBH-II(gl)	e	40			
Clostridium cellulolyticum	EG-A	cloned	51		French,p-AS(30-60),IEC,IEC,HIC	*(279)*
C. cellulolyticum	EG-D	cloned	38,56,62		osmotic,p-AS(80),HAC,SEC	*(92)*
Clostridium josui	EG	e	45		IEC(urea),IEC	*(338)*
	EG	cloned	39		osmotic,IEC,SEC,IEC	*(339)*
Clostridium stercorarium	CBH	e	87	3.9	IEC,IEC,SEC	*(340)*
	BG	e	85	4.8	IEC,SEC	*(341)*
	EG-I	e	100	4.6	IEC,IEC	*(342)*
	EG-II	e	87			
	CBH-I	e	82	4.7		
	CBH-II	e	128	4.7		
	BG	e	85	4.8		
	EG-I	e	100		IEC,IEC,SEC	*(134)*
Clostridium thermocellum	BG	peri	50	4.68	IEC,HAC,IEC,SEC,IEF-col	*(343)*
C. thermocellum	BG	cloned			French,p-heat,p-strept,IEC,SEC	*(344)*
C. thermocellum	EG-I	cellulosome	64		lyo,IEC,SEC	*(138)*
	EG-II	cellulosome	110			

Continued on next page

Table 1. *Continued*

Organism	Enzyme	Location	M_r (kDa)	pI	Purification Methods	Reference
C. thermocellum	EG-5	cellulosome	35	4.6	son,p-heat(60/30),p-AS(30-60),IEC,IEC,IEC	(*231,345*)
C. thermocellum	EG-S	cellulosome	83	3.55	IEC,IEC,HAC,HIC	(*199*)
C. thermocellum	EG-D	cloned	65	5.4	son,p-strep,p-heat,p-AS	(*346*)
C. thermocellum	cellulosome				son,AC-cell,SEC	(*347*)
C. thermocellum	cellulosome				SEC, centr-suc	(*348*)
C. thermocellum	cellulosome				AC-cell	(*349*)
C. thermocellum	EG-7	cloned	49	4.3	son,p-AS(50),SEC,SEC,Chrofo	(*350*)
C. thermocellum	EG-I	cloned	100		son,p-strep,p-heat(60/10)p-AS(60),HIC,IEC	(*291*)
C. thermocellum	all	cellulosome			p-acet,IEC	(*19*)
C. thermocellum	EG-D	cloned			p-heat(60/10),AC-thiocellobiose	(*224,225*)
C. thermocellum	CBH-3	cloned	78	4.75	p-strep,p-heat(55/10),p-AS(60),SEC,IEC,IEC	(*351*)
C. thermocellum	EG(no gl)	cellulosome	76	5.05	p-AS(80),SEC,IEC,HIC	(*198*)
	BG-B	cloned	84	4.4	son,p-strep,p-AS(70-100),IEC,HIC,HAC	(*352*)
C. thermocellum	EG	cellulosome	56	6.2	p-etha,IEC(6 M urea)	(*353*)
C. thermocellum	EG(gl)	e	90	6.72	p-AS(80),IEC,PAGE-prep	(*200,354*)
C. thermocellum	cellulosome				affinity digestion	(*222*)
	CBH	cellulosome	78,68		IEC	(*355*)
C. thermocellum	CMCase	e			IEC,SEC	(*356*)
	CMCase	cellulos.	60,56,48		SEC,SEC,SEC	(*357*)
C. thermocellum	EG-G	cellulosome	50		French,p-strep,p-heat(60/15),p-AS(50),IEC,SEC	(*290*)
C. thermocellum	cellulosome				IEC,AC-jacalin	(*243*)
C. thermocopriae	BG	e	50		p-AS(45-90),IEC,SEC	(*358*)
	EG	e	51			
Coniophora puteana	CBH-I(gl)	e	52	3.6	IEC,SEC,IEC	(*359*)
	CBH-II(gl)	e	50	3.55		
	CBdehydro(gl)	e	2x110	3.9	IEC,SEC	(*360*)
Coriolus versicolor	EG-I	e	26		p-AS(70),IEC,SEC	(*361*)
	EG-II	e	29			
C. versicolor	BG(gl)	e	300		p-AS(80),SEC,SEC	(*101*)
Dichomitus squalens	EG	e	42	4.8	AS-prec(0-80),chrofo,HIC,IEC,SEC	(*362*)

Continued on next page

Table 1. *Continued*

Organism	Enzyme	Location	M_r (kDa)	pI	Purification Methods	Reference
	EG	e	56	4.3		
	EG	e	47	4.1		
	CBH	e	39	4.6		*(363)*
	CBH	e	36	4.5		
Dolabella auricularia	EG(gl)		44	8.6	SEC,IEC,IEC	*(364)*
	EG		51		SEC,IEC,AC-oligo	*(236)*
Erwinia chrysanthemi	EG-Z	cloned			AC-cell,IEC	*(70)*
E. chrysanthemi	EG-I	e	45	4.3	IEC	*(365)*
	EG-Y	cloned			son,p-AC(20-50),IEC	*(366)*
Erwinia herbicola	BG	e	122		p-AS(40-70),SEC,IEC,SEC	*(367)*
Eupenicillium javanicum	all	e	21-65		flash evap,p-AS(60),SEC,IEC,IEC,PAGE-prep	*(368,369)*
Euphausia superba	BG	i	155		p-AS(20-70),IEC,AC-conA,SEC	*(370)*
	BG(gl)	i	155			
Evernia prunastri	BG	i	4x70	3.12	French,p-AS(90),IEC,SEC,SEC	*(100)*
	EG	e	70	4.4	p-AS(90),IEC,SEC	*(155)*
Favolus arcularius	CMCase	e	28	5	lyo,p-AS(80),IEC,IEF	*(371)*
Fungus Y-94	EG	e	68	4.4	p-AS(60),IEC,HAC,IEC	*(372)*
	BG	e	2x120		IEC,Chrofo,SEC	*(373)*
	EG	e			p-AS(30-80),IEC,Chrofo,SEC,PAGE-prep	*(284)*
Fusarium lini	EG(gl)	e	28	8.3	p-AS(45),SEC	*(374)*
F. lini	CBH(gl)-I	e	57		p-AS,IEF-flat	*(375)*
	CBH(gl)-II	e	50			
F. lini	CMCase	e	40-65	6	p-AS(48),IEF-flat	*(265)*
Fusarium moniliforme	EG(gl)	e	25		p-AS(80),SEC,IEC,HAC,HAC	*(376)*
Fusarium solani	CBH	e		4.9	p-AS(20-80),IEC,IEF	*(377,378)*
Geotrichum candidum	EG	e	130		AC-cell,IEC,SEC,SEC	*(214)*
	CMCase	e	63			
Gliocladium sp.	all	e	10-70		IEF-flat	*(379,380)*
Helix pomatia	EG				SEC,IEC,SEC	*(381)*
Humicola grisea	EG	e	128		IEC,SEC,IEC,SEC	*(382)*
Humicola insolens	CBH(gl)	e	72		lyo,p-AS(60),p-acid,IEC,SEC	*(383,384)*

Continued on next page

Table 1. *Continued*

Organism	Enzyme	Location	M_r (kDa)	pI	Purification Methods	Reference
	EG(gl)	e	57			
	BG(gl)	e	250	4.23		
H. insolens	BG	e	45	8.1	p-AS(75),IEC,SEC,HAC	*(385)*
	EG	e	45			
Humicola lanuginosa	BG(gl)	e	110-135	~4	p-AS(85),p-acet,IEC,SEC	*(150)*
Irpex lacteus (Driselase)	EG-I	e	60		p-AS(20-60),IEC,IEC,SEC,IEF-col	*(386)*
	EG-II	e	35			
I. lacteus (Driselase)	CBH	e	65		p-AS(20-80),IEC,SEC,IEC,IEC,SEC,AC-starch	*(387-390)*
	EG(gl)	e	56			
	EG	e	15			
Lenzites saepiaria	EG	e	85		p-AS(80),p-strep,SEC	*(391)*
Lenzites trabea	EG	e	29		SEC,IEC,HAC,SEC	*(392)*
Macrophomina phaseolina	BG	e	320		p-AS(80),IEC,SEC	*(393)*
	BG	e	220			
Microbispora bispora	CBH-II	cloned	93	4.2	son,p-heat(60/10)	*(289)*
Myceliophthora thermophila	EG	e			p-AS(30,50,80),SEC	*(394)*
	BG	e				
M. thermophila	EG(no gl)	e	100		p-AS(50-80),IEC,IEC	*(395)*
Mucor miehei	BG	e			p-AS(60),SEC,IEC,SEC,IEC,SEC,IEC	*(396)*
Mucor racemosus	BG(gl)	e	91		p-AS(70-100),SEC,IEC	*(397)*
	BG(gl)	bound	91			
Neocallimastix frontalis	BG	e	120	3.85	p-AS(40-85),IEC,SEC,IEF-flat	*(398)*
N. frontalis	complex	e			HAC	*(399)*
	BG	e	85	6.95	PAGE	*(400)*
N. frontalis	cellulosome	e	670		AC-cell	*(401)*
Paecilomyces sp.	EG(gl)	e	2x43	4.15	IEC,IEC,IEC	*(285)*
Penicillium sp.	CBH-I(gl)	e	52		flash-evap,p-acet,IEC,IEC,IEC,Chrofo,SEC	*(402)*
Penicillium camemberti	BG	e	61		AC-cell,p-AS(20-80),SEC,IEC,IEC,SEC	*(212)*
	CBH	e	99			
	EG	e	87			
Penicillium capsulatum	EG-I(gl)	e	38	4.12	lyo,SEC,IEC	*(403)*

Continued on next page

Table 1. *Continued*

Organism	Enzyme	Location	M_r (kDa)	pI	Purification Methods	Reference
	EG-II(gl)	e	36	3.92		
	complex	e	2x135	4.02	lyo,SEC	*(404)*
Penicillium funiculosum	GG	e	65	4.65	p-AS(20-80),Chrofo	*(405)*
P. funiculosum	CBH(gl)	e	46	4.36	p-AS(20-80),IEC,IEF-col	*(260)*
Penicillium pinophilum	CBH(gl)	e	51	5.0	p-AS(20-80),IEC,IEC	*(260,406)*
P. pinophilum	CBH-I	e			p-AS(20-85),SEC,IEC,AC-thiocellobiose	*(224)*
	CBH-II	e				
Penicillium purpurogenum	CBH-I(gl)	e	84		SEC,IEC,IEC,SEC,Chrofo	*(407)*
	CBH-II(gl)	e	64			
Phanerochaete chrysosporium	CBO	e	90		p-AS,IEC,HIC,SEC,IEC	*(408)*
P. chrysosporium	CBQase	e			p-AS(40-50),IEC	*(409)*
P. chrysosporium	EG	e	44,38,36	4-5	p-AS(80),SEC,IEC	*(226)*
	CBH	e	60,62,50	4-5		
P. chrysosporium	CBQoxidored	e	75	6.4	p-AS(60),IEC,HAC,IEC	*(410-412)*
Phoracantha semipunctata	EG	e		4	AC-cell	*(413)*
Phytophthora infestans	EG	e	21	3.2	p-AC(50-90),SEC,IEF-col	*(414)*
	BG-I	e	160	3.3		
	BG-I	e	230	4.7		
Piromyces sp. E2	BG	e	45	4.15	SEC,AC-cellobiose,Chrofo,SEC,IEC	*(232)*
Pisolithus tinctorius	BG	i	3x150	3.8	SEC,IEC,IEC	*(415)*
Pseudomonas fluoresc. var. *cell.*	EG	cloned	30	7.5	IEC-HPLC	*(416)*
	EG	cloned	38	6.7		
P. fluorescens var. *cellulosa*	BG	i			SEC,SEC,IEC,SEC	*(417)*
	BG	bound				
P. fluorescens var. *cellulosa*	all	e			p-AS(90),SEC,IEC,SEC	*(139,418,419)*
Pyricularia oryzae	BG	e	240	4.15	flash,p-AS(100),IEC,SEC,IEF-col,SEC,centr-suc	*(420)*
Pyrococcus furiosus	BG	i	4x58	4.4	IEC,IEC,SEC	*(421)*
Rhizobium trifolii	BG	peri	74	5.4	IEC,SEC,AC-glyco	*(244)*
Robillarda sp. Y20	EG(no gl)	e	59	3.5	p-etha,IEC,Chrofo	*(422)*
	BG(gl)	e	76	7.5		

Continued on next page

Table 1. *Continued*

Organism	Enzyme	Location	M_r (kDa)	pI	Purification Methods	Reference
	BG(no gl)	e	54	3.8		
Ruminococcus sp.	EG	e	225,22,10		IEC,SEC	*(423)*
Ruminococcus albus	complex	e	1000		SEC,p-AS(80),IEF-col	*(424,425)*
		e	25			
R. albus	CMCase	e	50		IEC,SEC,IEC,IEF-col,SEC	*(426)*
	CBH	e	2x100		IEC,IEC,IEF-col	*(427)*
	BG(gl)	e	82-110		IEC,IEC,IEC,IEF-col	*(428)*
R. albus	EG	e	30	6.1	SEC,p-AS(80),SEC,IEF	*(424)*
Ruminococcus flavefaciens	EG-I	e	140		IEC-HPLC,SEC-HPLC	*(429)*
	EG-II	e	100			
	CBH-A(no gl)	e	2x118		IEC-HPLC,HAC-HPLC,SEC-HPLC	*(430)*
Sambucus nigra	EG		54		AC-cell,p-AS(85),AC-cell	*(431)*
Schizophyllum commune	EG-I	e	40	3.7	p-etha,IEC	*(432)*
	EG-II	e	38	3.5		
S. commune	BG	e	97		IEC,SEC	*(433)*
Sclerotium rolfsii	CBH(gl)	e	42	4.32	p-AS(90),SEC,IEC,PAGE-prep	*(434)*
S. rolfsii	EG-I(gl)	e	20+30	4.55	p-AS(90),IEC,SEC,PAGE-prep,IEF-col	*(435)*
	EG-II(gl)	e	27	4.2		
	EG-III(gl)	e	78	4.51		
	CBH(gl)	e	42	4.32	p-AS(100),SEC,IEC,PAGE-prep	*(436)*
	BG(gl)	e	90-110		p-AS(100),SEC,IEC,IEF-flat	*(437)*
S. rolfsii	BG	e	90-107	4.5	p-AS(90),SEC,IEC,IEF-col	*(438)*
Sclerotinia sclerotiorum	BG	e	4x65		p-AS(90),SEC,IEC	*(156)*
	EG	e	48	6.2	p-AS(90),SEC,IEF,IEC	*(157)*
	EG	e	34	3.7		
Scytalidium lignicola	BG-I	e	74		p-acet,SEC,IEC	*(170)*
	BG-II	e	69			
Sporocytophaga myxococcoides	EG-I	e	46	7.5	IEC,SEC	*(130)*
	EG-II	e	52	4.75		
Sporotrichum pulverulentum	CBQoxidored	e	58		p-AS(60),IEC,AC-conA	*(439)*
	CBO	e	74		p-AS(55),SEC,IEC,SEC,AC-dye	*(440,441)*

Continued on next page

Table 1. *Continued*

Organism	Enzyme	Location	M_r (kDa)	pI	Purification Methods	Reference
S. pulverulentum	BG	e	170	4.5	p-AS(0-90),IEF-slab,HIC,IEF-slab	*(442)*
S. pulverulentum	CBO	e	93		IEC,SEC,HIC,IEF-col	*(443)*
S. pulverulentum	EG-I(gl)	e	32	5.32	p-AS(90),IEC,SEC,AC-conA,IEC	*(94)*
	EG-II	e	37	4.72		
	EG-III(gl)	e	28	4.4		
	EG-IV(gl)	e	38	4.65		
	EG-V(gl)	e	37	4.2		
	CBH(no gl)	e	48	4.3		
Sporotrichum thermophile	CBH(gl)	e	64	4.52	p-AS(20-80),SEC,IEC,HAC,	*(444)*
S. thermophile	BG-A	i	440	4.48	son,p-AS(20-75),IEC,SEC	*(445)*
	BG-B	i	40	4.4		
Stachybotrys atra	BG(gl)	e	65	4.8	p-strep,p-acet,HAC,SEC,AC-thiogluc	*(167)*
Streptomyces A20	EG	e			PAGE-prep	*(446)*
Streptomyces KSM-9	CMCase	e	32,32,92		p-AS(70),IEC,IEC,SEC	*(447)*
	CMCase	cloned			p-AS(90),SEC,IEC,SEC	*(448)*
Streptomyces flavogriseus	CBH	e	45	4.15	IEC,SEC,IEF,AC-conA	*(449)*
Streptomyces lividans	BG(gl)	i	51	4.3	IEC,SEC	*(186)*
S. lividans	EG(gl)	e	46	3.3	p-etha,IEC-HPLC,SEC-HPLC	*(450)*
Streptomyces reticuli	EG	e	82		IEC,SEC	*(451-453)*
S. reticuli	EG	e, bound	120+42 (300)		French,SEC,IEC	*(453)*
	CMCase	e	40-48			
S. reticuli	BG	i	50	4.5	son,IEC,IEC,IEC,HIC,Chrofo	*(454)*
Talaromyces emersonii	EG-I(gl)	e	35	3.19	p-AS(90),SEC,IEC,IEC,PAGE-prep	*(135,455,456)*
	EG-II(gl)	e	35	3.08		
	EG-III(gl)	e	35	2.93		
	EG-IV(gl)	e	35	2.68		
	BG-I	e	135		p-AS,SEC,IEC,PAGE-prep	*(457,458)*
	BG-II	e	100			
	BG-III	e	46			
	BG-IV	i	58			

Continued on next page

Table 1. *Continued*

Organism	Enzyme	Location	M_r (kDa)	pI	Purification Methods	Reference
T. emersonii	BG	e and i			PAGE	*(458)*
	all	e			p-AS(20-80),SEC,IEC	*(29)*
Termitomyces sp.	BG	e			IEC,p-AC,SEC	*(459)*
Termitomyces clypeatus	BG(gl)	c	4x110	4.5	p-AS(60-100),IEC,SEC,SEC	*(143)*
Thermoascus aurantiacus	BG(gl)	e	87		p-AS(10,20,30,40,50,60),SEC,SEC	*(159,460)*
	EG-I(gl)	e	78			
	EG-II(gl)	e	49			
	EG-III(gl)	e	34			
T. aurantiacus	BG(gl)	i	2x79	4.5	IEC,IEC,HAC,p-heat,SEC	*(99)*
Thermomonospora curvata	BG	bound	66		p-AS(50-70),IEC-HPLC	*(461)*
Thermomonospora fusca	EG-I(gl)	e	46	4.5	p-AS(80),SEC,SEC,HA	*(197)*
	EG-II	e	94	3.5		
T. fusca	all	e	42-108	3-5	p-AS(55),SEC,HAC	*(196)*
T. fusca	EG-II	cloned	2x42		IEC	*(462)*
Thermophilic anaerobe	EG(no gl)	cloned	40		p-heat(70/5),AC-cell,IEC	*(221)*
Thermotoga sp.	BG	bound	75	4.2	IEC,IEC	*(463)*
	CBH	e	36		IEC,IEC,SEC	*(464)*
Thermus sp.	BG	i	48		lysozyme,son,HIC,SEC	*(465,466)*
Torulopsis wickerhamii	GG(gl)	e	2x75		p-AS(100),IEC,SEC,centr-CsCl	*(281)*
Trametes gibbosa	BG	e	640		p-AS(50),SEC	*(160)*
Trichosporon adeninovorans	BG	e	570		p-acet,HAC,SEC	*(172)*
		e	525			
Trichoderma koningii	BG,EG,CBH	e			p-AS(20-80),AC-conA,SEC,IEC,IEF-col	(142,259,262,467)
T. koningii	EG	e			AC-cell	*(468)*
	all	e			AC-cell	*(211)*
Trichoderma reesei	CBH-I	e	65	3.9	AC-thiocellobiose	*(88,224)*
	CBH-II	e	56	5.9		
T. reesei	CBH-I	e			AC-thiooligo	*(228)*
	CBH-II					
T. reesei	EG-III	e	48	5.1	p-AS(40-60),IEC,SEC,IEC	*(469)*
T. reesei	BG	i	98	4.4	p-AS(28-57),SEC,IEF-col	*(470)*

Continued on next page

Table 1. *Continued*

Organism	Enzyme	Location	M_r (kDa)	pI	Purification Methods	Reference
T. reesei	EG-I	e			AC-immuno	*(247)*
T. reesei	CBH-II(gl)	e	58	6.3	AC-immuno	*(12)*
T. reesei	CBH(gl)	e	65	4.2	AC-cell,AC-immuno	*(216)*
T. reesei	CBH-I	e	64	4.2	AC-immuno	*(246)*
T. reesei	CBH-I	e	67	3.62	SEC,PAGE-prep,IEC	*(91)*
	CBH-I(core)		55	3.23		
T. reesei	EG	all			AC-immuno,SEC,Chrofo,AC-conA,	*(245)*
T. reesei	CBH	e	64	4.75	IEC,IEC,IEF-rotofor,AC-conA	*(471)*
T. viride	EG	e			IEC,IEC,SEC	*(86)*
T. reesei	EG(no gl)	e	25	7.5	SEC,IEF-flat	*(64)*
T. reesei	BG(gl)	e	2x35		p-AS(20-40),AC-cell,IEC,SEC	*(304)*
T. reesei	BG-I(gl)	e	71	8.7	p-acet,IEC,IEC,SEC	*(169)*
	BG-II(gl)	e	114	4.8		
	CHB-I	e	65,58,54		IEC,IEC,IEC	*(93)*
T. reesei	EG-II(gl)	e	55	4.5	SEC,IEC,IEC,IEC	*(288)*
	EG-III(gl)	e	48	5.5		
	CBH-I(gl)	e	64	3.9		
	CBH-II(gl)	e	53	5.9		
T. reesei	all	e			IEC-HPLC	*(472)*
T. reesei	all	e			Chrofo	*(264)*
T. reesei	EG-I(gl)	e			AC-conA	*(238)*
	CBH-I	e				
T. reesei	BG,CBH,EG	e			p-AS(20-80),IEC,Chrofo	*(473)*
T. reesei QM9414	BG,CBH,EG	e			p-AS(70),SEC,IEC,IEC,SEC	*(106)*
T. reesei QM9414	BG,CBH,EG	e			IEF	*(267)*
T. reesei QM9414	all	e			AC-cell,IEC,PAGE-prep	*(474)*
T. reesei (Celluclast)	CBH-I	e			AC-thiooligo	*(227)*
	CBH-II	e				
T. reesei (Celluclast)	CBH-I	e	65	4	IEC,SEC,SEC-HPLC	*(475)*
	EG-I	e	52	4.5		
	EG-II	e	48	7		

Continued on next page

Table 1. *Continued*

Organism	Enzyme	Location	M_r (kDa)	pI	Purification Methods	Reference
T. reesei (Celluclast)	CBH-I	e		4.1	p-AS(90),SEC,HAC,HIC,Chrofo	(*476*)
	CBH-II	e		6.1		
T. reesei (Maxazyme)	EG	e	23-57	3-8	SEC,IEC,IEC,SEC,IEC	(*477*)
T. reesei (technical)	EG	e	26	7.3	IEC,SEC	(*478*)
	CBH	e	36	3.2		
T. viride	all	e			AC-cell	(*215*)
T. viride	CBH-I	e	66	4.2	IEC,IEC,IEC	(*479*)
	EG-I	e	54	4.7		
	EG-II	e	115	5.5		
T. viride QM9414	EG	e	20	7.52	SEC,IEC,IEF-col	(*480*)
	EG	e	51	4.66	SEC,IEC,IEF-col,AC-immuno	(*65*)
T. viride (Maxazyme)	CBH-I(gl)	e	53	5.3	SEC,IEC,SEC,SEC,IEC,AC-cell	(*481*)
	CBH-II(gl)	e	60	3.5		
	CBH-III(gl)	e	62	3.8		
	EG-I(gl)	e	50	5.3		
	EG-II(gl)	e	45	6.9		
	EG-III(gl)	e	58	6.5		
	EG-IV	e	23	7.7		
	EG-V(gl)	e	57	4.4		
	EG-VI(gl)	e	52	3.5		
	BG(gl)	e	76	3.9		
T. viride (Meicelase)	CBH-II(gl)	e	48		AC-cell,IEC	(*209*)
T. viride (Onozuka SS)	EG-I(gl)		12	4.6	SEC,IEC,AC-cell,IEC-dipolar,IEF-col	(*280*)
	EG-II(gl)		50	3.39		
	BG(no gl)	e	47	5.74	p-AS(90),IEC,IEF-col,SEC	(*277*)
T. viride (Panacellase SS)	EG-(gl)	e	37,52,50		AC-cell,IEC,IEC	(*210*)
T. viride (technical)	BG	e	76		IEC,AC-cell,IEC,IEC,SEC	(*482*)
Trinervitermes trinervoides	CMCase				p-acet,IEC,IEC	(*483*)
Tropaeolum majus L.	EG(gl)	seed	29	5.0	p-AS(90),IEC,IEC,SEC	(*484*)
Yeast from soil	CMCase	e	34		p-AS(60-90),p-etha,IEC	(*485*)

Continued on next page

Table 1. *Continued*

Footnotes to Table 1

Abbreviations used:

Enzymes

BG, ß-glucosidase; BG(gl), ß-glucosidase (glycoprotein); CBdehydro, cellobiose dehydrogenase; CBH, cellobiohydrolase; CBH(gl), cellobiohydrolase (glycoprotein); CBO, cellobiose oxidase; CBQoxidored, cellobiose-quinone-oxidoreductase; CD, cellodextrinase; CMCase, carboxymethylcellulase; EG, endoglucanase; EG(gl), endoglucanase (glycoprotein); GG, glucano glucohydrolase.

Chromatography, affinity

AC-cell, on cellulose; AC-cellobiamine, on cellobiamine; AC-cellobiose, on cellobiose; AC-conA, on concanavalin A; AC-dye, on triazine dye, reactive red; AC-glucit, on glucitol; AC-glyc, on a glycoprotein; AC-immuno, on antibodies; AC-jac, Jacalin; AC-oligo, cello-oligomer; AC-salicine, on salicine; AC-seph, Sephadex; AC-starch, on starch; AC-thiocellobiose, on thio-cellobiose; AC-thiogluc; thio-glucoside; AC-thiooligo, thio-cello-oligomer.

Chromatography, general

HAC, on hydroxyapatite; HIC, hydrophobic interaction; IEC, ion exchange; IEC-dipolar, ion exchange on dipolar matrix (arginine); SEC, size exclusion.

Electrophoresis

IEF, in cylindrical PAGE; IEF-col, in sucrose gradient stabilized columns; IEF-rotofor, Rotofor (Bio-Rad); PAGE-prep, preparative PAGE.

Precipitation

p-acet, with acetone; p-acid, with acid; p-AS, with ammonium sulfate (no specifications); p-AS(80), with ammonium sulfate, 0 to 80% of saturation; p-AS(20-70), with ammonium sulfate, 20 to 70% of saturation; p-AS(20,30,40), with ammonium sulfate, fractionated cuts; p-etha; with ethanol; p-PEI, with polyethyleneimine; p-heat(60/15), heated to 60°C for 15 min; p-prot, with protamine sulfate; p-strep, with streptomycin sulfate.

Miscellaneous abbreviations

bound, cell-wall bound; centr-suc, sucrose gradient centrifugation; centr-CsCl, isopycnic banding centrifugation; e, extracellular; French, French press; GH-extr, guanidinium hydrochloride extraction; i, intracellular; lyo, lyophilization; osmotic, osmotic shock; peri, periplasmic activity staining protocols (*151,152*). For example, from the cell-free culture broth of *Chaetomium thermophile* the glycosylated EG and CBHs (molecular mass 36, 40 and 60 kDa, respectively) were bulk-precipitated with 70 to 80% ammonium sulfate (*153,154*). The complexity of the subsequent ion-exchange column profile indicates that fractionated precipitation might have provided a much cleaner starting material. Other examples include the purification of ß-glucosidase and endoglucanase from *Evernia prunastri* (*100,155*) and from *Sclerotinia sclerotiorum* (*156,157*). In both cases a single ammonium sulfate cut of 0 to 90% of saturation was performed, although differences in molecular mass (260 to 270 for BGs, and 38 to 70 for EGs) and in isoelectric point (3.7 and 6.2 in the case of *Sclerotinia sclerotiorum*) might have proven large enough for successful fractionation with salt.

Fractionated precipitations have not always met with success. While an EG of *Aspergillus niger* emerged from a 20 to 40% ammonium sulfate cut at 83% yield and a purification factor of 2, the BG and CBH were recovered at only 50% yield and no improvement in purity (*158*). Likewise fractionated precipitation of cellulases from *Thermoascus aurantiacus* did not result in any purification (*159*). In both cases, a small adjustments to pH and ionic strength of the starting buffer might have improved the situation. In contrast, extremely successful ammonium sulfate fractionations have been reported, too. A single 0 to 50% cut separated BG and EG from *Trametes gibbosa*, and BG was recovered with a purification factor of 10 (*160*).

Salts other than ammonium sulfate, such as NaCl, KCl, $MgSO_4$, and Na_2SO_4, are also in use for protein precipitation. However, only in specialized cases, e.g. protein crystallization, do they offer advantages over ammonium sulfate (*161*). A recently described procedure using $ZnCl_2$ (*162*) appears appealing because the salt load in the precipitated protein is very low. This method has not been used with cellulases but in view of occasional sensitivity to metal ions (*163*) compatibility with Zn^{2+} must be tested prior to application.

Organic Solvent Precipitation. A widely known and commercially practiced technique (*164*), protein precipitation with organic solvents is as convenient as salt precipitation. However, the application of this technique requires closer control because organic solvents in large excess or at temperatures approaching ambient begin to irreversibly denature proteins, while ammonium sulfate in most cases tends to stabilize them. Lately, environmental concerns seem to disfavor the use of organic solvents. Nevertheless, precipitation with acetone and ethanol, the most frequently used organic solvents, remains a viable alternative to salt precipitation, as both protein fractionation and removal of lipophilic contaminants are possible. On the other hand, non-proteinaceous components, such as polysaccharides may co-precipitate with proteins. This is a welcome feature, though, when co-precipitation is needed for the recovery of protein from very dilute solutions.

The process of protein precipitation by organic solvents must be monitored much more carefully than that of salt precipitation, because the hydration of organic solvent molecules is exothermic and heat development is greatest during the initial phase of solvent addition. Another reason is that protein solubility in aqueous/organic solutions is much more temperature dependent than in salt solutions, which means, that for organic solvent precipitation to be effective, the temperature of the protein solution must initially be as close as possible to 0°C. While this may be conveniently accomplished by placing the container in an ice/salt mixture, care must be taken that the aqueous solution does not freeze! As the organic solvent (pre-cooled to at least -20°C) is added drop-wise its cryoprotective function permits dropping the temperature of the mixture slowly below 0°C, which, in turn, enhances protein precipitation. An interesting variant of ethanol precipitation involves the presence of guanidinium hydrochloride which adds considerable selectivity to the process (*165*).

Most proteins precipitate at 20 to 50% (v/v) solvent although the majority of cellulases has been precipitated at 60 to 80% (v/v) ethanol or acetone. Precipitation with acetone and ethanol has given good yields (65 to 90%) and purification factors of 8 to 10 have been achieved (*166-168*). In other instances almost no purification resulted (factor 1 to 1.4) although yields were substantial (*169-171*). And there are cases, where both yield and purification factor were low (*150,172*), possibly because heat-induced denaturation and temperature-dependent solubility were underestimated.

Polyethylene Glycol Precipitation. Introduced in 1964 (*173*) polyethylene glycol (PEG) has become a valuable reagent for protein precipitation (*174*). In combination with pH and divalent metal ions (*175*) PEG is a very selective protein precipitant. As early as 1976, preliminary tests documented the feasibility of precipitating cellulases with PEG (*176*), and yet, there is a conspicuous absence of publications that report the use of PEG for cellulase purification. One of the reasons might be the lower cost of both ammonium sulfate and organic solvents compared to PEG since many purification protocols serve as pilot experiment for large scale operations.

Heat and pH Precipitation. Precipitation of unwanted proteins by heat is a valuable entry into protein purification (*6,141*). Although largely an empirical method, there are parameters that can (and carefully must) be controlled. The pK_a of buffer substances is temperature dependent and some compounds experience rather large dpH/dT values (*177*) which will shift the pH of the heated extract into a range where the desired enzyme is no longer stable. Traces of proteolytic enzymes, at 4°C of no significant threat to the investigated enzyme, may become a nightmare at elevated temperatures. The addition of salt and, better yet, of protease inhibitors will remedy this problem to a certain extent.

Heat denaturation is generally useful only under two conditions: (a) the desired protein has an intrinsically higher heat resistance than most other proteins (which is rare) or acquires this differential property by binding of ligands (substrates, but also unrelated compounds found empirically (*11*)); and (b) there is sufficient extraneous protein to warrant the treatment in the first place. Neither condition is really met with cellulases. They are either low in concentration and adding cellulose might already precipitate them (affinity precipitation) or the culture filtrate contains only low levels of contaminating proteins. However, for thermostable and genetically overexpressed proteins, the method has been put to good use (*178,179*).

Precipitation of unwanted proteins by denaturation at extremes of pH has proven useful as a preliminary purification step for recombinant proteins. Prerequisite for this technique is that the protein of interest is stable and soluble at the chosen pH (acid or alkaline). Because cellulases are for the most part slightly acidic proteins (pI 3 to 6) and are only moderately stable below pH 3 and above pH 10, pH dependant precipitation has rarely been used in cellulase purifications. In one instance the culture broth was adjusted to pH 3 with 1 N HCl and allowed to stand for 24 h. A precipitate was removed by centrifugation but neither the enzyme's stability nor a purification factor was recorded (*180*).

Miscellaneous Precipitations. A borderline case between precipitation and affinity chromatography is the separation of cellulases of a commercial product, Meicelase, by chitosan. BG remains in solution while EGs and CBHs form conjugates with chitosan which can be removed by centrifugation. The pellet is soluble upon change in pH and makes the trapped enzymes accessible to further purification (*181*).

Ion Exchange Chromatography. As probably the most versatile of all separation methods ion exchange chromatography (IEC) commands a central position in nearly all purification protocols for cellulases. In the early days cellulose-based ion exchange matrices were viewed as particularly suitable to cellulase purification because they combined ion exchange with affinity chromatography (*182,183*). A subsequent study showed that the affinity aspect was limited to cellulases that have the ability to degrade crystalline cellulose (*184*). The availability of agarose, acrylamide, dextran-acrylamide, and styrene based ion exchange supports as well as biospecific affinity matrices has ended efforts to combine ion exchange with affinity chromatography, because the two methods in sequence provide better resolution than their combination in one step. Nevertheless, there may be special situations where IEC on a cellulose matrix offers advantages (*185*) although one has to be prepared for much lower yields than with the use of trisacryl supports (*138*).

Anion exchangers are the most versatile chromatography supports for cellulase purification for many cellulases are neutral to slightly acidic proteins. Choice of buffer type, pH range and displacing ion open an almost unlimited number of options for separation. Most cellulases bind within a pH range of 5 to 9 at low ionic strength (10 to 25 mM) and are eluted at moderate salt concentrations, but neither very shallow (0.01 to 0.025 M) nor very steep gradients (0 to 1 M) are unusual. In only few cases have pH gradients been explored, possibly because they are more difficult to conceptualize and harder to vary than salt gradients. Although a gentle method, IEC of cellulases may lead to considerable loss in protein and activity, and also to unacceptable band broadening (trailing). Stabilization of cellulases has been accomplished by mixing 10% (v/v) glycerol to the running buffer (*186*). Band trailing at very shallow salt gradients (where unspecific adsorption to the IEC matrix becomes a significant factor) has been suppressed by adding methyl-glucoside to the running buffer (Paech, unpublished observation).

Ion exchangers are used in column and in batch mode. The latter is preferred when the viscosity of the protein solution would prevent acceptable flow rates in a column. With occasional shaking or very gentle stirring a small quantity of the ion exchange material is equilibrated with the extract for several hours, the mixture is allowed to settle under gravity, decanted, and then loaded onto a chromatography column. Release of bound proteins is accomplished in gradient or batch mode by salt or pH. Alternatively, the ion exchanger can be filtered off, washed first with buffer, and then with a salt solution to release the bound enzyme(s).

There are too many examples of successful ion exchange chromatography in cellulase purification to allow their individual listing here. Suffice to say that all protocols include commonly available reagents and no unusual buffer systems. As mentioned above, cellulose-based chromatography supports have been abandoned, but when used in a variation of an earlier method apparently no disadvantage was noticed (*187,188*).

Adsorption Chromatography on Hydroxyapatite. Among inorganic adsorbents (*6*) hydroxyapatite (HA) is by far the most frequently used chromatographic support for cellulase purification. Introduced to protein chemistry (*189*) long before cellulose-based ion exchangers and Sephadex, HA with predictable performance characteristics has become commercially available only recently. The difficulty of preparing HA resides with controlling crystal growth during mixing of dibasic sodium phosphate and calcium chloride and maintaining crystallinity during treatment of the resulting dibasic calcium phosphate (brushite) with sodium hydroxide. The entire procedure is rather involved but can produce a separation matrix of superior quality (*190*). This is why many scientists still prefer to prepare (and cherish) their own hydroxyapatite. However, properties of HA differ widely among preparations (which, consequently, extends to commercial suppliers) and the user is advised to test different batches for a particular separation problem. Once a suitable preparation is found it should be stored and handled with great care if subsequent protein separations are to be repeated with some degree of predictability. Several factors may contribute to the decay of separation efficiency. HA crystals are physically not very stable and tend to disintegrate under pressure and shear forces. Their crystallinity is also affected by the buffer chemistry. Therefore, HA must always be stored under consistent conditions with respect to buffer type and strength, and temperature (e.g., 1 mM sodium phosphate, pH 7.0, at 4°C). Although said to be unnecessary, addition of 0.03% sodium azide as preservative should be made routine practice when working with cellulases. Another drawback of HA is its slightly basic nature. HA eagerly absorbs atmospheric and dissolved CO_2 which alters surface charges of the crystals and, more importantly, aggregates the crystals. This results in drastically reduced column flow rates. The use of boiled water for preparing buffer solutions, and sparging of buffers with nitrogen reduces the problem and increases the longevity of a HA batch. Further improvement on flow rates may be achieved by mixing HA with Sephadex G-25 (*191*), but the merit of this approach has not been evaluated in detail.

Hydroxyapatite chromatography (HAC) is most useful for the separation of proteins with neutral to acidic isoelectric point (*190,192,193*) which happens to be the range for many cellulases. Loaded at low ionic strength (1 to 20 mM) in buffered phosphate solutions (pH 6.5 to 7) or in unbuffered solutions of NaCl, $CaCl_2$, or $MgCl_2$ onto HA, proteins are eluted in batch or gradient mode with the same buffer or salt at up to 1 M strength. Particularly the group of Wilson (*194-197*) has refined HAC to great performance. Recovery of enzyme activity is in the range of 70 to 90% and purification factors vary from 2 to 5 (*198*). However, due to the large differences in properties of HA, one has to be prepared for lower recoveries and purification factors (*199*) and should be reminded that application of HAC by analogy requires more trial-and-error experiments than with most other chromatography techniques. Low recoveries and marginal improvement in specific activity (*92,199*) indicate an expired hydroxyapatite more than the inapplicability of the method to the separation problem.

Size Exclusion Chromatography. Size exclusion chromatography (SEC) can produce substantial purification factors because individual components of a cellulolytic system often differ significantly in size or are accompanied by extraneous proteins of largely different size. SEC is a 'must' in almost any purification scheme. Those who master the art of pouring a SEC column may wish to continue doing that although precast columns for FPLC systems surpass in convenience and sometimes resolution open-end columns. Others, who have never seen a manually poured SEC column will not wish to acquire that skill but rather continue the use of FPLC columns. In any event, with open-end columns and with FPLC columns, there is little that can go wrong with SEC. Cellulase may show low mobility in SEC (*200-202*). However, surprises cannot be avoided and unusual usage of SEC material is still possible. The unexpected presence of a dextrinase shrank the bed volume of a Sephadex column by 25% (*203*). Affinity of Sephadex to cellulases has also been exploited in recovering EG-C from fermentation broth (*204,205*).

Affinity Chromatography. Affinity chromatography (AC) has been lauded as the most powerful of all protein separation techniques (*206-208*). Applied to enzymes AC takes advantage of their biospecific interaction with substrate analogs, inhibitors, or cofactors. In a broader sense group-specific (e.g., binding of lectins to glycoproteins) and structure-specific interactions (immunoaffinity) are also exploited in affinity chromatography. When immobilized on a suitable support the biospecific ligand will retain only the enzyme of interest and release it when presented with a displacing reagent. This ideal case has rarely been realized (*206*). Besides weaknesses in the chemistry of the ligand/matrix there is always the trade-off between selective binding and successful release. Increase in ligand selectivity diminishes the likelihood of specific desorption. Eventually, the enzyme must be liberated by non-specific techniques (temperature, pH and ionic strength change, chaotropic and denaturing reagents) which frequently result in drastically reduced yields.

The application of cellulose to the purification of cellulases has contributed as much to the understanding of their mechanism of action as it has to their separation (*191,209-217*). An appraisal of the affinity of *Trichoderma reesei* cellulases for cellulose and the means of their recovery by Reese (*218*) is quite instructive in this context. AC on cellulosic material (in batch or column mode) mainly accomplishes a group separation: BGs do not adsorb to cellulose (they remain in the supernatant or run through the column), EGs adsorb with moderate affinity to cellulose (they are released with buffers at low or at high ionic strength), and CBHs bind with high affinity to cellulose (they are displaced with distilled water or with a chaotropic reagent). Lowering of the ionic strength (from 20 mM buffer to distilled water) gave excellent separation of the enzymes from *T. viride* (*210*) and *T. koningii* (*211*), while increasing the ionic strength was advantageous for the release of EGs from *Geotrichum candidum* (*214*) and *Penicillium camembertii* (212). The latter example shows that batch mode is more convenient but may result in inferior separation compared to column mode. Elution from cross-linked cellulose was also afforded with CM-cellulose (*215*). AC on cellulose sometimes does not accomplish a group separation (Avicelase and CMCase activity of *Bacillus subtilis* largely elute together (*219*)) but always excels in the purification of cloned EGs (*220,221*).

Resolution of a cellulase-cellulose complex by allowing the enzymes to completely degrade the cellulosic substrate has been suggested by Brown's group more than a decade ago. Applied to the cellulases of *Cellulomonas fimi* in a diafiltering flow-through reactor (to remove inhibitory di- and oligosaccharides), even in the presence of an auxiliary cellulase system (*Trichoderma reesei*) none of the enzymes were released (Paech, unpublished observation). Others, however, succeeded with this technique which they named affinity digestion when they recovered almost quantitatively the cellulosome from *Clostridium thermocellum* (*222*).

The introduction of immobilized *p*-aminobenzyl 1-thio-ß-D-cellobioside as biospecific adsorbent for CBHs was a significant step towards a complete separation of cellulolytic enzyme mixtures by AC (*223,224*). When applied to cellulases from *Trichoderma reesei* CBH I was released with 100 mM lactose and CHB II with 10 mM cellobiose, while EGs did not bind at all (*223*). These adsorption characteristics are not necessarily applicable to all cellulase systems. For example, the EG of *Clostridium thermocellum* cloned into *E. coli* has been successfully recovered from cell-free extracts by adsorption to Sepharose 4B-supported thio-cellobioside (*225*). On the other hand, this matrix failed with extracts from *Phanerochaete chrysosporium* (*226*). A drawback of the use of thio-cellobioside is its potential destruction by BGs, although addition of 1 mM glucono-lactone, an effective BG inhibitor, to all buffers affords some protection of the AC matrix. Replacing all interglycosidic oxygen atoms in immobilized cellobioside and cellotrioside with sulfur has significantly improved the longevity of these AC columns (*227,228*). Progress in synthesis of thio-oligosaccharides (*229*) may soon provide more specific AC matrices. The general usefulness of a recently introduced commercial cellobiose affinity gel (Sigma) could not be verified (Paech, unpublished observation).

Effective AC for the purification of BGs is still in its infancy. In an early report *p*-aminobenzyl 1-thio-ß-D-glucopyranoside coupled to Sepharose 2B separated 2 BGs from *Stachybotris atra* but the potential of this technique was not explored in detail because enzyme recovery was too low (*167*). 1-Thio-ß-D-glucopyranoside attached to chitosan through an aliphatic linker seems to bind BG from *Alcaligenes faecalis* quite well (*230*), but no further application of this method has been reported. AC of BG from *Candida guilliermondii* on an immobilized salicin derivative provided good separation and yields, but the life span of the column was not mentioned (*231*). Cellobiose coupled to epoxy-activated Sepharose 6B apparently provides also a reasonable AC matrix for the purification of BG from *Piromyces* sp. E2 (*232*). Salicin or cellobiose coupled to controlled porous glass served as suitable matrix for isolating BG and aryl-ß-glucosidase from culture broths of *Trichoderma reesei* QM9414 (*233*). In this case, enzymes were released from the columns by pH shifts or salt gradients, rather than disaccharides. A new dimension in BG purification opened when methacrylamide-N-methylene-bis-methacrylamide supported cellobiamine was used for resolving crude extracts of *Aspergillus niger*. The purity of a BG obtained by this method surpassed by a factor of 10 that isolated by chromatofocusing (*97*). BGs unrelated to cellulases were purified on N-(9-carboxynonyl)-deoxynojirimycin-AH-Sepharose with a factor of over 30 (*234*) and on gluconate-Sepharose with a factor of 550 (*235*).

Affinity chromatography of EGs on specific supports is mechanistically easy to envisage but synthetically difficult to realize. However, the salt-promoted binding of EGs from *Dolabella auricularia* to a cellooligomer ligand on Sepharose CL-2B (which resulted in a respectable purification factor) must be considered a step in this direction (*236*).

Non-specific affinity chromatography on immobilized lectins (*237*), specific for the carbohydrate moiety of glycoproteins, has been used in cellulase purification with variable success. Agarose-immobilized concanavalin A (ConA), a ligand that responds to α-D-manno- and α-D-glucopyranosyl residues, separated non-glycosylated BG from glycosylated EG of a partially purified sample from a commercial broth of *Trichoderma viride* (*238*). A ConA column was also used to recover only the non-glycosylated BG from *T. viride* culture fluids (*239*). In the latter case it was mentioned that methyl-α-glucoside, typically used to release bound glycoproteins from ConA, did not recover any enzyme and had to be replaced with 0.1 M borate buffer. Nothing was mentioned on the capacity and effectiveness of the column after repetitive use. In view of the cost for a ConA column this is an important consideration as another example shows. Although endowed with a 14% carbohydrate moiety, the induced BG of *Stachybotrys atra* did not bind to ConA, while the constitutive BG was retained very strongly (*167*). In fact, the liganding was so tight that a regeneration of the column was not possible and the technique was abandoned. The apparent tenacity with which ConA binds certain glycoproteins has been exploited in both purification and immobilization of BG from *Aspergillus niger* (*240*). Examples of successful applications of ConA include the purification of EGs from *Thermomonospora fusca* (*26*), *Aspergillus fumigatus* (*203*), *Bacillus* (*241*), and *Chaetomium cellulolyticum* (*242*).

Other lectins and glycoproteins as affinity ligand have also been used. A subcellulosome fraction from *Clostridium thermocellum* was purified on a Jacalin column (*243*). A somewhat exotic approach to isolate a BG from *Rhizobium trifolii* by AC was taken with the immobilization of a cell-wall glycoprotein from *Saccharomyces cerevisiae* (*244*).

Structure-specific AC relies on the selectivity of antibodies for their antigen. Given the rapid progress in monoclonal antibody technology and the fact that relatively impure protein mixtures can be used for immunization, affinity chromatography on immunoadsorbents will become more commonplace in purification schedules. The selection of a suitable antibody from a panel of monoclonal antibodies requires careful consideration of and screening for the desired properties. The most useful antibody combines binding selectivity with ease of antigen release. Change in ionic strength or pH is preferred. Organic solvents such as isopropanol or ethylene glycol, and nonionic surfactants may be tolerated. Chaotropic reagents, extreme pH, and temperature jumps should be avoided.

The first successful purification by immunochromatography was accomplished with polyclonal antibodies directed towards CBH-I of *Trichoderma reesei* (*216*). The low yield of this preparation did not discourage the authors to also purify CBH-II (*12*) and an EG (*245*) from *Trichoderma reesei*, but prompted other groups to raise monoclonal antibodies against CBH-I (*246*) and EG-I (*247*) and re-assess the properties of these enzymes. Through all of these studies (including work on cellulases of other organisms (*248*)) it became clear that low yield and cross-reactivity of polyclonal antibodies related to the high degree of homology among different cellulases are a shortcoming of immunochromatography, and that resolution of multiple enzymes derived from the same parent protein is still not possible. However, progress in this field can be anticipated as monoclonal antibodies directed towards individual domains of cellulases are being developed (*249-253*).

Hydrophobic Interaction Chromatography. In this method the non-covalent interaction of a hydrophobic stationary phase with a hydrophobic segment of the analyte in a polar liquid phase accomplishes molecular partition so essential to chromatography. The retained molecule is released by decreasing the ionic strength of the application buffer (negative gradients of ammonium sulfate, sodium chloride, or other salts), by decreasing the dielectric constant of the liquid phase (increasing gradients of ethylene glycol, isopropanol, acetonitrile), or by a simultaneous or sequential application of both regimes. The immediate concern in protein purification, of course, how well the enzyme will withstand the rigors of any of these procedures. Thus, loss due to denaturation must always be anticipated in hydrophobic interaction chromatography (HIC).

Hydrophobic interaction forces have been implied in the binding of cellulases to cellulose for many years (*218,254*). And although the exact nature of these forces is not known, researchers were quick in exploiting this feature for chromatographic purification as soon as suitable separation matrices became available. Recoveries of activity are modest (around 50 to 65%) and purification factors from 2 to 5 can be expected. A particularly interesting case of HIC is the purification of an exo-ß-(1,3)-glucanase from *Candida albicans* on phenyl-Sepharose eluted with 50% acetonitrile (recovery 45%, purification factor 40).

The advantage of HIC is that prior 'desalting' of the protein solution to be applied is unnecessary, and so is concentration. Therefore, HIC is the ideal step after IEC or a salt precipitation.

Preparative Isoelectric Focusing and Chromatofocusing. These two techniques have been used extensively in cellulase purifications because they reveal small differences in protein structure (isoelectric point) of seemingly identical enzymes with respect to size and catalytic function. Over the years access to these technologies has improved, but caution must be exercised in the interpretation of the data. An apparently homogeneous protein after preparative isoelectric focusing (IEF) showed different enzymic activities. Subsequent analyses confirmed that the isolated material was an enzyme complex rather than a single protein (*255,256*). Another weak point of IEF and chromatofocusing (Chrofo) is that complete removal of ampholytes from glycoproteins is difficult to accomplish and to ascertain (one promising procedure described is not applicable for many cellulases because it involves the use of dialysis membranes (*257*)). Residual ampholytes, however, may interfere with activity assays and synergy experiments. For example, an effect of residual ampholytes on the binding ability of EGs from *Aspergillus niger* (*258*) cannot be excluded. It is perhaps in recognition of this notion that preparative IEF and Chrofo in recent purification protocols appear as earlier steps (e.g., (*232*)) while in earlier publications these techniques assumed the final step in the preparation of a protein (*245,259-261*). IEF in vertical columns with sucrose stabilized pH gradients used in earlier days with great skill (*259,262*) has been replaced by the more convenient chromatofocusing on solid supports in either open-end (*97,263*) or in FPLC columns (*91,258,264*). IEF on horizontal agarose gels has rarely been used in preparative mode (*265*) but served well for monitoring the appearance of various cellulolytic enzymes during the fermentation of *Trichoderma reesei* (*266,267*).

Preparative Gel Electrophoresis. Over the years many devices were described in the literature to accomplish electrophoretic separation of proteins on a preparative scale (*6*). Only recently, though, true preparative electrophoretic separation of a cellulolytic system from a thermotolerant bacterium was accomplished (*268*). All other reported uses of electrophoresis are 'preparative' only in the sense that the protein was recovered for further analyses (*158,269*). For preparing substantial quantities of enzyme methods other than flat-bed agarose or vertical polyacrylamide gel electrophoresis must be selected.

Unusual Techniques. From the previous paragraphs it is clear that resolving the cellulolytic enzyme system and assigning catalytic functions is hampered by structural and functional multiplicity of the proteins involved. Skilled application of separation and detection techniques has shed much light on soluble cellulase enzymes, but failed to a large extent with bacterial enzymes that associate in tight protein complexes (cellulosome from *Clostridium thermocellum*), or display an extremely high affinity for cellulose (e.g. cellulases from *Cellulomonas fimi*).

The cellulases from *C. fimi* are produced at an estimated quantity of 1 mg per liter of fermentation broth and were partially resolved by extracting the enzyme-substrate complex with 8 M guanidinium hydrochloride (GH). The enzymes apparently refold to their native structure upon removal of GH and can be purified by conventional techniques (*248*). Since a GH extraction is a cumbersome process and results in large and unpredictable loss of protein, characterization of *C. fimi* cellulases was approached by molecular biology techniques as soon as they became available. Indeed, the first cellulase cloned was an endoglucanase from *C. fimi* (*270*). Nevertheless, the problem of tight substrate binding remains as the purification of a cloned *C. fimi* exoglucanase (*Cex*) by GH gradient elution from an Avicel affinity column shows (*271*).

In an attempt to improve on the isolation and purification of native *C. fimi* cellulases, particularly to circumvent the folding/unfolding process with GH, a new electrophoretic method was developed (*272*). *C. fimi* exposed Avicel was freed of biomass and culture fluid, suspended in 1% (w/v) low melting agarose at 26°C (close to the gelling point) and cast into a cylinder that was sealed off on one side with a 0.62 μm Durapore membrane. This cup was mounted into a specially designed electrophoresis chamber such molecules migrating in the electric field to the anode were retained by an ultrafiltration membrane (PM-10) in a separate compartment. The content of this compartment was collected after one hour of electroelution, concentrated by ultrafiltration and separated by anion exchange chromatography. In two peaks, endoglucanase D (*273*) and a mixture of two endoglucanases with molecular masses of 115 to 125 kDa (tentatively assigned to endoglucanase B (*274*) and C (*275*)) emerge in near homogeneous form (*272*). In agreement with their high affinity for cellulose, EG-A and the exoglucanase Cex did not elute under these conditions. It is expected that, upon buffer change and prolonged electroelution, both Cex and EG-A will be recovered. This, then, is an elegant technique not only for resolving the enzymes of *C. fimi*, but also investigating solids where cellulases would be expected but have not been found so far, such as sewage sludge (D.E. Eveleigh, personal communication).

Purification in the denatured state has been reviewed (*276*) but is not new to the cellulase field (*277,278*). This technique appeared particularly fortuitous in the purification of cloned EG-A from *Clostridium cellulolyticum* (*279*). Ultracentrifugation was used in earlier years for establishing protein homogeneity (*210,277,280*), but recently was applied to free a glucan glucohydrolase from *Torulopsis wickerhamii* from bound polysaccharides in a cesium chloride gradient (*281*), the argument being that attached carbohydrates reduce enzymic activity (*282*). Various techniques have been used to free cellulases of interfering ligands: ion exchange on Amberlite (*86,283,284*) and Dowex to remove glucono-lactone (*94*) and treatment with guanidinium hydrochloride (*285*).

The use of HPLC technology is still limited due to the known difficulties in recovering proteins with biological activity. There are instances, where analytical HPLC (both in SEC and IEC mode) has been applied to cellulases, and examples are given in Table I.

No purification procedure is complete without recommendation for storage of the enzyme. Buffers in the neutral range are adequate, but must contain sodium azide (0.03% w/v) to prevent microbial growth. Refrigeration suffices for short-term storage, but freezing is preferred. This should be done quickly, possibly in the presence of cryoprotectants (glycerol, sucrose, 1,2-propanediol). Alternatively, an ammonium sulfate slurry is slowly dripped from a separatory funnel into liquid nitrogen (*286*). The beads thus formed are stored at -70°C (or lower) and can later be conveniently weighed out in small portions if so desired.

Conclusion

Purification of cellulases presents multi-faceted problems. On one side, one can apply the entire gamut of purification techniques and obtain pure enzyme, on the other side, one is not quite sure whether the right enzyme was purified. This is because of the notorious multiplicity of cellulases either designed by nature or produced post-translationally by proteases and de-glycosylating enzymes. One way to reduce this ambiguity is to simplify the starting material, as shown with enzymes from *Trichoderma reesei* QM9414 (*287,288*). Another option is clone the genes and express the enzymes in a suitable host. Frequently their purification schedule is simple (*289*). However, neither absence of protein multiplicity (*290*) nor ease of purification is guaranteed (*291*). In spite of knowledge and access to technology, choice of methods will remain a matter of intuition and rationalization of observations, and procurement of pure cellulase proteins a challenge for many years to come.

Literature Cited

1. Kornberg, A. *Methods Enzymol.* **1990**, *182*, 1-5.

2. Guide to Protein Purification; Deutscher, M.P., Ed.; Academic Press: New York, 1990.
3. *Protein Purification Applications*; Harris, E.L.V.; Angal, S., Eds.; Oxford University Press: Oxford, 1990.
4. *Protein Purification Methods*; Harris, E.L.V.; Angal, S., Eds.; Oxford University Press: Oxford, 1989.
5. *Protein Purification*; Janson, J.-C.; Rydén, L., Eds.; VCH: Weinheim, 1990.
6. Scopes, R. *Protein Purification*; 2nd Edition, Springer-Verlag: New York, NY, 1987.
7. *Protein Purification: Micro to Macro*; Burgess, R., Ed.; Alan R. Liss: New York, 1987.
8. *Practical Protein Chromatography*; Kenney, A.; Fowell, S., Eds.; Humana Press: Totowa, NJ, 1992.
9. Goddette, D.W.; Paech, C.; Yang, S.S.; Mielenz, J.R.; Bystroff, C.; Wilke, M.; Fletterick, R.J. *J. Mol. Biol.* **1992**, *228*, 580-595.
10. Berhow, M.A.; McFadden, B.A. *FEMS Microbiol. Lett.* **1983**, *17*, 269-272.
11. Paech, C.; Dybing, C.D. *Plant Physiol.* **1986**, *81*, 97-102.
12. Niku-Paavola, M.-L.; Lappalainen, A.; Enari, T.-M.; Nummi, M. *Biotechnol. Appl. Biochem.* **1986**, *8*, 449-458.
13. Lamed, R.; Setter, E.; Kenig, R.; Bayer, E.A. *Biotechnol. Bioeng. Symp.* **1983**, *13*, 163-181.
14. Lamed, R.; Setter, E.; Bayer, E.A. *J. Bacteriol.* **1983**, *156*, 828-836.
15. Lamed, R.; Bayer, E.A. *Adv. Appl. Microbiol.* **1988**, *33*, 1-46.
16. Lamed, R.; Bayer, E.A. In: *Biochemistry and Genetics of Cellulose Degradation*; Aubert, J.-P., Béguin, P., Millet, J., Eds.; Academic Press: New York, 1988, pp 101-116.
17. Reese, E.T.; Siu, R.G.H.; Levinson, H.S. *J. Bacteriol.* **1950**, *59*, 485-497.
18. Gilligan, W.; Reese, E.T. *Can. J. Microbiol.* **1954**, *1*, 90-107.
19. Wu, J.H.D.; Orme-Johnson, W.H.; Demain, A.L. *Biochemistry* **1988**, *27*, 1703-1709.
20. Fujii, M.; Homma, T.; Ooshima, K.; Taniguchi, M. *Appl. Biochem. Biotechnol.* **1991**, *28/29*, 145-156.
21. Nidetzky, B.; Hayn, M.; Macarron, R.; Steiner, W. *Biotechnol. Lett.* **1993**, *15*, 71-76.
22. Woodward, J. *Bioresour. Technol.* **1991**, *36*, 67-75.
23. Henrissat, B.; Driguez, H.; Viet, C.; Schülein, M. *Bio/Technology* **1985**, *3*, 722-726.
24. Coughlan, M.P.; Ljungdahl, L.G. In: *Biochemistry and Genetics of Cellulose Degradation*; Aubert, J.-P., Béguin, P., Millet, J., Eds.; Academic Press: New York, 1988, pp 11-30.
25. Coughlan, M.P. *Biotechnol. Genet. Eng. Rev.* **1985**, *3*, 39-109.
26. Irwin, D.C.; Spezio, M.; Walker, L.P. *Biotechnol. Bioeng.* **1993**, *42*, 1002-1013.
27. Ryu, D.D.Y.; Kim, C.; Mandels, M. *Biotechnol. Bioeng.* **1984**, *26*, 488-496.
28. Fägerstam, L.G.; Pettersson, G. *FEBS Lett.* **1980**, *119*, 97-100.
29. McHale, A.; Coughlan, M.P. *FEBS Lett.* **1980**, *117*, 319-322.
30. Wood, T.M.; McCrae, S.I.; Bhat, K.M. *Biochem. J.* **1989**, *260*, 37-43.
31. Poulsen, O.M.; Petersen, L.W. *Biotechnol. Bioeng.* **1992**, *39*, 121-123.

32. Wood, T.M.; McCrae, S.I. In: *Hydrolysis of Cellulose: Mechanisms of Enzymatic and Acid Catalysis (Advances in Chemistry Series 181)*; Brown, R.D.,Jr., Jurasek, L., Eds.; American Chemical Society: Washington, DC, 1979, pp 181-209.
33. Woodward, J.; Lima, M.; Lee, N.E. *Biochem. J.* **1988**, *255*, 895-899.
34. Gow, L.A.; Wood, T.M. *FEMS Microbiol. Lett.* **1988**, *50*, 247-252.
35. Beldman, G.; Voragen, A.G.J.; Rombouts, F.M.; Pilnik, W. *Biotechnol. Bioeng.* **1988**, *31*, 173-178.
36. Tuka, K.; Zverlov, V.V.; Velikodvorskaya, G.A. *Appl. Biochem. Biotechnol.* **1992**, *37*, 201-207.
37. Walker, L.P.; Wilson, D.B.; Irwin, D.C.; McQuire, C.; Price, M. *Biotechnol. Bioeng.* **1992**, *40*, 1019-1026.
38. Gascoigne, J.A.; Gascoigne, M.M. *Biological Degradation of Cellulose*; Butterworth: London, 1960.
39. Warzywoda, M.; Larbre, E.; Pourquié, J. *Bioresour. Technol.* **1992**, *39*, 125-130.
40. Montenecourt, B.S. *Trends Biotechnol.* **1983**, *1*, 156-161.
41. Woodward, J.; Hayes, M.K.; Lee, N.E. *Bio/Technology* **1988**, *6*, 301-304.
42. van Tilbeurgh, H.; Claeyssens, M.; De Bruyne, C.K. *FEBS Lett.* **1982**, *149*, 152-156.
43. Betz, S.F. *Protein Sci.* **1993**, *2*, 1551-1558.
44. Sharrock, K.R. *J. Biochem. Biophys. Methods* **1988**, *17*, 81-106.
45. Kremer, S.M.; Wood, P.M. *Appl. Microbiol. Biotechnol.* **1992**, *37*, 750-755.
46. Ghose, T.K. *Pure Appl. Chem.* **1987**, *59*, 257-268.
47. Wood, T.M.; Bhat, K.M. *Methods Enzymol.* **1988**, *160*, 87-112.
48. Hulme, M.A. *Methods Enzymol.* **1988**, *160*, 130-135.
49. Deshpande, M.V.; Pettersson, G.; Eriksson, K.-E. *Methods Enzymol.* **1988**, *160*, 126-130.
50. van Tilbeurgh, H.; Loontiens, F.G.; De Bruyne, C.K.; Claeyssens, M. *Methods Enzymol.* **1988**, *160*, 45-59.
51. Canevascini, G. *Methods Enzymol.* **1988**, *160*, 112-116.
52. Kremer, S.M.; Wood, P.M. *FEMS Microbiol. Lett.* **1992**, *92*, 187-192.
53. Tomme, P.; Heriban, V.; van Tilbeurgh, H.; Claeyssens, M. In: *Plant Cell Wall Polymers (ACS Symposium Series 399)*; Lewis, N.G., Paice, M.G., Eds.; American Chemical Society: Washington, DC, 1989, pp 570-586.
54. Biely, P. In: *Trichoderma reesei Cellulases: Biochemistry, Genetics, Physiology, and Application*; Kubicek, C.P., Eveleigh, D.E., Esterbauer, H., Steiner, W., Kubicek-Pranz, E.M., Eds.; Royal Society of Chemistry: Cambridge, 1990, pp 30-46.
55. Breuil, C.; Saddler, J.N. *Biochem. Soc. Trans.* **1985**, *13*, 449-450.
56. Claeyssens, M.; Aerts, G. *Bioresour. Technol.* **1992**, *39*, 143-146.
57. Claeyssens, M. In: *Biochemistry and Genetics of Cellulose Degradation*; Aubert, J.-P., Béguin, P., Millet, J., Eds.; Academic Press: New York, 1988, pp 393-397.
58. Linder, W.A.; Dennison, C.; Quicke, G.V. *Biotechnol. Bioeng.* **1983**, *25*, 377-385.
59. Baker, W.L.; Panow, A. *J. Biochem. Biophys. Methods* **1991**, *23*, 265-273.
60. Kolev, D.; Witte, K.; Wartenberg, A. *Acta Biotechnol.* **1991**, *11*, 359-365.

61. Safarík, I.; Safaríková, M. *J. Biochem. Biophys. Methods* **1991**, *23*, 301-306.
62. Bühler, R. *Appl. Environ. Microbiol.* **1991**, *57*, 3317-3321.
63. *Enzyme Nomenclature*; Academic Press: San Diego, CA, 1992.
64. Uelker, A.; Sprey, B. *FEMS Microbiol. Lett.* **1990**, *69*, 215-219.
65. Håkansson, U.; Fägerstam, L.G.; Pettersson, G.; Andersson, L. *Biochem. J.* **1979**, *179*, 141-149.
66. Sprey, B.; Uelker, A. *FEMS Microbiol. Lett.* **1992**, *92*, 253-258.
67. Béguin, P. *Annu. Rev. Microbiol.* **1990**, *44*, 219-248.
68. Lao, G.; Ghangas, G.S.; Jung, E.D.; Wilson, D.B. *J. Bacteriol.* **1991**, *173*, 3397-3407.
69. Gebler, J.; Gilkes, N.R.; Claeyssens, M.; Wilson, D.B.; Béguin, P.; Wakarchuk, W.W.; Kilburn, D.G.; Miller, R.C.,Jr.; Warren, R.A.J.; Withers, S.G. *J. Biol. Chem.* **1992**, *267*, 12559-12561.
70. Barras, F.; Bortoli-German, I.; Bauzan, M.; Rouvier, J.; Gey, C.; Heyraud, A.; Henrissat, B. *FEBS Lett.* **1992**, *300*, 145-148.
71. Damude, H.G.; Gilkes, N.R.; Kilburn, D.G.; Miller, R.C.,Jr.; Warren, R.A.J. *Gene* **1993**, *123*, 105-107.
72. Henrissat, B.; Claeyssens, M.; Tomme, P.; Lemesle, L.; Mornon, J.-P. *Gene* **1989**, *81*, 83-95.
73. Henrissat, B. *Gene* **1993**, *125*, 199-204.
74. Henrissat, B.; Bairoch, A. *Biochem. J.* **1993**, *293*, 781-788.
75. Henrissat, B. *Biochem. J.* **1991**, *280*, 309-316.
76. Bedarkar, S.; Gilkes, N.R.; Kilburn, D.G.; Kwan, E.; Rose, D.R.; Miller, R.C.,Jr.; Warren, R.A.J.; Withers, S.G. *J. Mol. Biol.* **1992**, *228*, 693-695.
77. Rouvinen, J.; Bergfors, T.; Teeri, T.; Knowles, J.K.C.; Jones, T.A. *Science* **1990**, *249*, 380-386.
78. Knowles, J.K.C.; Lentovaara, P.; Murray, M.; Sinnott, M.L. *J. Chem. Soc. Chem. Commun.* **1988**, 1401-1402.
79. Wood, T.M. In: *The Ekman-Days International Symposium on Wood Pulping Chemistry, Volume 3*; SPCI: Stockholm, 1981, pp 31-38.
80. Claeyssens, M.; Tomme, P.; Brewer, C.F.; Hehre, E.J. *FEBS Lett.* **1990**, *263*, 89-92.
81. Withers, S.G.; Dombroski, D.; Berven, L.A.; Kilburn, D.G.; Miller, R.C.,Jr.; Warren, R.A.J.; Gilkes, N.R. *Biochem. Biophys. Res. Commun.* **1986**, *139*, 487-494.
82. Fägerstam, L.G.; Pettersson, L.G.; Engström, J.Å. *FEBS Lett.* **1984**, *167*, 309-315.
83. Ahn, D.H.; Kim, H.; Pack, M.Y. *Biotechnol. Lett.* **1993**, *15*, 127-132.
84. Gilkes, N.R.; Warren, R.A.J.; Miller, R.C.,Jr.; Kilburn, D.G. *J. Biol. Chem.* **1988**, *263*, 10401-10407.
85. Moormann, M.; Schlochtermeier, A.; Schrempf, H. *Appl. Environ. Microbiol.* **1993**, *59*, 1573-1578.
86. Nakayama, M.; Tomita, Y.; Suzuki, H.; Nisizawa, K. *J. Biochem.* **1976**, *79*, 955-966.
87. Henrikson, G.; Pettersson, G.; Johansson, G.; Ruiz, A.; Uzcategui, E. *Eur. J. Biochem.* **1991**, *196*, 101-106.

88. Tomme, P.; van Tilbeurgh, H.; Pettersson, G.; Van Damme, J.; Vandekerckhove, J.; Knowles, J.; Teeri, T.; Claeyssens, M. *Eur. J. Biochem.* **1988**, *170*, 575-581.
89. Mohammad, J.; Eriksson, K.-O.; Johansson, G. *J. Biochem. Biophys. Methods* **1993**, *26*, 41-49.
90. van Tilbeurgh, H.; Tomme, P.; Claeyssens, M.; Bhikhabhai, R.; Pettersson, G. *FEBS Lett.* **1986**, *204*, 223-227.
91. Offord, D.A.; Lee, N.E.; Woodward, J. *Appl. Biochem. Biotechnol.* **1991**, *28/29*, 377-386.
92. Shima, S.; Igarashi, Y.; Kodama, T. *Appl. Microbiol. Biotechnol.* **1993**, *38*, 750-754.
93. Chen, H.; Hayn, M.; Esterbauer, H. *Biochem. Mol. Biol. Int.* **1993**, *30*, 901-910.
94. Eriksson, K.-E.; Pettersson, B. *Eur. J. Biochem.* **1975**, *51*, 213-218.
95. Rudick, M.J.; Elbein, A.D. *J. Bacteriol.* **1975**, *124*, 534-541.
96. Umezurike, G.M. *Biochem. J.* **1975**, *145*, 361-368.
97. Watanabe, T.; Sato, T.; Yoshioka, S.; Koshijima, T.; Kuwahara, M. *Eur. J. Biochem.* **1992**, *209*, 651-659.
98. Mega, T.; Matsushima, Y. *J. Biochem.* **1979**, *85*, 335-341.
99. Bedino, S.; Testore, G.; Obert, F. *Ital. J. Biochem.* **1985**, *34*, 341-355.
100. Yagüe, E.; Estevez, M.P. *Eur. J. Biochem.* **1988**, *175*, 627-632.
101. Evans, C.S. *Appl. Microbiol. Biotechnol.* **1985**, *22*, 128-131.
102. Bonnerjea, J.; Oh, S.; Hoare, M.; Dunnill, P. *Bio/Technology* **1986**, *4*, 954-958.
103. Lee, Y.-H.; Fan, L.T. *Adv. Biochem. Eng.* **1980**, *17*, 101-129.
104. Ghosh, B.K.; Ghosh, A. In: *Microbial Degradation of Natural Products*; Winkelmann, G., Ed.; VCH: Weinheim, 1992, pp 83-126.
105. Rapp, P.; Beermann, A. In: *Biosynthesis and Biodegradation of Cellulose*; Haigler, C.H., Weimer, P.J., Eds.; Marcel Dekker: New York, 1991, pp 535-597.
106. Gong, C.-S.; Ladisch, M.R.; Tsao, G.T. In: *Hydrolysis of Cellulose: mechanisms of Enzymatic and Acid Catalysis (ACS Series 181)*; Brown, R.D.,Jr., Jurasek, L., Eds.; American Chemical Society: Washington, DC, 1979, pp 261-287.
107. Enari, T.-M.; Niku-Paavola, M.-L. *CRC Crit. Rev. Biotechnol.* **1987**, *5*, 67-87.
108. Eriksson, K.-E.; Wood, T.M. In: *Biosynthesis and Biodegradation of Wood Components*; Higuchi, T., Ed.; Academic Press: Orlando, 1985, pp 469-503.
109. Coughlan, M.P. *Biochem. Soc. Trans.* **1985**, *13*, 405-406.
110. Béguin, P.; Gilkes, N.R.; Kilburn, D.G.; Miller, R.C.,Jr.; O'Neill, G.P.; Warren, R.A.J. *CRC Crit. Rev. Biotechnol.* **1987**, *6*, 129-162.
111. Marsden, W.L.; Gray, P.P. *CRC Crit. Rev. Biotechnol.* **1986**, *3*, 235-276.
112. Robson, L.M.; Chambliss, G.H. *Enzyme Microb. Technol.* **1989**, *11*, 626-644.
113. Gilkes, N.R.; Henrissat, B.; Kilburn, D.G.; Miller, R.C.,Jr.; Warren, R.A.J. *Microbiol. Rev.* **1991**, *55*, 303-315.
114. Gilkes, N.R.; Kilburn, D.G.; Miller, R.C.,Jr.; Warren, R.A.J. *Bioresour. Technol.* **1991**, *36*, 21-35.
115. Goyal, A.; Ghosh, B.; Eveleigh, D. *Bioresour. Technol.* **1991**, *36*, 37-50.
116. Persson, I.; Tjerneld, F.; Hahn-Hägerdahl, B. *Process Biochem.* **1991**, *26*, 65-74.
117. Wood, T.M. In: *Biosynthesis and Biodegradation of Cellulose*; Haigler, C.H., Weimer, P.J., Eds.; Marcel Dekker: New York, 1991, pp 491-553.

118. Béguin, P.; Millet, J.; Chauvaux, S.; Salamitou, S.; Tokatlidis, K.; Navas, J.; Fujino, T.; Lemaire, M.; Raynaud, O.; Daniel, M.-K.; Aubert, J.-P. *Biochem. Soc. Trans.* **1992**, *20*, 42-46.
119. Kubicek, C.P. *Adv. Biochem. Eng. Biotechnol.* **1992**, *45*, 1-27.
120. Wood, T.M. *Biochem. Soc. Trans.* **1992**, *20*, 46-53.
121. Wilson, D.B. *CRC Crit. Rev. Biotechnol.* **1992**, *12*, 45-63.
122. Gilbert, H.J.; Hazlewood, G.P. *J. Gen. Microbiol.* **1993**, *139*, 187-194.
123. Teather, R.M.; Gilbert, H.J.; Hazlewood, G.P. In: *Genetics and Molecular Biology of Anaerobic Bacteria*; Sebald, M., Ed.; Springer Verlag: Berlin, Heidelberg, New York, 1993, pp 569-585.
124. *Hydrolysis of Cellulose: Mechanisms of Enzymatic and Acid Catalysis (Advances in Chemistry Series 181)*; Brown, R.D.,Jr.; Jurasek, L., Eds.; American Chemical Society: Washington, DC, 1979.
125. *Biomass (Methods in Enzymology 160)*; Wood, T.M., Kellogg, S.T., Eds.; Academic Press: San Diego, 1988.
126. *Trichoderma reesei Cellulases: Biochemistry, Genetics, Physiology, and Application*; Kubicek, C.P., Eveleigh, D.E., Esterbauer, H., Steiner, W., Kubicek-Pranz, E.M., Eds.; Royal Society of Chemistry: Cambridge, 1990.
127. Teeri, T.T.; Penttilä, M.; Keränen, S.; Nevalainen, H.; Knowles, J.K.C. In: *Biotechnology of Filamentous Fungi. Technology and Products*; Finkelstein, D.B., Ball, C., Eds.; Butterworth-Heinemann: Boston, 1992, pp 417-443.
128. Gum, E.K.,Jr.; Brown, R.D.,Jr. *Biochim. Biophys. Acta* **1977**, *492*, 225-231.
129. Verma, J.P.; Martin, H.H. *Arch. Mikrobiol.* **1967**, *59*, 355-380.
130. Goksoeyr, J. *Methods Enzymol.* **1988**, *160*, 338-342.
131. Hill, J.P.; Dickinson, F.M. *Biochem. Soc. Trans.* **1989**, *17*, 1079-1080.
132. Vartak, H.G.; Rele, M.V.; Rao, M.; Deshpande, V.V. *Anal. Biochem.* **1983**, *133*, 260-263.
133. Park, C.-H.; Orozco-Avila, I. *Biotechnol. Prog.* **1992**, *8*, 521-526.
134. Bronnenmeier, K.; Staudenbauer, W.L. *Enzyme Microb. Technol.* **1990**, *12*, 431-436.
135. Moloney, A.P.; McCrae, S.I.; Wood, T.M.; Coughlan, M.P. *Biochem. J.* **1985**, *225*, 365-374.
136. Pizzichini, M.; Fabiani, C. *Sep. Sci. Technol.* **1991**, *26*, 175-187.
137. Pokorny, M.; Zupancic, S.; Steiner, T.; Kreiner, W. In: *Trichoderma reesei Cellulases: Biochemistry, Genetics, Physiology, and Application*; Kubicek, C.P., Eveleigh, D.E., Esterbauer, H., Steiner, W., Kubicek-Pranz, E.M., Eds.; Royal Society of Chemistry: Cambridge, 1990, pp 168-184.
138. Halliwell, G.; Halliwell, N. *Biochim. Biophys. Acta* **1989**, *992*, 223-229.
139. Yamane, K.; Suzuki, H. *Methods Enzymol.* **1988**, *160*, 200-210.
140. Himmel, M.E.; Oh, K.; Tucker, M.P.; Grohmann, K.; Rivard, C.J. *Biotechnol. Bioeng. Symp.* **1986**, *17*, 413-423.
141. Englard, S.; Seifter, S. *Methods Enzymol.* **1990**, *182*, 285-300.
142. Wood, T.M. *Methods Enzymol.* **1988**, *160*, 221-233.
143. Sengupta, S.; Ghosh, A.K.; Sengupta, Su. *Biochim. Biophys. Acta* **1991**, *1076*, 215-220.

144. Gibbs, R.A.; Posner, B.A.; Filpula, D.R.; Dodd, S.W.; Finkelman, M.A.J.; Lee, T.K.; Wroble, M.; Whitlow, M.; Benkovic, S.J. *Proc. Natl. Acad. Sci. USA* **1991**, *88*, 4001-4004.
145. Green, A.A.; Hughes, W.L. *Methods Enzymol.* **1955**, *1*, 67-90.
146. Dixon, M.; Webb, E.C. *Adv. Protein Chem.* **1961**, *16*, 197-216.
147. Arakawa, T.; Timasheff, S.N. *Methods Enzymol.* **1985**, *114*, 49-77.
148. Richardson, P.; Hoare, M.; Dunnill, P. *Biotechnol. Bioeng.* **1990**, *36*, 354-366.
149. Przybycien, T.M.; Bailey, J.E. *Enzyme Microb. Technol.* **1989**, *11*, 264-276.
150. Anand, L.; Vithayathil, P.J. *J. Ferment. Bioeng.* **1989**, *67*, 380-386.
151. Coughlan, M.P. *Methods Enzymol.* **1988**, *160*, 135-144.
152. Mathew, R.; Rao, K.K. *Anal. Biochem.* **1992**, *206*, 50-52.
153. Ganju, R.K.; Murthy, S.K.; Vithayathil, P.J. *Biochim. Biophys. Acta* **1989**, *993*, 266-274.
154. Ganju, R.K.; Murthy, S.K.; Vithayathil, P.J. *Carbohydr. Res.* **1990**, *197*, 245-255.
155. Yagüe, E.; Estevez, M.P. *Biochem. Int.* **1989**, *18*, 141-147.
156. Waksman, G. *Biochim. Biophys. Acta* **1988**, *967*, 82-86.
157. Waksman, G. *Biochim. Biophys. Acta* **1991**, *1073*, 49-55.
158. Bagga, P.S.; Sandhu, D.K.; Sharma, S. *J. Appl. Bacteriol.* **1990**, *68*, 61-68.
159. Tong, C.C.; Cole, A.L.; Shepherd, M.G. *Biochem. J.* **1980**, *191*, 83-94.
160. Bhattacharjee, B.; Roy, A.; Majumder, A.L. *Biochem. Int.* **1992**, *28*, 783-793.
161. McPherson, A. *Eur. J. Biochem.* **1990**, *189*, 1-23.
162. Zaworski, P.G.; Gill, G.S. *Anal. Biochem.* **1988**, *173*, 440-444.
163. Prasertsan, P.; Doelle, H.W. *Appl. Microbiol. Biotechnol.* **1986**, *24*, 326-333.
164. Cohn, E.J.; Strong, L.E.; Hughes, W.L.,Jr.; Mulford, D.J.; Ashworth, J.N.; Melin, M.; Taylor, H.L. *J. Am. Chem. Soc.* **1946**, *68*, 459-475.
165. Pepinsky, R.B. *Anal. Biochem.* **1991**, *195*, 177-181.
166. Freer, S.N. *Arch. Biochem. Biophys.* **1985**, *243*, 515-522.
167. De Gussem, R.L.; Aerts, G.M.; Claeyssens, M.; De Bruyne, C.K. *Biochim. Biophys. Acta* **1978**, *525*, 142-153.
168. Murao, S.; Sakamoto, R.; Arai, M. *Methods Enzymol.* **1988**, *160*, 274-299.
169. Chen, H.; Hayn, M.; Esterbauer, H. *Biochim. Biophys. Acta* **1992**, *1121*, 54-60.
170. Desai, J.D.; Ray, R.M.; Patel, N.P. *Biotechnol. Bioeng.* **1983**, *25*, 307-313.
171. Gondé, P.; Ratomahenina, R.; Arnaud, A.; Galzy, P. *Can. J. Biochem. Cell Biol.* **1985**, *63*, 1160-1166.
172. Büttner, R.; Bode, R.; Scheidt, A.; Birnbaum, D. *Acta Biotechnol.* **1988**, *8*, 517-525.
173. Polson, A.; Potgieter, G.M.; Largier, J.F.; Mears, G.E.F.; Joubert, F.J. *Biochim. Biophys. Acta* **1964**, *82*, 463-475.
174. Ingham, K.C. *Methods Enzymol.* **1990**, *182*, 301-306.
175. Hall, N.P.; Tolbert, N.E. *FEBS Lett.* **1978**, *96*, 167-169.
176. Hönig, W.; Kula, M.-R. *Anal. Biochem.* **1976**, *72*, 502-512.
177. Ellis, K.J.; Morrison, J.F. *Methods Enzymol.* **1982**, *87*, 405-426.
178. Patchett, M.L.; Neal, T.L.; Schofield, L.R.; Strange, R.C.; Daniel, R.M.; Morgan, H.W. *Enzyme Microb. Technol.* **1989**, *11*, 113-115.
179. Joliff, G.; Béguin, P.; Juy, M.; Millet, J.; Ryter, A.; Poljak, R.; Aubert, J.-P. *Bio/Technology* **1986**, *4*, 896-900.

180. Hayashida, S.; Mo, K.; Hosoda, A. *Appl. Environ. Microbiol.* **1988**, *54*, 1523-1529.
181. Homma, T.; Fujii, M.; Mori, J.; Kawakami, T.; Kuroda, K.; Taniguchi, M. *Biotechnol. Bioeng.* **1993**, *41*, 405-410.
182. Marshall, J.J. *Anal. Biochem.* **1973**, *53*, 191-198.
183. Marshall, J.J. *Biochem. Soc. Trans.* **1973**, *1*, 198-200.
184. Boyer, R.F.; Redmond, M.A. *Biotechnol. Bioeng.* **1983**, *25*, 1311-1319.
185. Boyer, R.F.; Allen, T.L.; Dykema, P.A. *Biotechnol. Bioeng.* **1987**, *29*, 176-179.
186. Mihoc, A.; Kluepfel, D. *Can. J. Microbiol.* **1990**, *36*, 53-56.
187. McGavin, M.; Forsberg, C.W. *J. Bacteriol.* **1988**, *170*, 2914-2922.
188. McGavin, M.J.; Forsberg, C.W.; Crosby, B.; Bell, A.W.; Dignard, D.; Thomas, D.Y. *J. Bacteriol.* **1989**, *171*, 5587-5595.
189. Swingle, S.M.; Tiselius, A. *Biochem. J.* **1951**, *48*, 171-174.
190. Bernardi, G. *Methods Enzymol.* **1971**, *22*, 325-339.
191. Hurst, P.L.; Nielsen, J.; Sullivan, P.A.; Shepherd, M.G. *Biochem. J.* **1977**, *165*, 33-41.
192. Gorbunoff, M.J. *Methods Enzymol.* **1985**, *182*, 329-339.
193. Roe, S. In: *Protein Purification Methods - A Practical Approach*; Harris, E.L.V., Angal, S., Eds.; Oxford University Press: Oxford, 1989, pp 175-244.
194. Matsushita, O.; Russell, J.B.; Wilson, D.B. *J. Bacteriol.* **1991**, *173*, 6919-6926.
195. Maglione, G.; Matsushita, O.; Russell, J.B.; Wilson, D.B. *Appl. Environ. Microbiol.* **1992**, *58*, 3593-3597.
196. Wilson, D.B. *Methods Enzymol.* **1988**, *160*, 314-323.
197. Calza, R.E.; Irwin, D.C.; Wilson, D.B. *Biochemistry* **1985**, *24*, 7797-7804.
198. Romaniec, M.P.M.; Fauth, U.; Kobayashi, T.; Huskisson, N.S.; Barker, P.J.; Demain, A.L. *Biochem. J.* **1992**, *283*, 69-73.
199. Fauth, U.; Romaniec, M.P.M.; Kobayashi, T.; Demain, A.L. *Biochem. J.* **1991**, *279*, 67-73.
200. Ng, T.K.; Zeikus, J.G. *Biochem. J.* **1981**, *199*, 341-350.
201. Ait, N.; Creuzet, N.; Forget, P. *J. Gen. Microbiol.* **1979**, *113*, 399-402.
202. Bagnara, C.; Gaudin, C.; Belaich, J.-P. *Biochem. Biophys. Res. Commun.* **1986**, *140*, 219-229.
203. Stewart, J.C.; Heptinstall, J. *Methods Enzymol.* **1988**, *160*, 264-274.
204. Béguin, P.; Eisen, H. *Eur. J. Biochem.* **1978**, *87*, 525-531.
205. Coutinho, J.B.; Gilkes, N.R.; Warren, R.A.J.; Kilburn, D.G.; Miller, R.C.,Jr. *Mol. Microbiol.* **1992**, *6*, 1243-1252.
206. Wilchek, M.; Miron, T.; Kohn, J. *Methods Enzymol.* **1984**, *104*, 3-55.
207. Ostrove, S. *Methods Enzymol.* **1990**, *182*, 357-371.
208. Carlsson, J.; Janson, J.-C.; Sparrman, M. In: *Protein Purification*; Janson, J.-C., Rydén, L., Eds.; VCH: Weinheim, 1989, pp 275-329.
209. Gum, E.K.,Jr.; Brown, R.D.,Jr. *Biochim. Biophys. Acta* **1976**, *446*, 371-386.
210. Shoemaker, S.; Brown, R.D.,Jr. *Biochim. Biophys. Acta* **1978**, *523*, 147-161.
211. Halliwell, G.; Griffin, M. *Biochem. J.* **1978**, *169*, 713-715.
212. Jun, C.W.; Min, M.Z.; Sel, K.M. *Folia Microbiol.* **1992**, *37*, 199-204.
213. Nummi, M.; Niku-Paavola, M.-L.; Enari, T.-M.; Raunio, V. *Anal. Biochem.* **1981**, *116*, 137-141.
214. Mo, K.; Hayashida, S. *Agric. Biol. Chem.* **1988**, *52*, 1675-1682.

215. Weber, M.; Foglietti, M.J.; Percheron, F. *J. Chromatogr.* **1980**, *188*, 377-382.
216. Nummi, M.; Niku-Paavola, M.-L.; Lappalainen, A.; Enari, T.-M.; Raunio, V. *Biochem. J.* **1983**, *215*, 677-683.
217. Poulsen, O.M.; Petersen, L.W. *Biotechnol. Bioeng.* **1987**, *29*, 799-804.
218. Reese, E.T. *Process Biochem.* **1982**, *17(5/6)*, 2-6.
219. Au, K.-S.; Chan, K.-Y. *J. Gen. Microbiol.* **1987**, *133*, 2155-2162.
220. Owolabi, J.B.; Béguin, P.; Kilburn, D.G.; Miller, R.C.,Jr.; Warren, R.A.J. *Appl. Environ. Microbiol.* **1988**, *54*, 518-523.
221. Saito, T.; Suzuki, T.; Hayashi, A.; Honda, H.; Taya, M.; Iijima, S.; Kobayashi, T. *J. Ferment. Bioeng.* **1990**, *69*, 282-286.
222. Morag, E.; Bayer, E.A.; Lamed, R. *Enzyme Microb. Technol.* **1992**, *14*, 289-292.
223. van Tilbeurgh, H.; Bhikhabhai, R.; Pettersson, L.G.; Claeyssens, M. *FEBS Lett.* **1984**, *169*, 215-218.
224. Tomme, P.; McCrae, S.; Wood, T.M.; Claeyssens, M. *Methods Enzymol.* **1988**, *160*, 187-193.
225. Tomme, P.; Chauvaux, S.; Béguin, P.; Millet, J.; Aubert, J.-P.; Claeyssens, M. *J. Biol. Chem.* **1991**, *266*, 10313-10318.
226. Uzcategui, E.; Raices, M.; Montesino, R.; Johansson, G.; Pettersson, G.; Eriksson, K.-E. *Biotechnol. Appl. Biochem.* **1991**, *13*, 323-334.
227. Driguez, H. In: *Biochemistry and Genetics of Cellulose Degradation*; Aubert, J.-P., Béguin, P., Millet, J., Eds.; Academic Press: New York, 1988, pp 399-402.
228. Orgeret, C.; Seillier, E.; Gautier, C.; Defaye, J.; Driguez, H. *Carbohydr. Res.* **1992**, *224*, 29-40.
229. Schou, C.; Rasmussen, G.; Schülein, M.; Henrissat, B.; Driguez, H. *J. Carbohydr. Chem.* **1993**, *12*, 743-752.
230. Holme, K.R.; Hall, L.D.; Armstrong, C.R.; Withers, S.G. *Carbohydr. Res.* **1988**, *173*, 285-291.
231. Roth, W.W.; Srinivasan, V.R. *Prep. Biochem.* **1978**, *8*, 57-71.
232. Teunissen, M.J.; Lahaye, D.H.T.P.; Huis in't Veld, J.H.J.; Vogels, G.D. *Arch. Microbiol.* **1992**, *158*, 276-281.
233. Rogalski, J.; Wojtas-Wasilewska, M.; Leonowicz, A. *Acta Biotechnol.* **1991**, *11*, 485-494.
234. Legler, G.; Bieberich, E. *Arch. Biochem. Biophys.* **1988**, *260*, 427-436.
235. Kanfer, J.N.; Mumford, R.A.; Raghavan, S.S.; Byrd, J. *Anal. Biochem.* **1974**, *60*, 200-205.
236. Anzai, H.; Uchida, N.; Nishide, E.; Nisizawa, K.; Matsuda, K. *Agric. Biol. Chem.* **1988**, *52*, 633-640.
237. Sutton, C. In: *Protein Purification Methods - A Practical Approach*; Harris, E.L.V., Angal, S., Eds.; Oxford University Press: Oxford, 1989, pp 268-282.
238. Gong, C.-S.; Chen, L.F.; Tsao, G.T. *Biotechnol. Bioeng.* **1979**, *21*, 167-171.
239. Kminkova, M.; Kucera, J. *J. Chromatogr.* **1982**, *244*, 166-168.
240. Woodward, J.; Marquess, H.J.; Picker, C.S. *Prep. Biochem.* **1986**, *16*, 337-352.
241. Paul, J.; Varma, A.K. *Biotechnol. Lett.* **1992**, *14*, 207-212.
242. Fähnrich, P.; Irrgang, K. *Biotechnol. Lett.* **1984**, *6*, 251-256.
243. Kobayashi, T.; Romaniec, M.P.M.; Fauth, U.; Demain, A.L. *Appl. Environ. Microbiol.* **1990**, *56*, 3040-3046.

244. Abe, M.; Higashi, S. *J. Gen. Appl. Microbiol.* **1982**, *28*, 551-562.
245. Niku-Paavola, M.-L.; Lappalainen, A.; Enari, T.-M.; Nummi, M. *Biochem. J.* **1985**, *231*, 75-81.
246. Riske, F.J.; Eveleigh, D.E.; Macmillan, J.D. *J. Ind. Microbiol.* **1986**, *1*, 259-264.
247. Nieves, R.A.; Ellis, R.P.; Himmel, M.E. *Appl. Biochem. Biotechnol.* **1990**, *24-25*, 397-406.
248. Langsford, M.L.; Gilkes, N.R.; Wakarchuk, W.W.; Kilburn, D.G.; Miller, R.C.,Jr.; Warren, R.A.J. *J. Gen. Microbiol.* **1984**, *130*, 1367-1376.
249. Aho, S.; Olkkonen, V.; Jalava, T.; Paloheimo, M.; Bühler, R.; Niku-Paavola, M.-L.; Bamford, D.H.; Korhola, M. *Eur. J. Biochem.* **1991**, *200*, 643-649.
250. Aho, S.; Paloheimo, M. *Biochim. Biophys. Acta* **1990**, *1087*, 137-141.
251. Luderer, M.E.H.; Hofer, F.; Hagspiel, K.; Allmaier, G.; Blaas, D.; Kubicek, C.P. *Biochim. Biophys. Acta* **1991**, *1076*, 427-434.
252. Mischak, H.; Hofer, F.; Messner, R.; Weissinger, E.; Hayn, M.; Tomme, P.; Esterbauer, H.; Küchler, E.; Claeyssens, M.; Kubicek, C.P. *Biochim. Biophys. Acta* **1989**, *990*, 1-7.
253. Uemura, S.; Ishihara, M.; Jellison, J. *Appl. Microbiol. Biotechnol.* **1993**, *39*, 788-794.
254. Golovchenko, N.P.; Kataeva, I.A.; Akimenko, V.K. *Enzyme Microb. Technol.* **1992**, *14*, 327-331.
255. Sprey, B.; Lambert, C. *FEMS Microbiol. Lett.* **1983**, *18*, 217-222.
256. Sprey, B. *FEMS Microbiol. Lett.* **1987**, *43*, 25-32.
257. Allen, R.C.; Saravis, C.A.; Maurer, H.R. *Gel Electrophoresis and Isoelectric Focusing of Proteins*; Walter de Gruyter: Berlin, 1984.
258. Lee, N.E.; Lima, M.; Woodward, J. *Biochim. Biophys. Acta* **1988**, *967*, 437-440.
259. Wood, T.M.; McCrae, S.I. *Biochem. J.* **1978**, *171*, 61-72.
260. Wood, T.M.; McCrae, S.I.; Macfarlane, C.C. *Biochem. J.* **1980**, *189*, 51-65.
261. Sakamoto, R.; Kanamoto, J.; Arai, M.; Murao, S. *Agric. Biol. Chem.* **1985**, *49*, 1275-1281.
262. Wood, T.M.; McCrae, S.I. *Biochem. J.* **1972**, *128*, 1183-1192.
263. Huang, L.; Forsberg, C.W. *Appl. Environ. Microbiol.* **1988**, *54*, 1488-1493.
264. Hayn, M.; Esterbauer, H. *J. Chromatogr.* **1985**, *329*, 379-387.
265. Rao, M.; Deshpande, V.; Mishra, C. *Biotechnol. Bioeng.* **1986**, *28*, 1100-1105.
266. Labudova, I.; Farkas, V. *Biochim. Biophys. Acta* **1983**, *744*, 135-140.
267. Farkas, V.; Jalanko, A.; Kolarova, N. *Biochim. Biophys. Acta* **1982**, *706*, 105-110.
268. Nieves, R.A.; Himmel, M.E. *Bio-Rad US/EG Bulletin* **1993**, *1830*, 1-3.
269. Sami, A.J.; Akhtar, M.W. *Enzyme Microb. Technol.* **1993**, *15*, 586-592.
270. Whittle, D.J.; Kilburn, D.G.; Warren, R.A.J.; Miller, R.C.,Jr. *Gene* **1982**, *17*, 139-145.
271. MacLeod, A.M.; Gilkes, N.R.; Escote-Carlson, L.; Warren, R.A.J.; Kilburn, D.G.; Miller, R.C.,Jr. *Gene* **1992**, *121*, 143-147.
272. Paech, C.; Christianson, T.; Paech, S.; Baums, C. *Biochem. Biophys. Res. Commun.* **1994**, (submitted).
273. Meinke, A.; Gilkes, N.R.; Kilburn, D.G.; Miller, R.C.,Jr.; Warren, R.A.J. *J. Bacteriol.* **1993**, *175*, 1910-1918.

274. Meinke, A.; Braun, C.; Gilkes, N.R.; Kilburn, D.G.; Miller, R.C.,Jr.; Warren, R.A.J. *J. Bacteriol.* **1991**, *173*, 308-314.
275. Coutinho, J.B.; Moser, B.; Kilburn, D.G.; Warren, R.A.J.; Miller, R.C.,Jr. *Mol. Microbiol.* **1991**, *5*, 1221-1233.
276. Knuth, M.W.; Burgess, R.R. In: *Protein Purification: Micro to Macro*; Burgess, R., Ed.; Alan R. Liss: New York, 1987, pp 279-305.
277. Berghem, L.E.R.; Pettersson, L.G. *Eur. J. Biochem.* **1974**, *46*, 295-305.
278. Håkansson, U.; Fägerstam, L.; Pettersson, G.; Andersson, L. *Biochim. Biophys. Acta* **1978**, *524*, 385-392.
279. Fierobe, H.-P.; Gaudin, C.; Belaich, A.; Loutfi, M.; Faure, E.; Bagnara, C.; Baty, D.; Belaich, J.-P. *J. Bacteriol.* **1991**, *173*, 7956-7962.
280. Berghem, L.E.R.; Pettersson, L.G.; Axio-Fredriksson, U.B. *Eur. J. Biochem.* **1976**, *61*, 621-630.
281. Himmel, M.E.; Tucker, M.P.; Lastick, S.M.; Oh, K.K.; Fox, J.W.; Spindler, D.D.; Grohmann, K. *J. Biol. Chem.* **1986**, *261*, 12948-12955.
282. Alurralde, J.L.; Ellenrieder, G. *Enzyme Microb. Technol.* **1984**, *6*, 467-470.
283. Ikeda, R.; Yamamoto, T.; Funatsu, M. *Agric. Biol. Chem.* **1973**, *37*, 1153-1159.
284. Yamanobe, T.; Mitsuishi, Y. *Agric. Biol. Chem.* **1990**, *54*, 301-307.
285. Liu, X.; Gao, P.; Cheng, Y. *Biotechnol. Appl. Biochem.* **1992**, *16*, 134-143.
286. McCurry, S.D.; Gee, R.; Tolbert, N.E. *Methods Enzymol.* **1982**, *90*, 515-521.
287. Fägerstam, L.G.; Pettersson, G. *FEBS Lett.* **1979**, *98*, 363-367.
288. Bhikhabhai, R.; Johansson, G.; Pettersson, G. *J. Appl. Biochem.* **1984**, *6*, 336-345.
289. Hu, P.; Chase, T.,Jr.; Eveleigh, D.E. *Appl. Microbiol. Biotechnol.* **1993**, *38*, 631-637.
290. Lemaire, M.; Béguin, P. *J. Bacteriol.* **1993**, *175*, 3353-3360.
291. Hazlewood, G.P.; Davidson, K.; Laurie, J.I.; Huskisson, N.S.; Gilbert, H.J. *J. Gen. Microbiol.* **1993**, *139*, 307-316.
292. Han, Y.W.; Srinivasan, V.R. *J. Bacteriol.* **1969**, *100*, 1355-1363.
293. Day, A.G.; Withers, S.G. *Biochem. Cell Biol.* **1986**, *64*, 914-922.
294. Büttner, R.; Bode, R.; Birnbaum, D. *J. Basic Microbiol.* **1991**, *31*, 423-428.
295. Murao, S.; Sakamoto, R.; Arai, M. *Agric. Biol. Chem.* **1985**, *49*, 3511-3517.
296. Murao, S.; Sakamoto, R. *Agric. Biol. Chem.* **1979**, *43*, 1791-1792.
297. Murao, S.; Nomura, Y.; Yoshikawa, M.; Shin, T.; Oyama, H.; Arai, M. *Biosci. Biotechnol. Biochem.* **1992**, *56*, 1366-1367.
298. Parry, J.B.; Stewart, J.C.; Heptistall, J. *Biochem. J.* **1983**, *213*, 437-444.
299. Kitpreechavanich, V.; Hayashi, M.; Nagai, S. *Agric. Biol. Chem.* **1986**, *50*, 1703-1711.
300. Sanyal, A.; Kundu, R.K.; Dube, S.; Dube, D.K. *Enzyme Microb. Technol.* **1988**, *10*, 91-99.
301. Kwon, K.-S.; Kang, H.G.; Hah, Y.C. *FEMS Microbiol. Lett.* **1992**, *97*, 149-153.
302. Okada, G. *Methods Enzymol.* **1988**, *160*, 259-264.
303. Okada, G. *Agric. Biol. Chem.* **1985**, *49*, 1257-1265.
304. Enari, T.-M.; Niku-Paavola, M.-L.; Harju, L.; Lappalainen, A.; Nummi, M. *J. Appl. Biochem.* **1981**, *3*, 157-163.
305. Hoh, Y.K.; Yeoh, H.-H.; Tan, T.K. *Appl. Microbiol. Biotechnol.* **1992**, *37*, 590-593.

306. Witte, K.; Wartenberg, A. *Acta Biotechnol.* **1989**, *9*, 179-190.
307. Singh, A.; Agrawal, A.K.; Abidi, A.B.; Darmwal, N.S. *FEMS Microbiol. Lett.* **1990**, *71*, 221-224.
308. Araujo, A.; D'Souza, J. *J. Ferment. Technol.* **1986**, *64*, 463-467.
309. Awad, M.; Lewis, L.N. *J. Food Sci.* **1980**, *45*, 1625-1628.
310. Ozaki, K.; Ito, S. *J. Gen. Microbiol.* **1991**, *137*, 41-48.
311. Ozaki, K.; Sumitomo, N.; Ito, S. *J. Gen. Microbiol.* **1991**, *137*, 2299-2305.
312. Yoshimatsu, T.; Ozaki, K.; Shikata, S.; Ohta, Y.-I.; Koike, K.; Kawai, S.; Ito, S. *J. Gen. Microbiol.* **1990**, *136*, 1973-1979.
313. Lusterio, D.D.; Suizo, F.G.; Labunos, N.M.; Valledor, M.N.; Ueda, S.; Kawai, S.; Koike, K.; Shikata, S.; Yoshimatsu, T.; Ito, S. *Biosci. Biotechnol. Biochem.* **1992**, *56*, 1671-1672.
314. Fukumori, F.; Kudo, T.; Horikoshi, K. *J. Gen. Microbiol.* **1985**, *131*, 3339-3345.
315. Hansen, C.K.; Diderichsen, B.; Jorgensen, P.L. *J. Bacteriol.* **1992**, *174*, 3522-3531.
316. Planas, A.; Juncosa, M.; Cayetano, A.; Querol, E. *Appl. Microbiol. Biotechnol.* **1992**, *37*, 583-589.
317. Robson, L.M.; Chambliss, G.H. *J. Bacteriol.* **1987**, *169*, 2017-2025.
318. Greene, R.V.; Griffin, H.L.; Freer, S.N. *Arch. Biochem. Biophys.* **1988**, *267*, 334-341.
319. Patchett, M.L.; Daniel, R.M.; Morgan, H.W. *Biochem. J.* **1987**, *243*, 779-787.
320. Plant, A.R.; Oliver, J.E.; Patchett, M.L.; Daniel, R.M.; Morgan, H.W. *Arch. Biochem. Biophys.* **1988**, *262*, 181-188.
321. Creuzet, N.; Frixon, C. *Biochimie* **1983**, *65*, 149-156.
322. Groleau, D.; Forsberg, C.W. *Can. J. Microbiol.* **1983**, *29*, 504-517.
323. Huang, L.; Forsberg, C.W.; Thomas, D.Y. *J. Bacteriol.* **1988**, *170*, 2923-2932.
324. Huang, L.; Forsberg, C.W. *Appl. Environ. Microbiol.* **1987**, *53*, 1034-1041.
325. Sakai, K.; Tachiki, T.; Kumagai, H.; Tochikura, T. *Agric. Biol. Chem.* **1986**, *50*, 2287-2293.
326. Umezurike, G.M. *Biochem. J.* **1979**, *177*, 9-19.
327. Lin, L.-L.; Thomson, J.A. *FEMS Microbiol. Lett.* **1991**, *84*, 197-203.
328. Hazlewood, G.P.; Davidson, K.; Laurie, J.I.; Romaniec, M.P.M.; Gilbert, H.J. *J. Gen. Microbiol.* **1990**, *136*, 2089-2097.
329. Luna-Arias, J.P.; Andaluz, E.; Ridruejo, J.C.; Olivero, I.; Larriba, G. *Yeast* **1991**, *7*, 833-841.
330. Kilian, S.G.; Prior, B.A.; Venter, J.J.; Lategan, P.M. *Appl. Microbiol. Biotechnol.* **1985**, *21*, 148-153.
331. Poulsen, O.M.; Petersen, L.W. *Biotechnol. Bioeng.* **1989**, *34*, 65-71.
332. Meinke, A.; Schmuck, M.; Gilkes, N.R.; Kilburn, D.G.; Miller, R.C.,Jr.; Warren, R.A.J. *Glycobiology* **1992**, *2*, 321-326.
333. Moser, B.; Gilkes, N.R.; Kilburn, D.G.; Warren, R.A.J.; Miller, R.C.,Jr. *Appl. Environ. Microbiol.* **1989**, *55*, 2480-2487.
334. Nakamura, K.; Kitamura, K. *Methods Enzymol.* **1988**, *160*, 211-216.
335. Nakamura, K.; Kitamura, K. *J. Ferment. Technol.* **1983**, *61*, 379-382.
336. Kitaoka, M.; Sasaki, T.; Taniguchi, H. *Biosci. Biotechnol. Biochem.* **1992**, *56*, 652-655.
337. Lusis, A.J.; Becker, R.R. *Biochim. Biophys. Acta* **1973**, *329*, 5-16.

338. Fujino, T.; Sukhumavasi, J.; Sasaki, T.; Ohmiya, K.; Shimizu, S. *J. Bacteriol.* **1989**, *171*, 4076-4079.
339. Fujino, T.; Sasaki, T.; Ohmiya, K.; Shimizu, S. *Appl. Environ. Microbiol.* **1990**, *56*, 1175-1178.
340. Bronnenmeier, K.; Rücknagel, K.P.; Staudenbauer, W.L. *Eur. J. Biochem.* **1991**, *200*, 379-385.
341. Bronnenmeier, K.; Staudenbauer, W.L. *Appl. Microbiol. Biotechnol.* **1988**, *28*, 380-386.
342. Bronnenmeier, K.; Staudenbauer, W.L. *Appl. Microbiol. Biotechnol.* **1988**, *27*, 432-436.
343. Ait, N.; Creuzet, N.; Cattaneo, J. *Biochem. Biophys. Res. Commun.* **1979**, *90*, 537-546.
344. Gräbnitz, F.; Seiss, M.; Rücknagel, K.P.; Staudenbauer, W.L. *Eur. J. Biochem.* **1991**, *200*, 301-309.
345. Mosolova, T.P.; Kalyuzhnyi, S.V.; Varfolomeyev, S.D.; Velikodvorskaya, G.A. *Appl. Biochem. Biotechnol.* **1993**, *42*, 9-18.
346. Chaffotte, A.F.; Guillou, Y.; Goldberg, M.E. *Eur. J. Biochem.* **1992**, *205*, 369-373.
347. Lamed, R.; Bayer, E.A. *Methods Enzymol.* **1988**, *160*, 472-482.
348. Hon-nami, K.; Coughlan, M.P.; Hon-nami, H.; Ljungdahl, L.G. *Arch. Microbiol.* **1986**, *145*, 13-19.
349. Bhat, K.M.; Wood, T.M. *Carbohydr. Res.* **1992**, *227*, 293-300.
350. Golovchenko, N.P.; Singh, R.N.; Velikodvorskaya, G.A.; Akimenko, V.K. *Appl. Microbiol. Biotechnol.* **1993**, *39*, 74-79.
351. Singh, R.N.; Akimenko, V.K. *Biochem. Biophys. Res. Commun.* **1993**, *192*, 1123-1130.
352. Romaniec, M.P.M.; Huskisson, N.; Barker, P.; Demain, A.L. *Enzyme Microb. Technol.* **1993**, *15*, 393-400.
353. Petre, J.; Longin, R.; Millet, J. *Biochimie* **1981**, *63*, 629-639.
354. Ng, T.K.; Zeikus, J.G. *Methods Enzymol.* **1988**, *160*, 351-355.
355. Morag, E.; Halevy, I.; Bayer, E.A.; Lamed, R. *J. Bacteriol.* **1991**, *173*, 4155-4162.
356. Mori, Y. *Biotechnol. Lett.* **1992**, *14*, 131-136.
357. Mori, Y. *Biosci. Biotechnol. Biochem.* **1992**, *56*, 1198-1203.
358. Jin, F.; Toda, K. *J. Ferment. Bioeng.* **1989**, *67*, 8-13.
359. Schmidhalter, D.R.; Canevascini, G. *Arch. Biochem. Biophys.* **1993**, *300*, 551-558.
360. Schmidhalter, D.R.; Canevascini, G. *Arch. Biochem. Biophys.* **1993**, *300*, 559-563.
361. Idogaki, H.; Kitamoto, Y. *Biosci. Biotechnol. Biochem.* **1992**, *56*, 970-971.
362. Rouau, X.; Foglietti, M.-J. *Biochem. Soc. Trans.* **1985**, *13*, 451-452.
363. Rouau, X.; Odier, E. *Carbohydr. Res.* **1986**, *145*, 279-292.
364. Anzai, H.; Nisizawa, K.; Matsuda, K. *J. Biochem.* **1984**, *96*, 1381-1390.
365. Boyer, M.H.; Chambost, J.-P.; Magnan, M.; Cattanéo, J. *J. Biotechnol.* **1984**, *1*, 241-252.
366. Boyer, M.-H.; Cami, B.; Chambost, J.-P.; Magnan, M.; Cattanéo, J. *Eur. J. Biochem.* **1987**, *162*, 311-316.

367. Garibaldi, A.; Gibbins, L.N. *Can. J. Microbiol.* **1975**, *21*, 513-520.
368. Tanaka, M.; Taniguchi, M.; Matsuno, R.; Kamikubo, T. *Methods Enzymol.* **1988**, *160*, 251-259.
369. Tanaka, M.; Taniguchi, M.; Matsuno, R.; Kamikubo, T. *J. Ferment. Technol.* **1981**, *59*, 177-183.
370. Chen, C.-S.; Lian, K.-T. *Agric. Biol. Chem.* **1986**, *50*, 1229-1238.
371. Enokibara, S.; Mori, N.; Kitamoto, Y. *J. Ferment. Bioeng.* **1992**, *73*, 230-232.
372. Yamanobe, T.; Mitsuishi, Y.; Yagisawa, M. *Agric. Biol. Chem.* **1988**, *52*, 2493-2501.
373. Yamanobe, T.; Mitsuishi, Y. *Agric. Biol. Chem.* **1989**, *53*, 3359-3360.
374. Vaidya, M.; Seeta, R.; Mishra, C.; Deshpande, V.; Rao, M. *Biotechnol. Bioeng.* **1984**, *26*, 41-45.
375. Mishra, C.; Vaidya, M.; Rao, M.; Deshpande, V. *Enzyme Microb. Technol.* **1983**, *5*, 430-434.
376. Matsumoto, K.; Endo, Y.; Tamiya, N.; Kano, M.; Miyauchi, K.; Abe, J. *J. Biochem.* **1974**, *76*, 563-572.
377. Wood, T.M. *Biochem. J.* **1971**, *121*, 353-362.
378. Wood, T.M.; McCrae, S.I. *Carbohydr. Res.* **1977**, *57*, 117-133.
379. Szakács-Dobozi, M.; Halász, A. *J. Chromatogr.* **1986**, *365*, 51-55.
380. Szakács-Dobozi, M.; Halász, A.; Vámos-Vigyázó, L. *J. Chromatogr.* **1985**, *347*, 310-315.
381. Marshall, J.J. *Comp. Biochem. Physiol. B Comp. Biochem.* **1973**, *44*, 981-988.
382. Hayashida, S.; Mo, K. *Appl. Environ. Microbiol.* **1986**, *51*, 1041-1046.
383. Hayashida, S.; Yoshioka, H. *Agric. Biol. Chem.* **1980**, *44*, 1721-1728.
384. Hayashida, S.; Ota, K.; Mo, K. *Methods Enzymol.* **1988**, *160*, 323-332.
385. Rao, U.S.; Murthy, S.K. *Indian J. Biochem. Biophys.* **1988**, *25*, 687-694.
386. Kubo, K.; Nisizawa, K. *J. Ferment. Technol.* **1983**, *61*, 383-389.
387. Kanda, T.; Wakabayashi, K.; Nisizawa, K. *J. Biochem.* **1976**, *79*, 977-988.
388. Kanda, T.; Nakakubo, S.; Wakabayashi, K.; Nisizawa, K. *J. Biochem.* **1978**, *84*, 1217-1226.
389. Kanda, T.; Nakakubo, S.; Wakabayashi, K.; Nisizawa, K. In: *Hydrolysis of Cellulose: Mechanisms of Enzymatic and Acid Catalysis (Advances in Chemistry Series 181)*; Brown, R.D.,Jr., Jurasek, L., Eds.; American Chemical Society: Washington, DC, 1979, pp 211-236.
390. Kanda, T.; Wakabayashi, K.; Nisizawa, K. *J. Biochem.* **1980**, *87*, 1625-1634.
391. Bhattacharjee, B.; Roy, A.; Lahiri Majumder, A. *Biochem. Mol. Biol. Int.* **1993**, *30*, 1143-1152.
392. Herr, D.; Baumer, F.; Dellweg, H. *Arch. Microbiol.* **1978**, *117*, 287-292.
393. Saha, S.C.; Sanyal, A.; Kundu, R.K.; Dube, S.; Dube, D.K. *Biochim. Biophys. Acta* **1981**, *662*, 22-29.
394. Sen, S.; Abraham, T.K.; Chakrabarty, S.L. *Can. J. Microbiol.* **1982**, *28*, 271-277.
395. Roy, S.K.; Dey, S.K.; Raha, S.K.; Chakrabarty, S.L. *J. Gen. Microbiol.* **1990**, *136*, 1967-1971.
396. Yoshioka, H.; Hayashida, S. *Agric. Biol. Chem.* **1980**, *44*, 2817-2824.
397. Borgia, P.I.; Mehnert, D.W. *J. Bacteriol.* **1982**, *149*, 515-522.
398. Hebraud, M.; Fevre, M. *Appl. Environ. Microbiol.* **1990**, *56*, 3164-3169.

399. Li, X.; Calza, R.E. *Biochim. Biophys. Acta* **1991**, *1080*, 148-154.
400. Li, X.; Calza, R.E. *Appl. Environ. Microbiol.* **1991**, *57*, 3331-3336.
401. Wilson, C.A.; Wood, T.M. *Appl. Microbiol. Biotechnol.* **1992**, *37*, 125-129.
402. Funaguma, T.; Tsuji, H.; Hara, A. *J. Ferment. Technol.* **1986**, *64*, 77-80.
403. Connelly, I.C.; Coughlan, M.P. *Enzyme Microb. Technol.* **1991**, *13*, 462-469.
404. Connelly, I.C.; Filho, E.X.F.; Healy, A.-M.; Fleming, M.; Griffin, T.O.; Mayer, F.; Coughlan, M.P. *Enzyme Microb. Technol.* **1991**, *13*, 470-477.
405. Wood, T.M.; McCrae, S.I. *Carbohydr. Res.* **1982**, *110*, 291-303.
406. Wood, T.M.; McCrae, S.I. *Carbohydr. Res.* **1986**, *148*, 331-344.
407. Kamagata, Y.; Sasaki, H.; Takao, S. *J. Ferment. Technol.* **1986**, *64*, 211-217.
408. Bao, W.; Renganathan, V. *FEBS Lett.* **1992**, *302*, 77-80.
409. Kelleher, T.J.; Montenecourt, B.S.; Eveleigh, D.E. *Appl. Microbiol. Biotechnol.* **1987**, *27*, 299-305.
410. Samejima, M.; Eriksson, K.-E.L. *Eur. J. Biochem.* **1992**, *207*, 103-107.
411. Samejima, M.; Eriksson, K.-E.L. *FEBS Lett.* **1991**, *292*, 151-153.
412. Westermark, U.; Eriksson, K.-E. *Methods Enzymol.* **1988**, *160*, 463-468.
413. Weber, M.; Coulombel, C.; Darzens, D.; Foglietti, M.J.; Percheron, F. *J. Chromatogr.* **1986**, *355*, 456-462.
414. Bodenmann, J.; Heiniger, U.; Hohl, H.R. *Can. J. Microbiol.* **1985**, *31*, 75-82.
415. Cao, W.; Crawford, D.L. *Can. J. Microbiol.* **1993**, *39*, 125-129.
416. Hong, Y.; Pasternak, J.J.; Glick, B.R. *Curr. Microbiol.* **1990**, *20*, 339-342.
417. Hwang, J.-T.; Suzuki, H. *Agric. Biol. Chem.* **1976**, *40*, 2169-2176.
418. Yamane, K.; Suzuki, H.; Nisizawa, K. *J. Biochem.* **1970**, *67*, 19-35.
419. Yoshikawa, T.; Suzuki, H.; Nisizawa, K. *J. Biochem.* **1974**, *75*, 531-540.
420. Hirayama, T.; Horie, S.; Nagayama, H.; Matsuda, K. *J. Biochem.* **1978**, *84*, 27-37.
421. Kengen, S.W.M.; Luesink, E.J.; Stams, A.J.M.; Zehnder, A.J.B. *Eur. J. Biochem.* **1993**, *213*, 305-312.
422. Uziie, M.; Sasaki, T. *Enzyme Microb. Technol.* **1987**, *9*, 459-465.
423. Srivastava, S.K.; Ali, A.; Khanna, S. *Biotechnol. Lett.* **1991**, *13*, 907-912.
424. Wood, T.M. *Methods Enzymol.* **1988**, *160*, 216-221.
425. Wood, T.M.; Wilson, C.A.; Stewart, C.S. *Biochem. J.* **1982**, *205*, 129-137.
426. Ohmiya, K.; Maeda, K.; Shimizu, S. *Carbohydr. Res.* **1987**, *166*, 145-155.
427. Ohmiya, K.; Shimizu, M.; Taya, M.; Shimizu, S. *J. Bacteriol.* **1982**, *150*, 407-409.
428. Ohmiya, K.; Shirai, M.; Kurachi, Y.; Shimizu, S. *J. Bacteriol.* **1985**, *161*, 432-434.
429. Doerner, K.C.; White, B.A. *Appl. Environ. Microbiol.* **1990**, *56*, 1844-1850.
430. Gardner, R.M.; Doerner, K.C.; White, B.A. *J. Bacteriol.* **1987**, *169*, 4581-4588.
431. Webb, S.T.J.; Taylor, J.E.; Coupe, S.A.; Ferrarese, L.; Roberts, J.A. *Plant Cell Environ.* **1993**, *16*, 329-333.
432. Paice, M.G.; Desrochers, M.; Rho, D.; Roy, C.; Rollin, C.F.; De Miguel, E.; Yaguchi, M. *Bio/Technology* **1984**, *2*, 535-539.
433. Desrochers, M.; Jurasek, L.; Paice, M.G. *Appl. Environ. Microbiol.* **1981**, *41*, 222-228.
434. Patil, R.V.; Sadana, J.C. *Can. J. Biochem. Cell Biol.* **1984**, *62*, 920-926.
435. Sadana, J.C.; Lachke, A.H.; Patil, R.V. *Carbohydr. Res.* **1984**, *133*, 297-312.

436. Sadana, J.C.; Patil, R.V.; Shewale, J.G. *Methods Enzymol.* **1988**, *160*, 424-431.
437. Sadana, J.C.; Patil, R.V. *Methods Enzymol.* **1988**, *160*, 307-314.
438. Shewale, J.G.; Sadana, J. *Arch. Biochem. Biophys.* **1981**, *207*, 185-196.
439. Morpeth, F.F.; Jones, G.D. *Biochem. J.* **1986**, *236*, 221-226.
440. Morpeth, F.F. *Biochem. J.* **1985**, *228*, 557-564.
441. Jones, G.D.; Wilson, M.T. *Biochem. J.* **1988**, *256*, 713-718.
442. Deshpande, V.; Eriksson, K.-E.; Pettersson, B. *Eur. J. Biochem.* **1978**, *90*, 191-198.
443. Ayers, A.R.; Ayers, S.B.; Eriksson, K.-E. *Eur. J. Biochem.* **1978**, *90*, 171-181.
444. Fracheboud, D.; Canevascini, G. *Enzyme Microb. Technol.* **1989**, *11*, 220-229.
445. Meyer, H.-P.; Canevascini, G. *Appl. Environ. Microbiol.* **1981**, *41*, 924-931.
446. Curioni, A.; Dal Belin Peruffo, A.; Nuti, M.P. *Electrophoresis* **1988**, *9*, 327-330.
447. Nakai, R.; Horinouchi, S.; Uozumi, T.; Beppu, T. *Agric. Biol. Chem.* **1987**, *51*, 3061-3065.
448. Nakamura, A.; Fukumori, F.; Horinouchi, S.; Masaki, H.; Kudo, T.; Uozumi, T.; Horikoshi, K.; Beppu, T. *J. Biol. Chem.* **1991**, *266*, 1579-1583.
449. MacKenzie, C.R.; Bilous, D.; Johnson, K.G. *Can. J. Microbiol.* **1984**, *30*, 1171-1178.
450. Théberge, M.; Lacaze, P.; Shareck, F.; Morosoli, R.; Kluepfel, D. *Appl. Environ. Microbiol.* **1992**, *58*, 815-820.
451. Schlochtermeier, A.; Niemeyer, F.; Schrempf, H. *Appl. Environ. Microbiol.* **1992**, *58*, 3240-3248.
452. Schlochtermeier, A.; Walter, S.; Schröder, J.; Moorman, M.; Schrempf, H. *Mol. Microbiol.* **1992**, *6*, 3611-3621.
453. Wachinger, G.; Bronnenmeier, K.; Staudenbauer, W.L.; Schrempf, H. *Appl. Environ. Microbiol.* **1989**, *55*, 2653-2657.
454. Heupel, C.; Schlochtermeier, A.; Schrempf, H. *Enzyme Microb. Technol.* **1993**, *15*, 127-132.
455. Coughlan, M.P.; Moloney, A.P. *Methods Enzymol.* **1988**, *160*, 363-368.
456. Moloney, A.P.; Coughlan, M.P. *Biochem. Soc. Trans.* **1985**, *13*, 457.
457. Coughlan, M.P.; McHale, A. *Methods Enzymol.* **1988**, *160*, 437-443.
458. McHale, A.; Coughlan, M.P. *Biochim. Biophys. Acta* **1981**, *662*, 152-159.
459. Osore, H.; Okech, M.A. *J. Appl. Biochem.* **1983**, *5*, 172-179.
460. Shepherd, M.G.; Cole, A.L.; Tong, C.C. *Methods Enzymol.* **1988**, *160*, 300-307.
461. Bernier, R.; Stutzenberger, F. *Letters in Applied Microbiology* **1988**, *7*, 103-107.
462. McGinnis, K.; Kroupis, C.; Wilson, D.B. *Biochemistry* **1993**, *32*, 8146-8150.
463. Ruttersmith, L.D.; Daniel, R.M. *Biochim. Biophys. Acta* **1993**, *1156*, 167-172.
464. Ruttersmith, L.D.; Daniel, R.M. *Biochem. J.* **1991**, *277*, 887-890.
465. Takase, M.; Horikoshi, K. *Agric. Biol. Chem.* **1989**, *53*, 559-560.
466. Takase, K.; Horikoshi, K. *Appl. Microbiol. Biotechnol.* **1988**, *29*, 55-60.
467. Wood, T.M. *Biochem. J.* **1968**, *109*, 217-227.
468. Halliwell, G.; Vincent, R. *Biochem. J.* **1981**, *199*, 409-417.
469. Macarrón, R.; Acebal, C.; Castillón, M.P.; Domínguez, J.M.; De la Mata, I.; Pettersson, G.; Tomme, P.; Claeyssens, M. *Biochem. J.* **1993**, *289*, 867-873.
470. Inglin, M.; Feinberg, B.A.; Loewenberg, J.R. *Biochem. J.* **1980**, *185*, 515-519.
471. Ogawa, K.; Toyama, D.; Fujii, N. *J. Gen. Appl. Microbiol.* **1991**, *37*, 249-259.

472. Bissett, F.H. *J. Chromatogr.* **1979**, *178*, 515-523.
473. Evans, E.T.; Wales, D.S.; Bratt, R.P.; Sagar, B.F. *J. Gen. Microbiol.* **1992**, *138*, 1639-1646.
474. Gritzali, M.; Brown, R.D.,Jr. In: *Hydrolysis of Cellulose: Mechanisms of Enzymatic and Acid Catalysis (ACS Series 181)*; Brown, R.D.,Jr., Jurasek, L., Eds.; American Chemical Society: Washington, DC, 1979, pp 237-260.
475. Schülein, M. *Methods Enzymol.* **1988**, *160*, 234-242.
476. Witte, K.; Heitz, H.J.; Wartenberg, A. *Acta Biotechnol.* **1990**, *10*, 41-48.
477. Voragen, A.G.J.; Beldman, G.; Rombouts, F.M. *Methods Enzymol.* **1988**, *160*, 243-251.
478. Massiot, P. *Lebensm. -Wiss. Technol.* **1992**, *25*, 120-125.
479. Shoemaker, S.; Watt, K.; Tsitovsky, G.; Cox, R. *Bio/Technology* **1983**, *1*, 687-690.
480. Håkansson, U.; Fägerstam, L.G.; Pettersson, G.; Andersson, L. *Biochim. Biophys. Acta* **1978**, *524*, 385-392.
481. Beldman, G.; Searle-Van Leeuwen, M.F.; Rombouts, F.M.; Voragen, F.G.J. *Eur. J. Biochem.* **1985**, *146*, 301-308.
482. Gong, C.-S.; Ladisch, M.R.; Tsao, G.T. *Biotechnol. Bioeng.* **1977**, *19*, 959-981.
483. Potts, R.C.; Hewitt, P.H. *Comp. Biochem. Physiol. B Comp. Biochem.* **1974**, *47*, 317-326.
484. Edwards, M.; Dea, I.C.M.; Bulpin, P.V.; Reid, J.S.G. *J. Biol. Chem.* **1986**, *261*, 9489-9494.
485. Hatano, T.; Kosaka, M.; Cui, Z.; Kawaguchi, M.; Miyakawa, T.; Fukui, S. *J. Ferment. Bioeng.* **1991**, *71*, 313-317.

RECEIVED February 17, 1994

Chapter 8

Cellobiose Dehydrogenase

A Hemoflavoenzyme from *Phanerochaete chrysosporium*

V. Renganathan and Wenjun Bao

Department of Chemistry, Biochemistry, and Molecular Biology, Oregon Graduate Institute of Science and Technology, 20000 Northwest Walker Road, P.O. Box 91000, Portland, OR 97291–1000

Cellobiose dehydrogenase (CDH) is an extracellular hemoflavoenzyme produced by cellulose-degrading cultures of the lignocellulose-degrading basidiomycete *Phanerochaete chrysosporium.* CDH is monomeric and contains one heme *b* and one FAD on two distinct domains. In addition, CDH may also have a cellulose-binding domain. The heme iron of CDH is ferric, low spin, and is hexacoordinate with a histidine and a methionine as the fifth and sixth ligands. CDH oxidizes cellobiose to cellobionolactone in the presence of electron acceptors such as cytochrome *c*, dichlorophenol-indophenol, and ferricyanide. In the CDH reaction, flavin is the primary acceptor of electrons from cellobiose; the reduced flavin then transfers the electrons to the heme iron, one electron at a time. The reduced heme in turn transfers its electrons to ferricytochrome *c*. Whereas electron transfer to cytochrome *c* occurs only from the heme, electron transfer to other acceptors can occur directly from the reduced flavin. CDH enhances crystalline cellulose degradation by cellulases and this may be one of its physiological functions. This report reviews the progress made in understanding the structure, function, and mechanism of CDH, and its potential role in lignocellulose degradation.

Many cellulolytic fungi, including *P. chrysosporium* (*1,2*), *Sporotrichum thermophile* (*3*), *Coniophora puteana* (*4*), and *Sclerotium rolfsii* (*5*), produce extracellular cellobiose oxidizing enzymes, in addition to cellulases. All of these dehydrogenases appear to oxidize cellobiose to cellobionolactone. Cellobiose dehydrogenases (CDH) from *P. chrysosporium* (*1,6,7*), *S. thermophile* (*3,8*), and *C. puteana* (*4,9*), a brown-rot fungus, are characterized as hemoflavoenzymes containing one heme *b* and one FAD or FMN per monomer. In addition to CDH, *P. chrysosporium* produces another cellobiose-oxidizing flavoenzyme, cellobiose:quinone oxidoreductase (CBQase) (*2*).

0097–6156/94/0566–0179$08.00/0

The nature of cellobiose-oxidizing enzymes produced by *S. rolfsii*, *Monilia* sp., *Chaetomium cellulolyticum*, and *Fomes annosus* is not known (*5,10-12*).

Only a few enzymes are known to contain a heme and a flavin on a monomeric subunit. Apart from CDH, the following enzymes are characterized as hemoflavoenzymes: flavocytochrome b_2 or lactate dehydrogenase from *Saccharomyces cerevisiae* (*13*), spermidine dehydrogenase from *Serratia marcescens* (*14*), rubredoxin oxidase from *Desulfovibrio gigas* (*15*), nitric oxide synthetase from murine macrophages (*16*), and a fatty acid monooxygenase from *Bacillus megaterium* (*17*). Among these, only flavocytochrome b_2, which dehydrogenates lactate to pyruvate, is well characterized and this enzyme may serve as a model for CDH (*13,18,19*). Flavocytochrome b_2 is a tetramer; each monomer contains one FMN and one heme *b*. The polypeptide chain is organized into two domains, a cytochrome domain to which the heme binds, and a flavodomain to which the FMN binds (*13,18*). The two domains are linked by a protease-sensitive segment (*18*). The heme iron is in the ferric oxidation state and is hexacoordinate. Histidine residues 43 and 66 of the apoprotein function as the fifth and the sixth ligands to the heme iron (*13*). The FMN group is the primary acceptor of electrons from lactate. The reduced flavin then transfers the electrons to the heme iron in two single steps. The heme, in turn, transfers the electrons to cytochrome oxidase via ferricytochrome *c* (*13*).

CDH (also known as cellobiose oxidase) from *P. chrysosporium* was discovered by Eriksson and coworkers (*1*). Several methods for the purification of CDH from *P. chrysosporium* have been reported (*1,6,7,19,20*); however, a procedure that we recently developed provides CDH in high yield and with high specific activity compared to other methods (*6*). CDH from *P. chrysosporium* is a monomer with a molecular mass of approximately 90,000. It contains one heme *b* and one FAD per monomer. It is a glycoprotein with a neutral carbohydrate content of 9.4% (*6*). CDH is very stable between pH 3 and 10.5, and below 50°C. However, below pH 2 or above 50°C, the activity is lost due to release of the flavin from the active site. The heme appears to be tightly bound to the apoprotein under these conditions (*6*).

Spectral Studies

The heme iron of CDH exists in the ferric state (*1,6,7*). The UV-visible spectrum of oxidized CDH shows absorbances at 420, 529, and 570 nm; on the addition of cellobiose or dithionite, absorbances shift to 428, 534, and 564 nm (Figure 1) (*1,6,7*). These spectral characteristics of CDH are very similar to the those of the hemoflavoenzymes, flavocytochrome b_2 and spermidine dehydrogenase (*6,14,23*). Our preliminary studies indicate that the resonance Raman spectral characteristics of CDH are also very similar to those of flavocytochrome b_2 suggesting similar heme structures for both of these enzymes (*23*; Cohen, Bao, Renganathan, and Loehr, unpublished results). Ferric CDH does not bind azide or cyanide; this is indicative of a hexacoordinate heme iron as is found in flavocytochrome b_2 (*6*). Two histidines serve as the fifth and sixth ligands to the heme iron of flavocytochrome b_2 (*13*). However, based on nuclear magnetic resonance and infrared-magnetic circular dichroism studies, Cox et al. have concluded that CDH has a histidine and a methionine as the fifth and sixth coordinations to the heme iron (*24*).

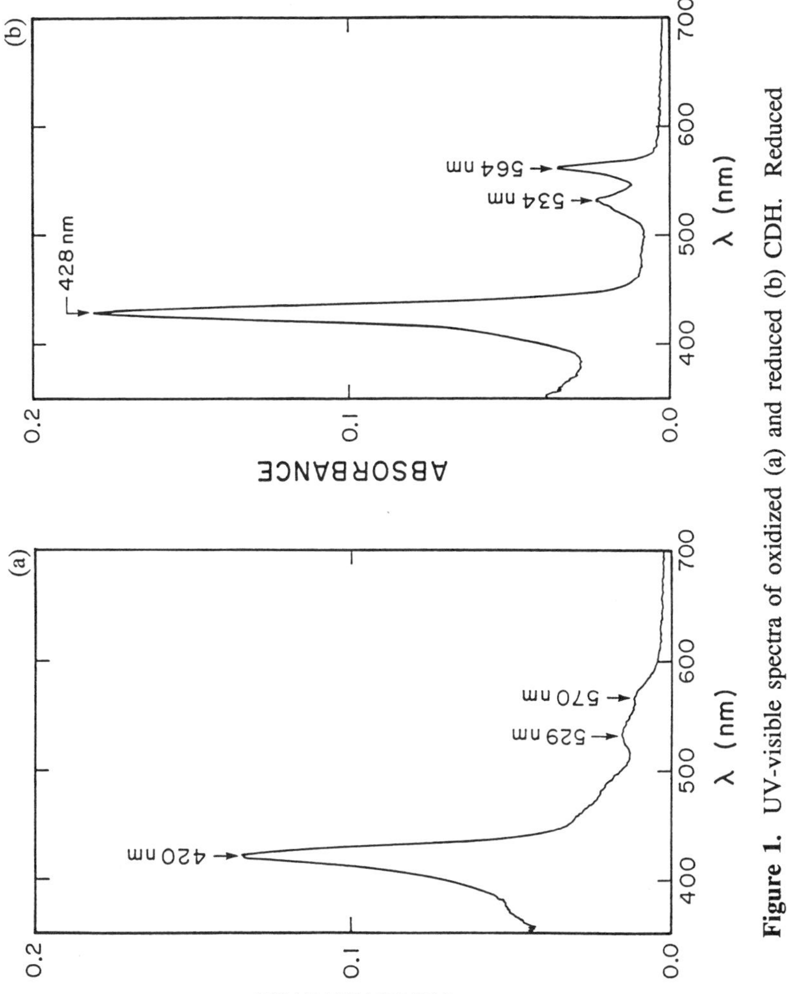

Figure 1. UV-visible spectra of oxidized (a) and reduced (b) CDH. Reduced CDH was generated by the addition of cellobiose (50 μM) to the oxidized form. (Reproduced with permission from ref. *6*, copyright Academic Press, 1993.)

Structural Organization

The heme and the FAD of CDH are bound noncovalently to different domains (*20,21*). On treatment with a protease such as papain, CDH is hydrolyzed into two peptide fragments: a larger peptide ($M_r \approx 55{,}000$) containing the flavin, and a smaller peptide ($M_r \approx 35{,}000$) containing the heme (*20,21*). The flavin-containing peptide is catalytically active and oxidizes cellobiose to cellobionolactone in the presence of quinones or dichlorophenol-indophenol. However, unlike native CDH, the flavodomain cannot transfer its electrons to cytochrome *c* (*20,21*). These characteristics of the flavodomain are very similar to those of CBQase, the second cellobiose-oxidizing extracellular enzyme identified from *P. chrysosporium* (*2*). It has been postulated that CBQase is formed via the proteolytic hydrolysis of CDH (*20,21*). Our recent observations--that culture conditions favoring protease production increase the yield of CBQase and decrease the yield of CDH--support this hypothesis (Bao, Lymar, and Renganathan, unpublished results). Our laboratory first reported (*25*) that CDH and CBQase bind to microcrystalline cellulose, and we proposed that these enzymes may have a specific cellulose-binding domain similar to that of cellulases. Later, Henriksson et al. (*19*) demonstrated that the cellulose-binding domain of CDH is located on the flavodomain.

Kinetic Studies

Cellobiose, cellotriose, cellotetraose, cellopentaose, and lactose function as substrates for CDH (*1*). CDH oxidizes these sugars in the presence of electron acceptors such as cytochrome *c*, dichlorophenol-indophenol, quinones, ferricyanide, Mn(III)-malonate complex, and even ferric iron and its complexes (*1,6,7,20,22,26*). Our recent steady-state kinetic studies suggest that cellobiose, with a K_m of 25 µM, is the preferred substrate, whereas lactose, with a K_m of 630 µM, is the least preferred substrate. However, all the sugar substrates exhibited similar k_{cat} values (*6*). Our kinetic investigation also suggests that cytochrome *c* is the preferred electron acceptor for CDH (*6*). Cytochrome *c* is the physiological electron acceptor for the hemoflavoenzymes flavocytochrome b_2 and spermidine dehydrogenase (*13,14*); however, unlike CDH, these enzymes are intracellular and membrane associated.

As in the flavocytochrome b_2 reaction, the flavin moiety of CDH appears to receive the electrons from cellobiose first and then transfer them to the heme iron, one electron at a time. The reduced heme, in turn, transfers the electrons to ferricytochrome *c* (Scheme I). Jones and Wilson (*27*) observed that the rate constant for heme iron reduction in CDH increases with enzyme concentration and interpreted

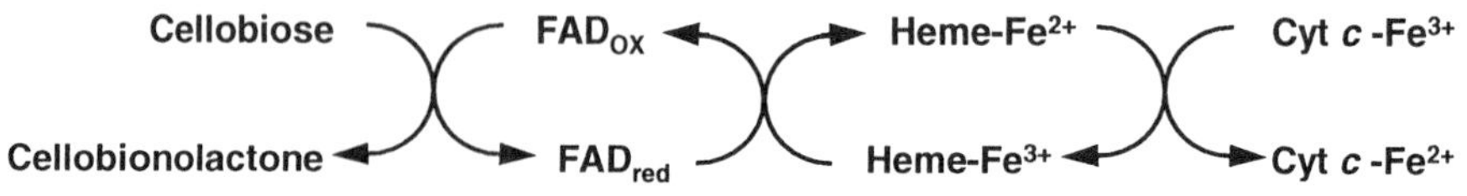

Cellobiose → Cellobionolactone; FAD_{ox} → FAD_{red}; FAD_{red} + Heme-Fe^{3+} → FAD_{ox} + Heme-Fe^{2+}; Heme-Fe^{2+} + Cyt *c*-Fe^{3+} → Heme-Fe^{3+} + Cyt *c*-Fe^{2+}

Scheme I. Catalytic cycle of CDH

this as due to the intermolecular reduction of the heme center by flavin groups; whereas, in the case of flavocytochrome b_2, electron transfer from flavin to heme is intramolecular (*19*). Samejima et al. (*28*) studied the effect of pH on the reduction of cytochrome *c* and dichlorophenol-indophenol by CDH. Although both of these electron acceptors are active at pH 4.2, only dichlorophenol-indophenol is active at pH 5.9. Furthermore, a stopped-flow kinetic study indicated that reduction of the flavin at pH 4.2 and 5.9 is fast, whereas the heme reduction is fast only at pH 4.2. These observations have led to the proposal that reduction of heme by the flavin decreases at a higher pH. Since the heme transfers electrons to cytochrome *c*, the reaction rate for cytochrome *c* reduction also has been suggested to decrease at a higher pH (*28*).

CDH Reaction with Oxygen

CDH was first characterized as an oxidase because it consumes oxygen during its reaction (*1*). However, recent studies clearly indicate that oxygen is a very poor acceptor of electrons from CDH (*6,22,29*). Accordingly, we renamed the enzyme cellobiose dehydrogenase to correctly reflect its nature. In the absence of a suitable electron acceptor, O_2 serves as the electron acceptor and is reduced to H_2O_2. We have demonstrated the stoichiometric production of H_2O_2 from cellobiose oxidation (*6*). Based on pre-steady-state kinetic studies, Wilson et al. identified the heme group as the site for reaction with oxygen (*29*). CDH can reduce O_2 by one electron to produce a superoxide radical, or by two electrons to produce H_2O_2. However, superoxide radicals are unstable and readily disproportionate to H_2O_2 and O_2. Thus, both one-electron and two-electron reduction of oxygen will lead to H_2O_2 generation. It has been hypothesized that a superoxide radical is formed in the CDH reaction; however, formation of superoxide radical has not yet been established (*26*).

Interaction of CDH with Lignin Peroxidase

The lignin degradative system of *P. chrysosporium* contains two important extracellular peroxidases: lignin peroxidase (LiP) and manganese peroxidase (MnP) (*30,31*). CDH, in the presence of cellobiose, inhibits a variety of reactions catalyzed by LiP, MnP, and horseradish peroxidase (HRP) (*6,22*). Oxidation of a phenol by the peroxidases proceeds through the corresponding phenoxy radical, and produces quinone and polymerized phenol as the final products. CDH inhibits phenol oxidation by reducing the phenoxy radicals (*22*). Phenoxy radicals are also produced during lignin depolymerization by phenol oxidases, and these radicals tend to condense with themselves and with the lignin substrate (*32,33*). It has been proposed that reduction of phenoxy radicals by cellobiose-oxidizing enzymes prevents these polymerization reactions and thus increases the rate of depolymerization (*34*).

In addition to phenols, nonphenolic compounds such as 3,4-dimethoxybenzyl alcohol and methoxybenzenes serve as LiP substrates. LiP oxidation of these compounds proceeds through the corresponding aromatic cation radical intermediate (*30,31*). CDH is suggested to inhibit these LiP reactions by reducing the cation radical intermediates (*22*). In the LiP reaction, H_2O_2 oxidizes the ferric native enzyme by two electrons to produce compound I. Suitable peroxidase substrates then reduce compound I by one electron to generate compound II. Donation of one more electron from the substrate to compound II completes the catalytic cycle and regenerates the ferric peroxidase (*30,31,35*). Our recent studies suggest electron transfer from reduced CDH to H_2O_2-oxidized forms of LiP and HRP (Bao and Renganathan, unpublished observations). Also, in the presence of cellobiose and very low concentrations of veratryl alcohol, CDH protected LiP from H_2O_2-dependent inactivation (*36,37*,Bao and Renganathan, unpublished results).

Role of CDH in Cellulose Degradation

Our laboratory examined the contribution of CDH to cellulose degradation by studying the effect of CDH addition on microcrystalline cellulose hydrolysis by *Trichoderma viride* cellulase (*38*). At low concentrations (5-10 μg/mL), CDH enhances both glucose and cellobiose production (Figure 2). However, at higher levels, production of reducing sugar decreases and cellobionolactone increases. The decrease observed

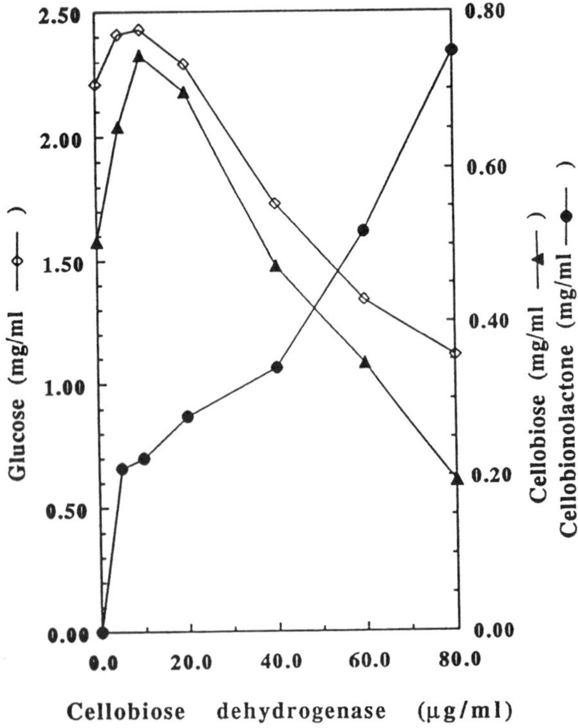

Figure 2. Effect of *P. chrysosporium* CDH on *T. viride* cellulase-catalyzed conversion of microcrystalline cellulose to glucose (□), cellobiose (▲), and cellobionolactone (•). (Reproduced with permission from ref. *38*, copyright Elsevier Science Publishers, 1992.)

at higher CDH concentrations is attributed to the increased oxidation of cellobiose to cellobionolactone. Cellulose weight loss, in the presence of 10 μg mL^{-1} CDH, increases by 19% compared to the control cellulase incubations (Table I). However, at 60 μg mL^{-1}, CDH weight loss decreases by 24% compared to that of the control. In these reactions, oxygen functions as the electron acceptor for CDH leading to the production of H_2O_2. We suspected that oxy-radicals generated from H_2O_2 might be inactivating the cellulases, particularly at high CDH concentrations. In support of this hypothesis, we demonstrated that catalase addition increases the cellulose weight loss (Table I) and reducing sugar production. Hydrolysis of crystalline cellulose is recognized as the rate-limiting step in the saccharification of cellulose to glucose (*39*). Our findings suggest that CDH increases the velocity of this rate-limiting step.

Kremer and Wood (*26*) proposed that hydroxyl radical is produced by the reaction of ferrous iron, generated through the one-electron reduction of ferric iron by CDH, and H_2O_2, generated from the CDH reaction with oxygen. Hydroxyl radicals generated in this fashion are suggested to disrupt the crystalline structure of cellulose

Table I.

T. viride Cellulase Catalyzed Hydrolysis of Microcrystalline Cellulose. Effect of CDH and/or Catalase Addition on Cellulose Weight Loss.[a]

Enzymes	Weight loss (mg)	Percentage weight loss relative to control[b]
1. Cellulase (0.2 mg mL^{-1}) (control)	91.5 ± 1.0	100 ± 1.1
2. Cellulase (0.2 mg mL^{-1}) + CDH (10 μg mL^{-1})	108.5 ± 0.6	118.6 ± 0.6
3. Cellulase (0.2 mg mL^{-1}) + CDH (60 μg mL^{-1})	69.5 ± 0.1	76.0 ± 0.1
4. Cellulase (0.2 mg mL^{-1}) + CDH (10 μg mL^{-1}) + catalase (5 μg mL^{-1})	109.2 ± 2.3	119.3 ± 2.1
5. Cellulase (0.2 mg mL^{-1}) + CDH (60 μg mL^{-1}) + catalase (5 μg mL^{-1})	86.7 ± 0.2	94.8 ± 0.2

[a] Enzymes were incubated with 600 mg microcrystalline cellulose in 50 mM acetate, pH 5, in a total volume of 30 mL. Weight loss was determined as described (*36*).

[b] Cellulose weight loss observed in the absence of CDH and catalase was considered to be 100%.

SOURCE: Reproduced with permission from ref. *38*, copyright Elsevier Science Publishers, 1992.

(*26*). We observed very similar enhancement of cellulose hydrolysis by CDH in the presence or absence of catalase. In the presence of catalase, H_2O_2 is decomposed, severely inhibiting hydroxyl radical formation. Therefore, hydroxyl radicals are not responsible for the enhancement of cellulose hydrolysis observed in our study. An alternate explanation for enhanced cellulose hydrolysis is that cellobiose is an inhibitor of cellulases (*40*), and by dehydrogenating cellobiose to cellobionolactone CDH is removing this inhibition. In addition, CDH appears to be capable of oxidizing cellulose directly (*8,26*); however, the nature of the degradation products and the possible significance of this reaction is not known. Further studies are necessary to fully understand the mechanism by which CDH enhances cellulose degradation.

Though CDH was discovered in 1978, most information on this enzyme has accumulated over the past five years through the efforts of several laboratories. In the next few years, further research should provide a more thorough understanding of the structure, function, and mechanism of CDH, and the role of CDH in lignocellulose degradation.

Acknowledgments

Our research was supported by grants DE-FG06-87ER 13715 and DE-PR06-92ER 20065 from the U.S. Department of Energy, Office of Basic Energy Sciences.

Literature Cited

1. Ayers, A.R.; Ayers, S.B.; Eriksson, K.-E. *Eur. J. Biochem.* **1978,** *90,* 171.
2. Westermark, U.; Eriksson, K.-E. *Acta Chim. Scan.* **1975,** *B29,* 419.
3. Coudray, M.-R.; Canevascini, G.; Meier, H. *Biochem. J.* **1982,** *203,* 277.
4. Schmidhalter, D.R.; Canevascini, G. *Appl. Microbiol. Biotechnol.* **1992,** *37,* 431.
5. Sadana, J.C.; Patil, R.V.J. *Gen. Microbiol.* **1985,** *131,* 1917.
6. Bao, W.; Usha, S.N.; Renganathan, V. *Arch. Biochem. Biophys.* **1993,** *300,* 705.
7. Morpeth, F.F. *Biochem. J.* **1985,** *228,* 557.
8. Canevascini, G.; Borer, P.; Dreyer, J.-L. *Eur. J. Biochem.* **1991,** *198,* 43.
9. Schmidhalter, D.; Canevascini, G. *Arch. Biochem. Biophys.* **1993,** *300,* 559.
10. Dekker, R.F.H. *J. Gen. Microbiol.* **1980,** *120,* 309.
11. Fahnrich, P.; Irrgang, K. *Biotechnol. Lett.* **1982,** *4,* 775.
12. Hutterman, A.; Noelle, A. *Holzforschung* **1982,** *36,* 283.
13. Xia, Z.-X.; Shamala, N.; Bethege, P.H.; Lim, L.W.; Bellamy, H.D.; Xuong, N. H.; Lederer, F.; Mathews, F.S. *Proc. Natl. Acad. Sci. USA* **1987,** *84,* 2629.
14. Tabor, C.W.; Kellogg, P.D. *J. Biol. Chem.* **1970,** *245,* 5424.
15. Chen, L.; Liu, M.-Y.; LeGall, J.; Fareleira, P.; Santos, H.; Xavier, A.V. *Biochem. Biophys. Res. Commun.* **1993,** *193,* 100.
16. Marletta, M.A. *J. Biol. Chem.* **1993,** *268,* 12231.
17. Narhi, L.O.; Fulco, A.J. *J. Biol. Chem.* **1986,** *261,* 7160.
18. Celerier, J.; Risler, Y.; Schwenke, J.; Janot, J.-M.; Gervais, M. *Eur. J. Biochem.* **1989,** *182,* 67.

19. Walker, M.C.; Tollin, G. *Biochemistry* **1992,** *31,* 2798.
20. Henriksson, G.; Pettersson, G.; Johansson, G.; Ruiz, A.; Uzcategui, E. *Eur. J. Biochem.* **1991,** *196,* 101.
21. Wood, J.D.; Wood, P.M. *Biochim. Biophys. Acta* **1992,** *119,* 90.
22. Samejima, M.; Eriksson, K.-E. *Eur. J. Biochem.* **1992,** *207,* 103.
23. Desbois, A.; Tegoni, M.; Gervais, M.; Lutz, M. *Biochemistry* **1989,** *28,* 8011.
24. Cox, M.C.; Rogers, M.S.; Cheesman, M.; Jones, J.D.; Thomson, A.J.; Wilson, M.T.; Moore, G.R. *FEBS Lett.* **1992,** *307,* 233.
25. Renganathan, V.; Usha, S.N.; Lindenburg, F. *Appl. Microbiol. Biotechnol.* **1990,** *32,* 609.
26. Kremer, S.M.; Wood, P.M. *Eur. J. Biochem.* **1992,** *208,* 807.
27. Jones, J.D.; Wilson, M. *Biochem. J.* **1988,** *256,* 713.
28. Samejima, M.; Phillips, R.S.; Eriksson, K.-E. *FEBS Lett.* **1992,** *306,* 165.
29. Wilson, M.T.; Hogg, N.; Jones, G.D. *Biochem. J.* **1990,** *270,* 265.
30. Kirk, T.K.; Farrell, R.F. *Ann. Rev. Microbiol.* **1987,** *41,* 465.
31. Gold, M.H.; Wariishi, H.; Valli, K. In *Biocatalysis in Agricultural Biotechnology;* Whitaker, J.R.; Sonnet, P.E., Eds.; American Chemical Society: Washington, DC, 1989; pp 127-140.
32. Hammel, K.E.; Moen, M.A. *Enzyme Microb. Technol.* **1991,** *13,* 15.
33. Wariishi, H.; Valli, K.; Gold, M.H. *Biochem. Biophys. Res. Commun.* **1991,** *176,* 269.
34. Ander, P.; Mishra, C.; Farrell, R.L.; Eriksson, K.-E.L. *J. Biotechnol.* **1990,** *13,* 190.
35. Renganathan, V.; Gold, M.H. *Biochemistry* **1986,** *25,* 1626.
36. Wariishi, H.; Gold, M.H. *J. Biol. Chem.* **1990,** *265,* 2070.
37. Cai, D.; Tien, M. *J. Biol. Chem.* **1992,** *267,* 11149.
38. Bao, W.; Renganathan, V. *FEBS Lett.* **1992,** *302,* 77.
39. Ohmine, K.; Ooshima, H.; Harano, Y. *Biotechnol. Bioeng.* **1983,** *25,* 2041.
40. Gritzali, M.; Brown, R.D. In *Hydrolysis of Cellulose: Mechanisms of Enzymatic and Acidic Catalysis;* Brown, R.D.; Jurasek, L., Eds.; American Chemical Society: Washington, DC, 1979; pp 237-260.

RECEIVED January 19, 1994

Chapter 9

Cellulase Production Technology

Evaluation of Current Status

George P. Philippidis

Bioprocessing Branch, National Renewable Energy Laboratory, 1617 Cole Boulevard, Golden, CO 80401–3393

The production of cellulase, the enzyme complex that can hydrolyze cellulose, is an integral part of the biochemical technology designed to convert lignocellulosic biomass to chemicals such as ethanol, a leading alternative renewable transportation fuel. The cost of cellulase is a significant factor in the economics of biomass-to-ethanol conversion technology. As a result, it has a direct impact on the cost of the ethanol produced and therefore needs to be minimized. The key to producing highly active cellulase inexpensively lies in a combination of critical factors: improved enzyme quality, prolonged enzyme lifetime, enhanced enzyme productivity and yield, and minimal cost of growth medium and substrate. To date, both mesophilic and thermophilic organisms have been employed in cellulase synthesis, using soluble and insoluble carbohydrates as substrates. Defined and undefined media have been developed, and operating parameters, such as temperature, pH, inducer supplementation, and substrate feeding rate, have been manipulated in various patterns to optimize enzyme productivity in batch, fed-batch, and continuous modes of operation. So far, a fed-batch process using the mesophilic fungus *Trichoderma reesei* Rut C30 and ball-milled hardwood pulp has led to the highest performance reported in the literature, an enzyme concentration of 57,000 International Units (IU)/L at a production rate of 427 IU/L/h. Various *T. reesei* strains have been successfully cultivated in pilot (15,000-L) and commercial (200,000-L) scale fermenters. This paper presents a critical evaluation of the current status of research and development in cellulase enzyme production. It focuses on examination of crucial issues and identification of strategies that will lead to a technically and economically feasible process for cellulase synthesis and utilization in the context of biomass conversion.

0097–6156/94/0566–0188$09.80/0

The dependence of the U.S. economy on imported fossil fuels renders the country vulnerable to fluctuations in the oil supply. The continued instability in the Middle East underlines the need for development of alternative energy resources so that our economic growth can be uncoupled from unpredictable energy markets. Utilization of cellulose for synthesis of alternative transportation fuels, such as ethanol, to replace gasoline seems to be the most promising technology that can provide a permanent solution to our energy needs. Lignocellulosic materials, as a source of cellulose, are the most abundant renewable resource on earth with an annual production of approximately 1.8 x 10^{15} kg (*1*). Cellulose is readily available from agricultural residues, herbaceous crops, forestry by-products, pulp and paper industry wastes, and municipal solid waste. In addition to ethanol, other commodity and specialty chemicals can also be produced from cellulose, such as xylose, acetone, acetate, glycine, organic acids, glycerol, ethylene glycol, furfural, and animal feed (*2*).

Lignocellulosic biomass consists primarily of cellulose, hemicellulose, and lignin. Cellulose is the major biomass component, typically in the 40%-50% range on a dry basis. Although rich in carbohydrates (cellulose and hemicellulose), lignocellulose is an insoluble substrate with a complex structure, as outlined in the next section. This structure makes its conversion to fermentable sugars and subsequently to ethanol difficult. In a simplified representation, lignocellulose can be depicted as a network of cellulose fibers embedded in a sheath of hemicellulose and lignin. Bonding among cellulose chains presents additional constraints to the hydrolysis of cellulose.

Despite its compact structure, cellulose can be converted first to fermentable sugars (primarily glucose) and eventually to ethanol or other chemicals using cellulose-hydrolyzing enzymes called cellulases and ethanol-producing microorganisms. The major cellulase-producing organisms that have reached pilot plant or industrial scale are *T. reesei*, *Aspergillus niger*, and *Penicillium funiculosum*. Cellulases have been used for several years in food processing, feed preparation, waste water treatment, detergent formulation, textile production, and other areas. Additional potential applications include the production of wine, beer, and fruit juice (*3*). Nevertheless, all these uses are of a rather small magnitude compared with the cellulase requirements for bioconversion of lignocellulosic biomass to fuel ethanol, an alternative to gasoline with a market potential exceeding 150 billion gallons (5.7 x 10^{11} L) annually. Extensive research in the past decade has shown that the simultaneous saccharification and fermentation (SSF) process in a single step is the most effective and economic way to convert cellulose into ethanol using cellulases and fermentative organisms (*4*).

The current status of the lignocellulose bioconversion technology, as developed at NREL (*5*), consists of five major unit operations (Figure 1): (1) pretreatment of biomass using chemical, physical, or mechanical means to disrupt its structure and render it more accessible to enzymatic attack; (2) enzyme production using organisms that synthesize and secrete cellulase, such as fungi or bacteria; (3) fermentation of xylose, derived from hemicellulose during pretreatment, into ethanol; (4) SSF to initially convert cellulose to cellobiose and glucose by cellulase and then ferment the sugars to ethanol by yeasts or bacteria; and (5) ethanol recovery from the SSF effluent. According to a recent economic analysis of this design (*6*), the cellulase production unit contributes 6% of the total capital cost of an ethanol plant (Figure 2) and 4.5% of the final cost of ethanol (Figure 3). Despite its diverse applicability, cellulase remains an expensive hydrolytic enzyme because of the unfavorable process economics of its production from expensive soluble sugars or inexpensive solid biomass. In addition, cellulase production influences the kinetics and economics of SSF, the most costly unit in the overall process (Figures 2 and 3). Because the commercialization of the technology depends on its economic viability, it is imperative to minimize the cost of cellulase. There are, in general, two approaches to that goal: (1) increasing the enzyme productivity of the employed microorganism, and (2) decreasing the enzyme needs of the SSF process. Increase of enzyme productivity requires identification of microorganisms that grow fast and synthesize significant amounts of cellulase or, alternatively, secrete cellulases of high specific activity and stability. Decrease of enzyme input into the SSF process, on the other hand, requires understanding of the factors that affect the catalytic properties of cellulase and improvement of the SSF conditions with regard to both enzyme activity (faster hydrolysis) and stability (long-term utilization).

The scope of this chapter is to assess the current status of research and development in cellulase enzyme production, present critical issues of the technology, and identify strategies that will lead to optimization of cellulase synthesis and utilization.

The Substrate: Cellulosic Biomass

Cellulose is an insoluble, high-molecular-weight, linear polymer of D-glucose residues linked by β-1,4-glucosidic bonds. It is the most abundant natural polymer with a degree of polymerization usually between 3,500 and 10,000 (*7*). Cellulose chains, held together by hydrogen bonds and van der Waals forces, form insoluble elementary fibrils, which, in turn, are held together by hemicellulose to form microfibrils that are approximately 25 nm wide (*8*). Lamellae of microfibrils are surrounded by layers of lignin (a phenylpropane polymer) and hemicellulose; both protect cellulose from enzymatic attack. Microfibril regions of dense interchain bonding form crystalline areas that render cellulose resistant to enzymes and chemical reagents. In contrast, the looser structure of amorphous cellulose can be hydrolyzed more readily.

As a means of increasing the digestibility of cellulose, cellulosic biomass is pretreated mechanically, physically, and/or chemically prior to being exposed to cellulases. Physical pretreatment methods include the use of hammer mills, ball mills, and roll mills to increase the surface area and reactivity of the substrate. Although effective at rendering cellulose digestible, these methods are energy intensive. The major physical pretreatment techniques are steam, ammonia, and carbon dioxide explosion, which result in significant particle size reduction and therefore an increase of the substrate surface area. Chemical pretreatments include exposure of cellulose to alkali or dilute acid. The use of dilute sulfuric acid has proven to be efficient at significantly improving the accessibility of cellulose to cellulase (*9*). Unfortunately, most cost-effective pretreatment methods leave lignin

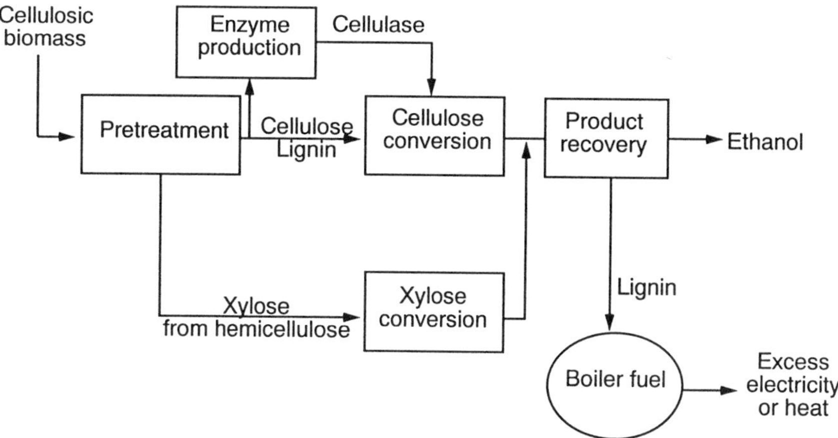

Figure 1. Schematic overview of the biomass-to-ethanol conversion process.

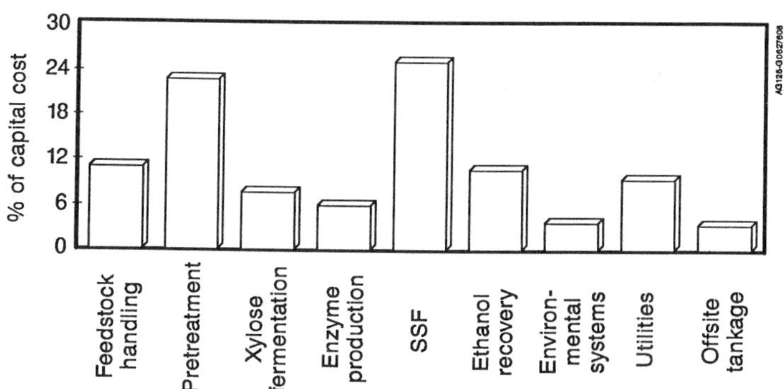

Figure 2. Breakdown of the fixed capital cost of a biomass-to-ethanol conversion plant.

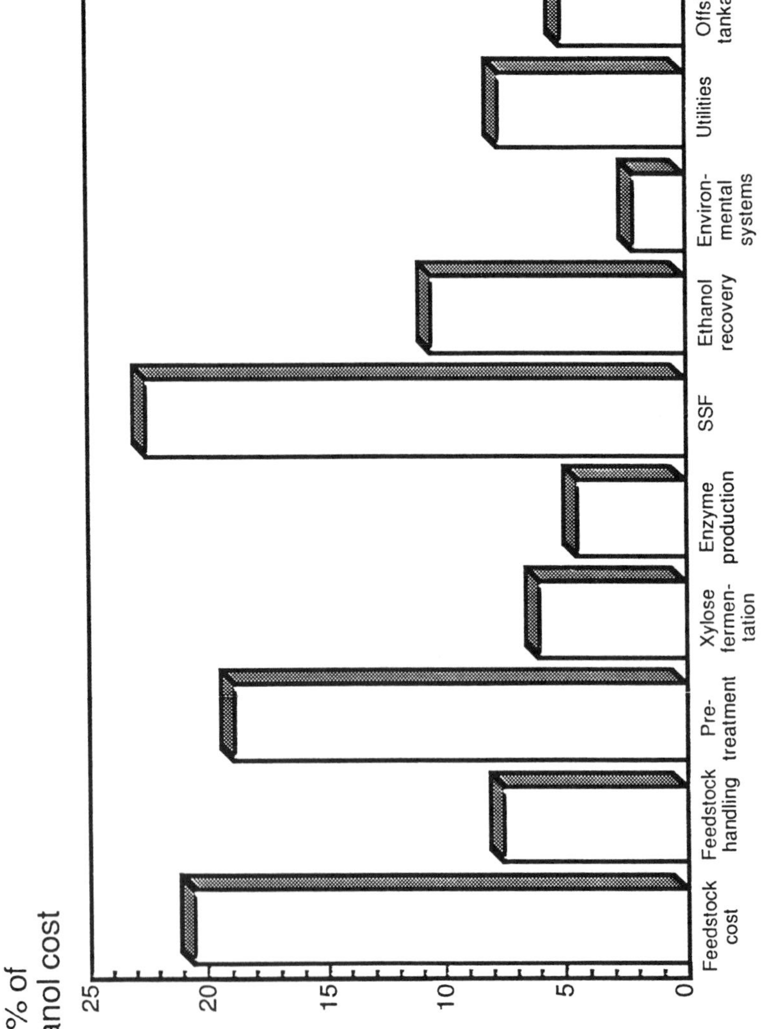

Figure 3. Breakdown of the final ethanol cost of a biomass-to-ethanol conversion plant.

largely intact. The presence of lignin is a major concern to the biomass conversion process because it not only prevents cellulase from reaching cellulose, but also adsorbs cellulase components, making them unavailable for cellulose hydrolysis (*10*).

The Enzyme: Cellulase

Sources, Composition, and Properties. Cellulases, the enzymes that are able to degrade cellulose, perform a crucial task in the SSF process by catalyzing the hydrolysis of cellulose to soluble, fermentable carbohydrates. They are synthesized by fungi, bacteria, and plants, but research and development has focused primarily on fungal and bacterial cellulases produced both aerobically and anaerobically. The aerobic mesophilic fungus *T. reesei* QM 6a and its mutants have been the most intensely studied sources of cellulases (*11,12*). Other fungal cellulase producers include *T. viride, T. lignorum, T. koningii, Penicillium spp., Fusarium spp., Aspergillus spp., Chrysosporium pannorum*, and *Sclerotium rolfsii* (*13-15*). Fungal cellulases are glycoproteins that require no cofactor for enzymatic activity. They can be refrigerated or frozen for years with no significant loss of activity (*2*). However, incubation of mesophilic cellulases at optimal reaction conditions (50°C and pH 4.8) results in significant activity losses (*2*). In contrast, thermophilic cellulases, such as those from *Thermoascus aurantiacus*, remain stable and active at temperatures exceeding 50°C (*16*).

In reality, cellulase is not a single enzyme but a multicomponent entity with variable composition depending on its source. In general, cellulases secreted by fungi are believed to consist of three major classes of components: (1) 1,4-β-D-glucan glucanohydrolases (**endoglucanases**; EC 3.2.1.4); (2) 1,4-β-D-glucan cellobiohydrolases and 1,4-β-D-glucan glucohydrolases (**exoglucanases**; EC 3.2.1.91 and 3.2.1.74, respectively); and (3) β-D-glucoside glucohydrolases (**β-glucosidases**; EC 3.2.1.21). Often, cellulases also encompass other glucanase and hemicellulase activities. The reported molecular weights of cellulase components range from 5,600 to 89,000 (*16*). Endoglucanases and exoglucanases adsorb to the surface of cellulose particles in order to initiate hydrolysis, whereas β-glucosidases are soluble enzymes. Even within the same organism, multiple forms of the cellulase components with differing substrate specificities have been identified. For example, the *T. reesei* cellulase consists of two endoglucanases (EG I and II) and two exoglucanases (CBH I and II), with CBH I being the most abundant component (*17*). Among the numerous fungal cellulases, those synthesized by *Trichoderma* have the advantage of possessing all three cellulase components, being more resistant to chemical inhibitors, and exhibiting short-term stability at 50°C. Unfortunately, they have rather low specific activity and β-glucosidase levels, are sensitive to product inhibition, and are slowly inactivated at their optimal temperature of 50°C. At optimal reaction conditions (50°C and pH 4.8), endoglucanases are much more stable and resistant to chemicals than exoglucanases (*18*).

Several bacteria have also been identified that produce extracellular cellulases, such as *Acidothermus cellulolyticus, Micromonospora bispora, Bacillus* sp.,

Cytophaga sp., *Streptomyces flavogriseus*, *Thermomonospora fusca* and *curvata*, *Cellulomonas uda*, *Clostridium stercorarium* and *thermocellum*, *Acetivibrio cellulolyticus*, and *Ruminococcus albus*, with the majority being aerobic (*19,20*). Bacterial cellulases have a different composition than their fungal counterparts. Cellulolytic bacteria produce endoglucanases and either cellobiose phosphorylase, which catalyzes the conversion of cellobiose to glucose and glucose-1-phosphate, or β-glucosidase, or both (*21*). Based on the ability of the thermophilic bacterium *C. thermocellum* to produce cellulase and ferment sugars to ethanol, the direct microbial conversion (DMC) scheme has been proposed as an alternative to SSF (*22*). However, problems of low ethanol yield and high by-product formation (*23*) have shadowed the potential of DMC.

Enzyme Synthesis. Cellulase is an inducible enzyme complex with cellulose, cellobiose and other oligosaccharides, lactose, sophorose, and other readily metabolizable carbohydrates acting as apparent inducers for virtually all cellulase-synthesizing organisms studied to date (*24,25*). It is believed that in *Trichoderma* conidial cellobiohydrolases initiate the degradation of cellulose to cellobiose, which is taken up by the fungus and then induces enzyme synthesis and secretion (*26*). However, the actual inducing molecule remains unknown, possibly a β-glycoside. Interestingly, some inducers like sophorose are required at extremely low concentrations, whereas others are needed at high levels. Furthermore, the required induction time for some, like sophorose, is only about 2 h, in contrast to other inducers that need to be present in the medium for about 30 h before cellulase synthesis occurs (*2*). Some bacterial cellulases are cell-associated or form multiple-unit aggregates termed "cellulosomes" (*27*).

Cellulase expression is subject to catabolite repression by glucose at concentrations higher than approximately 0.1 g/L (*2*). The repression is believed to occur at the pretranslational level (*28*). When *T. reesei* was cultivated in 5 g/L glucose, cellulase was synthesized only after glucose was almost completely consumed (*29*). In the case of the thermophilic bacterium *A. cellulolyticus*, repression caused by readily metabolizable sugars, such as glucose, is relieved by addition of exogenous cAMP (*30*). Derepressed mutants of *T. reesei* have been developed and cultivated in both batch and continuous modes (*31*). Catabolite repression has been also observed with other soluble sugars, such as glycerol (*32,33*). Moreover, these researchers report differential effects of the repressor on the expression of the various cellulase components. The presence of such multiple endoglucanase and exoglucanase forms may be due to several factors, such as the existence of multiple genes; post-translational and proteolytic modifications of the gene products; variable cellulase composition; the presence of other proteins, glycoproteins, and polysaccharides in the cellulase complex; and variable glycosylation of the cellulase components (*1*). Studies of cellulase synthesis in *Trichoderma spp.* have shown that the induction, expression, and secretion of cellulase are closely associated processes (*34*).

In general, the endoglucanase and exoglucanase activities of fungal cellulases are extracellular, whereas β-glucosidase activity seems to be cell-associated (*1*). As a result, cell-free broths of fungi have low β-glucosidase activity. In order to address this deficiency, hypersecretive mutants have been isolated. Such a mutant of *P. funicusolum*, when cultivated on 3% rice bran at pH 4.5, yielded β-glucosidase activity as high as 30,000-36,000 IU/L (*35*). In some bacteria the enzyme is cell-bound (*Cytophaga* sp.), while in others it is extracellular (*Cellvibrio vulgaris*) or in both forms (*Cellvibrio fulvus*). In some cases the localization of cellulase is reportedly affected by the growth conditions and age of the culture (*36,37*).

Mechanism of Enzymatic Action. Recent biochemical and molecular studies of fungal cellulases indicate the existence of three main domains in their structure. These domains, a cellulose binding site, an active site, and a glycosylated hinge connecting these two sites, seem to play a crucial role in the expression and catalytic activity of cellulase (*38*). Furthermore, the three components of fungal cellulases (endoglucanase, exoglucanase, and β-glucosidase) are believed to function synergistically. A generally accepted enzymatic mechanism assumes that endoglucanases initiate the attack on cellulose by randomly cleaving β-1,4-glucosidic bonds and thus creating shorter-length cellulose chains; neighboring exoglucanases start degrading these chains at the non-reducing termini, thus generating cellobiose and glucose residues; finally, β-glucosidases cleave cellobiose to form glucose units (*1*). However, as cellulose hydrolysis proceeds, its rate decreases and eventually ceases well before all of the substrate is hydrolyzed (*39*).

Based on observations that amorphous cellulose is more readily hydrolyzed by cellulase than crystalline, it has been postulated that as hydrolysis progresses, the cellulosic biomass becomes enriched in crystalline cellulose, and hence the hydrolysis rate diminishes (*5*). Other factors that may also be contributing to the decrease in the hydrolysis rate include enzyme adsorptive loss to lignin, thermal, mechanical, and chemical enzyme deactivation, and enzyme inhibition by the hydrolysis products cellobiose and glucose. Exo- and endoglucanases are primarily inhibited by cellobiose, whereas glucose inhibits mainly β-glucosidase (*39*). The SSF process circumvents the inhibition problems to a significant extent by allowing continuous removal of glucose by the fermentative microorganism. Compared with their fungal counterparts, bacterial cellulases are in general more resistant to product inhibition by about one order of magnitude (*20*).

Although useful, the proposed enzymatic mechanism (*1*) does not account for the various specificities of the multiple forms of endoglucanase and exoglucanase and their interactions. The case of *T. reesei* cellulase illustrates the shortcomings of the model: optimal synergism between CBH II and EG II or EG I has been observed at a mass ratio of 95:1, unlike a 1:1 optimal ratio reported for CBH I and EG I, as well as CBH I and EG II (*17*). Surprisingly, even cooperation between CBH I and CBH II has been described (*40*). It should be noted, however, that the degree of cooperation among the components varied depending on the substrate used for the study.

The rate and extent of cellulose hydrolysis catalyzed by cellulase can be considered a function of primarily the concentration, composition, and quality of the substrate and cellulase enzyme (*5*). The activity of cellulase varies widely among the various cellulosic substrates. The crystallinity (*41*) and surface area (*42,43*) seem to be the two major characteristics of cellulose that determine its degradation by cellulase. Dyed substrates have been used for enzyme detection and activity measurement with increased sensitivity (*44,45*). Carboxymethyl cellulose (CMC) is used as substrate for endoglucanases exclusively, whereas the β-glucosidase activity of cellulase is determined by hydrolysis of p-nitrophenyl-β-glucopyranoside. The overall activity of cellulase is measured using filter paper cellulose as substrate. It is expressed in international units (IU), where one IU of cellulase activity on filter paper is the amount of enzyme that yields 1 μmol of glucose per minute (*46*). This activity unit is used throughout this chapter to describe volumetric productivity in IU/L/h, yield on substrate in IU/g, and concentration in IU/L for cellulases.

Enzyme Production

Table 1 summarizes representative data from various cellulase production processes, ranging from shake-flask experiments to 15,000-L fermentations. The operational conditions of each process and the corresponding cellulase productivities and concentrations are reported to offer a concise overview of the current status of the cellulase production technology. In addition, using the substrate and cellulase concentration data, the yield of enzyme with respect to the carbon source or substrate (in the case of lignocellulosic biomass with unspecified composition) has been calculated and included in the table.

Substrate Selection and Media Optimization. The cost of substrate plays a crucial role in the economics of enzyme production. It is therefore imperative to identify inexpensive substrates that lead to high levels of cellulase productivity and yield. Cellulase production and activity are influenced by several factors, such as the carbon, nitrogen, and phosphorus sources, the ratio of carbon to nitrogen provided, trace elements, pH, and aeration rate (*29,47-49*). The identity of the carbon source may affect not only the production of cellulase, but also its composition. Cellulase from *T. reesei* grown on cellulose or lactose has a better balanced composition with regard to its overall activity than on other carbon sources, such as sophorose (*34*). Several substrates have been evaluated as carbon sources for the production of cellulase, both insoluble (cellulose, hemicellulose, agricultural residues, whey, corn stover, wheat straw, steam-exploded wood, rice bran, waste newsprint, potato pulp) and soluble (lactose, cellobiose, xylose) ones. Soluble carbohydrates promote faster cell growth and therefore decrease the fermentation time needed for cellulase production. For instance, lactose has been reported to enhance cellulase productivity and yield, when compared with cellulose (*50*), presumably because of the higher cell growth rate achieved in the presence of this soluble substrate. Some saccharides, such as sophorose and sorbose, reportedly act as inducers of cellulase synthesis. As an example, when *T. reesei* was cultivated on 10 g/L xylose, cellulase production was half that observed with 10 g/L of Solka Floc cellulose (*51*). Supplementation of the xylose with 3 g/L sorbose doubled enzyme production without affecting cell growth. Similarly, sorbose induced cellulase production from partially enzymatically hydrolyzed newspaper (*52*).

However, soluble sugars are expensive, raising the cost of the enzyme. For this reason, less expensive cellulosic substrates have been examined. Among them,

Table 1. Representative List of Data from Cellulase Production Fermentations Carried Out by Cellulase-Secreting Microorganisms under Various Operational Modes and Conditions (N/A denotes unavailable data)[1].

Organism	Mode[2]	Conditions[3]	Time[4] (days)	Media[5]	Substrate[6] (carbon source)	Productivity[7] (IU/L/h)	Concentration[8] (IU/L)	Yield[9] (IU/g)	Reference
T. reesei MCG-77	Continuous 2-stage; 15 L	1:T=32, pH=4.5 2:T=28, pH=3.5	0.027	Defined	Lactose (50-100 g/L)	90	N/A	N/A	*(79)*
T. reesei QM 9414	Batch 150 L	T=28, pH=3.2	8	Peptone	Cellulose (60 g/L)	57	11,000	183	*(71)*
A. cellulolyticus	Batch 1-2.5 L	T=55, pH=5.2	3.3	Yeast extract	Cellobiose (5 g/L) & Cellulose (15 g/L)	2.5	69	3.5	*(78)*
T. reesei QM 9414	Batch (flasks)	T=25, pH=5.0	21	Defined	Cellulose (10 g/L) Xylose (10 g/L)	N/A	2,000	100	*(51)*
T. reesei Rut C30	Fed-batch 5 L	T=28, pH=4.8	14	Corn steep liquor	Cellulose (230 g/L) Xylose (100 g/L)	45.4	12,500	38	*(67)*
T. reesei L27	Batch 5 L	T=28, pH=4.8	8	Corn steep liquor	Cellulose (50 g/L)	N/A	6,200	124	*(81)*
T. pseudokoningii	Batch (flasks)	T=27, pH=5.0	10	Defined	Cellulose (10 g/L)	N/A	110	11	*(88)*
S. flavogriseus	Batch (flasks)	T=29, pH=6.8	3	Yeast extract Peptone	Cellulose (10 g/L)	N/A	480	48	*(32)*
T. reesei Rut C30	Batch (flasks)	T=27, pH=5.5	10	Peptone	Aspen Wood (60 g/L) Cellulose (8 g/L)	N/A	3,900	57	*(53)*

Continued on next page

Table 1 continued. Representative List of Data from Cellulase Production Fermentations Carried Out by Cellulase-Secreting Microorganisms under Various Operational Modes and Conditions (N/A denotes unavailable data)[1].

Organism	Mode[2]	Conditions[3]	Time[4] (days)	Media[5]	Substrate[6] (carbon source)	Productivity[7] (IU/L/h)	Concentration[8] (IU/L)	Yield[9] (IU/g)	Reference
P. pinophilum	Batch 16 L	T=35, pH=3.5-5.0 (no pH control)	10	Peptone	Cellulose (60 g/L) or Straw (60 g/L)	N/A	6,000	100	*(59)*
P. pinophilum	Batch 16 L	T=35, pH=3.5-6.0 (no pH control)	3	Corn steep liquor	Cellulose (30 g/L) Straw (30 g/L)	137	10,000	167	*(49)*
T. reesei MCG-77	Batch 5 L	T=30, pH=5.0	5	Soya meal (20 g/L)	Straw (steamed)	31	1,900	95	*(77)*
T. reesei NG14	Batch 10 L	T=28, pH>3.0	14	N/A (60 g/L)	Cellulose	45	15,000	250	*(2)*
T. reesei Rut C30	Batch 14 L	T=25, pH>5.0	8	Peptone (50 g/L)	Cellulose	N/A	14,400	288	*(47)*
T. reesei Rut C30	Batch 14 L	T=28, pH=5.0	7	Defined (20 g/L)	Cellulose	29	4,200	210	*(54)*
T. reesei Rut C30	Continuous 1-stage; 14 L	"	0.046	"	"	97	2,100	105	*(54)*
T. reesei Rut C30	Continuous 2-stage; 14 L	"	0.017	"	"	58	3,300	165	*(54)*
T. reesei Rut C30	Fed-batch 14 L	"	8.3	"	Cellulose (100 g/L)	247	22,400	224	*(54)*
T. reesei Rut C30	Fed-batch 20 L	T=25, pH>4.0	11.4	Defined (232 g/L)	Pulp cellulose	427	57,000	246	*(57)*

Continued on next page

Table 1 continued. Representative List of Data from Cellulase Production Fermentations Carried Out by Cellulase-Secreting Microorganisms under Various Operational Modes and Conditions (N/A denotes unavailable data)[1].

Organism	Mode[2]	Conditions[3]	Time[4] (days)	Media[5]	Substrate[6] (carbon source)	Productivity[7] (IU/L/h)	Concentration[8] (IU/L)	Yield[9] (IU/g)	Reference
T. reesei Rut C30	Batch 5 L	T=28, pH=5.0	10	Corn steep liquor	aspen wood (50 g/L)	50	4,500	90	*(89)*
T. thermo-monospora	Continuous 1-stage; 70 L	T=55, pH=7.2	0.125	Yeast extract	cellulose (10g/L) or cellobiose (5 g/L)	39	310	31	*(90)*
T. reesei CL-847	Batch 3,000 L	T=25, pH=5.0	6	Yeast extract	lactose (60 g/L) & cellulose (5 g/L)	N/A	10,500	162	*(76)*
T. reesei L27	Batch	N/A	8	N/A	cellulose (80 g/L)	94	18,000	225	*(69)*
T. reesei SVG-17	Fed-batch 15,000 L	T=30, pH=4.5-5.0	5.4	N/A	wheat straw (60 g/L)	61	6,500	N/A	*(3)*

[1]Cellulase activity is given as international units (IU); one IU of cellulase activity is that amount of enzyme that yields 1 µmol of glucose per minute *(46)*.
[2]Temperature (T) is expressed in °C.
[3]Refers to the operational mode of the cultivation (batch, fed-batch, continuous).
[4]Duration of run; for continuous operations, the dilution rate is given (h^{-1}).
[5]"Defined media" refers to media consisting of defined ingredients; otherwise, the rich (undefined) ingredient is given.
[6]For fed-batch operations, the total amount (per unit working volume) of carbon source provided is given. For continuous operations, the concentration of the carbon source in the feed is given.
[7]Cellulase productivity (volumetric production rate).
[8]Cellulase concentration.
[9]Yield of cellulase (IU) per unit mass (g) of substrate consumed. The yield has been calculated by dividing the cellulase concentration by the substrate concentration. It is an approximate value because it has been assumed in the calculation that all the provided substrate has been completely consumed.

steam-exploded aspen wood is an attractive choice because of its good digestibility. Unfortunately, when aspen wood was used as the sole carbon source of *T. reesei* Rut C30, cellulase activity was low (1,600 IU/L) and the enzyme was deficient in exoglucanase and β-glucosidase (*53*). Supplementation of the wood with pure cellulose enhanced the cellulase production and composition (Table 2). In a comparative study of carbon sources, Solka Floc cellulose, Avicel cellulose, hammer-milled barley straw, ball-milled barley straw, wheat bran, and combinations were tested (*49*). Avicel cellulose yielded the highest enzyme concentration at 4,000 IU/L, a five-fold improvement over wheat bran. A mixture of corn steep liquor and inorganic salts served as the nitrogen source in that study. Enzyme yield also depends on substrate concentration. When the substrate (a cellulose and ball-milled straw mixture) concentration was raised from 15 to 60 g/L, the *P. pinophilum* cellulase concentration increased five-fold after three days of cultivation (*49*). In the same study, the pH of the cultivation medium was left uncontrolled in the 3.5-5.0 range and the temperature was set at 35°C (Table 1). An increase in cellulase concentration proportional to substrate loading has been also reported for *T. reesei* up to 50 g/L of cellulose, reaching a maximum of 8,000 IU/L at a production rate of 55 IU/L/h (*54*). Above that cellulose concentration, however, little enhancement in cellulase productivity was observed, as the viscosity of the medium increased significantly and the cell growth rate slowed.

Numerous media have been developed and evaluated for use in cellulase production. A typical medium composition is shown in Table 3. Some metallic ions (Ca^{2+}, Fe^{2+}, Co^{2+}, and Zn^{2+}) have been found necessary for cellulase synthesis, although they did not enhance cell growth; it is postulated that they affect enzyme secretion or prevent an inducer molecule from diffusing out of the cell (*29*). The presence of Tween-80 at 0.1-0.2 mL/L reportedly results in enhanced cellulase production by increasing the permeability of the cell membrane and thus improving enzyme secretion (*47*). In general, *T. reesei* grows well in defined media containing lactose, ammonium salts, and minerals (*55*), although it grows faster in rich (undefined) media with fructose or glucose as the carbon source and peptone as the nitrogen source (*2*). The use of amino acids (*2*) and yeast extract (*56*), as nitrogen sources seems to stimulate cell growth and cellulase synthesis. The maximal specific growth rate of *T. reesei* mutants cultivated in batch mode in defined media ranges from 0.1 to 0.25 h^{-1}, depending on the identity of the carbon and nitrogen sources employed (*2*).

In a recent study, the importance of media optimization was evaluated. Various nitrogen and carbon sources were examined for the production of cellulase by the fungus *P. pinophilum* (*49*). The nitrogen sources under consideration were casein hydrolyzate, peptone, yeast extract, corn steep liquor, urea, and inorganic nitrogen salts. Among them, corn steep liquor yielded the highest cellulase concentration, more than 2,000 IU/L, whereas $NaNO_3$ led to the highest cellulase specific activity, 0.78 IU/mg protein. The carbon source was 10 g/L of hammer-milled barley straw. Hence, use of the optimal nitrogen source in that study resulted in a five-fold improvement in enzyme production. However, use of rich nitrogen sources does not always result in enzyme concentration or productivity enhancement (*57*). In another report, nitrogen supplied in the form of $Ca(NO_3)_2$ yielded the highest cell mass and

Table 2. Cellulase Production by *Trichoderma reesei* Rut C30 in Peptone Medium Containing Steam-Exploded Aspen Wood and Various Amounts of Pure Solka Floc Cellulose (compiled from *53*).[1]

Cellulosic Substrate[2]	Enzymatic Activity (IU/mL)[3]			
	Cellulase	Exo	Endo	β-G
Wood (Aspen)	1.6	1.0	7.7	0.2
Wood + Cellulose (95/5)	3.1	2.2	7.9	0.3
Wood + Cellulose (90/10)	3.6	2.5	8.2	0.4
Wood + Cellulose (80/20)	3.9	2.6	8.7	0.4
Cellulose	4.1	2.7	9.2	0.5

[1]Exo, Endo, and β-G are the exoglucanase, endoglucanase, and β-glucosidase components of cellulase, respectively.
[2]The percentage of each component (on a weight basis) in the mixed substrate is given in parenthesis.
[3]Although all activities are expressed in IU/mL, the substrate of reference for each cellulase component is different (see *53*). Therefore, for meaningful comparisons, the table should be read only vertically.

Table 3. Typical Medium Composition used for Cellulase Production (compiled from *57*)[1].

Component	Concentration
Cellulose	10.0 g/L (variable)
$(NH_4)_2SO_4$	1.4 g/L
KH_2PO_4	2.0 g/L
$MgSO_4.7H_2O$	0.3 g/L
$CaCl_2.2H_2O$	0.4 g/L
$FeSO_4.7H_2O$	9.2 mg/L
$MnSO_4.H_2O$	2.0 mg/L
$ZnSO_4.7H_2O$	2.7 mg/L
$CoCl_2.6H_2O$	2.0 mg/L
Urea (optional)	0.3 g/L
Tween-80 (optional)	0.1-0.2 mL/L

[1]Undefined media usually include 1 g/L of peptone, yeast extract, or corn steep liquor.

cellulase concentrations for the fungus *Phoma herbarum* cultivated on 10 g/L cellulase (*48*). Finally, the use of urea has also been suggested (*47*) because it increases cellulase productivity, thus shortening the duration of the cellulase fermentation (Table 3).

The effect of lignin on *T. reesei* cellulase production and activity has also been studied (*58*). Although lignin and some of its derivatives (vanillin, protocatechuic acid, and verulic acid) had no effect on enzyme activity, in general they negatively influenced the growth of *T. reesei* and the production of cellulase components. Addition of 0.3 g/L conifer lignin to the growth medium resulted in a two-fold decrease in cell mass, eight-fold decrease in exoglucanase activity, and five-fold decrease in endoglucanase and β-glucosidase activity. Supplementation of the growth medium with 0.5 g/L of lignin led to complete inhibition of cellulase synthesis. This potential effect of lignin on cellulase production should be investigated because it is unknown whether the lignin present in lignocellulosic biomass will have a similar effect on cellulase production as purified lignin.

Fermentation Conditions for Cellulase Production. As shown in Table 1, the optimal temperature for fungal cellulase production is around 28°C, since these fungi are mesophilic organisms. In contrast, thermophilic bacteria prefer temperatures around 55°C. More variability is found in the optimal pH value, which ranges between 3.2 and 6.8. In a study of the pH effect on cellulase production by *T. reesei* Rut C30, although a pH of 5.0 enhanced overall cellulase synthesis, the productivity of a particular cellulase component, β-glucosidase, was maximal at 6.0 (*47*). It seems that by manipulating the temperature and pH conditions of the fermentation, the composition of cellulase can be influenced. As noted earlier, in the case of *P. pinophilum*, maximal cellulase production was obtained when the pH was allowed to fluctuate naturally between 3.5 and 5.0 (*59*).

Oxygen is another element that plays an important role in cellulase production. Although the oxygen uptake rate during the stage of cell growth is four-fold higher than during enzyme synthesis, the provision of adequate oxygen during this latter stage is imperative for enzyme synthesis (2). A minimal dissolved oxygen concentration of 20% of the saturation level reportedly satisfies the requirements of cellulase synthesis both in small-scale (*47,60*) and large-scale (*3*) operations. However, the formation of foam during cellulase synthesis is associated with oxygen supplementation to the fermenter. To circumvent the problem, various antifoam agents have been employed in the fermentation (*47,56*).

As noted earlier, *T. reesei* cellulases have low β-glucosidase activity, which limits their ability to convert cellulose to glucose. In order to remedy this disadvantage, cocultivation of *T. reesei* has been suggested with organisms producing high levels of β-glucosidase, such as *Aspergillus niger* (*61*), *Aspergillus phoenicis* (*62*), *Schizophyllum commune* (*63*), and even immobilized *Alcaligenes faecalis* (*64*). The β-glucosidase synthesized by these organisms complemented the high endoglucanase and exoglucanase activities of *T. reesei* cellulases. Similarly, the use of mixed cultures has successfully complemented the high cellulase production levels of *Trichoderma harzianum* with the high β-glucosidase activity of *Aspergillus ustus* (*65*). As an alternative to β-glucosidase supplementation, cellobiose-fermenting organisms, such as *Hansenula spp.* (*66*), *Streptomyces spp.* (*67*), and *Brettanomyces custersii* (*68*), could be used in monocultures or mixed cultures with *T. reesei.*

Construction of improved cellulase-producing organisms. In an effort to reduce the production cost of cellulase, hypersynthesizing mutant strains of cellulolytic microorganisms have been created using ultraviolet (UV) light and/or mutagenic agents. As a result, enzyme productivities as high as 427 IU/L/h have been attained during fed-batch fermentations of *T. reesei* (*57*). Enzyme yields reaching 288 IU/g cellulose have been reported for a batch cultivation of the same organism (Table 1). A mutant of *T. reesei* QM 9414, morphologically different from the parent strain (higher branching degree), produced two-fold more cellulase than the parent strain (*44*). The cellulase of another mutant had a three-fold higher β-glucosidase specific activity than the parent strain. Moreover, mutant strains of *T. reesei* have been isolated that produce cellulase at two- to six-fold higher rates without any effect on enzyme composition (30% endoglucanase, 70% exoglucanase, and less than 1% β-glucosidase) (*69,70*). Mutagenesis of the *T. reesei* strain QM 9414 with UV light and nitrosoguanidine yielded a hyperproducing and catabolite repression-resistant mutant (*71*), which produced 90% more enzyme than the parent strain. The method of mutation and selection has been also used to improve strains of cellulolytic bacteria, such as *T. fusca* (*56*). Furthermore, recombinant DNA techniques have been employed to develop cellulase overproducing systems. Overexpression of *Bacillus subtilis* endoglucanase from a cloned strong, native promoter reached 60% of the total amount of secreted protein (*72*). *Escherichia coli* is a popular host for foreign proteins, but in the case of cellulase components the inability of the bacterium to glycosylate the gene products may result in the formation of less active or inactive proteins. Glycosylation of cellulase, although not fully comprehended, seems to play an important role in enzyme secretion and stability (*26*).

Cellulases, in general, have low specific activities, typically 100-fold less than amylases. Thus, research is being conducted to increase activity via strain mutation to produce more active enzymes, site-specific mutagenesis to improve the catalytic turnover rate of the active site, and isolation of novel microorganisms (*1*). As an example, a cellulase isolated from the anaerobic fungus *Neocallimastix frontalis* was several times more active than the *T. reesei* Rut-C 30 cellulase (*73*). Unfortunately, these optimization methods usually do not lead to the same extent of improvement for the endoglucanase and exoglucanase activities of cellulase. More efficient mutant selection methods and improved enzyme characterization techniques are needed to help optimize cellulase production. The development of monoclonal antibodies specific for each cellulase component is expected to offer insight into the synergism between endoglucanases and exoglucanases and elucidate the CBH to EG ratio that leads to optimal cellulose hydrolysis rates.

Thermostable cellulases are of particular interest because of their longer activity, stability, and resistance to chemical, mechanical, and thermal deactivation. The use of thermophilic organisms, primarily bacteria, offers several advantages: higher growth and metabolic rates at elevated temperatures, increased mass transfer rates, low cooling requirements for the fermentation, and lower contamination risk.

Disadvantages include more expensive construction materials, water evaporation, and low oxygen solubility in the fermentation media at high temperatures. Several thermophilic bacteria (50-60°C) secrete thermostable cellulases resistant to proteolysis and chemical or mechanical denaturation. As an example, *Acidothermus cellulolyticus* has temperature and pH optima of 55°C and 5.2, respectively (*74*). When grown on cellulose or cellobiose, *A. cellulolyticus* produces enzyme at the late exponential and early stationary phase of its growth in a mode similar to that of *T. reesei*, with D-cellobiose and D-xylose acting as inducers (*75*). According to the same study, cellulase synthesis is apparently inducible by soluble derivatives of cellulose, like cellobiose and other low-molecular-weight carbohydrates, such as lactose, sophorose, and xylose.

The main drawback in the use of bacteria is their low cellulase productivities and yields compared with those of *T. reesei* (Table 4). In an effort to remedy this disadvantage, various compounds in addition to cellulose have been screened for their ability to induce bacterial cellulase synthesis. As noted earlier, the thermophile *A. cellulolyticus* is a potential source of thermostable cellulases, especially when cultivated in the presence of cellulose. A review of the international literature suggests that the bacteria *Cellulomonas fimi*, *Thermomonospora fusca*, *Pseudomonas fluorescens* and *cellulosa*, *Acetovibrio cellulolyticus*, and *Acidothermus cellulolyticus* synthesize highly active cellulases that are highly resistant to end-product inhibition and therefore deserve closer attention.

Engineering Aspects of Cellulase Production

During the batch cultivation of *T. reesei* for production of cellulase (*2,56*), two phases can be identified: first, cell growth occurs, followed by enzyme synthesis and secretion, which increase significantly as cell growth enters the stationary phase. The two phases of cell growth and cellulase synthesis exhibit different temperature and pH optima. Growth is favored by temperatures in the range of 32°-35°C, whereas enzyme production occurs optimally at 25°-28°C (*2*). Similarly, cell growth is optimal at a pH of approximately 4.0, while cellulase synthesis needs a pH of 3.0. Other researchers have reported different optima (Table 1). The different requirements of the two stages of cellulase production can serve as a guide for developing a strategy aimed at the optimization of enzyme productivity. In batch or fed-batch mode, the optimal profiles of pH and temperature should be applied to the system, whereas in continuous operations a multistage process would be desirable, where the first stages will bring cell concentration to high levels and the later ones will allow only cellulase synthesis to take place.

Batch Cellulase Production. Batch is the most common operational mode for cellulase production, mainly because it is the easiest to start up and control. Cellulase concentrations as high as 15,000 IU/L and enzyme yields of 288 IU/g cellulose have been reported (Table 1). The highest enzyme activity was obtained with *T. reesei* cultivated in the presence of 60 g/L roll-milled cotton cellulose (*2*). A pilot plant study of *T. reesei* CL-847 cellulase production was carried out in a 3,000-L fermenter using 60 g/L of lactose as the carbon source and 1 g/L of yeast

Table 4. Cellulase Production by Thermophilic Bacteria and Mesophilic Fungi Grown on Cellulosic Substrates (compiled from *15* and *78*).

Organism	Substrate	Enzyme Concentration (IU/L)	Productivity (IU/L/h)	Fermentation Time (h)
Thermomonospora fusca	Avicel	150	5.1	29
Thermomonospora curvata	Cellulose	100	1.4	72
Clostridium thermocellum	Solka Floc	140	1.9	72
Acidothermus cellulolyticus	Solka Floc	105	1.5	70
Thielavia terrestris	Solka Floc	110	3.4	32
Trichoderma reesei QM 9414	Solka Floc	2,600	21.7	120
Trichoderma reesei QM 6a	Solka Floc	5,000	15.0	333
Trichoderma reesei L27	Avicel	18,000	93.7	192

extract as the nitrogen source (*76*). Cellulose was also provided at a concentration of 5 g/L to induce cellulase production. After 8 days of cultivation, the cellulase concentration reached 10,500 IU/L with an overall enzyme yield of 162 IU/g carbon source.

The feasibility of cellulase production using inexpensive lignocellulosic biomass as the carbon source was investigated in a batch 5-L fermenter with *T. reesei* MCG-77 (*77*). Several substrates were examined, ranging from municipal waste to pretreated wheat straw. The results are shown in Table 5. Bleached sulfite pulp was used as a reference substrate because it is a good although relatively more expensive material. Steam treatment improved the yield of cellulase from wheat straw almost two-fold. Furthermore, by increasing the straw concentration from 20 to 50 g/L, the cellulase and β-glucosidase activities doubled to 3,800 and 1,950 IU/L, respectively. Controlling the pH at 6.0 was beneficial. Interestingly, considerable production was exhibited by already enzymatically hydrolyzed wheat straw; the enzyme productivity was 31 IU/L/h and the concentration 2,400 IU/L. Similar results were obtained with NaOH-treated municipal waste and steam-treated straw, which yielded cellulase at rates of 24 and 31 IU/L/h, respectively. Wastepaper yielded half the enzyme activity of municipal waste. At the pilot plant level, *T. reesei* SVG-17 produced 2,200 IU/L from 50 g/L of thermally pretreated wheat straw in a 15,000-L fermenter (*3*). The data from these studies indicate that inexpensive lignocellulosic biomass, such as municipal waste and fresh or hydrolyzed cellulosic biomass, can support production of cellulase at considerable rates.

The batch production of cellulase by *A. cellulolyticus* has also been investigated (*78*). Among various carbon sources tested, this thermophile synthesized cellulase only when it grew on cellulose, xylose, and cellobiose, with specific growth rates of 0.10, 0.14, and 0.18 h^{-1}, respectively. When cultivated in yeast extract supplemented with salts, minerals, and 2.5-10 g/L of cellobiose as the carbon source at 55°C and pH 5.2, *A. cellulolyticus* produced cellulase at a rate of 0.92-2.49 IU/L/h yielding a final concentration of 21-69 IU/L. Both the productivity and the concentration of cellulase increased proportionally to the cellobiose concentration. Supplementation of the media with Avicel cellulose or Solka Floc cellulose increased the enzyme concentration further up to a maximum of 105 IU/L. Still, this performance lags considerably behind the fungal systems.

Although batch fermentations lead to high cellulase concentrations and yields, their productivities usually suffer. In a *T. reesei* study, the productivity did not exceed 45 IU/L/h when cultivated on cotton cellulose (*2*), a relatively small value compared to the productivities of fed-batch and continuous operations (Table 1). Another fungus, *P. pinophilum*, yielded cellulase at a rate of 137 IU/L/h, but the enzyme concentration did not exceed 10,000 IU/L (*49*).

Fed-Batch Cellulase Production. Another popular mode of cellulase synthesis is fed-batch operation. Its main advantages, when compared with batch or continuous operations, are higher cellulase productivity, concentration, and yield per unit mass of utilized substrate. These are primarily due to the more efficient use of the substrate achieved in a fed-batch system by manipulating the feeding rates of media

Table 5. Cellulase Production on Various Lignocellulosic Substrates in a 5-L Batch Fermentation (compiled from *77*).

Cellulosic Substrate	**Substrate Concentration** (g/L)	**Enzyme Concentration** (IU/ml)	**Productivity** (IU/L/h)	**Enzyme Yield** (IU/g)[1]
Spruce sulfite pulp	20	3.7	61	185
Wheat straw (untreated)	20	1.1	13	79
Wheat straw (steamed)	20	1.9	31	151
Wheat straw (NaOH treated)	20	2.0	25	133
Wheat straw (8-day stored)	20	1.6	27	97
Wheat straw (hydrolyzed)	30	2.4	31	147
Printed newspaper	30	0.7	19	48
Mixed wastepaper	35	1.1	17	64
Municipal waste	30	2.5	24	130

[1]Enzyme yield is expressed in IU cellulase/gram carbohydrate (cellulose and hemicellulose) content of the substrate.

and substrate. Furthermore, the long lag periods of several hours usually observed in batch fermentations at high biomass concentrations can be significantly reduced by slowly feeding the substrate to the fermenter. A three-stage inoculum preparation starting at 800 mL and ending at 220 L was applied in a 15,000-L cellulase production unit to optimize enzyme production at 2,200 IU/L (*3*). In another study, using a mixed substrate feed of cellulose (20 g/L) and xylose (30 g/L) and feeding rates of 15 g/L/day and 5 g/L/day, respectively, an enzyme concentration of 12,500 IU/L at a productivity of 45.4 IU/L/h was attained (*74*). In that work, it was observed that substitution of up to 25% (w/w) of cellulose by xylose improved enzyme production. Xylose, a soluble carbohydrate, enhanced the slow growth of the fungus on cellulose. In a biomass conversion plant, xylose will be readily available from the hydrolysis of hemicellulose (Figure 1), which occurs during the biomass pretreatment step. Furthermore, the amount of cellulose spared from consumption in the cellulase fermenter will be available for ethanol production in the SSF process.

In a comparative study of a batch, fed-batch, one-stage continuous, and two-stage continuous production of cellulase by *T. reesei* Rut C30 using Solka Floc cellulose, the fed-batch operation was reportedly superior to the other modes with regard to enzyme concentration, yield, and productivity (*54*). Cellulose was fed whenever a decrease in cell growth rate was detected. Cellulase concentration was found to be proportional to the substrate feeding rate at the expense, however, of enzyme productivity. After 8.3 days of cultivation in 100 g/L Solka Floc cellulose, the cellulase concentration reached 22,400 IU/L at an average production rate of 247 IU/L/h. During another fed-batch study in a 14-L fermenter with the same organism, but at 232 g/L of ball-milled cellulose, the productivity reached 427 IU/L/h and the concentration 57,000 IU/L (*57*). These values are the highest ones published to date. In that study, the temperature was set at 27°C during the first 42 h of operation in order to promote cell growth and then switched to 25°C to enhance cellulase production. At the pilot plant scale, fed-batch cultivation of *T. reesei* SVG-17 on 60 g/L of thermally pretreated wheat straw produced 6,500 IU/L, a significant improvement on the 2,200 IU/L obtained in batch mode (*3*). Because of its high productivity, the fed-batch mode is also used in commercial operations with fermenter sizes in the 200,000-L range.

Continuous Cellulase Production. Depending on the kinetics of the biochemical process of interest, continuous operations may attain higher productivity levels than batch ones. Using a two-stage continuous system, researchers investigated the production of cellulase by *T. reesei* MCG-77 cultivated in a defined medium of lactose (50-100 g/L), ammonium sulphate, and mineral salts in 15-L fermenters (*79*). Different sets of conditions were used for the two stages of cellulase production: (1) 32°C, pH 4.5, 600-800 rpm agitation, and 0.3-0.5 vvm aeration for cell growth; and (2) 28°C, pH 3.5, 400-600 rpm agitation, and 0.2-0.3 vvm aeration for cellulase synthesis. A silicone agent was used for foam control. The flow rates of the carbon, nitrogen, and oxygen sources were used as process control variables to maximize enzyme productivity, based on the consideration that the three nutrients were utilized for biomass formation, maintenance requirements, and cellulase biosynthesis. The specific growth rate maintained during cell growth was 0.1 h^{-1}. In the second stage, maximal cellulase productivity was obtained at a dilution rate of about 0.03 h^{-1}. Enzyme productivity decreased at higher dilution rates. At steady state, the cell mass concentration was maintained around 10 g/L, bringing the enzyme productivity of the continuous process to 90 IU/L/h (8 IU/g cell mass/h).

Continuous cellulase production has also been studied using *T. reesei* Rut C30 in one- and two-stage operations (*54*). Unlike the previous study (*79*), this time both stages were operated at the same conditions (28°C and pH 5.0), but the working volume of the first stage was half the volume of the second one. At a cellulose feed concentration of 20 g/L, the one-stage system yielded 2,100 IU/L of cellulase at a rate of 97 IU/L/h and a yield of 105 IU/g, whereas the two-stage operation resulted in a higher cellulase concentration (3,300 IU/L) and enzyme yield (165 IU/g), but significantly lower productivity (58 IU/L/h). In the one-stage process, increasing the dilution rate beyond 0.046 h^{-1} led to lower enzyme productivities. In the two-stage system, optimal productivity was achieved at a dilution rate of 0.017 h^{-1}; productivity decreased at higher dilution rates. In still another study, continuous one-stage cellulase production by thermophilic *Thermomonospora* sp. exhibited its highest productivity (39 IU/L/h) at a dilution rate of 0.125 h^{-1} (*56*). Cell mass concentration also peaked at the same dilution rate value.

It should be mentioned that an alternative means of improving cellulase productivity is through cell recycle. For the one-stage and two-stage continuous systems described above, cell recycle resulted in high productivities at high dilution rates, 0.064 and 0.027, respectively (*54*). Operating at high dilution rates decreases the residence time of the process and its cost.

Downstream Processing in Cellulase Production. Today, commercial cellulase manufacturing is conducted throughout the world primarily in suspension cultures, while solid state fermentations are gradually being phased out (*3,80*). In large-scale batch and fed-batch operations, the fermentation is terminated shortly after maximal productivity has been reached, and the broth is rapidly chilled to 5°-10°C to preserve cellulase activity (*3*). The type of downstream processing applied to the cellulase stream depends on the intended application for the enzyme. For immediate or relatively near-term use of cellulase, as is the case for the SSF process, it is advantageous to introduce the whole broth in the SSF units, as has been demonstrated for *T. reesei* L27 (*81*). That study showed that whole broth exhibits higher cellulase activity than the culture filtrate, thus leading to higher ethanol yields in the SSF process. The improvement in enzyme activity is due to the additional cellulase present in the cell mass of the mycelia. In addition to providing higher enzyme activity, the use of whole broth eliminates the need for costly separation of solids from the effluent of the cellulase production fermenter. It should be noted that the mycelia cannot survive under the SSF conditions (38°-40°C, pH 5.0) and thus do not interfere with the SSF kinetics. The use of whole broth is also useful for simple cellulose hydrolysis to soluble sugars (*80*). If necessary, a filtration step can be introduced to separate the cell mass from the enzyme-rich liquid. Vacuum drum filters coated with aluminosilicates, as filter aid, allow 80% of the enzyme activity to pass into the filtrate (*3*).

For longer term enzyme use, cellulase is subjected to purification. Hence, after filtration, the fermentation effluent is concentrated by ultrafiltration using membranes with a cut-off size of 10 kDa (*80*) or 20-30 kDa (*3*). Addition of lactose at a concentration of 15-20 g/L significantly improves the activity yield of these membranes, whereas small amounts of $CuSO_4$ remove foul odors (*3*). Other enzyme stabilizers, such as glycerol, are also added to commercial preparations to extend the life of the enzyme. Food-grade biocides, like sodium benzoate at 2 g/L (*3*), are introduced to prevent microbial contamination. Finally, spray drying, lyophilization, or simple precipitation are employed to obtain cellulase powder. Spray drying is carried out at 120°-150°C to preserve cellulase activity (*3,82*). Precipitation is performed by adding a solvent (isopropanol, acetone, methanol, ethanol) to the enzyme ultrafiltration concentrate at 8°-10°C, followed by drying at 50°C or lyophilization (*3,83*). In all cases the presence of lactose was found to be beneficial for enzyme recovery (*3*). Depending on the type of medium employed in cellulase production, diafiltration with salt solutions to remove medium components may be necessary prior to lyophilization (*84*).

Recommendations for Research and Development

As stated earlier, the cost of cellulase is a significant factor in the economics of the biomass-to-ethanol conversion technology and has a direct impact on the cost of the ethanol produced (Figure 3). Recent technical evaluations indicate that enzyme concentrations and productivities exceeding 20,000 IU/L and 200 IU/L/h, respectively, need to be reached in the cellulase production operation (*12*). The key to producing inexpensive (per unit of activity) and highly active cellulase lies in a combination of critical factors: improved enzyme quality, prolonged enzyme utilization, enhanced enzyme productivity and yield, and minimal cost of media and substrate.

Enhancement of Cellulase Quality. The ideal cellulase should have:

- Balanced composition of endoglucanase, exoglucanase, and β-glucosidase activities to convert cellulose into glucose efficiently and to the maximal extent; this is especially desirable for the fungal cellulases that contain low β-glucosidase levels

- High specific activity to minimize the amount of enzyme needed in the SSF process, thus alleviating the need for high enzyme productivity and reducing the size of both the cellulase and SSF fermenters

- Resistance to inhibition by hydrolysis (cellobiose, glucose) and fermentation (ethanol) products in order to retain its activity during the course of SSF and reduce the duration of the process

- Resistance to thermal, physical, and chemical deactivation to perform well under conditions that may accelerate the rate and yield of SSF

- Little affinity for lignin to minimize the adsorptive loss and inactivation of cellulase components

- High activity on various forms of lignocellulosic biomass to accommodate the use of multiple substrates for ethanol production.

In the search for better cellulases with the above features, it is essential to examine all possible sources, both bacterial and fungal. Cellulases from these microorganisms need to be screened according to strict assay procedures that remain to be established, so that the properties of all the enzymes are compared on a consistent basis. In addition to aerobes, the screening should also include anaerobic microorganisms. For instance, cellulase from the anaerobic fungus *Neocallimastix frontalis* has higher specific activity than *T. reesei* Rut C30 cellulase (*73*). A significant advantage of anaerobes is the elimination of the aeration needs during cellulase production, which greatly improves the economics of the process.

Enhancement of Cellulase Stability. Reduction of the cellulase cost can also be achieved by extending the life of the enzyme, which requires that the activity and stability of cellulase be prolonged. Then cellulase could be used repeatedly, either through recycling or immobilization in a reactor, thus decreasing the needs of SSF for enzyme. However, cell or enzyme recycle may require a costly means of separation. Alternatively, immobilization or attachment of the enzyme, enzyme components, or whole cells on an appropriate matrix may extend the activity and stability of cellulases. This benefit, however, usually comes at the expense of lower hydrolysis rates, because of external and internal diffusion limitations realized with immobilization or attachment matrices (*64*). The limitations can be overcome to some degree by reducing the particle size of the immobilized biocatalyst and increasing the agitation speed in the fermenter. It therefore needs to be carefully investigated whether the prolonged use of the enzyme, which results in lower enzyme cost, can compensate for the slower rates of enzyme synthesis and the cost of immobilization or biocatalyst separation.

An alternative to this engineering approach could be the modification of the binding and active sites of cellulase through genetic manipulations. Recombinant DNA techniques (*85*) and crystallographic studies (*11*) have started providing researchers with useful insights into the cellulase structure and its interaction with the substrate and are expected to aid in the enhancement of enzyme stability in the near future.

Enhancement of Cellulase Yield and Productivity. Enzyme yield and productivity directly affect the cost of cellulase. Improved enzyme yield reflects a more efficient utilization of the carbon source with respect to cellulase synthesis and hence reduces the cost of the substrate. Enhanced enzyme productivity diminishes the duration of the cellulase cultivation and thus decreases the number and cost of fermenters required for a certain enzyme output. Higher cellulase yields and productivities can be achieved by (1) using a hyperproducing strain, (2) using an organism that grows fast and attains high cell concentrations, and (3) improving the configuration or conditions of the fermentation. As mentioned earlier, fungi have exhibited the highest enzyme productivities to date (Table 1), and possibly even higher productivities and yields may be reached in the future. In contrast, bacterial cellulase production is low. For example, the thermophile *T. fusca*, a promising bacterial producer, synthesized in a continuous operation only 39 IU/L/h of cellulase at an enzyme yield of 31 IU/g cellulose (*56*). However, bacteria can reach high cell mass concentrations, such as 100-200 g/L, at high growth rates, when cultivated appropriately. Moreover, cellulases from thermophilic bacteria function and remain active at temperatures exceeding 50°C, where reaction rates are considerably faster than at the optimal temperature for most fungal cellulases. These features, further improved through genetic manipulations, can make bacteria an attractive source of cellulases that should be pursued. By the same token, thermophilic fungi also need to be examined.

The configuration of the fermenter, the mode of operation, and the operating conditions of the fermentation offer opportunities for improvement in the engineering design of cellulase production. The initial lag phase, common to all fermentations, can be minimized by identifying the optimal age and size of inoculum. As indicated earlier, cellulase production consists of two phases, cell growth and enzyme synthesis. The optimal pH and temperature conditions of the two phases may differ, so it will be advantageous to manipulate these two variables first to maximize cell mass production and then to boost cellulase synthesis. For batch and fed-batch operations, the main disadvantage of this option is the increase in capital cost because of the greater number of vessels needed. However, for continuous operation in a cascade of reactors, it will be advantageous to use different stages for each phase. Because each stage will operate under optimal conditions, high enzyme productivities and yields are to be expected, as experimental results have shown (*79*). However, achieving steady state in a multiple bioreactor system is not a trivial task; in one study, 75-100 h passed before steady state was reached (*79*). Finally, the potential benefits of cell recycle, such as operation at higher dilution rates and increased productivity, need to be investigated.

The two-stage continuous setup and the fed-batch operation show promise for enzyme synthesis from inexpensive lignocellulosic biomass. Potential sources of biomass include wastepaper, municipal waste, newspaper, and agricultural residues because they all can reportedly support cellulase synthesis (Table 5). It should be emphasized, however, that the kinetic behavior of the employed microorganism has to be understood, before an attempt is made to optimize cellulase production. Investigation of the cellulase production kinetics is a priority and should be used as a source of information for the formulation of mathematical models that simulate the production of cellulase and can serve as tools in optimizing the process and addressing scale-up concerns. Such a model for enzyme production by *T. reesei* Rut C30, which seems to correspond well with experimental data, can identify means for improvement in cellulase productivity and yield (*86*).

Substrate Selection and Media Optimization. The cost of the media and carbon source plays a major role in the economics of cellulase production. Soluble carbohydrates, such as lactose, xylose, and cellobiose, are popular carbon sources at the bench-scale level because they allow fast growth of the cells and induce cellulase synthesis. Their high cost, however, makes them prohibitive for a large-scale ethanol production operation. Lignocellulosic biomass, on the other hand, provides a considerably less expensive carbon source and cellulase inducer, which is and will continue to be available in large quantities. The drawback to the use of cellulose is the slow growth rate of the organisms, because of the need to first hydrolyze cellulose to glucose before it can be assimilated. As reported in the literature, this problem can be alleviated by supplementing cellulose with small amounts of soluble sugars (e.g. cellobiose, xylose) or ball-milled biomass to enhance cell growth. Unfortunately, as in the case of soluble sugars, preparation of ball-milled biomass is costly. The area of substrate pretreatment needs to be studied further. A pretreatment method that can inexpensively make the substrate readily digestible by cellulase will benefit both the cellulase production unit and the SSF process.

Similarly to the carbon source, the cost of the medium employed in cellulase production is important. It is essential to search for media that are both inexpensive and available in quantities that will satisfy the projected needs of ethanol production. The medium includes nitrogen, phosphorus, sulfur, and magnesium sources, as well as trace elements. Usually peptone, yeast extract, or corn steep liquor are provided as rich sources of nitrogen, vitamins, and trace elements (Table 1). However, the cost of these ingredients may be prohibitive for the economics of cellulase production. Hence, media, such as those shown in Table 3, need to be evaluated for their effect on cell growth and enzyme synthesis. The use of pretreated lignocellulosic biomass as substrate for the cellulose production unit and the SSF process may provide some of the essential nutrients and therefore reduce the complexity and cost of the needed medium. Another likely source of nutrients is the cell mass present in the effluent streams of the SSF and xylose fermentation units. As previous studies have shown (*47,49,87*), media optimization can have a great impact on cellulase productivity. The criteria for medium and substrate selection are optimal enzyme productivity, adequate availability, and low cost.

Conclusion

The purpose of this paper was to assess the current status of research and development in cellulase production through a thorough review of the pertinent literature, which exceeded 300 publications. The ultimate goal is to recommend research directions that need to be pursued in order to establish cellulase production technology, which will provide low-cost cellulase enzymes for the biomass conversion process.

From the overview presented, it is evident that further improvements in cellulase production require a better understanding of the kinetics of cell growth and enzyme synthesis. *T. reesei* strains, currently the most promising cellulase-producing organisms, can synthesize cellulase at concentrations usually exceeding 10,000 IU/L and productivities higher than 100 IU/L/h, as demonstrated in Table 1. However, the best bacterial candidates should also be examined because they can reach high cell concentrations and therefore high enzyme titers. Such candidates include *Cellulomonas, Thermonospora, Pseudomonas, Acetovibrio, and Acidothermus spp.* Moreover, it is recommended that anaerobic fungi, such as *Neocallimastix frontalis*, also be studied, because of their high specific activity and compatibility with SSF, which would eliminate any aeration requirements for enzyme production. Finally, alternative schemes of enzyme and ethanol production, such as the DMC process, should be examined more thoroughly.

The formulation of simple, descriptive models will be of great importance because they can serve as tools in determining the combined effect of several process variables on enzyme productivity. Based on such models, the crucial issues of cellulase production can be addressed: microorganism and substrate selection; media optimization; investigation of various modes of operation; and engineering means to enhance cellulase productivity and yields and prolong enzyme activity. The technical feasibility and performance of fed-batch and continuous enzyme production processes should be examined extensively, in light of their high productivity potential and compatibility with the other continuous unit operations of an integrated biomass-to-ethanol conversion process.

Acknowledgments. This work was funded by the Ethanol from Biomass Program of the Biofuels Systems Division of the U.S. Department of Energy.

Literature Cited

1. Eveleigh, D.E. *Phil. Trans. R. Soc. Lond.* **1987,** *321*, 435-447.
2. Ryu, D.D.Y; Mandels, M. *Enzyme Microb. Technol.* **1980,** *2*, 91-102.
3. Pokorny, M.; Zupancic, S.; Steiner, Th.; Kreiner, W. In *Trichoderma reesei Cellulases: Biochemistry, Genetics, Physiology and Application*; Kubicek, C.P.; Eveleigh, D.E.; Esterbauer, H.; Steiner, W.; Kubicek-Pranz, E.M., Eds.; The Royal Society of Chemistry: Cambridge, UK, 1990; pp 168-184.
4. Philippidis, G.P.; Wyman, C.E. In *Recent Advances in Biotechnology*; Vardar-Sukan, F.; Sukan, S.S., Eds.; Kluwer Academic Publishers: Dordrecht, The Netherlands, 1992; pp 405-411.
5. Philippidis, G.P.; Spindler, D.D.; Wyman, C.E. *Appl. Biochem. Biotechnol.* **1992,** 34/35, 543-556.
6. Hinman, N.D.; Schell, D.J.; Riley, C.J.; Bergeron, P.; Wyman, C.E. *Appl. Biochem. Biotechnol.* **1992,** 34/35, 639-649.
7. Reese, E.T.; Mandels, M.; Weiss, A.H. *Adv. Biochem. Eng.* **1972,** *2*, 181-200.
8. Fan, L.T.; Lee, Y.-H.; Beardmore, D.H. *Adv. Biochem. Eng.* **1980,** *14*, 101-117.
9. Grohmann, K.; Torget, R.; Himmel, M. *Biotechnol. Bioeng. Symp.* **1985,** *15*, 59-80.
10. Ooshima, H.; Burns, D.S.; Converse, A.O. *Biotechnol. Bioeng.* **1990,** *36*, 446-452.
11. Esterbauer, H; Hayn, M.; Abuja, P.M.; Claeyssens, M. In *Enzymes in Biomass Conversion*; Leatham, G.F.; Himmel, M.E., Eds.; American Chemical Society: Washington, DC, 1991; pp 301-312.
12. Persson, I.; Tjerneld, F.; Hahn-Hägerdal, B. *Process Biochem.* **1991,** *26*, 65-74.

13. Enari, T.M.; Markannen, P. *Adv. Biochem. Eng.* **1977**, *5*, 1-24.
14. Augustin, J.; Zemek, J.; Kuniak, L.; Fassatiova, O. *Folia Microbiol.* **1981**, *26*, 14-18.
15. Margaritis, A.; Merchant, R.F.J. *CRC Crit. Rev. Biotechnol.* **1986**, *4*, 327-367.
16. Feldman, K.A.; Lovett, J.S.; Tsao, G.T. *Enzyme Microb. Technol.* **1988**, *10*, 262-272.
17. Henrissat, B.; Driguez, H.; Viet, C.; Schülein, M. *Biotechnology* **1985**, *3*, 722-726.
18. Reese, E.T.; Mandels, M. *Biotechnol. Bioeng.* **1980**, *22*, 323-335.
19. Coughlan, M.P.; Ljungdahl, L.G. In *Biochemistry and Genetics of Cellulose Degradation*; Academic Press: New York, NY, 1988; pp 11-30.
20. Robson, L.M.; Chambliss, G.H. *Enzyme Microb. Technol.* **1989**, *11*, 626-644.
21. Park, W.S.; Ryu, D.D.Y. *J. Ferment. Technol.* **1983**, *61*, 563-571.
22. Lynd, L. R.; Grethlein, H.E.; Wolkin, R.H. *Appl. Environ. Microbiol.* **1989**, *55*, 3131-3139.
23. Hogsett, D.A.; Ahn, H.-J.; Bernardez, T.D.; South, C.R.; Lynd, L.R. *Appl. Biochem. Biotechnol.* **1992,** *34/35*, 527-541.
24. Canevascini, G.; Coudray, M.R.; Rey, J.P.; Southgate, R.J.G.; Meier, H.J. *J. Gen. Microbiol.* **1979,** *110*, 261-303.
25. Szakmary, K.; Wottawa, A.; Kubicek, C.P. *J. Gen. Microbiol.* **1991,** *137*, 2873-2878.
26. Kubicek, C.P.; Messner, R.; Gruber, F.; Mach, R.L.; Kubicek-Pranz, E.M. *Enzyme Microb. Technol.* **1993,** *15*, 90-99.
27. Lamed, R; Setter, E.; Kenig, R. *Biotechnol. Bioeng. Symp.* **1983**, *13*, 163-181.
28. El Gogary, S.; Leite, A.; Crivellaro, O.; Eveleigh, D.E.; El Dorry, H. *Proc. Natl. Acad. Sci. USA* **1989,** *86*, 6138-6142.
29. Mandels, M.; Reese, E.T. *J. Bacteriol.* **1957**, *73*, 269-278.
30. Shiang, M.; Linden, J.C; Mohagheghi, A.; Grohmann, K.; Himmel, M.E. *Biotechnol. Prog.* **1991,** *7*, 315-322.
31. Chaudhuri, B.K.; Sahai, V. *Enzyme Microb. Technol.* **1993,** *15*, 513-518.
32. Montenecourt, R.M.; Eveleigh, D.E. *Appl. Environ. Microbiol.* **1977**, *34*, 81-85.
33. Hoffman, R.M.; Wood, T.M. *Biotechnol. Bioeng.* **1985**, *27*, 81-85.
34. Sternberg, D.; Mandels, G. *J. Bacteriol.* **1979,** *139*, 761-769.
35. Lachke, A.H.; Bastawde, K.B.; Powar, V.K.; Srinivasan, M.C. *Biotechnol. Lett.* **1983,** *5*, 649-652.
36. Yamane, K.; Yoshikawa, T.; Suzuki, H.; Nisizawa, K. *J. Biochem.* **1971**, *69*, 771-780.
37. Berg, B. *Can. J. Microbiol.* **1975**, *21*, 51-57.
38. Claeyssens, M.; Tomme, P. In *Trichoderma reesei Cellulases: Biochemistry, Genetics, Physiology and Application*; Kubicek, C.P.; Eveleigh, D.E.; Esterbauer, H.; Steiner, W.; Kubicek-Pranz, E.M., Eds.; The Royal Society of Chemistry: Oxford, UK, 1990; pp 1-11.
39. Philippidis, G.P.; Smith, T.K.; Wyman, C.E. *Biotechnol. Bioeng.* **1993,** *41*, 846-853.
40. Fägerstam, L.G.; Pettersson, L.S. *FEBS Lett.* **1980**, *119*, 97-110.

41. Lee, Y.-H.; Fan, L.T. *Biotechnol. Bioeng.* **1983**, *25*, 939-966.
42. Grethlein, H.E. *Biotechnology* **1985**, *3*, 155-160.
43. Converse, A.O.; Ooshima, H.; Burns, D.S. *Appl. Biochem. Biotechnol.* **1990**, *24*, 67-73.
44. Farkas, V.; Labudova, I.; Bauer, S.; Ferenczy, L. *Folia Microbiol.* **1981**, *26*, 129-132.
45. Holtzapple, M.T.; Caram, H.S.; Humphrey, A.E. *Biotechnol. Bioeng.* **1984**, *26*, 753-757.
46. Ghose, T.K. *Pure Appl. Chem.* **1987**, *59*, 257-268.
47. Tangnu, S.K.; Blanch, H.W.; Wilke, C.R. *Biotechnol. Bioeng.* **1981**, *23*, 1837-1849.
48. Shinde, S.R.; Gangawane, L.V. *Indian bot. Reptr.* **1986**, *5*, 57-60.
49. Brown, J.A.; Collin, S.A.; Wood, T.M. *Enzyme Microb. Technol.* **1987**, *9*, 355-360.
50. Frein, E.M. Ph.D. Dissertation, Rutgers University, New Brunswick, NJ, 1986.
51. Schaffner, D.W.; Toledo, R.T. *Biotechnol. Bioeng.* **1991**, *37*, 12-16.
52. Chen, S.; Wayman, M. *Process Biochem.* **1991**, *26*, 93-100.
53. Khan, A.W.; Lamb, K.A. *Biotechnol. Lett.* **1984**, *6*, 663-666.
54. Hendy, N.A.; Wilke, C.R.; Blanch, H.W. *Enzyme Microb. Technol.* **1984**, *6*, 73-77.
55. Sternberg, D.; Dorval, S. *Biotechnol. Bioeng.* **1979**, *21*, 181-191.
56. Meyer, H.P.; Humphrey, A.E. *Chem. Eng. Commun.* **1982**, *19*, 149-154.
57. Watson, T.G.; Nelligan, I.; Lessing, L. *Biotechnol. Lett.* **1984**, *6*, 667-672.
58. Vohra, R.M.; Shirkot, C.K.; Dhawan, S.; Gupta, K.G. *Biotechnol. Bioeng.* **1980**, *22*, 1497-1500.
59. Brown, J.A.; Collin, S.A.; Wood, T.M. *Enzyme Microb. Technol.* **1987**, *9*, 176-180.
60. Robison, P.D. *Biotechnol. Lett.* **1984**, *6*, 119-122.
61. Knappert, D.; Grethlein, H.; Converse, A. *Biotechnol. Bioeng.* **1980**, *22*, 1449-1463.
62. Allen, A.; Sternberg, D. *Biotechnol. Bioeng. Symp.* **1980**, *10*, 189-197.
63. Desrochers, M.; Jurasek, L.; Paice, M.G. *Appl. Environ. Microbiol.* **1981**, *41*, 222-228.
64. Wheatley, M.A.; Phillips, C.R. *Biotechnol. Lett.* **1983**, *5*, 79-84.
65. Macris, B.J.; Paspaliari, M.; Kekos, D. *Biotechnol. Lett.* **1985**, *7*, 369-372.
66. Cavazzoni, V.; Adami, A. *Ann. Microbiol.* **1987**, *37*, 127-133.
67. Moldoveanu, N.; Kluepfel, D. *Appl. Environ. Microbiol.* **1983**, *46*, 17-21.
68. Spindler, D.S.; Wyman, C.E.; Grohmann, K.; Philippidis, G.P. *Biotechnol. Lett.* **1992**, *14*, 403-407.
69. Shoemaker, S.P.; Raymond, J.C.; Bruner, R. In *Trends in the Biology of Fermentations for Fuels and Chemicals*; Hollaender, A; Rabson, R.; Rogers, P.; San Pietro, A.; Valentine, R.; Wolfe, R., Eds.; Plenum Press: New York, NY, 1981, pp 89-109.
70. Creese, E. M.S. Dissertation, University of Western Ontario, London, Ontario, Canada, 1983.

71. Ghosh, V.K.; Ghose, T.K.; Gopalkrishnan, K.S. *Biotechnol. Bioeng.* **1982**, *24*, 241-243.
72. Kim, J.H.; Pack, M.Y. *Biotechnol. Lett.* **1993,** *15*, 133-138.
73. Wood, T.M.; Wilson, C.A.; McCrae, S.I.; Joblin, K.N. *FEMS Microbiol. Lett.* **1986**, *34*, 37-40.
74. Mohagheghi, A.; Grohmann, K.; Wyman, C.E. *Biotechnol. Bioeng.* **1990**, *35*, 211-216.
75. Shiang, M.; Linden, J.C.; Mohagheghi, A.; Tucker, M; Grohmann, K.; Himmel, M.E. *Biotechnol. Appl. Biochem.* **1991,** *14*, 30-40.
76. Warzywoda, M.; Chevron, F.; Ferre, V.; Pourquie, J. *Biotechnol. Bioeng. Symp.* **1983**, *13*, 577-580.
77. Doppelbauer, R.; Esterbauer, H.; Steiner, W.; Lafferty, R.M.; Steinmüller, H. *Appl. Microbiol. Biotechnol.* **1987**, *26*, 485-494.
78. Shiang, M.; Linden, J.C.; Mohagheghi, A.; Rivard, C.J.; Grohmann, K.; Himmel, M.E. *Appl. Biochem. Biotechnol.* **1990,** *24/25*, 223-35.
79. Ryu, D.; Andreotti, R.; Mandels, M.; Gallo, B.; Reese, E.T. *Biotechnol. Bioeng.* **1979,** *21*, 1887-1903.
80. Esterbauer, H.; Steiner, W.; Labudova, I.; Hermann, A.; Hayn, M. *Bioresource Technol.* **1991,** *36*, 51-65.
81. Schell, D.J.; Hinman, N.D.; Wyman, C.E.; Werdene, P.J. *Appl. Biochem. Biotechnol.* **1990,** *24/25*, 287-297.
82. Himmel, M; Oh, K.; Tucker, M.; Rivard, C.; Grohmann, K. *Biotechnol. Bioeng. Symp.* **1987**, *17*, 413-423.
83. Esterbauer, H.; Hayn, M.; Jungschaffer, G.; Taufratzhofer, E; Schurz, J. *J. Wood Chem. Technol.* **1983,** *3*, 261-287.
84. Klingspohn, U.; Schügerl, K. *J. Biotechnol.* **1993,** *29*, 109-119.
85. Elliston, K.O.; Yablonsky, M.D.; Eveleigh, D.E. In *Enzymes in Biomass Conversion*; Leatham, G.F., Himmel, M.E., Eds.; American Chemical Society: Washington, DC, 1991; pp 290-300.
86. Bader, J.; Klingspohn, U.; Bellgardt, K.-H.; Schügerl, K. *J. Biotechnol.* **1993,** *29*, 121-135.
87. Duff, S.J.B.; Cooper, D.G.; Fuller, O.M. *Enzyme Microb. Technol.* **1987**, *9*, 47-52.
88. Kalra, M.K.; Sandhu, D.K. *Acta Biotechnol.* **1986,** *6*, 161-166.
89. San Martin, R.; Blanch, H.W.; Wilke, C.R.; Sciamanna, A.F. *Biotechnol. Bioeng.* **1986**, *28*, 564-569.
90. Mohagheghi, A.; Grohmann, K.; Himmel, M.E.; Leighton, L.; Updegraff, D.M. *Int. J. of Systematic Bacteriol.* **1986**, *36*, 435-443.

RECEIVED March 17, 1994

Chapter 10

Cellulase Assays

Methods from Empirical Mathematical Models

William S. Adney, Christine I. Ehrman, John O. Baker, Steven R. Thomas, and Michael E. Himmel

Alternative Fuels Division, Applied Biological Sciences Branch, National Renewable Energy Laboratory, 1617 Cole Boulevard, Golden, CO 80401–3393

Various kinetic models have been developed to determine the effectiveness of cellulase enzyme preparations on the rate and extent of cellulose hydrolysis. The assay method proposed by Sattler et al. (*1*) was used to compare five dilute acid treated wood sawdust samples and a control microcrystalline cellulose, Sigmacell 50. This study confirmed that where kinetic information is required for a specific biomass conversion process application, such as simultaneous saccharification and fermentation (SSF), hydrolysis kinetics must be evaluated using the actual substrate to acquire meaningful results.

Although it is an abundant biopolymer, cellulose is unique in that it is highly crystalline, insoluble in water, and highly resistant to depolymerization. The enzymatic degradation of cellulose to small reducing sugars is generally considered to be accomplished by the synergistic action of three classes of enzymes: the 1,4-β-D-glucan 4-glucanohydrolases (EC 3.2.1.4), the exo-1,4-β-glucosidases, including both 1,4-β-D-glucan glucohydrolases (EC 3.2.1.74) and 1,4-β-D-glucan cellobiohydrolase (EC 3.2.1.91), and the β-D-glucosidases (EC 3.2.1.21).

The inconsistent structural features of cellulosic substrates, such as surface area, size distribution, crystallinity, and composition, affect the proportion of enzyme adsorbed and the relative rate and extent of hydrolysis (*2-4*). The use of pretreated, rather than ideal, biomass substrates for the production of fuels (ethanol) and other chemicals complicates most models not only because of inconsistent physical properties between individual pretreated substrates, but also because of variability in the effectiveness of a certain pretreatment (*5*). It becomes evident that universal kinetic models can be extrapolated for use on any substrate, if the key operational parameters for specific substrate-enzyme combinations are determined.

0097–6156/94/0566–0218$08.00/0

Background

Recent advances made in the assay of cellulase preparations by Sattler et al. (*1*) describe a relationship between the extent of hydrolysis, reaction time, and enzyme concentration. This procedure permits the ranking of the effectiveness of different enzymes and different pretreatment methods. In this study, Sattler's procedure was used in the comparison of different pretreated biomass substrates. With this approach, data are collected during the hydrolysis of a cellulosic substrate at various enzyme loadings and different digestion times. For a given digestion time, cellulose hydrolysis is described by a hyperbolic curve of glucose yield versus cellulase enzyme loading (FPU/g). A double reciprocal plot of hydrolysis yield as a function of enzyme loading can be obtained using the relationship proposed by Sattler et al. (*1*):

$$(Y/C_o)^{-1} = (K\ C_o/Y_{max})[E]^{-1} + (Y_{max}/C_o)^{-1},$$

where Y/C_o is the fraction of substrate hydrolyzed; [E] is given in FPU/g substrate initially added; and Y_{max}/C_o is the maximum fraction of substrate that could theoretically be hydrolyzed in a given time period at an infinite enzyme concentration. The y-axis intercept in the double reciprocal plot, $(Y_{max}/C_o)^{-1}$, may be used to estimate the relative enzyme-saturated rate at a particular time for the enzyme preparation on a given substrate. Ideally, an enzyme should have a high Y_{max} and a low value for $K\ C_o/Y_{max}$, because a shallow slope indicates that the enzyme concentration does not strongly affect the degree of hydrolysis. Thus, a low value for K is desirable because it is indicative of a higher enzyme-substrate affinity, less significant saturation effects, and higher rates at low enzyme loadings. Using this information, a prediction can be made, for a given pretreated substrate, of the enzyme loading required to achieve a particular level of hydrolysis in a given length of time. This information is potentially useful in not only ranking biomass with regard to ease and extent of hydrolysis, but in modeling the performance of specific biomass substrates proposed for use in conversion processes to produce fuels and chemicals.

In this study, this method of kinetic analysis was applied to the hydrolysis of microcrystalline cellulose (Sigmacell 50) as well as samples of pretreated sawdust from five species of hardwoods and softwoods (maple, sycamore, walnut, red oak, and pine) that were collected from sawmills located in the Appalachian region of the southern United States. Waste material from the lumbering of these species of woods represents a significant source of raw material feedstock for conversion into fuels and chemicals. These wood samples were air dried to 8%-10% moisture, pretreated using a dilute sulfuric acid cooking scheme at 160°C, exhaustively washed, and then used as substrates for the enzymatic digestion studies. Although the glucan content of each wood was found to be relatively invariant throughout the samples tested, hemicellulosic sugar and lignin contents were unique to each wood as reported by Vinzant et al. (*7*). In this study, these and other differences in chemical composition were related to the resulting kinetic performance.

Materials and Methods

Biomass Handling and Pretreatment. Samples of wood sawdust were obtained from green trees harvested in Ohio, Kentucky, and West Virginia in January 1992. The samples of wood sawdust were dried and milled according to the procedures described by Vinzant et al. (*7*). Dilute sulfuric acid pretreatment and composition analysis was conducted on 500-g milled wood samples as described by Torget et al. (*6*) and Vinzant et al. (*7*), respectively.

Standard Reagents, Enzyme, and Substrate. Laminex (Code #6-5950, Lot #13-90091-01) from Genencor, International (South San Francisco, CA) was the cellulase mixture used in all hydrolysis studies. The filter paper activity in this preparation was determined using IUPAC methods (8) and was found to contain 63.8 ± 0.5 FPU/mL (n=4) and 82.1 U/mL β-glucosidase activity. The enzyme preparation contained 126.6 ± 2.9 mg/mL protein as determined by the Pierce Micro BCA method (Pierce, Rockford, IL) with bovine serum albumin as standard. Sigmacell 50 (Sigma Chemical, Lot 113F-0187) was used as the model substrate for the hydrolysis of a clean, microcrystalline cellulose. All other chemicals were reagent grade and obtained from Sigma unless otherwise noted.

Enzymatic Hydrolysis. Enzymatic hydrolysis of the standard cellulose preparation was initiated by the addition of 2 g of Sigmacell 50 to 90-mL of 50 mM citrate buffer, pH 4.8 containing 0.02% sodium azide as a preservative, in a 250-mL Erlenmeyer flasks. The mixture was pre-equilibrated to 50°C in a New Brunswick Scientific (Model G-26) shaking incubator at 200 rpm for 1 h before the addition of 10-mL of appropriately diluted (in the same buffer) enzyme mixture, which was also pre-incubated for 1 h.

Enzymatic hydrolyses of the pretreated sawdust samples were carried out exactly as in the case of the Sigmacell 50 model substrate, with a substrate loading of 2 g dry weight pretreated biomass per 100-mL total digestion mixture. In contrast to Sigmacell 50, which is approximately 98% cellulose, these pretreated substrates have been found to contain from 55% to 63% cellulose on a dry weight basis (*7*). In the analysis of the results, the enzyme loadings for each of the pretreated substrates are given in terms of FPU per gram of cellulose content, rather than FPU per gram of substrate.

Products From Enzymatic Digestion. One-milliliter aliquots were removed from well-dispersed reaction slurries in such a way as to ensure that representative mixtures of enzyme and biomass were obtained for various time points during the hydrolysis. The aliquots were placed in 1.5-mL microcentrifuge tubes, sealed with Gripper/1.5-mL tube closures (Rainin Instrument Co., Inc., Woburn, MA), and boiled for 10 min in a boiling water bath to terminate the enzyme reactions. The aliquots were then neutralized with calcium carbonate and the debris removed by centrifugation for 10 min using an Eppendorf (Model 5415C) microcentrifuge. After filtration through a 0.2 µm syringe filter, the aliquots were analyzed for neutral sugars by ion-moderated partition (IMP) chromatography using a Fast Carbohydrate column (BioRad) with distilled water as the mobile phase and refractive index detection. Glucose concentrations were confirmed using a YSI Model 2700 Select Biochemistry Analyzer with attached Model 2710 autosampler (Yellow Springs Instrument Co., Yellow Springs, OH). The concentrations of glucose and cellobiose contained in the cellulase digestion supernatant were calculated according to the relationship:

$$Y = [Glu]0.9 + [Cell]0.95$$

where [Glu] and [Cell] are concentrations in g per L by HPLC. The hydrolysis yield was expressed as the fraction (Y/C_0), where C_0 is the total cellulose concentration. Double reciprocal plots of yield as a function of time were used to predict the maximum digestibility of individual substrates at infinite enzyme concentrations (Y_{max}) and to predict finite enzyme loadings required to achieve a given level of hydrolysis in various reaction times.

Results

Hydrolysis of Sigmacell 50. Enzymatic hydrolysis of Sigmacell 50 was monitored over a period of 96 h with different enzyme-substrate ratios ranging from 5 to 100 FPU/g (Figure 1). An initial rapid reaction phase lasting about 10 h was observed, followed by a relatively slow reaction phase with the maximum enzyme loading used (100 FPU/g cellulose), hydrolyzing roughly 70 % of the Sigmacell 50 in 96 h. The relationship of enzyme dosage to hydrolysis yields at different reaction times, as shown in Figure 2, demonstrates the manner in which the fraction of cellulose hydrolyzed increases, in an apparently hyperbolic fashion, as the enzyme-substrate ratio is increased for each incubation time. For a given enzyme-substrate ratio, the level of cellulose conversion was observed to increase, again in apparently hyperbolic fashion, as incubation times were extended. These results are consistent with those reported by Sattler et al. (*1*) for Sigmacell 50. A double reciprocal plot of cellulose hydrolysis yield versus enzyme-substrate ratio and linear least-squares fitting of the data (Figure 3) allows for determination of the maximum theoretical value for cellulose conversion at infinite enzyme loading (Y_{max}), and a given incubation time. With correlation coefficients for the fits being generally greater than 0.95, this method allows for convenient and reliable interpolation to obtain estimates of the extent of conversion expected at given digestion times for enzyme loadings intermediate between those actually tested.

The proper level of β-glucosidase is known to be required in cellulase mixtures for achievement of maximal rates and extents of hydrolysis (*9*). No additional β-glucosidase was added to any of the digestion mixtures because the starting enzyme mixture contained 81.2 U/mL β-glucosidase activity as measured on *p*-nitrophenyl-β-D-glucopyranoside under the conditions of the digestions. This level of β-glucosidase activity was adequate to prevent buildup of high concentrations of cellobiose, because only low levels of cellobiose (less than 1.5 mg/mL) were observed early in the digestion and no more than trace amounts were present after 24 h.

Hydrolysis of Pretreated Wood Samples. Hydrolysis data for the pretreated substrates maple, sycamore, pine, walnut, and red oak were analyzed in the same fashion as for Sigmacell 50 using the assumptions outlined by Sattler et al. (*1*). Enzyme loadings were calculated based on cellulose content as determined previously (*6*). Double reciprocal plots of 1/(percent cellulose conversion) versus 1/(FPU/g cellulose) were used to predict conversion yields for given enzyme loadings in FPU/g

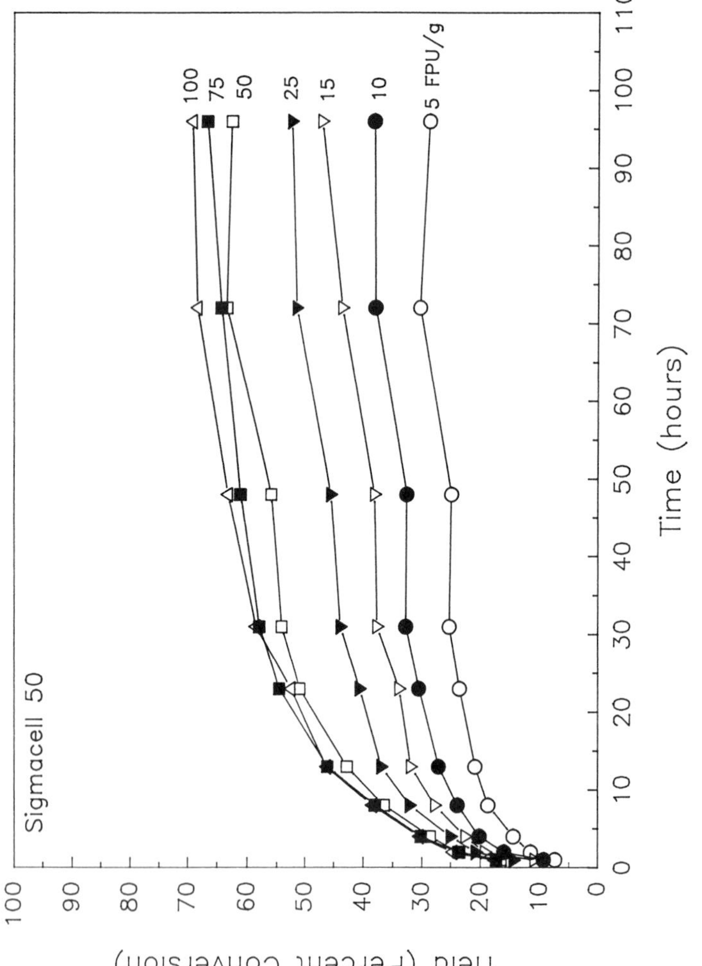

Figure 1. Relationship of increasing time on the hydrolytic yield from Sigmacell 50 expressed as a percentage of cellulose conversion (Y/C_0) using Genencor Laminex at various loadings.

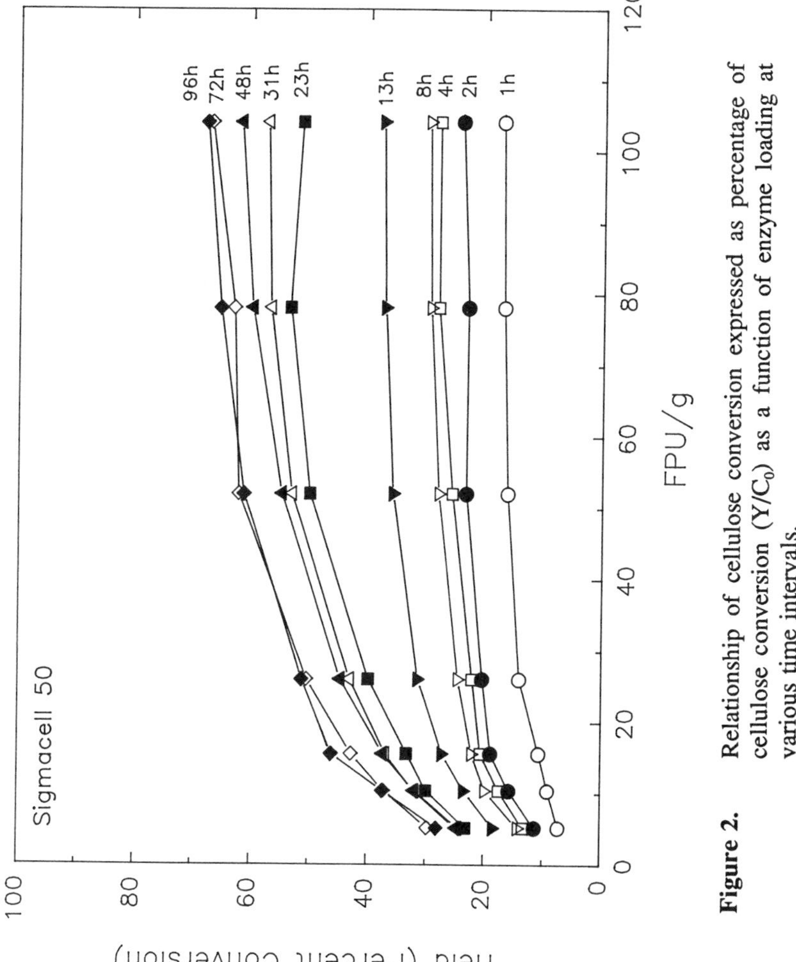

Figure 2. Relationship of cellulose conversion expressed as percentage of cellulose conversion (Y/C_0) as a function of enzyme loading at various time intervals.

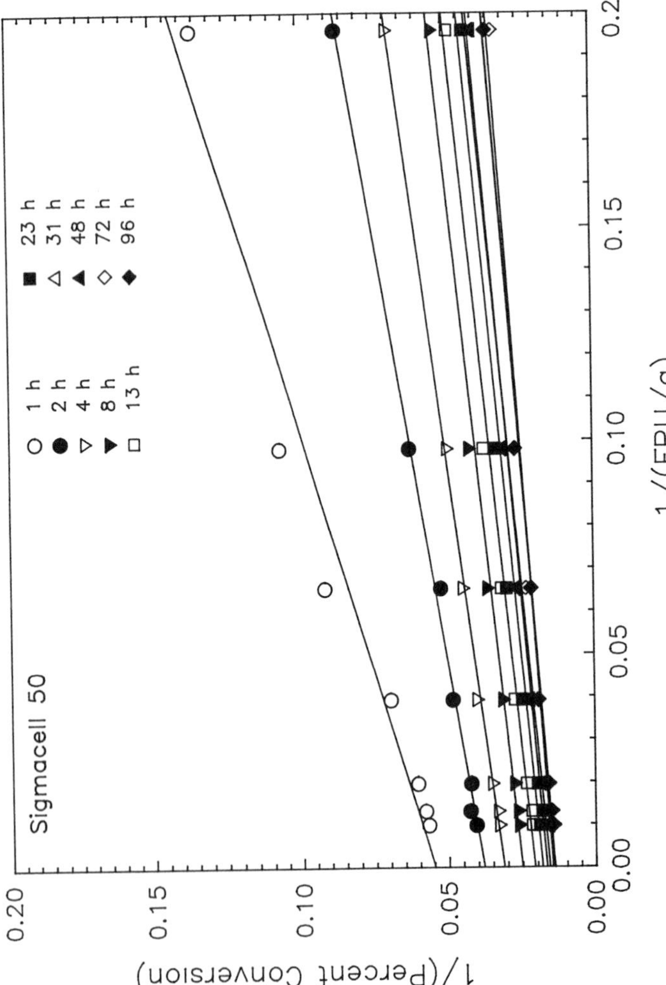

Figure 3. Double reciprocal plots (1/percent conversion (Y/C_0) versus 1/enzyme loading in FPU/g of cellulose) for the hydrolysis of Sigmacell 50 at various time intervals during the digestions. Lines represent the least squares linear regression fit of data.

cellulose and reaction times. As in the case of the double reciprocal plots for the hydrolysis of Sigmacell 50, the correlation coefficient for the linear least-squares fit of hydrolysis data of pretreated woods was found to be greater than 0.95 in most cases, except for incubation times of less than 2 h. Double reciprocal plots shown in Figure 4 for sycamore are typical of the relationship observed for cellulose conversion as a function of enzyme loading for all pretreated substrates used in this study. Linear-least squares analyses from the double reciprocal plots for pretreated wood samples were used to generate the curves in Figures 5-7. Figures 5-7 provide a direct means of comparing the overall conversion levels of the pretreated substrates used in this study at 24, 48, and 96 h. At all time points, maple was easiest to convert, followed by sycamore, red oak, walnut, and pine. A comparison of double reciprocal plots at 24 h (Figure 8) shows that Y_{max}/C_o values for all of the pretreated substrates with the exception of pine are clustered between 75% and 100% conversion of the cellulose contents (and are thus well above the Y_{max}/C_o value of 74.6% for the model compound Sigmacell 50), whereas for pretreated white pine, only 12.9% of the total cellulose would appear to be available for hydrolysis in 24 h by even an infinite concentration of cellulase.

The slopes of the double reciprocal plots in Figures 4 and 8 indicate significant differences among all of the pretreated lignocellulosics in terms of the relationship between enzyme loading and the percentage of the theoretical maximum yield (100 x Y/Y_{max}) that is actually achieved.

The activity of Laminex against pine is much more strongly dependent on loading (large value for slope) than are the activities of the cellulase mixture against the other substrates. The curves for maple and Sigmacell 50 have the smallest slopes in Figure 8, which is to say that at any finite cellulase loading, the extent of cellulose solubilization at 24 h will be a larger fraction of the conversion level theoretically achievable at infinite cellulase loading, than will be the case for the other substrates. For the pretreated pine, on the other hand, the extent of conversion falls off very rapidly as the enzyme loading is decreased (Figure 8); obtaining 24-h yields anywhere near the Y_{max} for pine will therefore require quite high enzyme loadings.

In Table 1, the six substrates are ranked according to the theoretical hydrolyzability of their cellulose contents in a 24-h digestion (i.e., according to the values of 100 x Y_{max}/C_o in the first column). It will be noted that ranking the substrates on the basis of the preferred (large) values of Y_{max}/C_o places the pretreated wood substrates in the same order that would be observed, were they ranked on the basis of the preferred (small) values of slope of the double reciprocal plots for 24 h digestions (second column, Table 1). The sole exception to this correlation is the microcrystalline-cellulose model substrate, Sigmacell 50, which has the second-smallest (and therefore second-best) slope among all of the substrates, but which also has the second-smallest (and in this case, next-to-worst) Y_{max}/C_o value for a 24 h digestion. Basically similar rankings can be established using the Y_{max}/C_o values at 48 h and 96 h digestion times. (The confusion seen between the "top three" substrates at 48 and 96 h, with the extent of conversion shown as exceeding 100%, is probably traceable to underestimates of the glucan content of the substrates in an analytical procedure involving hydrolysis catalyzed by sulfuric acid.)

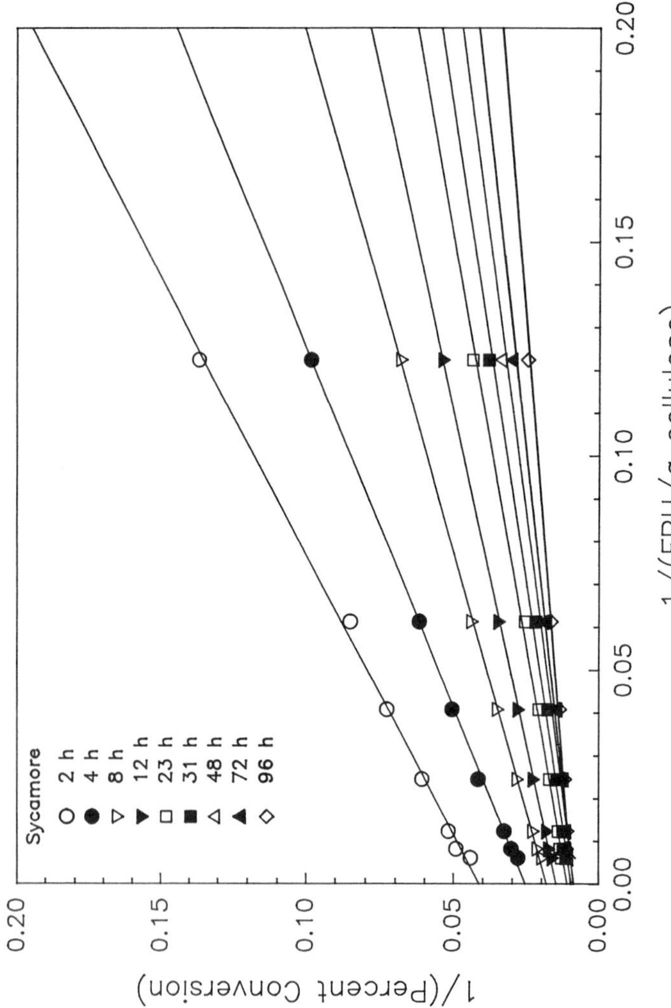

Figure 4. Double reciprocal plots (1/percent conversion (Y/C_0) versus 1/enzyme loading in FPU/g of cellulose) for the hydrolysis of pretreated sycamore at various time intervals during the digestions. Lines represent the least squares linear regression fit of data.

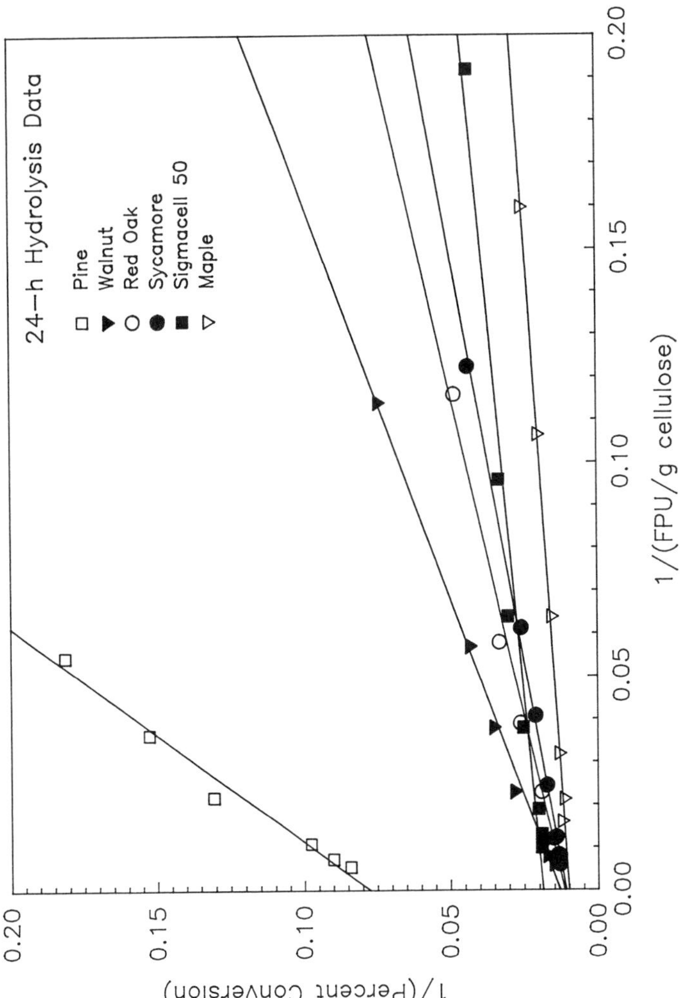

Figure 5. Comparison of the double reciprocal plots of 24-h digestions data from the enzymatic hydrolysis of Sigmacell 50 and pine, walnut, red oak, sycamore and maple.

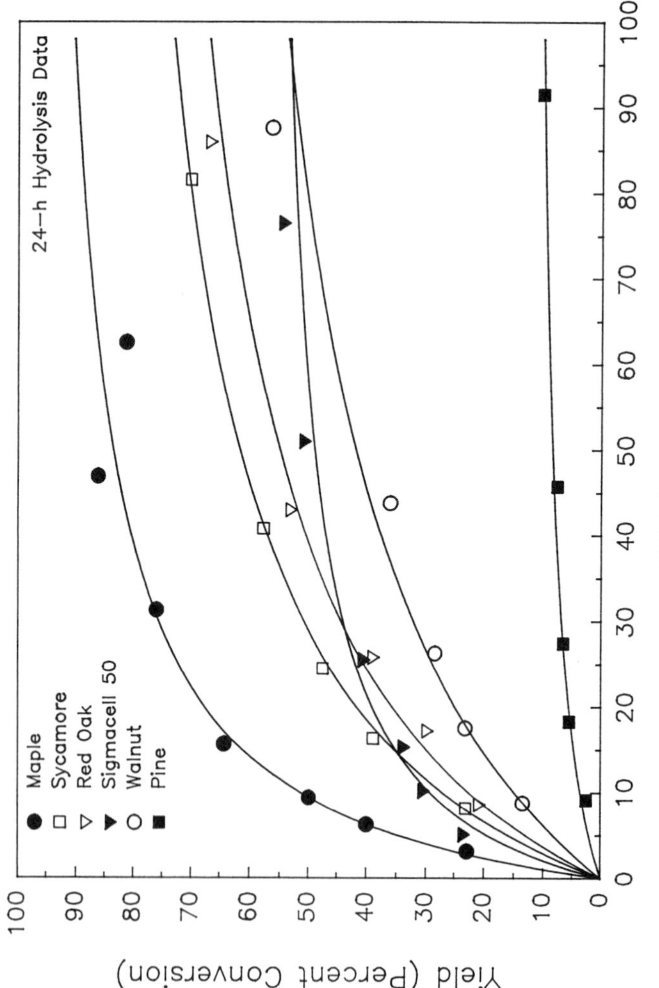

Figure 6. Comparison of 24-h hydrolysis data for Sigmacell 50, pine, walnut, red oak, sycamore, and maple. Lines represent best fit of data from linear least squares regression analysis extrapolated from double reciprocal plots.

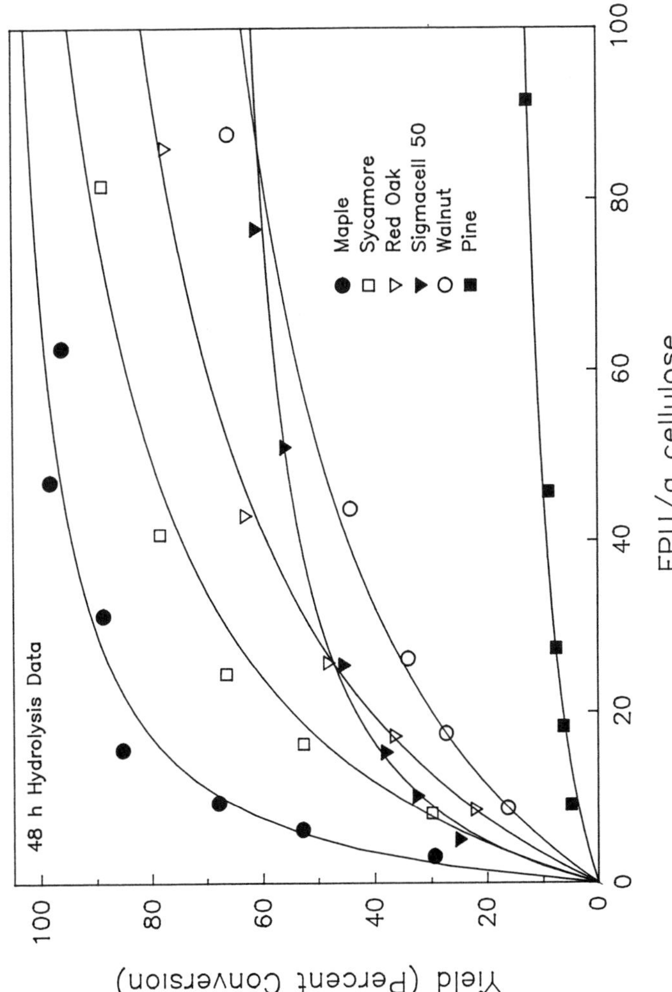

Figure 7. Comparison of 48-h hydrolysis data for Sigmacell 50, pine, walnut, red oak, sycamore, and maple. Lines represent best fit of data from linear least squares regression analysis extrapolated from double reciprocal plots.

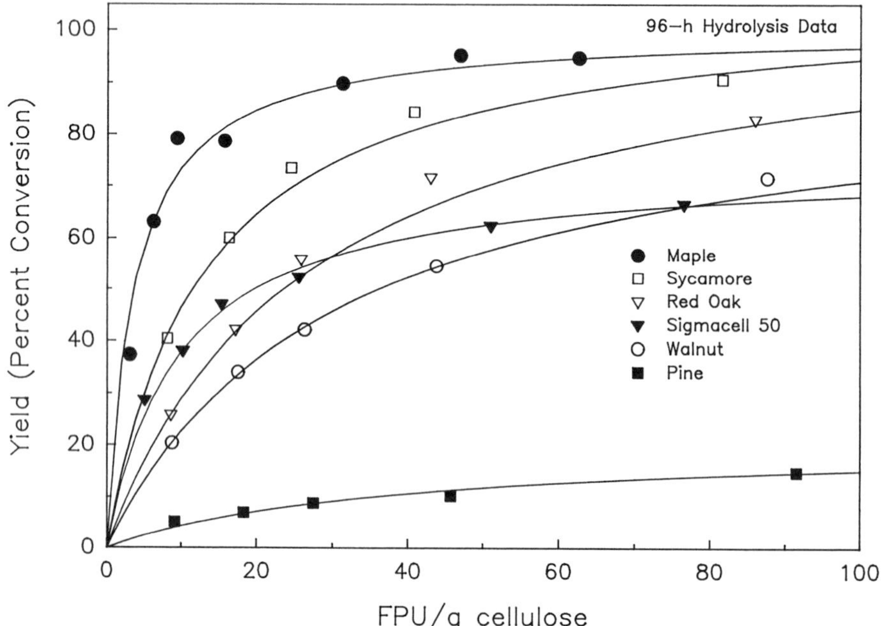

Figure 8. Comparison of 96-h hydrolysis data for Sigmacell 50, pine, walnut, red oak, sycamore, and maple. Lines represent best fit of data from linear least squares regression analysis extrapolated from double reciprocal plots.

Table 1.
Comparison of Y_{max} and Slopes for Different Pretreated Substrates at Selected Sampling Times.

	24-h	24-h slope	48-h	48-h slope	96-h	96-h slope
Maple	99.0	0.0933	108.7	0.0567	101.0	0.0364
Sycamore	90.9	0.2582	116.3	0.1934	106.4	0.1218
Red Oak	86.2	0.3271	108.8	0.3077	108.2	0.2530
Walnut	76.3	0.5376	87.7	0.4411	92.6	0.3330
Sigmacell 50	58.8	0.1719	68.9	0.1714	74.6	0.1301
Pine	12.9	1.9923	16.9	1.9430	20.7	1.7972

Discussion

Filter paper assays, as described by the IUPAC procedure (*8*), are the most commonly used method of measuring saccharifying activity of an enzyme preparation on cellulose. The filter-paper assay, however, gives only a very limited indication of an enzyme's capability to hydrolyze diverse cellulosic substrates such as pretreated biomass, and provides inadequate information as to the extent and overall rate of hydrolysis. Data on the extent of cellulose conversion, the overall rate of conversion, and the minimum enzyme loading required to obtain specific hydrolysis yields is crucial information necessary for ranking feedstocks and pretreatments for a conversion process. Correlations between hydrolysis levels, reaction time, and enzyme-substrate ratio as observed in the present work tended to be readily analyzable in terms of the hyperbolic-kinetics model described by Sattler et al. (*1*).

The time-dependent hydrolysis of Sigmacell 50 for times of less than 100 h was described by Sattler et al. by two parallel pseudo-first-order reactions. The fraction most easily hydrolyzed from Sigmacell 50 was determined to be approximately 45%. In the current study, the delineation of easy- and difficult-to-hydrolyze portions of cellulose in pretreated substrates was not as well defined and may be more accurately described by more than two parallel pseudo-first-order reactions.

The ultimate goal for biomass pretreatment is to provide fermentable sugars in high yield and in the shortest period of time to the process. One measure of this goal is achieved by increasing the effectiveness of cellulase action. This may be accomplished by increasing the accessibility of cellulose to cellulases, perhaps by removal of hemicellulose and/or lignin, or by decreasing cellulose crystallinity. Different pretreatments must be considered if the rates and extents of cellulose conversion from different biomass substrates are to be enhanced, because the tissues and cells of various sources of biomass respond quite differently during pretreatment. Figures 9 and 10 illustrate one very direct, practical use to which the data resulting from this study can be put in the design and modeling of industrial processes. From the "percent of cellulose conversion" isolines shown in Figure 9 for the saccharification of soft maple, one can select either the enzyme loading required to convert a given percentage of the total cellulose to soluble sugar within a specified digestion time or, conversely, the digestion time required to achieve that extent of conversion using a specified enzyme loading. For a cellulose conversion level of 50%, Figure 10 compares the enzyme-loading/digestion-time trade-offs for saccharification of four different pretreated hardwoods, with Sigmacell 50 as a control. (Data for pretreated pine do not appear in Figure 10, because as for the highest enzyme loadings and longest digestion times studied here, conversion of the cellulose content of pine still fell far short of 50%).

In this study, ranking of substrates in terms of the enzyme loadings and digestion times required to achieve a specific extent of conversion was the most useful procedure. This assertion is supported by the observation that this approach (Figure 10) assigns Sigmacell 50 an unambiguous and meaningful fourth-place ranking for 24-h digestions (maple>sycamore>oak>Sigmacell 50>walnut>>pine), whereas rankings according to Y_{max}/C_o or double reciprocal slopes placed Sigmacell 50 either fifth or second, respectively. Neither of these rankings is particularly meaningful in terms of practical results. (Because as the curve for Sigmacell 50 is the only one in Figure 10

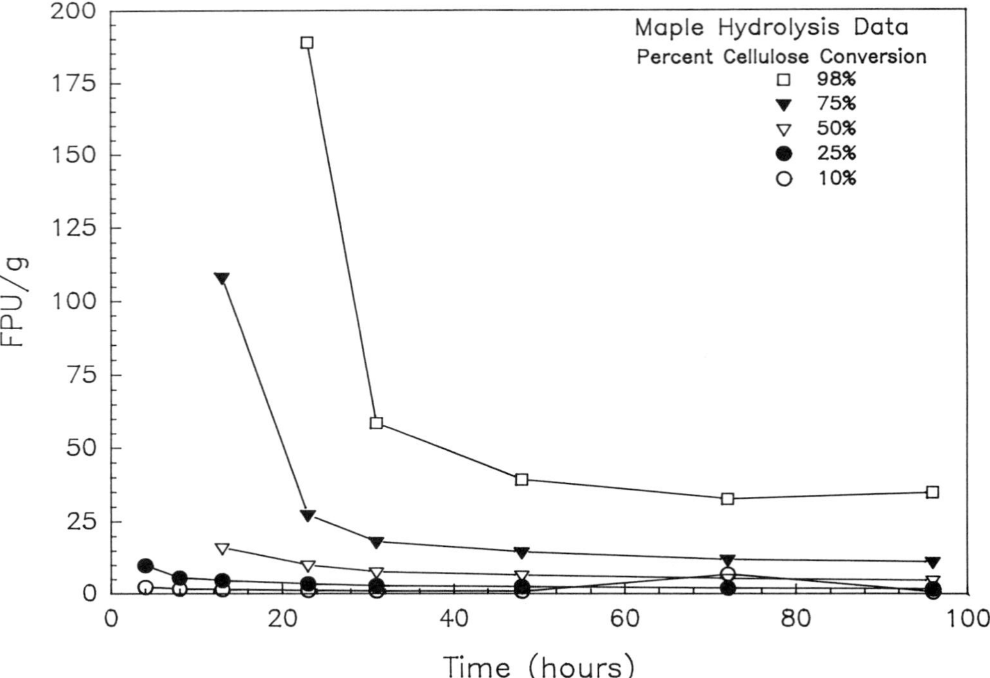

Figure 9. Relationship of enzyme loading and increasing incubation times on various cellulose conversion levels for maple, the "best" pretreated substrate. From this data, the cost in terms of enzyme loading can be determined for a given conversion time and level, information useful in modeling a conversion process.

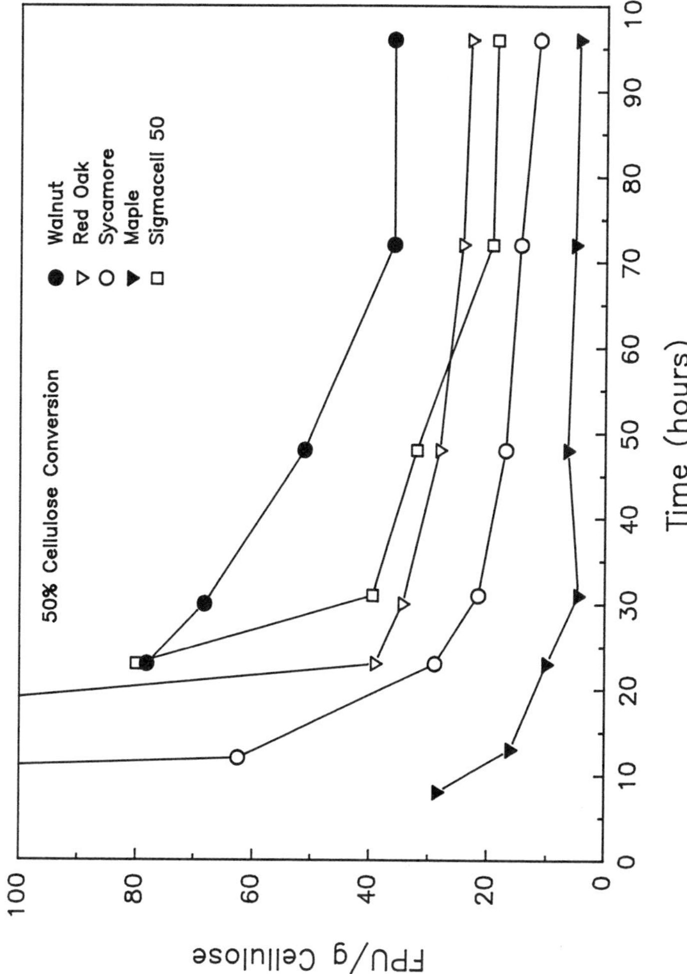

Figure 10. Comparison of enzyme loadings as a function of time for pretreated substrates at the 50% conversion level.

(and Figure 5) that intersects some of the other curves, the relative ranking of Sigmacell 50, based on Figure 10, will vary with digestion time).

The primary factors affecting the cost of saccharification of a cellulosic substrate include yield, the time required to obtain a target yield, and the amount of enzyme required, which is directly related to cost. The value of FPU, in these studies, for an enzyme preparation used can be directly interchanged for cost, or with milligram of protein loaded.

All of the results reported and analyzed here are for the simple batch hydrolysis of lignocellulosic materials (separate saccharification), but comparison with data obtained in other types of experiments strongly suggests that the predictive value of our results may extend to other experimental and process configurations as well. In a series of simultaneous saccharification and fermentation (SSF) experiments using the same pretreated wood samples used in the present study, Vinzant et al. (*7*) found similar conversion efficiencies. Vinzant et al. also reported that maple, sycamore, and red oak produced the highest theoretical yields of ethanol, followed by walnut and pine, which produced only low levels of ethanol. The average biomass conversion efficiency value for Sigmacell 50 from five fermentation runs was reported to be 78±3%, which is similar to the predicted Y_{max} value of 74.6% obtained in this study. It has been prominently noted in the literature (*10-11*) that saccharification and SSF experiments differ in at least one important respect: in SSF the ultimate saccharification product, glucose, is constantly utilized by the fermentative organisms, and thereby kept at a very low level, whereas in simple saccharification glucose accumulates, presenting the possibility of end-product inhibition of the cellulases. This difference notwithstanding, however, the close correspondence between the results obtained by empirical modeling based on saccharification data in this study and results of actual SSF runs clearly indicates that empirically analyzed enzyme hydrolysis studies such as the one presented here are able to demonstrate meaningful differences in availability to cellulases of the cellulose contents of different woods. Evaluation of hydrolysis data in this manner is potentially useful in identifying sources of biomass for conversion and pretreatment protocols.

Conclusions

The quality of the cellulosic substrate, the hydrolytic capability of the cellulase enzymes, and the mode of interaction between substrate and enzyme have been identified as the major factors that influence SSF performance in the conversion of biomass to ethanol (*12,13*). The utilization of various pretreated woody substrates for the production of ethanol using the SSF process has been found to be highly variable, depending primarily on the type of substrate used and the quality of the pretreatment. Accurate comparisons of the hydrolysis kinetics for different substrates and of different enzyme preparations are crucial for a designated process application.

It becomes evident from the data presented in this study that in cases where kinetic information is required for a designated process application, such as biomass conversion, hydrolysis kinetics must be evaluated using the actual substrate to acquire meaningful results. Utilization of a technique recently proposed by Sattler et al. (*1*) for assessing overall cellulase activity allows for the accurate prediction of enzyme loadings required to achieve given cellulose conversion levels using pretreated substrates. This method may also be useful in correlating enzymatic conversion of miscellaneous biomass substrates with SSF performance, even though the issues of feedback inhibition by glucose, and the possible existence of yeast toxins are not addressed in this approach.

Acknowledgments

The five different species of wood sawdust used in this study were supplied by Lannes Williamson Pallets, Inc., Southside, WV. This work was funded by the Ethanol from Biomass Program of the Biofuels Systems Division of the U.S. Department of Energy.

Literature Cited

1. Sattler, W.; Esterbauer, H.; Glatter, O.; Steiner, W. *Biotechnol. Bioeng.* **1989**, *33*, 1221-1234.
2. Atalla, R.H. In *Hydrolysis of Cellulose: Mechanisms of Enzymatic and Acid Catalysis*; Brown, R.D.; Jurasek, L., Eds.; American Chemical Society: Washington, DC, 1979; pp. 55-69.
3. Millett, M.A.; Effland, M.J.; Caulfield, D.F. In *Hydrolysis of Cellulose: Mechanisms of Enzymatic and Acid Catalysis*; Brown, R.D.; Jurasek, L., Eds.; American Chemical Society: Washington, DC, 1979; pp. 71-89.
4. Weimer, P.J. In *Biosynthesis and Biodegradation of Cellulose*; Haigler, C.H.; Weimer, P.J., Eds.; Marcel Dekker: New York, NY, 1991, pp. 263-291.
5. Grohmann, K.; Wyman, C. E.; and Himmel, M. E. In *Emerging Materials and Chemicals from Biomass*; Rowell, R.; Narayan, R.; Schultz, T., Eds.; ACS Series **476**, American Chemical Society: Washington, DC, 1991, pp. 354-392.
6. Torget, R.; Werdene, P.; Himmel, M. E.; Grohmann, K. *Appl. Biochem. Biotechnol.*, **1990**, *24/25*, 115-126.
7. Vinzant, T. B.; Ponfick, L.; Nagle, N. J.; Ehrman, C. I.; Reynolds, J. B.; and Himmel, M. E., *Appl. Biochem. Biotechnol.*, **1994**, In press.
8. Ghose, T. K. *Pure & Appl. Chem.* **1987**, *59*, 257-268.
9. Breuil, C.; Chan, M.; Gilbert M.; Saddler, J.N. *Bioresource Tech.* **1992**, *39*, 139-142.
10. Takagi, M.; Abe, S.; Suzuki, S.; Emert, G. H.; Yata, N. In *Bioconversion Symposium Proceedings*; Indian Institute of Technology: Delhi, 1977, pp. 551-571.
11. Philippidis, G. P.; Smith, T. K.; Wyman, C. E. *Biotech. Bioeng.* **1993**, *41*, 846-853.
12. Philippidis, G. P.; Spindler, D. D.; Wyman, C. E. *Appl. Biochem. Biotechnol.* **1992**, *34/35*, 543-556.
13. Hillis, W. E. In *Biosynthesis and Biodegradation of Wood Components*; Higuchi, T., Ed.; Academic Press: New York, NY, 1985, pp. 209-226.

RECEIVED March 17, 1994

Chapter 11

Components of *Trichoderma reesei* Cellulase Complex on Crystalline Cellulose

Three-Dimensional Visualization with Colloidal Gold

Rafael A. Nieves[1], Roberta J. Todd[2], Robert P. Ellis[2], and Michael E. Himmel[1]

[1]Applied Biological Sciences Branch, National Renewable Energy Laboratory, 1617 Cole Boulevard, Golden, CO 80401–3393
[2]Department of Microbiology, Colorado State University, Fort Collins, CO 80523

Bacterial cellulose and pretreated natural aspen cellulose were used as substrates for observation of cellulose-bound cellulases. Specific monoclonal antibodies which had previously been adsorbed to 10 nm and 15 nm gold spheres were used to detect bound endoglucanase and cellobiohydrolase via transmission electron microscopy. Three-dimensional electron micrographs demonstrated individually bound cellulases as well as clusters of bound enzymes. Significant changes in the interpretations of the micrographs were seen when these were observed in a three dimensional format as opposed to a two-dimensional view. The three dimensional electron micrographs indicated the sensitivity of this technique for these studies by revealing individual enzymes bound to individual cellulose microfibril(s).

Microbial cellulases have been of interest in the last few decades because of their potential for industrial applications. The most widely studied cellulases are those produced by the soft-rot fungus, *Trichoderma reesei*. These enzymes work synergistically to degrade crystalline cellulose to simple carbohydrates via hydrolysis. The glucose produced from this mechanism can be utilized as a primary feedstock for many fermentation processes. Hydrolysis of the cellulose is thought to be initiated by the 1,4-ß-D-glucan glucanohydrolases (endoglucanase I and endoglucanase II) which attack randomly along the cellulose chain exposing reducing and non-reducing ends. These hydrolyzed regions can then be attacked by the 1,4-ß-D-glucan cellobiohydrolases (CBH I and CBH II) which proceed with the subsequent degradation of cellulose. The cellobiose and small oligosaccharides produced from this synergistic effect are then converted to glucose by ß-D-glucosidase (*1-3*).

Electron microscopy studies of the action of cellulases on cellulosic substrates have been reported previously (*4-6*). These studies have relied on indirect visualization of hydrolysis as opposed to direct visualization and localization of the

0097–6156/94/0566–0236$08.00/0

enzymes (*4,6*). Another study demonstrated binding of cellobiohydrolases indirectly using gold colloids (*5*), but no studies have shown distinguishable, simultaneous binding of endoglucanases and exoglucanases. We have recently demonstrated the apparent association of endoglucanase and cellobiohydrolase on natural aspen cellulose by immunogold labelling and negative staining (*7*). Specific monoclonal antibodies (MoAb) directed against different cellulases were used to indirectly observe the enzymes bound to the solid substrate. Here we present further studies of cellulase binding using immunogold labels in three dimensional electron micrographs. Interpretation of these images becomes clearer as different planes of view are distinguished.

Materials and Methods

Preparation of Bacterial and Aspen Cellulose. Lyophilized cultures of *Acetobacter xylinum* were initially cultured at 30°C in 5 mL of Hestrin-Schramm medium until a visible pellicle at the air-liquid interface could be observed (*8*). The 5 ml culture was used to inoculate 100 mL of the medium and this was incubated for 3 days in stationary culture at 30°C. Using a sterile glass rod, the pellicle was transferred to 100 mL of cold 50 mM phosphate buffer pH 7.0. The pellicle was washed 3X in this manner and sliced into small squares. The squares were suspended in 25 mL of phosphate buffer, agitated for 2 m, and filtered through cheesecloth. The resulting cell suspension was washed 2X with phosphate buffer by centrifugation. Dilute sulfuric acid-pretreated aspen meal was prepared as described previously (*7*).

Ascites Production and Purification of Monoclonal Antibodies. Mice used for ascites production were injected with 0.5 ml of pristane (Sigma Chemical Co.) 1 to 4 weeks prior to injection of hybridoma cells. For each mouse, 5 x 10^6 cells of the specific hybridoma cell line was injected intraperitoneally. Ascites fluid was collected 5 days later and allowed to coagulate overnight at 4°C. The supernatant was clarified by centrifugation. Antibodies were purified using rProtein G Minicolumns (Gibco, Gaithersburg, MD). Ascites was diluted 1:10 with starting buffer (0.1 M sodium phosphate, pH 7.0, 0.15 M sodium chloride) and applied to columns which had been equilibrated with 10 bed volumes of starting buffer at a flow rate of 0.5 mL/min. The column was washed and the antibodies were eluted with 0.1 M glycine hydrochloride, pH 3.0. Thirty microliter aliquots of 1 M Tris buffer pH 10 were added to the fraction tubes prior to collection. Tubes containing the antibodies were pooled and analyzed for purity. Purified antibodies were bound to gold spheres as described previously (*7*). Purified CBH I antibodies were adsorbed to 10 nm gold colloids and EG I antibodies were adsorbed to 15 nm gold colloids.

Immunolabelling of Cellobiohydrolase I and Endoglucanase I Bound to Cellulose. EM grids were prepared by placing 200-mesh copper grids on floating films of 0.5% formvar in ethylene dichloride. The grids were carbon coated, air dried, and placed on top of drops of bacterial or aspen cellulose suspensions. The grids were stream washed with distilled water, blot dried, and stored until used.

EM grids to which aspen or bacterial cellulose had previously been adsorbed were incubated for 45 m with 1.3% milk-phosphate buffer pH 6.5. The grids were washed with distilled water and blot dried. The grids were placed face down on a drop of Genencor 150L cellulase (Genencor International, San Francisco, CA.) which contained *T. reesei* cellobiohydrolase I and endoglucanase I. Total protein concentration of the commercial cellulase was 1 mg/mL. The grids were washed, blot dried and then placed face down on a drop of an antibody mixture solution which contained anti-CBH I and/or anti-EG I MoAbs adsorbed on colloidal gold particles. MoAbs were diluted 1:10 with 1 part milk block and 9 parts immunolabel. After 30 m, the grids were stream washed, blot dried, and immediately stained with 1% phosphotungstic acid for 5 m. The grids were washed, blot dried, and stored for observation.

Electron Microscopy. Electron microscopy was performed on a Philips 400T transmission electron microscope at an accelerating voltage of 80 Kv. For three dimensional electron micrographs, fields of interest were initially photographed at a 0° angle on the specimen stage. A point of reference was identified, and the stage rotated approximately 4°C. The point of reference was then aligned to its original position on the screen and the second image was then photographed. The three dimensional effect could be observed by pairing stereo images and viewing with a stereoscope (Abrams Instrument Corp., Lansing, MI).

Results

Figure 1 demonstrated minimal background labelling of a negative control. Negative controls were performed by omitting the exposure of the grids to the commercial cellulase solution. This low magnification electron micrograph showed a single *A. xylinum* bacterial cell with protruding microfibrillar cellulose. All negative controls showed minimal background labelling as shown here, and previously (*7*). Figure 2 is a high magnification photomicrograph of bacterial cellulose. The width of the bacterial cellulose fibers ranged from 2.5 to 12.5 nm. When the bacterial cellulose was exposed to the commercial cellulase solution and then labelled with either anti-CBH I immunogold label (10 nm) or anti-EG I label (15 nm), sporadic labelling along the long axis of the fiber was observed (data not shown).

Aggregation of immunolabel was often seen on bacterial and pretreated aspen cellulosic substrates. Aggregation of the gold label had previously been observed in two dimensions and was not found to be an artifact (*7*). Figure 3 is a three dimensional photomicrograph of one of these aggregates on bacterial cellulose. CBH I was bound to the fiber but the majority of the enzyme appeared to be bound to non-ribbon cellulose (*9*). This was often seen on both cellulose substrates. Within the apparent non-ribbon cellulose there were also visible individual crystalline fibers to which the enzymes were bound at various planes on the viewing field.

When pretreated aspen cellulose was used, the conserved "wood" structure that is normally associated with cellulose of higher plants was observed (Figure 4). This electron micrograph demonstrated binding of both CBH I and EG I as seen by immunogold size differential. The convex deformation of the formvar (in the middle) demonstrated the depth perception that was obtained with this technique. When viewed in stereo, one wood fiber overlapped the other. At that junction some apparent amorphous cellulose was observed. The CBH molecules located at one end

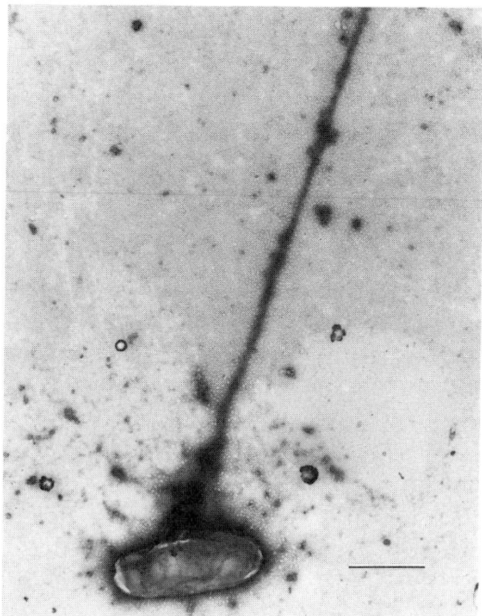

Figure 1. A negative control electron micrograph of anti-CBH I immunogold and bacterial cellulose. The adsorbed bacteria were reacted with the immunogold in the absence of cellulase. Minimal background labelling was observed on all negative controls. Bar = 1 micron.

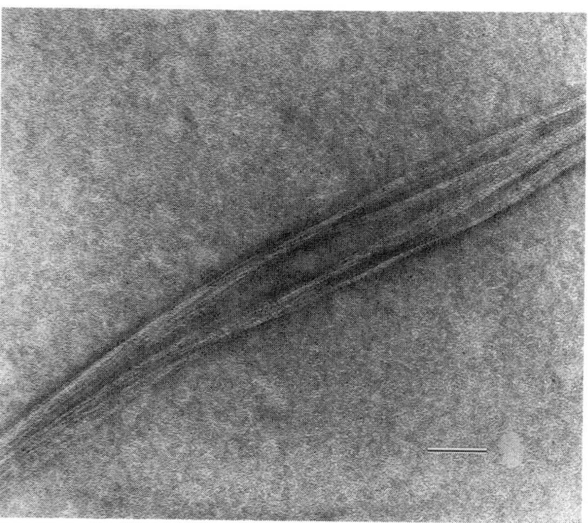

Figure 2. A high magnification photo electron micrograph of bacterial cellulose. The width of the bacterial microfibers ranged from 2.5 nm to 12.5 nm. Note the spiral formation and resolution of the fibers. Bar = 0.1 micron.

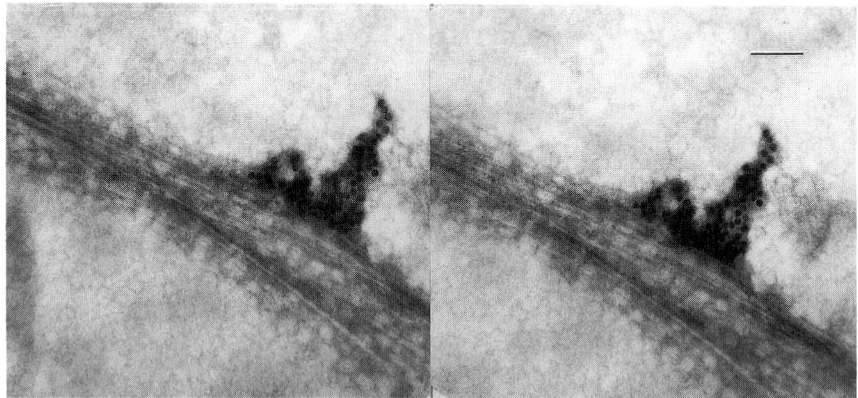

Figure 3. A stereo electron micrograph showing aggregated binding of CBH I to bacterial cellulose fibers. This aggregated binding may actually be to cellulose II which has been displaced. Bar = 0.1 micron.

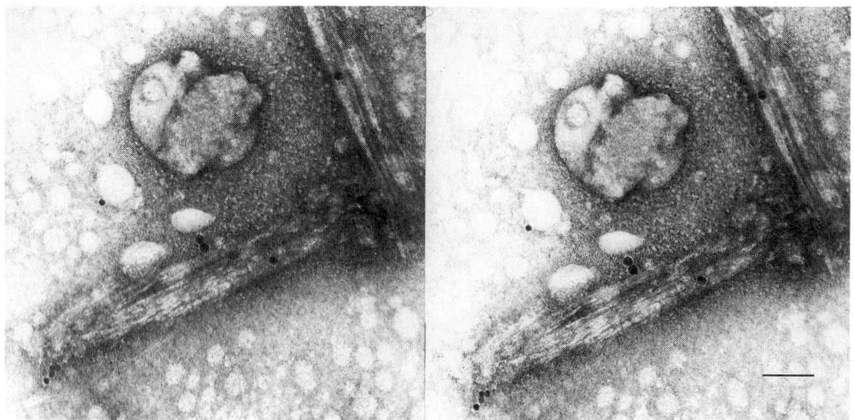

Figure 4. This stereo electron micrograph shows EG I (15 nm) and CBH I (10 nm) bound to intact aspen cellulose. The "woody" nature of the cellulose is still conserved. Bar = 0.1 micron.

of the cellulose fiber may indicate the terminal end containing the parallel non-reducing ends of the cellulose chains. This observation has been noted before in previous studies (*4,7*).

The advantage of obtaining a three dimensional representation as opposed to a two dimensional view was illustrated in the following figures. When Figure 5 was initially examined in two dimensions, it appeared as if an endoglucanase was present adjacent to eleven cellobiohydrolases bound to the terminal end of the aspen fibers. After careful observation in stereo it was revealed that the larger immunogold (EG I) was adjacent to eight cellobiohydrolases but these cellobiohydrolases were bound at different planes in the field of view. The three cellobiohydrolases bound to the terminal end of the microfiber were shown to be located at the plane closest to the grid surface. Logic suggested that the large immunogold label was not floating in space (a physical impossibility), but was actually bound to the terminal end of a bundle of fibers or to amorphous cellulose that projected towards the viewer. This was confirmed by noticing a single fiber(s) extending up from the surface to one of the 10 nm anti-CBH I immunogold spheres (arrow).

Figure 6 illustrated the importance of correct interpretation of electron photomicrographs. Initially, when this field was not observed in stereo, it appeared as if an endoglucanase were closely followed by two cellobiohydrolases along the axis of microfibers. When observed in stereo all three enzymes were at different planes. When this occurred, it was difficult to determine which label was actually bound to an endoglucanase or a cellobiohydrolase, because the difference in distance from one label to the other may have been significant enough to interpret a 10 nm sphere as a 15 nm sphere. Again the sensitivity of these images was revealed as the apparently larger immunogold label is actually bound to an upright fiber(s).

Discussion

Electron microscopy studies of biological materials has proven to be an invaluable tool. Indisputably, electron micrographs can sometimes be misinterpreted without careful examination. Three dimensional electron micrographs are useful for obtaining a more definitive interpretation of the observed field. The observations mentioned in this article are examples indicating the need for more useful visual information such as three-dimensional electron micrographs.

The aggregation of enzyme at a specific site, as observed on the bacterial cellulose was unique. In a previous report, overall binding of CBH I was observed along the long axis of the bacterial cellulose (*5*). The bacterial cellulose used in the present study was not pretreated and therefore may not have had as many non-reducing ends exposed as had the extracted cellulose used by Chanzy. The production of cellulose II by this particular *A. xylinum* strain was observed in other micrographs (data not shown). CBH I bound very well to this particular substrate in aggregates. CBH I binds so well to cellulose II (the amorphous substrate), that were it present at some point along the axis of the fiber, aggregated binding of the enzyme to this cellulose would be expected.

Observations made of immunogold labelling at different planes on aspen cellulose was not unexpected when working with natural cellulosic substrates. The rigidity and insolubility of cellulose could cause this type of observed binding. The fibers were

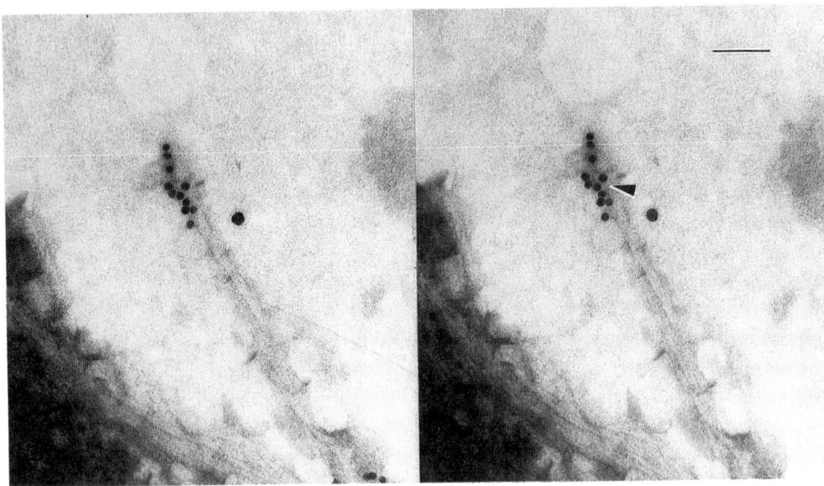

Figure 5. This figure demonstrates the appearance of immunogold in separate planes. On the lowest plane CBH I can be seen bound to the terminal ends of aspen cellulose. On a higher plane several cellobiohydrolases and one endoglucanase are seen. Close observation of this micrograph reveals a cellulose fiber extending from the lower plane of vision to one of the smaller gold particles (arrow). Bar = 0.1 micron.

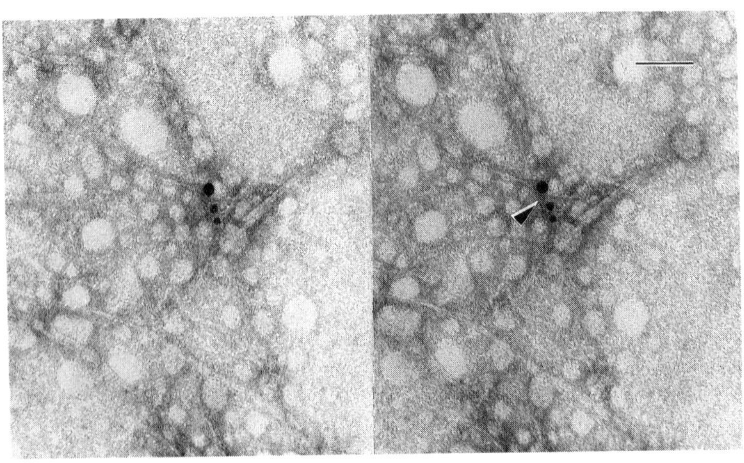

Figure 6. A stereo electron micrograph demonstrating a similar observation as that seen in figure 5. The arrow points to a single cellulose fiber extending from the grid, up to a large gold colloid (arrow). Bar = 0.1 micron.

extending out in all directions. Observations of the kind shown in Figures 5 and 6 were seen quite readily. The point must be made that an "apparent" endoglucanase was observed in Figure 6 since the difference in distance from the surface of the grid to the large immunogold may have been significant enough to have made a 10 nm gold sphere appear as a 15 nm sphere. The sensitivity of the immunolabels and this technique were demonstrated by Figures 5 and 6. The ability to be able to localize individual enzymes on the termini of individual fiber(s) was noteworthy. Without the use of three dimensionality, such interpretations could not have been made. Until three dimensional micrographs are utilized more readily for studies such as these, clarification of electron micrographs will not be complete.

Acknowledgments

This work was funded by the Biofuels Program Office at the National Renewable Energy Laboratory by the Biochemical Conversion Program at the Department of Energy Biofuels and Municipal Waste Technology Division.

Literature Cited

1. Coughlan, M.P. *Biotechnol. Genet. Eng. Rev.* **1985**, *31*, 39-109.
2. Wood, T.M.; McRae, S.I. *Adv. Chem. Ser.* **1979**, *181*, 181-209.
3. Goyal, A.; Ghosh, B.; Eveleigh, D. *Bioresource Technol.* **1991**, *36*, 37-50.
4. Chanzy, H.; Henrissat, B. *FEBS Lett.* **1985**, *184*, 285-288.
5. Chanzy, H.; Henrissat, B.; Vuong, R. *FEBS. Lett.* **1985**, *172*, 193-197.
6. White, A.R.; Brown, R.M. *Proc. Natl. Acad. Sci. USA* **1981**, *78*, 1047-1051.
7. Nieves, R.A.; Ellis, R.P.; Todd, R.J.; Johnson, T.J.A.; Grohmann, K.; Himmel, M.E. *Applied and Environ. Microbiol.* **1991**, *57*, 3163-3170.
8. Hestrin, S.; Schramm, M. *Biochem. J.* **1954**, *58*, 345-352.
9. Roberts, E.M.; Saxena, I.M.; Brown, R.M. In *Cellulose and Wood Chemistry Technology Proceedings of the Tenth Cellulose Conference,* Schuerch, C., Ed.; John Wiley & Sons: New York, NY, 1984; 689-704.

RECEIVED March 17, 1994

Bioenergy: Biodiesel and Electrolytic Cells

Chapter 12

Deposition of Metallic Platinum in Blue–Green Algae Cells

G. Duncan Hitchens[1], Tom D. Rogers[1], Oliver J. Murphy[1], Comer O. Patterson[2], and Ralph H. Hearn, Jr.[2]

[1]Lynntech, Inc., 7610 Eastmark Drive, Suite 105, College Station, TX 77840
[2]Department of Biology, Texas A&M University College Station, TX 77843

A method for placing metallic platinum in contact with the photosynthetic membranes of the unicellular blue green alga or cyanobacterium *Anacystis nidulans* (Synechococcus sp.) is described. It was found that cells treated in this way were capable of forming hydrogen gas when illuminated. The deposited platinum particles acted as a catalyst for the generation of hydrogen from photosynthetic light reactions in the absence of an added exogenous electron transfer agent. This exploratory work indicates that electron transfer can occur directly between the membrane-bound Photosystem I and the Pt particles. Electron micrographs of platinum treated algae show deposits of platinum at the surfaces of the internal photosynthetic membranes. The work has long-term implications for the use of cyanobacteria cells for the photoproduction of hydrogen fuel. The innovative aspect of the research has been to demonstrate a technique for placing metallic conductors in direct contact with the membrane structures of microorganisms. This approach can lead, for example, to new types of selective electrochemical biosensors.

Hydrogen has long been recognized as a future fuel and energy transfer medium (*1,2*). Hydrogen is obtained from abundantly available water; however, a primary source of energy is required to decompose water molecules. Biocatalytic processes are attractive for coupling solar energy to the decomposition of water because photosynthetic organisms already posses light harvesting structures (chlorophyll and accessory pigments) and associated redox capabilities of Photosystems I and II which are capable of carrying out dissociation of water and charge separation.

In numerous studies, photosynthetic electron transport in thylakoid membranes isolated from higher plants has been linked to hydrogen production (see reference *3* for a review of all aspects of photobiological hydrogen production). The coupling of photosynthesis to the production of hydrogen was first demonstrated in cell free

0097–6156/94/0566–0246$08.00/0

systems consisting of isolated spinach chloroplasts (*4-6*), with bacterial hydrogenase as the catalyst and ferredoxin as the electron transfer mediator (see Figure 1a). Instability of the hydrogen forming enzyme is a problem encountered in these systems. Platinum has been introduced as the hydrogen forming catalyst to provide greater stability (*7*), however, typically methyl viologen is required as the exogenous mediator (see Figure 1b) which itself is unstable in the presence of oxygen. Greenbaum (*8-13*) has demonstrated the photoproduction of hydrogen and oxygen using a platinum catalyst deposited directly onto isolated thylakoid membranes. Platinum was deposited from a dissolved platinum salt using hydrogen as the reductant. It was assumed that the metallic platinum formed was in direct contact with the reducing side of Photosystem I in such a way that electron transfer could occur *directly* between the membrane-bound Photosystem I and the platinum (see Figure 1c). The inorganic catalyst is relatively stable in the presence of oxygen and no other exogenous electron transfer mediator is required. Nevertheless systems based on isolated thylakoid membranes have short active lifetimes even when elaborate immobilization procedures are used (*14-16*). Blue green algae have photosystems that are much more stable than those of chloroplasts and can retain photosynthetic electron transport activities for many months (*17,18*). This paper presents a new approach to coupling the photosynthetic electron transport activities of the blue green alga *Anacystis nidulans* to metallic platinum as the hydrogen forming catalyst, without an exogenous electron transfer mediator (*19*).

Cell Growth and Platinum Treatment Procedures

Cells of *Anacystis nidulans* strain R2 (*Synechococcus* sp., PCC 7942) were grown on Medium Cs in a continuous culture apparatus described previously (*19*). After harvesting, the cells were washed and resuspended in 50 mM phosphate buffer (pH 7.1) containing 3 mM $MgCl_2$. The platinum deposition step was analogous to that described by Greenbaum for the platinization of thylakoid membranes (*8*). A solution containing 5.34 mg/mL of hexachloroplatinic acid in phosphate buffer was neutralized to pH 7 with NaOH. One mL of this solution was combined with 5 mL of cell suspension. The 6-mL volume was placed in a sealed glass vessel fitted with inlet and outlet ports. The cells were allowed to incubate in the platinum salt solution in the dark usually for 14 hours. After incubation, the deposition of platinum metal particles was carried out by passing molecular hydrogen (see Eq. 1) through the headspace above the sample in the sealed vessel.

$$Pt(Cl)_6{}^{2-}{}_{(aq)} + 2H_{2(g)} \dashrightarrow Pt_{(s)} + 6Cl^-{}_{(aq)} + 4H^+{}_{(aq)} \qquad (1)$$

Typically, hydrogen was passed into the vessel for 30 m. During this time, the color of the cell suspension darkened and metallic platinum (platinum black) particles were visible in the suspension and adhered to the walls of the vessel. The platinum treated cell suspension was filtered and washed before being used further.

Electron Micrograph Studies of Platinum Treated Cells

Electron micrographs (EMs) of intact cells of *Anacystis* that have been platinized are shown in Figures 2 and 3. After filtering and washing, the cells were fixed in glutaraldehyde-cacodylate solution, embedded in Spurr's resin and sectioned for EM. The cells were not stained. The optically dense regions are metallic platinum particles. In the frame is one cell that shows metal like particles distributed throughout its internal structure. Figure 2 also shows that not all the cells were platinized to the same degree. Large Pt particles can be seen in areas corresponding to the bulk solution and adhered to the outer surfaces of one of the cells. Examination of the close-up (i.e., Figure 3) shows the presence of platinum microparticles in close proximity to the intracellular photosynthetic membranes. Moreover, the deposited platinum particles follow the outline of the thylakoids which, in *A. nidulans,* appear to be arranged in concentric shells. Nierzwicki-Bauer et al (*20*) have shown that in a marine coccoid cyanobacterium, photosynthetic thylakoids are arranged in concentric shells. Significantly, these micrographs indicate that the cell membrane and the outer thylakoid layers do not act as a significant permeability barrier limiting the amount of intracellular $[Pt(Cl_6)]^{2-}$ if a sufficient incubation time in the platinum salt solution is used. Figure 4 was included for comparison purposes. It shows *A. nidulans* cells that had neither been incubated in the platinum salt solution or had been subject to hydrogen treatment. There were no indications of the presence of metallic platinum in these cells.

Hydrogen Production Measurements

Experiments were conducted to determine if platinum treated cells could produce hydrogen when illuminated. Measurements of light-dependent hydrogen formation were made amperometrically (*21,22*). This method is known to be sensitive to low aqueous phase hydrogen concentrations and responds rapidly. Before making the measurements, the platinization step was carried out as described except that samples of the cell suspension were withdrawn from the sealed vessel at intervals of 5, 15 and 30 m after starting hydrogen gas flow into the head space. By varying the exposure to hydrogen, the aim was to vary the amount of metallic platinum deposited in the cells. Two identical samples were taken at each time given above. Light-dependent hydrogen evolution was measured using one sample, and photosynthetic oxygen evolution was measured from the other.

Hydrogen measurements were carried out in a thermostated 0.6 mL volume reaction cell (Diamond General, Inc., Model 1271) described in reference *19*. Light from a projector lamp was passed through a cut-off filter (λ=above 600 nm) before being directed onto the front of the sample chamber and through an optical window. Prior to each hydrogen measurement, electrode conditioning was carried out by potential cycling between anodic and cathodic limits, typically +0.6 and -0.6V vs Ag/AgCl at 100 mV sec^{-1}, for 10 m. Careful choice of working electrode potential was also necessary for hydrogen detection. The sensitivity to hydrogen was measured over a range of potentials by making 50 μL additions of hydrogen-saturated phosphate buffer solution to the reaction cell. The system was most sensitive to hydrogen at 0.35 V vs Ag/AgCl which was used subsequently for all hydrogen

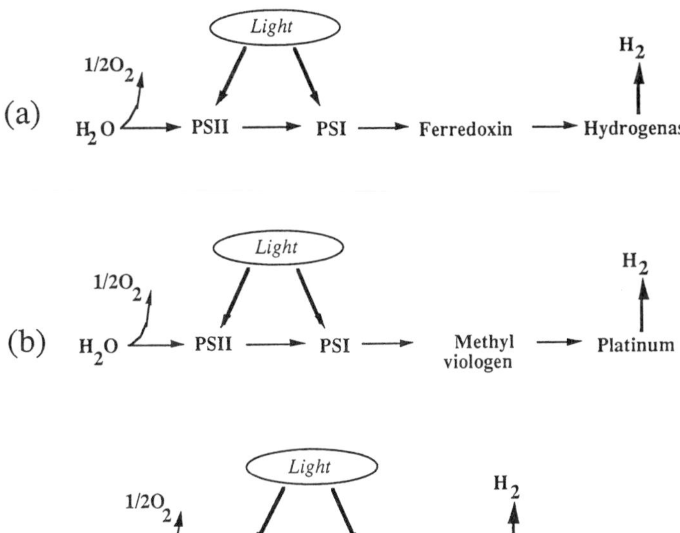

Figure 1. Schemes for Light-Driven Photobiological Hydrogen Formation from Water.

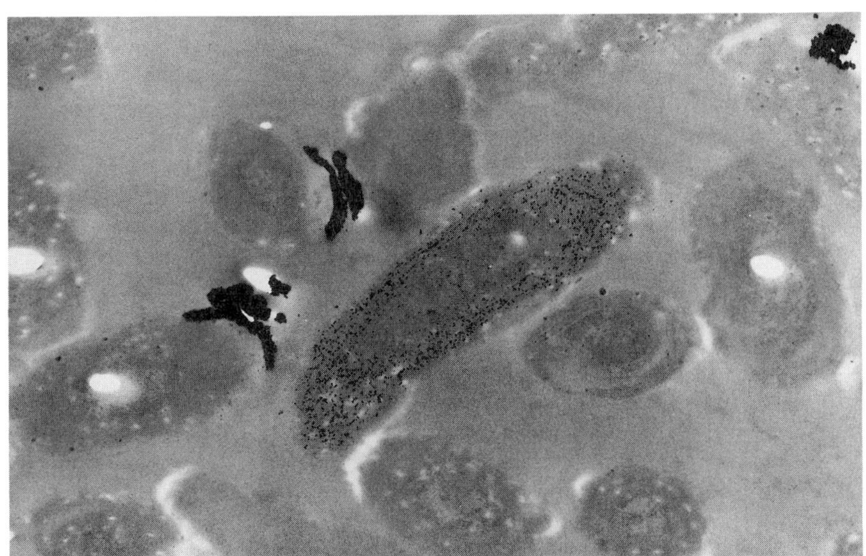

Figure 2. A Transmission Electron Micrograph of Platinized cells of *Anacystis nidulans*: (magnification x 30,000).

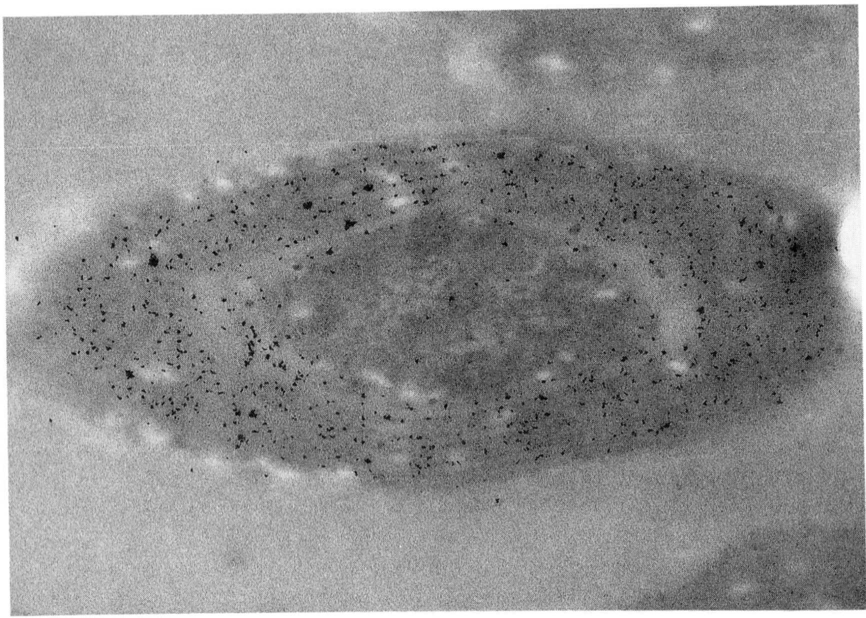

Figure 3. A Transmission Electron Micrograph of a Platinized *Anacystis nidulans* cell (magnification x 96,000).

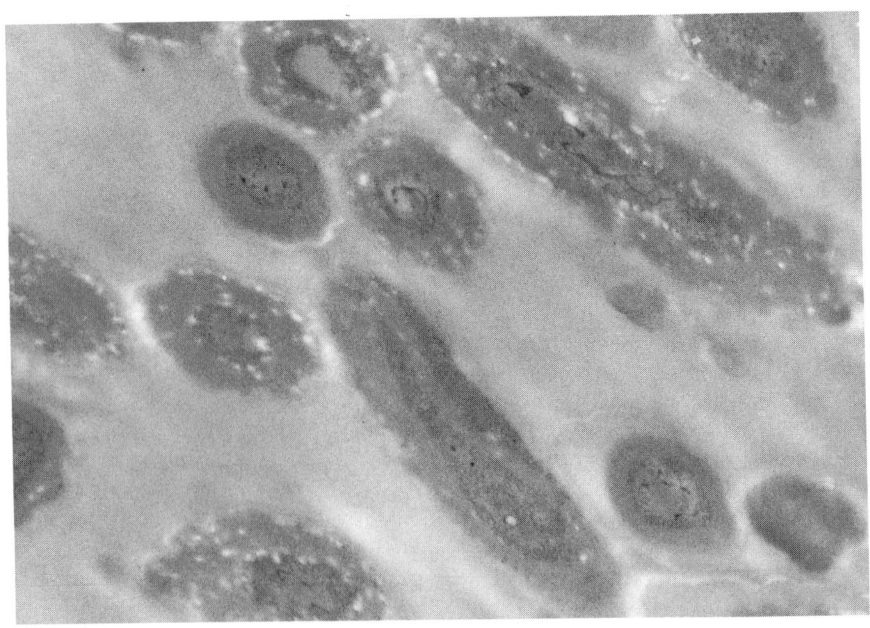

Figure 4. Untreated *Anacystis nidulans* Cells (magnification x 30,000)

determinations. Above this potential there was a pronounced decrease in sensitivity of the amperometric technique to hydrogen. This is expected since at neutral pH, the onset of oxide film formation on the platinum working electrode surface occurs at around 0.4 V vs Ag/AgCl. Platinum oxide is a less catalytic surface for hydrogen oxidation than metallic platinum. Before each run, the system was calibrated with additions of hydrogen-saturated buffer. The response time after injection of hydrogen-saturated buffer was rapid (10-20 s). Variation in sensitivity to hydrogen was observed between experimental runs. The current was specific to hydrogen since the electrochemical reactant must be a non-ionized small molecule in order to penetrate the membrane and oxidizable at 0.35 V vs Ag/AgCl. The same set-up was routinely used for oxygen evolution measurements except that a working electrode potential of -0.65 V vs Ag/AgCl was used.

The results of light-dependent hydrogen and oxygen production are given in Table I. The samples of the cell suspension taken immediately before beginning the platinum deposition step (platinum deposition time = 0 m) did not produce detectable amounts of hydrogen when illuminated. The sample taken after 5 m of platinum deposition produced measurable quantities of hydrogen. The amount of hydrogen generated increased with duration of the platinum deposition step until reaching a maximum at 30 m. To continue platinization of the cell suspension beyond 30 m was counter-productive in terms of hydrogen production. Oxygen production decreased in cells that had been subject to platinum deposition. The reduction in oxygen production could have been due to damage to the alga's photosynthetic apparatus by metallic platinum deposits or possibly the presence of the finely dispersed platinum catalyst encourages oxygen and hydrogen recombination to water.

It was assumed that since hydrogen formation occurred in the absence of an exogenous electron transfer agent, the metallic platinum formed was in direct contact with the reducing side of Photosystem I in such a way that electron transfer could occur *directly* between the membrane-bound Photosystem I and the platinum as illustrated in Figure 5 (see also *10*). The photosynthetic apparatus of blue green algae is functionally, structurally and molecularly similar to that of chloroplasts from eukaryotic organisms (*23-25*). Blue green algae have the ability to perform oxygenic photosynthesis and photosynthetic reactions take place in thylakoids. The thylakoids contain two reaction centers (Photosystems I and II) in which the photo-induced electron transfer reactions occur. The two photosystems act in series and encompass the redox span from $H_2O/^1/_2O_2$ to $NADP^+/NADPH$ (nicotinamide adenine dinucleotide phosphate). The electron transfer reactions are vectorial in nature with the reaction centers oriented such that the oxidation of water by PS II occurs on the inner face of the membrane (i.e., facing the thylakoid space) with electrons emerging from PS I on the opposite, or stroma, side of the membrane (*26,27*). Details of the organization of the integral membrane components making up PS I core complex are still unclear, but it appears that several component polypeptides are exposed on the stromal face of the thylakoid in higher plants (*28*). Still less information is available concerning organization of the PS I complex in cyanobacteria, but published data indicate similarities in organization between the cyanobacterial complex and that of higher plants (*29*). In cyanobacteria, the cytoplasmic face of thylakoids corresponds to the stromal face of thylakoids in higher plants. We therefore assume that some metallic platinum particles are deposited on the cytoplasmic faces of cyanobacterial thylakoids in immediate proximity to surface-exposed components of the PS I core complex.

Table I.
Hydrogen and Oxygen Formation Rates for Platinum Treated Cells

Time used for Pt Deposition (min)	Hydrogen (nmol h^{-1} mL^{-1})	Oxygen (nmol h^{-1} mL^{-1})
0	0.0	1730
5	12.3	1225
15	163.0	1170
30	655.0	1035

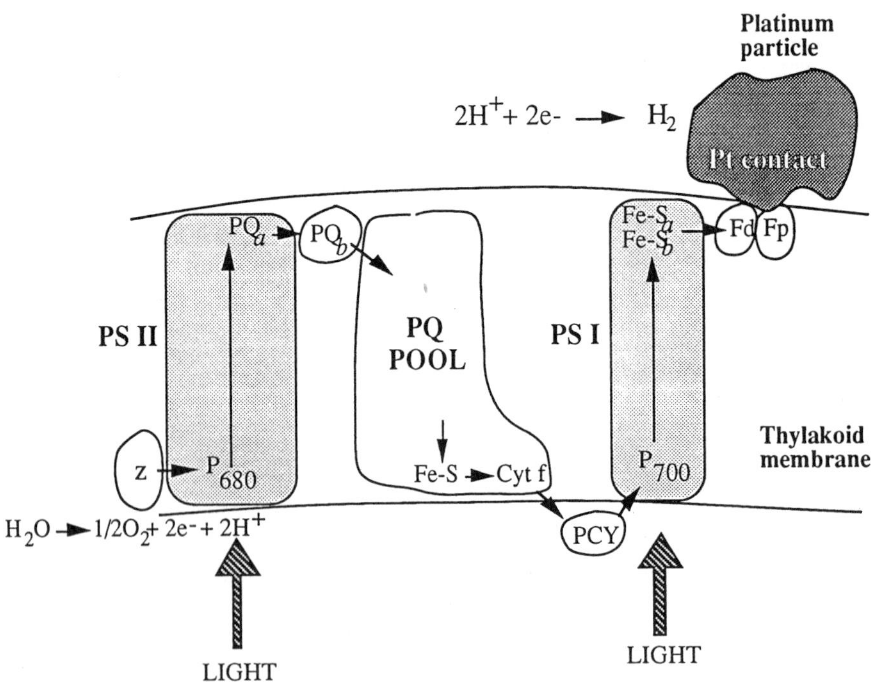

Figure 5. Photosynthetic Electron Transport in Blue Green Algae Showing Metallic Platinum in Electrical Contact with the Reducing Side of PS I (adapted from references *23-25*).

Table I shows that the highest hydrogen photoproduction rates attained were approximately 25% of the rate expected if all the electrons derived from the decomposition of water were used for hydrogen formation. In contrast, for platinized thylakoid membranes isolated from higher plants (*8*), steady state rates of oxygen and hydrogen production indicated that all electrons derived from PS II water splitting were used for hydrogen generation. Hydrogen production may have been limited because not all PS I reducing sites were in electronic contact with a hydrogen forming catalyst particle. The electron micrographs show that only selected cells contained platinum particles which can account for the low hydrogen yields compared to the oxygen yields.

Summary and Conclusions

Photobiological hydrogen production was achieved indicating that the platinum metal particles are in direct electronic contact with PS I. Whether this system offers the advantage of stable light-driven hydrogen production over long periods of time remains to be determined. In addition, many breakthroughs are required (*3*) to make photobiological hydrogen production systems both technically and economically viable as a means of fuel generation e.g., for hydrogen powered vehicles (*30*). The present approach has the cost disadvantage of requiring a noble metal catalyst. The work has, however, demonstrated a first step towards a viable means of making electrical contact to the charge transfer complexes contained within cells, particularly the redox complexes associated with membrane energy transduction. Usually, freely diffusing redox mediators are needed to relay electrons *indirectly* between the redox centers of proteins and electrodes (*31*). Methods that enable *direct electrical contact* between biological molecules and electrodes are being developed at a fast pace (*32-34*); however, these methods have largely focused on soluble redox proteins and enzymes. Direct electrochemical studies of intact multi-enzyme structures associated with photosynthetic (and respiratory) membranes is a new challenge. The technique described in this paper may have some long-term relevance to this problem. Being able to make direct electrical contact to membrane proteins can have practical benefits in several areas including: selective biosensors (*35,36*), biomolecular devices (10), and biofuel cells (*37*).

Acknowledgment

We wish to thank the National Science Foundation (Small Business Innovation Research Award ISI-8961216) for support of this work.

References

1. Bockris, J. O'M. *Energy Options*; Australia & New Zealand Book Company: Sydney, Australia, 1980.
2. Gutmann, F.; Murphy, O.J. In *Modern Aspects of Electrochemistry*; Bockris J. O'M; White R.E.; Conway B.E., Eds.; Plenum Press: New York, NY, 1983, Vol. 15; p 1.
3. Weaver, P.F.; Lien, S.; Seibert, M. *Solar Energy* **1980**, *24*, 3.
4. Arnon, D.; Mitsui, A.; Peneque, A. *Science* **1961**, *134*, 1425.

5. Beneman, J.R.; Berenson, J.A.; Kaplan, N.O.; Kamen, D. *Proc. Nat. Acad. Sci. U.S.A.* **1973**, *70*, 2317.
6. Greenbaum, E. *Biotechnol. Bioeng. Symp.* **1980**, *10*, 1.
7. Gisby, P.E.; Hall, D.O. *Nature* **1980**, *287*, 251.
8. Greenbaum, E. *Science* **1985**, *230*, 1373.
9. Greenbaum, E. *J. Phys. Chem.* **1988**, 92, 4571.
10. Greenbaum, E. In *Molecular Electronic Devices*; Carter, F.L.; Siatkowski, R.E.; Wohltjen, H., Eds.; Elsevier Science Publishers B.V.: North Holland, 1988; p 575.
11. Greenbaum, E. *Bioelectrochem. Bioenerg.* **1989**, *21*, 171.
12. Greenbaum, E. *J. Phys. Chem.* **1990**, *94*, 6151.
13. Greenbaum, E. *J. Phys Chem.* **1992**, *96*, 514.
14. Roa, K.K.; Rosa, L.; Hall, D.O. *Biochem. Biophys. Res. Comm.* **1976**, *68*, 21.
15. Woodward, J.; Greenbaum, E. *Biotechnology and Bioengineering Symp.* **1983**, *13*, 271.
16. Coquempot, M.F.; Thomasset, B.; Barbotin, N.J.; Gellf, G.; Thomas, D. *Eur. J. Appl. Microbiol. Biotechnol.* **1981**, *11*, 193.
17. Affolter, D.; Hall, D.O. *Photobiochem. Photobiophys.* **1986**, *12*, 193.
18. Gisby, P.E.; Rao, K.K.; Hall, D.O. *Method. Enzymol.* **1987**, *135*, 440.
19. Hitchens, G.D.; Rogers, T.D.; Murphy, O.J.; Patterson, C.O. *Biochim. Biophys. Res. Comm.* **1991**, *175*, 1029.
20. Nierzwicki-Bauer, S.A.; Balkwill, D.L.; Stevens, S.E. Jr. *J. Cell Biol.* **1983**, *97*, 713. 21.Wang, R.; Healey, H.L.; Myers, J. *Plant Physiol.* **1971**, *48*, 108.
22. Jones, L.W.; Bishop, N.I. *Plant Physiol.* **1976**, *57*, 659.
23. Cramer, W.A.; Crofts, A.R. In *Photosynthetic Energy Conversion in Plants and Bacteria*; Govindjee, Ed.; Academic Press: New York, NY, 1982; p 387.
24. Stewart, A.C.; Bendall, D.S. *Biochem. J.* **1981**, *194*, 877.
25. Guikema, J.; Sherman, L. *Biochim. Biophys. Acta*, **1982**, 440.
26. Harold, F.M. *The Vital Force: A Study of Bioenergetics*; W.H. Freeman: New York, NY, 1986.
27. *Topics in Photosynthesis*; Barber, J., Ed.; Elsevier: Amsterdam, Holland, 1987, Vol. 8.
28. Ortiz, W.; Lam, E.; Chollar, S.; Munt, D.; Malkin, R. *Plant Physiol.* **1985**, *77*, 389.
29. Wynn, R.M.; Omaha, J.; Malkin, R. *Biochemistry* **1989**, *28*, 5554.
30. Lemons, R.A. *J. Power Sources* **1990**, *29*, 251.
31. Turner, A.P.F. In *Analytical Uses of Immobilized Biological Compounds for Detection, Medical and Industrial Uses*; Guilbault, G.G.; Mascini, M., Eds.; Reidel Publishing Co.: Boston, MA, 1988; p 131.
32. Armstrong, F.A.; Cox, P.A.; Hill, H.A.O.; Lowe, V.J.; Oliver, B.N. *J. Electroanal. Chem.* **1987**, *217,* 331.
33. Frew, J.E.; Hill, H.A.O. *Eur. J. Biochem.* **1988**, *172*, 261.
34. Hitchens, G.D. *Trends Biochem. Sci.* **1989**, *14*, 152.
35. *Biosensors: Fundamentals and Applications*; Turner, A.P.F.; Karube, I; Wilson, G.S., Eds.; Oxford University Press: Oxford, Great Britain, 1987.
36. Yacynych, A.M. In *Bioinstrumentation, Research Developments and Applications*; Wise, D.L., Ed.; Butterworth Publishers: Stoneham, MA, 1990; p 1317.
37. Bennetto, H.P. *Life Chemistry Reports*; **1984**, *2*, 303.

RECEIVED January 19, 1994

Chapter 13

Genetic Engineering Approaches for Enhanced Production of Biodiesel Fuel from Microalgae

P. G. Roessler, L. M. Brown, T. G. Dunahay, D. A. Heacox, E. E. Jarvis, J. C. Schneider, S. G. Talbot, and K. G. Zeiler

National Renewable Energy Laboratory, 1617 Cole Boulevard, Golden, CO 80401–3393

Microscopic algae ("microalgae") have great potential as a future source of biological lipids for use in biodiesel fuel production. We are investigating the biochemistry and molecular biology of lipid biosynthesis in these organisms, and are developing genetic transformation systems that should allow us to genetically engineer microalgal strains for enhanced biodiesel production capabilities. As a result of these studies, we have isolated a full-length gene for the key lipid biosynthetic enzyme acetyl-CoA carboxylase. In addition, we have isolated portions of the genes that encode nitrate reductase and acetolactate synthase as a first step in the development of homologous selectable markers for microalgal transformation.

Diminishing global reserves of petroleum have prompted a search for alternative, renewable liquid fuels. One type of alternative fuel receiving consideration is derived from biological lipids. This fuel, referred to as "biodiesel," is produced by a simple transesterification process that converts glycerolipids into methyl or ethyl esters of fatty acids, along with glycerol as a byproduct. A small but burgeoning biodiesel industry is emerging, using oilseeds and waste animal fats as the primary lipid sources. Facilities with a total biodiesel production capacity of 32 million gallons (125 million liters) per year are currently operating in Europe, where high taxation of petroleum-derived fuels has stimulated the market for biodiesel (*1*). There are plans to increase the annual capacity of European facilities to more than 200 million gallons (800 million liters) in the next few years. On the other hand, the low cost of petroleum-derived diesel fuel in the United States has resulted in very limited commercial activity involving biodiesel in this country.

Environmental concerns are playing a major role in promoting the commercialization of biodiesel (*2*). Biodiesel is a cleaner burning fuel than conventional diesel. Because of the naturally low concentration of sulfur in biodiesel, the production of sulfur oxides during combustion is greatly reduced.

0097–6156/94/0566–0255$08.00/0

Emissions of carbon monoxide and particulates are also lower for biodiesel. Consequently, the use of biodiesel in urban areas of the United States that have poor air quality may help these areas to attain compliance with the Clean Air Act Amendments of 1990. Furthermore, unlike petroleum-derived diesel, biodiesel exhibits low toxicity and is biodegradable. This latter property has focused attention on the advantages of using lipid-based oils and hydraulic fluids in environmentally sensitive areas. Another benefit of biodiesel is that it can be produced photosynthetically, using carbon dioxide as the primary feedstock. This "greenhouse gas" is thereby recycled, resulting in a lower rate of atmospheric accumulation.

The small current market for biodiesel fuel can be satisfied by existing oilseed production capacities. As the market for biodiesel increases, however, additional sources of biological lipids will be needed (*3*). These new sources will have to be produced in large quantities, must not compete for resources (e.g., land and water) that are currently being used for conventional agriculture (or that will be needed for the future production of lignocellulosic energy crops), and must utilize inexpensive feedstock materials. With regard to this latter point, it is worth noting that the photosynthetic production of biodiesel from carbon dioxide has a distinct advantage over the heterotrophic production of biodiesel from expensive organic carbon sources.

Microscopic eukaryotic algae ("microalgae") are promising candidate organisms that may be able to fulfill the need for an additional lipid source. Certain strains of microalgae are able to produce up to 60% of their total cellular mass as lipid (*4*). Of equal importance is the fact that many microalgae grow extremely well in saline water that is unsuitable for crop irrigation. In addition, many microalgal strains are able to tolerate widely fluctuating temperatures and high light intensities. Consequently, the desert regions of the southwestern United States, which have large stores of saline groundwater, could be used for microalgal mass culture. These areas cannot be used for conventional agriculture, meaning that competing land uses would be minimal. Furthermore, because carbon dioxide is required in large quantities as a nutrient for microalgal mass culture, it is worth noting that coal-burning electric power plants are located in close proximity to some of the areas deemed suitable for microalgal mass culture. The carbon dioxide present in the flue gas produced by these facilities would therefore be available for utilization by microalgae for biodiesel production (*5*).

Massive accumulation of lipids in microalgae typically occurs only when the cells are environmentally stressed, but under these conditions the growth rates of cultures are greatly reduced, which consequently limits the overall rate of lipid production. Nutrient limitation is one type of stress that often leads to lipid accumulation coupled to reduced growth rates (*4*). It would be desirable to manipulate a microalgal strain so that it exhibits greater overall lipid productivity. One possible way to achieve this goal is through genetic engineering. This paper describes some of the research being carried out at the National Renewable Energy Laboratory (NREL) in an effort to develop microalgal strains that have enhanced biodiesel production capabilities. Current research efforts are focused primarily on the physiology, biochemistry, and molecular biology of oil-producing microalgae.

Biochemical Studies

The main goal of our biochemical research is to better understand the enzymatic pathways involved in microalgal lipid biosynthesis. A detailed knowledge of these pathways is necessary prior to any attempts to genetically engineer microalgae for altered lipid production capabilities. We are interested in identifying enzymes that influence the overall rate of lipid synthesis and the quality of the lipids produced. Identification of rate-controlling enzymes may eventually allow us to manipulate the activity of these enzymes in a manner that leads to an overall improvement in biodiesel production rates. Likewise, characterization of the key steps involved in fatty acid modification may allow us to control the nature of the lipids produced, which has implications regarding the quality of the final fuel product.

Fatty Acid Modification. We recently began a project designed to examine the biochemical processes that control the chain length and level of unsaturation of fatty acids in microalgae. Chain length and double bond number affect fuel quality factors such as oxidative stability, ignitability, and combustion performance (*6*). For instance, the thermal efficiency decreases with increasing carbon chain length (*7*). Likewise, although a single carbon-carbon double bond in the acyl chain helps to reduce the freezing point and increase the thermal efficiency of the fuel (*7*), greater numbers of double bonds impair the combustion performance (*6*). The presence of more than two double bonds also decreases the oxidative stability of biodiesel fuel. An ideal diesel fuel consists of medium to long unbranched hydrocarbon chains containing a single double bond (*7*).

Storage triacylglycerols (TAGs) from many microalgae are composed primarily of C_{14} to C_{18} fatty acids that are saturated or monounsaturated (*4*). The methyl or ethyl esters of these fatty acids would therefore function very well as a diesel fuel. Conversely, membrane lipids in most microalgae are rich in elongated and polyunsaturated fatty acids, and therefore are less desirable as a biodiesel feedstock. Because these less desirable fatty acids comprise from 20 to 80 mol % of the total (depending on the species and growth conditions), it would be advantageous to be able to manipulate the chain length and unsaturation level of membrane lipid fatty acids in order to improve fuel quality. In this respect, it is worth noting that mutants of the higher plant *Arabidopsis thaliana* have been isolated that are defective in certain fatty acid desaturation steps, and yet no obvious physiological effects of these mutations are exhibited under normal growth conditions. One such mutation reduces the proportion of linolenic acid (a fatty acid having eighteen carbon atoms and three double bonds, which in standardized fatty acid nomenclature is designated as 18:3) in leaves from 49% to 19%, with a concomitant increase in the level of linoleic acid (a fatty acid having eighteen carbon atoms and two double bonds, designated as 18:2) (*8*). There may be other situations, however, in which a severely altered fatty acid composition might negatively impact the growth of a mutant strain.

The unicellar marine alga *Nannochloropsis* is an excellent organism in which to study microalgal fatty acid elongation and desaturation. One strain of

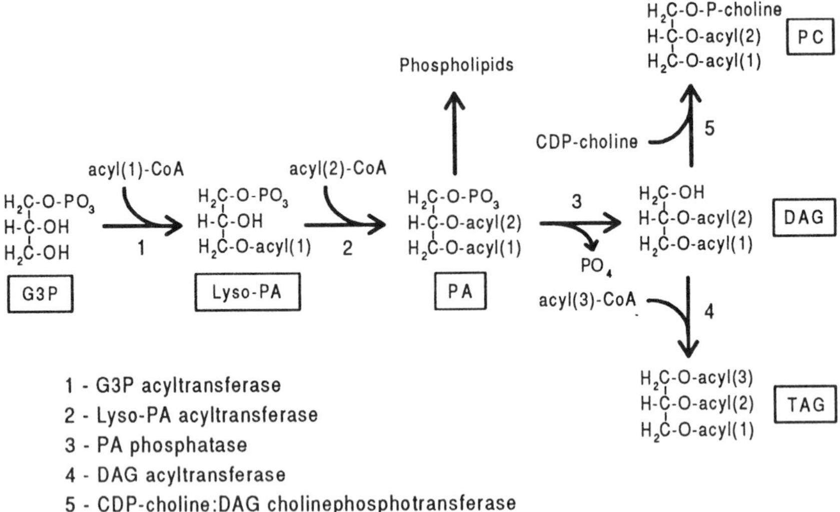

Figure 1. Pathways of glycerolipid biosynthesis.

Nannochloropsis has been shown to accumulate up to 55% of its cell mass as lipid under nitrogen-deficient conditions (*9*). Furthermore, light intensity affects the ratio of storage TAGs to photosynthetic membrane lipids in *Nannochloropsis*, thereby altering the ratio of medium/long chain, saturated/monounsaturated fatty acids to very long chain, polyunsaturated fatty acids (*10*). This model system should allow us to examine various biochemical aspects of fatty acid modification.

The synthesis of TAGs and many polar membrane lipids occurs within the microsomal membranes via the pathway shown in Figure 1. Two acyltransferases are required to produce phosphatidic acid (PA) from glycerol-3-P (G3P) and acyl-coenzyme A (acyl-CoA) molecules. This PA can be used directly as a substrate for polar lipid synthesis, or can be dephosphorylated to diacylglycerol (DAG), which is the substrate for the synthesis of TAG, phosphatidylcholine (PC), and galactolipids. It should be noted that PC can undergo an acyl exchange reaction at the *sn*-2 position with acyl-CoA; this reaction is catalyzed by lyso-PC acyltransferase. This latter point is important, in that fatty acids esterified to PC in this position are subject to desaturation by a PC-specific desaturase in higher plants.

The acyltransferases involved in microsomal lipid synthesis in higher plants have been shown to exhibit pronounced preferences for specific acyl-CoA molecules (e.g., *11, 12*). The acyl specificities exhibited by the lyso-PA and DAG acyltransferases are even more pronounced when the glycerolipid substrates contain certain fatty acids. These acyl specificities, along with specificities for particular glycerolipid substrates exhibited by enzymes involved in polar lipid biosynthesis, are believed to play a major role in determining the final acyl compositions of the various lipid classes. Thus, our initial studies involving the biochemistry of lipid biosynthesis in *Nannochloropsis* were designed to provide information on acyltransferase substrate specificities.

Microsomal membrane preparations were made from mid-log phase *Nannochloropsis* cells by a procedure that includes grinding the cells in the presence of glass beads, low-speed centrifugation to remove cellular debris, and ultracentrifugation at 100,000 x g to pellet the membranes. The membranes were incubated with [^{14}C]palmitoyl-CoA (16:0-CoA) or [^{14}C]oleoyl-CoA (18:1-CoA) in the presence of G3P for specific periods of time, after which the lipids were extracted with organic solvents and separated by one-dimensional thin layer chromatography (TLC) to separate neutral lipids or two-dimensional TLC to separate polar lipids. Areas of the TLC plates containing radioactivity (as determined by autoradiography) were scraped into vials for scintillation counting. These experiments indicated that 16:0-CoA is incorporated most readily into PA, while 18:1-CoA is incorporated most readily into PC (see Figure 2). The labeling of PC is probably due to acyl exchange with previously synthesized PC, rather than to *de novo* synthesis of PC, since CDP-choline (a substrate needed for PC synthesis) was not included in the assay mixture. In higher plants, 18:1 acyl groups incorporated into PC are usually subjected to further desaturation. PC may serve as a site for desaturation in *Nannochloropsis* as well, since the bulk of cellular unsaturated C_{18} fatty acids are found in PC. Preliminary results also suggest that 16:0-CoA is incorporated more strongly into PA than 18:1-CoA, leading to proportionally more label in DAG and TAG. The apparent difference in incorporation of 16:0-CoA and 18:1-CoA into PA implies that acyl chain specificity is exhibited for one or both of the acyltransferases that form this lipid. Although lyso-PA is present in very small quantities in membranes, a small amount of labeled compound that comigrates with this lipid intermediate was observed on TLC plates. This lipid was labeled to about the same extent regardless of whether 16:0-CoA or 18:1-CoA was provided (data not shown), suggesting that the enzyme that catalyzes the synthesis of lyso-PA (G3P acyltransferase) does not have a strong preference for one of these acyl chains over the other. If true, then the specificity for formation of PA must be exercised at the second step, lyso-PA acyltransferase. Indirect evidence has also suggested that the last enzyme of TAG biosynthesis, DAG acyltransferase, does not have a strong preference for particular acyl-CoA molecules.

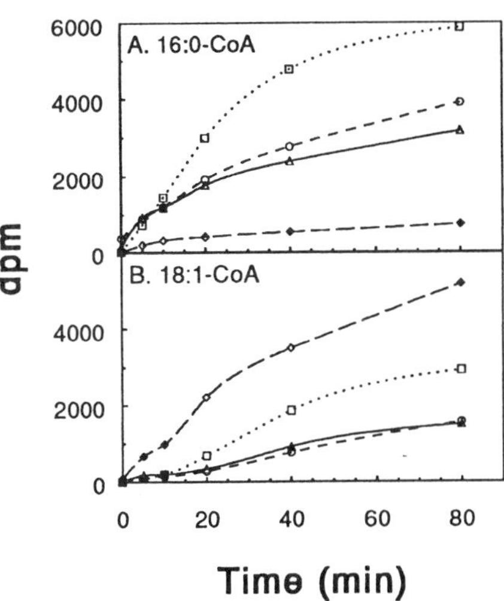

Figure 2. Incorporation of acyl-CoAs into lipids in *Nannochloropsis sp.* Symbols: (O) DAG; (Δ) TAG; (□) PA: and (◊) PC.

Our preliminary conclusions are that the enzyme that acylates the second position of the glycerol backbone, lyso-PA acyltransferase, has a high substrate specificity, whereas G3P acyltransferase and DAG acyltransferase are less discriminating. Additionally, lyso-PC acyltransferase prefers 18:1-CoA over 16:0-CoA. Confirmation of these conclusions will require substrate competition studies in which mixtures of acyl-CoAs (including acyl chain lengths greater than 18 carbons) are provided to the enzymes. *In vivo* pulse-chase experiments using various labeled substrates are also being carried out to analyze the pathways of lipid biosynthesis and fatty acid modification.

Acetyl-CoA Carboxylase. We have also carried out research to identify enzymes that play a role in controlling the overall rate of lipid biosynthesis in microalgae. For these studies, we have utilized the diatom *Cyclotella cryptica*. The cell walls of diatoms contain substantial amounts of silica, and thus this nutrient is required in large quantities. When silica becomes depleted from the medium, growth of *C. cryptica* cultures slows and lipids (primarily storage TAGs) accumulate rapidly (*13*). Within the first 4 hours of silicon deficiency, there is a doubling in the percentage of newly photoassimilated carbon that is partitioned into lipids, and a reduction in the percentage of carbon that is partitioned into storage carbohydrate. Using this model system, we have obtained evidence that acetyl-CoA carboxylase (ACCase) may play a role in this lipid accumulation process, in that the activity of this enzyme doubles after 4 hours of silicon deficiency and quadruples after 15 hours (*14*). This increase in activity can be blocked by the addition of protein synthesis inhibitors, suggesting control at the level of gene expression.

ACCase is a biotin-containing enzyme that catalyzes the ATP-dependent carboxylation of acetyl-CoA to form malonyl-CoA. Two partial reactions are involved in this process: 1) carboxylation of the enzyme-bound biotin molecule, and 2) transfer of the carboxyl group to acetyl-CoA. In the bacterium *Escherichia coli*, these reactions are catalyzed by four distinct, separable protein components: 1) biotin carboxylase, 2) biotin carboxyl carrier protein, and 3) two subunits of carboxyltransferase. In eukaryotic organisms, these entities are located on a single, multifunctional polypeptide typically having a molecular mass exceeding 200 kilodaltons (kDa) (*15*). The functional ACCase enzyme in eukaryotes is composed of multimers of this large polypeptide.

ACCase is believed to catalyze the rate-limiting step of fatty acid biosynthesis in animals (*16, 17*). The rate of fatty acid synthesis in higher plants also appears to be dependent on the activity of ACCase (*18, 19*). In addition, studies with *E. coli* have suggested that ACCase may catalyze the rate-limiting step of phospholipid biosynthesis (*20*). It is also worth noting that the expression of the ACCase gene in *E. coli* is correlated to the cellular growth rate (*21*).

In light of the regulatory significance of ACCase in lipid biosynthesis, we sought to purify this enzyme from *C. cryptica* and to determine some of its biochemical properties. The enzyme was purified 600-fold to near homogeneity by a combination of ammonium sulfate precipitation, gel filtration chromatography, and monomeric avidin affinity chromatography (*22*). The molecular mass of the native enzyme was determined to be between 700 and 800 kDa, while the subunit mass was determined to be approximately 200 kDa; this suggested that the functional enzyme was composed of four identical, biotin-containing subunits. Various cellular metabolites were tested for their ability to modulate the activity of the purified ACCase (*23*). ACCase activity was inhibited somewhat in the presence of the reaction products (malonyl-CoA, PO_4, and ADP), and inhibited more strongly in the presence of palmitic acid and palmitoyl-CoA. Pyruvate stimulated enzyme activity slightly (~50%) at high (1 mM) concentrations. The physiological significance of these effects is not known at this time.

Although we have determined some of the structural and catalytic properties of ACCase, characterization of various molecular aspects of the enzyme (e.g., primary structure, active site geometry, etc.) requires the isolation and cloning of the ACCase gene. A cloned copy of the gene is also required for research concerning the regulation of ACCase gene expression. Our efforts to clone this gene are discussed in the following section.

Genetic Engineering

As discussed earlier, genetic engineering is one approach for producing microalgal strains with enhanced biodiesel production capabilities. Two primary requirements must be fulfilled before a strain can be successfully engineered in this manner: 1) an expressible, recombinant gene that affects lipid metabolism must be available; and 2) methods must be developed to stably incorporate this cloned gene into a host cell in a manner that allows expression of the gene.

ACCase Gene Cloning. Because ACCase plays an important role in controlling the rate of fatty acid biosynthesis in many organisms, this enzyme has been receiving an increasing amount of attention as a target for manipulation via genetic engineering for the purpose of increasing the lipid production capabilities of various organisms. Therefore, we attempted to clone this gene from *C. cryptica*. When this project first started, the ACCase gene had only been isolated from rat (*24*) and chicken (*25*). The gene has now also been isolated from yeast (*26*) and all of the subunits of *E. coli* ACCase have been cloned (*27-30*). There have not been any previous reports of a full-length ACCase gene from eukaryotic algae or higher plants.

ACCase is relatively easy to purify from *C. cryptica*. Therefore, we chose to take a polymerase chain reaction (PCR)-based approach to clone the ACCase gene from this species (*31*). PCR primers were designed from partial amino acid sequences that were determined for various CNBr-generated fragments of the purified enzyme. These PCR primers were used to amplify a 146-base pair (bp) fragment of the ACCase gene, using DNA isolated from *C. cryptica* as a template. This PCR product was subsequently used to identify a full-length clone of the gene in a genomic *C. cryptica* library.

The complete DNA sequence of the ACCase gene from *C. cryptica* has now been determined. This sequence has been deposited in the GenBank genetic data base (accession number L20784). In addition, the sequences of several regions of

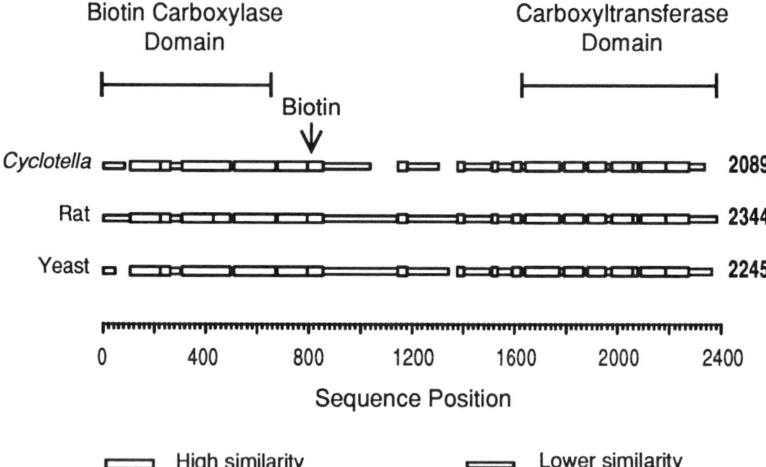

Figure 3. Sequence similarities among the ACCase enzymes from *C. cryptica*, rat, and yeast. Based on the MACAW program of Schuler et al. (32). The biotin attachment site is indicated by an arrow.

the ACCase gene transcript (i.e., messenger RNA) have been determined, which has allowed identification of the probable translation initiation codon, along with the precise localization of two introns in the gene. A 73-bp intron is present near the region of the gene that encodes the biotin-binding site of the enzyme, and a 447-bp intron is located near the start of the coding region of the gene. Upon removal of these introns, a coding sequence 6.3 kilobases long is obtained, which encodes a protein that contains 2089 amino acids and has a molecular mass of approximately 230 kDa. *C. cryptica* ACCase is therefore comparable in size to the multifunctional ACCases that have been isolated from various plants, animals, and yeast. The large size is necessary because of the three functional domains that are responsible for the different partial reactions involved in malonyl-CoA synthesis: biotin carboxylase, biotin carboxyl carrier protein, and carboxyltransferase.

Comparison of the deduced amino acid sequences of ACCase from *C. cryptica*, yeast, and rat and indicated strong similarities in the biotin carboxylase and carboxyltransferase domains (~50% identity), but less similarity in the middle region of the enzyme (~30% identity). This middle region, which includes the biotin carboxyl carrier protein domain, probably functions as a "swinging arm" to transfer the carboxylated biotin from the biotin carboxylase domain to the carboxyltransferase domain. As such, a fair amount of variability in the amino acid sequence of this region could be tolerated, thus explaining the comparatively lower sequence conservation. Regions with statistically significant similarity are shown in Figure 3, which is the graphical output generated by the MACAW sequence alignment computer program developed by Schuler et al. (*32*).

The amino-terminal sequence of the predicted ACCase polypeptide, shown below using the standard one-letter amino acid code, has characteristics of an endoplasmic reticulum (ER) "signal sequence" (*33*). Note the two positively charged arginine residues within the first few amino acids of the polypeptide, followed by a hydrophobic region:

MALRRGLYAAAATAILVTASVTAFAPQHSTFTPQSLSAAP. . .

Signal sequences direct polypeptides into the ER in eukaryotic cells. In diatoms, nuclear-encoded proteins destined for chloroplasts must be transported through the ER (*34*), which is consistent with the fact that diatom chloroplasts are completely enclosed by ER membranes (*35*). Because fatty acid biosynthesis occurs primarily in the chloroplasts of higher plants (*36*), we have assumed that ACCase is located in the chloroplasts of diatoms, and therefore a signal sequence would be necessary for chloroplast targeting. Alternatively, it is possible that the ACCase gene that we have cloned codes for an ER-localized ACCase that is responsible for producing the malonyl-CoA utilized for elongation of fatty acids from C_{16} or C_{18} to C_{20} and C_{22}. Because ACCase must pass through the ER before it can enter the chloroplast, it is also possible that a single ACCase isoform is functional in both of these cellular compartments. Further experimentation will be necessary to examine this possibility. The location of the cleavage site at the end of the putative ACCase signal sequence is not clear; attempts to determine the amino-terminal sequence of the mature enzyme have not been successful.

We have begun to manipulate the ACCase gene in order to produce forms that can be expressed in a variety of organisms. For example, to prevent possible problems with inefficient intron excision in heterologous hosts, we have removed the two introns by replacing fragments of the genomic DNA with corresponding fragments of cDNA. We have also performed site-specific mutagenesis of the gene to add certain restriction sites and eliminate others, thereby facilitating the construction of expression vectors. In addition, we are altering the 5' end of the coding region in order to change the sequence at the amino terminus of the expressed polypeptide; these changes will be useful for directing the enzyme into particular subcellular compartments of the host organisms. For example, to direct *C. cryptica* ACCase into the chloroplasts of higher plants, it will be necessary to add a gene fragment that encodes a chloroplast transit peptide. On the other hand, for host organisms in which fatty acid synthesis is cytoplasmic (e.g., yeast), it will be necessary to remove the signal sequence-encoding DNA from the gene.

Development of Genetic Transformation Systems for Microalgae. The isolation of genes that affect lipid accumulation in microalgae is the first step toward genetically engineering microalgal strains for enhanced biodiesel production. It is then necessary to incorporate these genes into the host cell in a stable manner that allows efficient expression of the gene. In recent years, the process of genetic transformation, defined here as the introduction and expression of exogenous genes in a host organism, has become routine for many bacterial, fungal, animal, and plant species (*37, 38*). However, stable transformation systems are not currently available for most microalgae. A major focus of our research is the development of genetic transformation systems for microalgae that have potential with respect to biodiesel fuel production. The two basic steps in this process, development of methods to introduce DNA into algal cells and detection of expression of the exogenous gene, are discussed in detail below.

Introduction of Foreign DNA. For both algae and higher plants, the cell wall is a significant barrier to the introduction of DNA or other macromolecules into the cell. For many plant species, the cell wall can be removed enzymatically to form protoplasts. Protoplasts can be induced to take up DNA through the cell membrane by a variety of methods, including electroporation (*39*) or treatment with polyethylene glycol (*40*). Previous work in our laboratory involving the transient expression of firefly luciferase has suggested that DNA can be introduced into algal cells by standard techniques if algal protoplasts can be formed (*41*). However, the production of protoplasts from most strains of microalgae is not straightforward. Currently, the only microalgal species for which an efficient transformation system exists is the green alga *Chlamydomonas reinhardtii*. Transformation of *C. reinhardtii* was facilitated by the availability of wall-less cells, produced either by genetic mutation (*42*) or through the use of autolysin, a species-specific cell wall-degrading enzyme produced by gametes during mating (*43*). Kindle and coworkers (*44*) found that DNA could be introduced into these wall-less cells simply by agitating the cells in the presence of glass beads. Unfortunately, wall-less mutants, or methods to generate wall-less cells, are not available for most microalgal species. The composition of microalgal cell walls is highly variable between species, and even between different strains of the same species. Furthermore, the cell walls often contain components, such as sporopollenin, that are resistant to degradation by the enzyme mixtures commonly used to degrade plant cell walls. To avoid the difficult work involved in developing protocols to generate viable protoplasts from each algal species of interest, a simple technique for introduction of DNA into intact (walled) algal cells is preferable.

Recently, several methods have been developed to introduce DNA into cells without prior removal of the cell wall. The technique of microprojectile bombardment, or biolistics, uses pressurized helium gas to propel DNA-coated gold particles into target cells or tissues, and has been used successfully to transform a large variety of cell types, including *C. reinhardtii* (*44, 45*). The glass bead agitation method described above has also been used successfully to introduce DNA into walled cells of yeast (*46*) and walled cells of *C. reinhardtii* (*43*), although the transformation frequency was low. Recently, two groups found that DNA uptake in higher plant cells could be facilitated by agitation of the cells with silicon carbide fibers (SiC) (*47, 48*). We have adapted the SiC protocol for the transformation of walled cells of *C. reinhardtii* (*49*) and found that the transformation efficiency was similar to that achieved using glass beads or particle bombardment. However, the SiC method may have advantages over the other protocols. For example, the cell survival rate was much higher using SiC as compared to glass beads, suggesting a more "gentle" mechanism. Furthermore, expensive equipment is not required for the SiC method, unlike the particle bombardment technique. We are currently working to adapt the SiC protocol for transformation of microalgal cells that have greater potential for fuel production.

Selectable Marker Genes. Even with the best DNA introduction technology, the fraction of cells in which gene entry and stable expression are achieved is very low. Therefore, marker genes are required to facilitate detection of these rare transformation events. These selectable markers also aid in the stabilization of introduced sequences, since cells that lose the marker gene can be selected against. In *C. reinhardtii*, successful transformation has required the use of homologous genes (*44, 50*). Although heterologous genes (e.g., bacterial marker genes) have been expressed at low levels in *C. reinhardtii* (*51*), they have not been effective as selectable markers in algae. This is probably due to poor expression of non-algal genes, possibly because of differences in algal promoter signals, codon bias, or DNA methylation. The DNA of *C. reinhardtii* has an unusually high guanosine plus cytosine (GC) content, which is reflected in an unusual bias toward these bases in codon usage (e.g., *52, 53*). Furthermore, a recent report suggests that DNA methylation may affect the expression of foreign genes in *C. reinhardtii* (*54*).

We are working to develop homologous genes as selectable markers for the transformation of certain species of microalgae that have potential utility for fuel production. The analysis of DNA from these organisms indicates that some microalgae, particularly the green algae, exhibit high GC contents and high degrees of cytosine methylation (*55*). The DNA of one of the target species, *Monoraphidium minutum*, has a GC content of 71%. This indicates that its codon usage may be biased toward G and C bases, as in *C. reinhardtii*, suggesting that heterologous selectable marker genes with a lower, more typical GC content may not be effective in *M. minutum*. By using homologous genes, problems with promoter specificity and codon bias will not inhibit gene expression, allowing successful selection of transformed cells. We are exploring the development of three different homologous selectable marker systems for the transformation of oil-producing microalgae. It should be noted that the availability of several marker genes for a given organism would provide a means for multiple rounds of genetic transformation in a single strain lineage.

Nitrate reductase. The nitrate reductase (NR) gene has been used in other laboratories as a transformation marker. In *C. reinhardtii*, NR-deficient mutant cells have been transformed successfully with the wild-type *C. reinhardtii* NR gene using the glass bead and SiC procedures described above. Progress has been made in the generation of mutants and in the cloning of the NR gene from *M. minutum* in hopes of developing a homologous transformation system similar to that used for *C. reinhardtii*.

Algal mutants lacking functional NR can be selected based on their resistance to chlorate (*56, 57*). Cells having a functional NR enzyme will take up chlorate (a nitrate analog) and reduce it to chlorite, which is toxic to the cells. The NR-deficient mutants do not reduce the chlorate and so are not subject to the cytotoxic effects of chlorite. Consequently, cells can be subjected to a positive selection regime by growing target organisms in the presence of chlorate and picking out resistant colonies. Using this protocol, we have isolated several putative NR-deficient mutants of *M. minutum*. These candidates grow in the presence of chlorate and are unable to utilize nitrate as a nitrogen source. Biochemical analyses are under way to confirm that this phenotype is in fact due to mutations in the NR structural gene.

The next step in the development of this particular transformation system for *M. minutum* is the isolation of the functional NR gene from the wild-type strain. This gene will be used to complement the NR-deficient strains; putative transformants will be selected by their ability to use nitrate as the sole nitrogen source. A genomic library of *M. minutum* DNA was constructed in a lambda vector. To obtain a homologous probe with which to screen the library, PCR primers were designed based on conserved sequences in the coding regions of NR genes from three other green algae: *C. reinhardtii* (Lefebvre, P., Univ. of Minn., personal comm., 1992), *Volvox carterii* (*58*), and *Chlorella vulgaris* (*59*). PCR amplification of total DNA from *M. minutum* produced a 750-bp fragment that was cloned and partially sequenced, confirming that the clone represents a portion of the NR gene. This NR gene fragment is currently being used to screen the *M. minutum* genomic library in order to obtain the full-length NR gene.

Orotidine-5'-phosphate decarboxylase. The orotidine-5'-phosphate decarboxylase (OPD) gene codes for an essential enzyme in the pyrimidine biosynthesis pathway. OPD-deficient mutants grow only if provided with an alternate source of pyrimidines, such as uracil, or if they have been transformed with a functional copy of an OPD gene. There is also a positive selection for OPD mutants; OPD converts the compound 5-fluoroorotic acid (FOA) into a toxic compound, killing wild-type cells, whereas OPD mutants can grow on FOA-containing media (*60*). A spectrophotometric assay can be used to measure OPD activity in cell extracts (*61*). The OPD gene has been cloned and sequenced from a number of organisms (e.g., *62-65*), and has been used successfully in several transformation systems (e.g., *66-68*).

Use of the FOA mutant selection method has allowed us to obtain several putative OPD mutants from ultraviolet-mutagenized *M. minutum* cells. These cells require uracil for growth and exhibit a low reversion rate, thus providing a powerful selection for cells that become transformed with the wild-type OPD gene. Work is in progress to isolate the wild-type OPD gene from *M. minutum*. Two approaches are being taken. The first relies on the PCR technique, using degenerate primers designed from regions of the gene that appear to be conserved between the OPD genes of various species. The second approach is to clone the gene directly by functional complementation of *E. coli* or *Saccharomyces cerevisiae* OPD mutants.

Herbicide resistance. Using herbicide resistance as a selectable marker has the advantage of traditional antibiotic selection in that resistance genes can be used as dominant selectable markers. In other words, wild-type organisms would be able to be transformed directly without the need to generate mutant strains, which is a process that can be particularly problematic in diploid cells such as diatoms. This method requires either 1) cloning and *in vitro* site-specific mutagenesis of the wild-type herbicide-sensitive gene to an herbicide-resistant form, or 2) cloning of the gene from a mutant strain that has become resistant to the herbicide. Herbicide resistance genes that have been developed in plant systems include 5-enol-pyruvylshikimate-3-phosphate, which is responsible for glyphosate resistance, and acetolactate synthase (ALS), which is involved in resistance to sulfonylurea and imidazolinone herbicides (*69*). These systems have potential as selectable markers for the transformation of microalgae as well. In fact, several species in the NREL Culture Collection, including *M. minutum* and *C. cryptica*, have been shown to be sensitive to such herbicides (*70*). Efforts are under way to clone and mutate the microalgal genes involved in resistance to these herbicides. For the ALS gene, a PCR-based approach similar to that described above for the NR gene was used to amplify a ~500-bp gene fragment from the DNA of *M. minutum;* this fragment exhibits strong sequence similarity to the ALS genes of other organisms. This partial gene will be used as a probe to isolate the full-length gene from a *M. minutum* genomic library.

Conclusions

Microalgae are a potential source of biological lipids for use in large-scale biodiesel production. We have made substantial progress in identifying some of the factors that are important in determining the quantity and quality of lipids produced by microalgae. This work has led to the cloning of ACCase, which is believed to play an important role in controlling the rate of fatty acid biosynthesis. Progress is also being made with respect to the development of methods to genetically engineer oil-producing microalgae, with the eventual goal of enhanced biodiesel production.

Acknowledgments

We are grateful to J. Bleibaum and G. Thompson of Calgene, Inc. for performing amino acid sequence analysis of ACCase. A significant portion of the ACCase gene cloning research was carried out in collaboration with J. Ohlrogge at Michigan State University; this support is gratefully acknowledged. This work was supported in part by the Biofuels Systems Division of the United States Department of Energy, the National Science Foundation, and the United States Department of Agriculture-Binational Agricultural Research and Development Fund.

Literature Cited

1. *Biodiesel Alert* **1993,** *7(1).*
2. Shay, E. G. *Biomass and Bioenergy* **1993,** *4*, 227-242.
3. Brown, L. M. In *Proceedings of the First Biomass Conference of the Americas, Vol II* (NREL/CP-200-5768); National Renewable Energy Laboratory, Golden, Colorado, 1993; pp. 902-909.
4. Roessler, P. G. *J. Phycol.* **1990,** *26*, 393-399.

5. Chelf, P.; Brown, L. M.; Wyman, C. E. *Biomass and Bioenergy* **1993,** *4*, 175-183.
6. Harrington, K. J. *Biomass* **1986,** *9*, 1-17.
7. Klopfenstein, W. E.; Walker, H. S. *J. Amer. Oil Chem. Soc.* **1983,** *60*, 1596-1598.
8. Browse, J.; McCourt, P.; Somerville, C. *Plant Physiol.* **1986,** *81*, 859-864.
9. Suen, Y.; Hubbard, J. S.; Holzer, G.; Tornabene, T. G. *J. Phycol.* **1987,** *23*, 289-296.
10. Sukenik, A.; Carmeli, Y.; Berner, T. *J. Phycol.* **1989,** *25*, 686-692.
11. Oo, K. -C.; Huang, A. H. C. *Plant Physiol.* **1989,** *91*, 1288-1295.
12. Griffiths, G.; Stobart, A. K.; Stymne, S. *Biochem. J.* **1985,** *230*, 379-388.
13. Roessler, P. G. *Arch. Biochem. Biophys.* **1988,** *267*, 521-528.
14. Roessler, P. G. *J. Phycol.* **1988,** *24*, 394-400.
15. Samols, D.; Thornton, C. G.; Murtif, V. L.; Kumar, G. K.; Haase, F. C.; Wood, H. G. *J. Biol. Chem.* **1988,** *263*, 6461-6464.
16. Kim, K. -H.; Lopez-Casillas, F.; and Bai, D. -H. *FASEB J.* **1989,** *3*, 2250-2256.
17. Lane, M. D.; Moss, J.; Polakis, S. E. In *Current Topics in Cellular Recognition*; Horecker, B. L.; Stadtman, E. R., Eds.; Academic Press, New York, NY, 1974, Vol. 8; pp 139-195.
18. Post-Beittenmiller, D.; Jaworski, J.G.; Ohlrogge, J.B. *J. Biol. Chem.* **1991,** *266*, 1858-1865.
19. Post-Beittenmiller, D.; Roughan, G.; Ohlrogge, J. B. *Plant Physiol.* **1992,** *100*, 923-930.
20. Jackowski, S.; Cronan Jr., J. E.; Rock, C. O. In *Biochemistry of Lipids, Lipoproteins and Membranes*; Vance, D. E.; Vance, J., Eds.; Elsevier Science Publishers B.V., Amsterdam, 1991; pp 43-85.
21. Li, S. -J.; Cronan Jr., J. E. *J. Bacteriol.* **1993,** *175*, 332-340.
22. Roessler, P. G. *Plant Physiol.* **1990,** *92*, 73-78.
23. Roessler, P. G.; Ohlrogge, J. B. In *The Biochemistry, Structure, and Utilization of Plant Lipids*; Harwood, J.; Quinn, P., Eds.; Portland Press, London, 1990; pp 108-110.
24. Lopez-Casillas, F.; Bai, D. -H.; Luo, X.; Kong, I. -S.; Hermodson, M. A.; Kim, K. -H. *Proc. Natl. Acad. Sci. U.S.A.* **1988,** *85*, 5784-5788.
25. Takai, T.; Yokoyama, C.; Wada, K.; Tanabe, T. *J. Biol. Chem.* **1988,** *263*, 2651-2657.
26. Al-Feel, W.; Chirala, S. S.; Wakil, S. J. *Proc. Natl. Acad. Sci. U.S.A.* **1992,** *89*, 4534-4538.
27. Li, S. -J.; Cronan Jr., J. E. *J. Biol. Chem.* **1992,** *267*, 855-863.
28. Li, S. -J.; Cronan Jr., J. E. *J. Biol. Chem.* **1992,** *267*, 16841-16847.
29. Kondo, H.; Shiratsuchi, K.; Yoshimoto, T.; Masuda, T.; Kitazono, A.; Tsuru, D.; Anai, M.; Sekiguchi, M.; Tanabe, T. *Proc. Natl. Acad. Sci. U.S.A.* **1991,** *88*, 9730-9733.
30. Alix, J.-H. *DNA* **1989,** *8*, 779-789.

31. Roessler, P. G.; Ohlrogge, J. B. *J. Biol. Chem.* **1993,** *268*, 19254-19259.
32. Schuler, G. D.; Altschul, S. F.; Lipman, D. J. *Proteins Struct. Funct. Genet.* **1991,** *9*, 180-190.
33. von Heijne, G. *J. Membrane Biol.* **1990,** *115*, 195-201.
34. Bhaya, D.; Grossman, A. *Mol. Gen. Genet.* **1991,** *229*, 400-404.
35. Gibbs, S. P. *J. Cell. Sci.* **1979,** *35*, 253-266.
36. Harwood, J. L. *Ann. Rev. Plant Physiol. Plant Mol. Biol.* **1988,** *39*, 101-138.
37. Fincham, J. R. S. *Microbiol. Reviews* **1989,** *53*, 148-170.
38. Potrykus, I. *Ann. Rev. Plant Physiol. Plant Mol. Biol.* **1991**, *42*, 205-225.
39. Fromm, M.; Taylor, L. P.; Walbot, V. *Proc. Natl. Acad. Sci. USA* **1985,** *82*, 5824-5828.
40. Krens, F. A.; Molendijk, L.; Wullems, G. J.; Schilperoort, R. A. *Nature* **1982**, *296*, 72-74.
41. Jarvis, E. E.; Brown, L. M. *Curr. Genet.* **1991**, *19*, 317-321.
42. Davis, D. R.; Plaskitt, A. *Genet. Res.* **1971**, *17*, 33-43.
43. Kindle, K. L. *Proc. Natl. Acad. Sci. USA* **1990**, *87*, 1228-1232.
44. Kindle, K. L.; Schnell, R. A.; Fernández, E.; Lefebvre, P. A. *J. Cell Biol.* **1989**, *109*, 2589-2601.
45. Klein, T. M.; Wolf, E. D.; Wu, R.; Sanford, J. C. *Nature* **1987**, *327*, 70-73.
46. Costanzo, M. C.; Fox, T. D. *Genetics* **1988**, *120*, 667-670.
47. Asano, Y.; Otsuki, Y.; Ugaki, M. *Plant Science* **1991**, *79*, 247-252.
48. Kaeppler, H. F.; Somers, D. A.; Rines, H. W.; Cockburn, A. F. *Theo. Appl. Genet.* **1992**, *84*, 560-566.
49. Dunahay, T. G. *Biotechniques* **1993,** *15*, 452-460.
50. Debuchy, R.; Purton, S.; Rochaix, J. -D. *EMBO J.* **1989**, *8*, 2803-2809.
51. Hall, L. M.; Taylor, K. B.; Jones, D. D. *Gene* **1993**, *124*, 75-81.
52. Goldschmidt-Clermont, M.; Rahire, M. *J. Mol. Biol.* **1986**, *191*, 421-432.
53. Silflow, C. D.; Youngblom, J. *Ann. N.Y. Acad. Sci.* **1986**, *466*, 18-30.
54. Blankenship, J. E.; Kindle, K. L. *Mol. Cell. Biol.* **1992**, *12*, 5268-5279.
55. Jarvis, E. E.; Dunahay, T. G.; Brown, L. M. *J. Phycol.* **1992**, *28*, 356-362.
56. Cove, D. J. *Mol. Gen. Genet.* **1976**, *146*, 147-159.
57. Toby, A. L.; Kemp, C. L. *J. Phycol.* **1977**, *13*, 368-372.
58. Gruber, H.; Goetinck, S. D.; Kirk, D. L., Schmitt, R. *Gene* **1992**, *120*, 75-83.
59. Cannons, A. C.; Iida, N.; Solomonson, L. P. *Biochem. J.* **1991**, *278*, 203-209.
60. Boeke, J. D.; LaCroute, F.; Fink, G. R. *Mol. Gen. Genet.* **1984**, *197*, 345-346.
61. Donovan, W. P.; Kushner, S. R. *J. Bacteriol.* **1983**, *156*, 620-624.
62. Rose, M.; Grisafi, P.; Botstein, D. *Gene* **1984**, *29*, 113-124.
63. Newbury, S. F.; Glazebrook, J. A.; Radford, A. *Gene* **1986**, *43*, 51-58.
64. Ohmstede, C.; Langdon, S. D., Chae, C.; Jones, M. E. *J. Biol. Chem.* **1986**, *261*, 4276-4282.

65. Turnbough, C. L.; Kerr, K. H.; Funderburg, W. R.; Donahue, J. P.; Powell, F. E. *J. Biol. Chem.* **1987**, *262*, 10239-10245.
66. Buxton, F. P.; Radford, A. *Mol. Gen. Genet.* **1983**, *190*, 403-405.
67. Boy-Marcotte, E.; Vilaine, F.; Camonis, J.; Jacquet, M. *Mol. Gen. Genet.* **1984**, *193*, 406-413.
68. van Hartingsveldt, W.; Mattern, I. E.; van Zeijl, C. M. J.; Pouwels, P. H.; van den Hondel, C. A. M. *Mol. Gen. Genet.* **1987**, *206*, 71-75.
69. Botterman, J.; Leemans, J. *Trends Genet.* **1988**, *4*, 219-222.
70. Galloway, R. E. *J. Phycol.* **1990**, *26*, 752-760.

RECEIVED March 17, 1994

Chapter 14

Microbial and Enzymatic Biofuel Cells

G. Tayhas R. Palmore and George M. Whitesides

Department of Chemistry, Harvard University, Cambridge, MA 02138

This review summarizes the literature published after 1985 relevant to microbial and enzymatic biofuel cells. It tabulates the experimental conditions used in operation, the characteristics, and the performance of the reported biofuel cells. The present state of research in biofuel cells is analyzed and suggestions for future research are included.

The purpose of this chapter is to review recent work in biofuel cells -- that is, fuel cells relying at least in part on enzymatic catalysis for their activity. The chapter is divided into three sections: (1) a brief review of the principles common to fuel cells (both biological and non-biological), supported with a bibliography of more comprehensive treatments, (2) an overview of the literature directly relevant to biofuel cells that encompasses both microbial and enzymatic cells, and (3) a summary of the obstacles that remain along the path to a practical biofuel cell, with suggestions for future research.

Fundamentals of Fuel Cells and the Distinction between Biofuel Cells and Non-Biological Fuel Cells

A fuel cell is an electrochemical device that manages the flow of electrons and charge-compensating positive ions (typically protons or alkali metal cations such as Li^+ or Na^+) in a redox reaction in such a way that the electrons move through an external circuit where they can do electrochemical work (Scheme 1). The only fuels now used in practical fuel cells are H_2 and O_2; other reducing fuels are converted to H_2 before they enter the fuel cell.

A biological fuel cell (abbreviated as *biofuel cell*) is the offspring of two parent technologies: fuel cells and biotechnology. Like conventional fuel cells, biofuel cells comprise an anode and a cathode separated by a barrier that is selective for the passage of positively charged ions; they require the addition of fuel to generate power (*1-9*). Unlike conventional fuel cells, which usually use precious metals as catalysts, biofuel cells utilize enzymatic catalysts, either as they occur in microorganisms, or as isolated proteins (*10*). There are two types of biofuel cells: direct and indirect (*11*). The *direct* biofuel cell is a configuration in which fuel is oxidized *at* the surface of the electrode. The role of the biological catalysts in this type of fuel cell is to catalyze these surface reactions. The *indirect* biofuel cell is a

0097–6156/94/0566–0271$08.00/0

configuration in which fuel reacts not at the electrode, but in solution or in a separate compartment, and a redox active *mediator* is added to shuttle electrons between the site of reaction and the electrode. Biological catalysts have been used in both direct and indirect configurations: typically, enzymes have been used as catalysts in direct biofuel cells, while microorganisms have often been used in indirect biofuel cells.

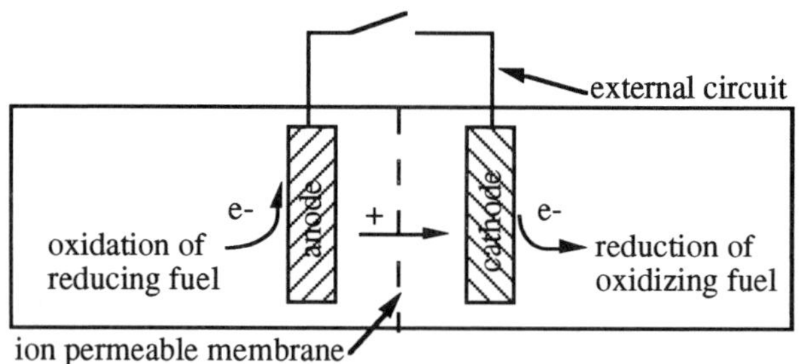

Scheme 1. Direction of electron and ion flow in fuel cell: external circuit open, no electrons or ions flow; external circuit closed, electrons move from the anode to the cathode and positively charged ions move from the anodic compartment to the cathodic compartment.

Power (i.e., electrochemical work, W) is abstracted from fuel cells via an external electrical circuit connecting the anode and the cathode. The power available from a fuel cell is the product of cell voltage (V_{cell}) and cell current (I_{cell}). Ideally, V_{cell} for a direct fuel cell is the difference between the formal potentials of the *fuels* in the anode and cathode compartments; V_{cell} for an indirect fuel cell is the difference between the formal potentials of the *mediators* in the anode and cathode compartments. Irreversible losses in voltage (overpotentials) reflect the influence of slow kinetics of heterogeneous electron transfer, ohmic resistances, and concentration gradients (*12*). I_{cell} is the rate at which charge moves through the fuel cell and is dependent on electrode size, the rates of the processes occurring in the fuel cell, and the rate of movement of ions across the membrane separating the fuel cell compartments. A range of different processes -- catalytic reactions in solution and at interfaces, mass transport, electron transfer, ion permeation, and leakage of fuels between compartments -- can limit the power extractable from a particular fuel cell, either individually or collectively. Power lost to irreversible processes usually appears as heat, and the management of heat is an issue in biofuel cells in which the enzymes are temperature sensitive. Maximizing both V_{cell} and I_{cell} are two important challenges in developing practical biofuel cells (*13*).

Dioxygen, either pure or as a component of air, is the most commonly used cathodic fuel: it is readily available; thermodynamically a good oxidant (although often kinetically poor); its reduction product (H_2O) is benign. The choice of the anodic fuel is more complicated. Dihydrogen has been used almost universally for its excellent electrode kinetics and potential. Because of issues of cost, engineering and safety associated with the generation, transport and storage of dihydrogen, methanol is now being considered seriously as an anodic fuel. Other fuels -- especially hydrocarbons or hydrocarbon mixtures such as methane, natural gas, gasoline and diesel fuel -- would be enormously interesting, but work on developing redox systems capable of using these fuels (other than by converting them to dihydrogen) is in its infancy.

Methanol is a non-explosive, water-soluble fuel that can be produced industrially in large quantities from coal, natural gas, or renewable resources. Ease in transport, storage, and safety are the advantages of methanol over dihydrogen as fuel.

Moreover, the theoretical cell voltage for a methanol/dioxygen fuel cell (1.19 V) is approximately the same as that of a dihydrogen/dioxygen fuel cell (1.23 V). Consequently, the development of catalysts that will oxidize methanol at low temperatures (< 200°C) is a high priority in fuel cell research (*14,15*).

One approach to the challenge of developing catalysts for the oxidation of methanol is to utilize biotechnology. Living organisms contain a range of enzymatic catalysts that oxidize methanol and other potential fuels (ethanol, glucose, glycerol, sucrose) to carbon dioxide under mild conditions (room temperature, atmospheric pressure, neutral pH). For example, the enzymatic oxidation of methanol to carbon dioxide has been studied using a combination of alcohol dehydrogenase (ADH), aldehyde dehydrogenase (AldDH), and formate dehydrogenase (FDH) with nicotinamide adenine dinucleotide (NAD^+) as cofactor (Scheme 2) (*16,17*). This method reduces three mol of NAD^+ to NADH for every mol of methanol oxidized. NADH must subsequently be reconverted to NAD^+ to continue the enzymatic reaction or to generate power.

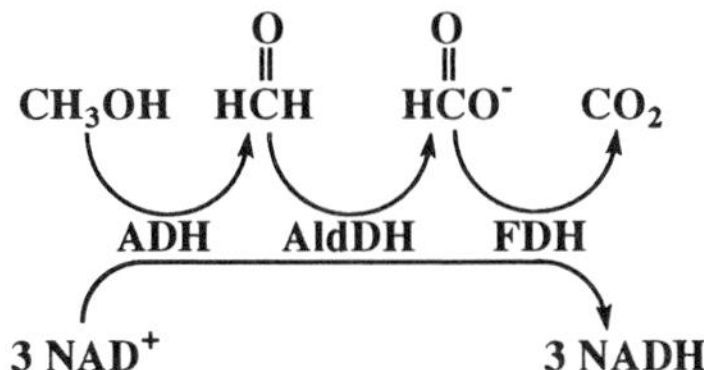

Scheme 2. Dehydrogenase/NAD^+ catalyzed oxidation of methanol to CO_2.

Both biochemical and electrochemical regeneration of NAD^+ have been the topic of many reports (18-27). The formal potential (E°′) of NAD^+ is -0.32 V vs. NHE, pH 7.0 (-0.62 V vs. SCE at pH 7.5) (*28*; The standard potential, E°, refers to the reduction potential of a given specie under standard conditions and unit activity for all reactants and products in the cell. Because it is difficult to construct a cell in which all activities are unity, one utilizes the Nernst equation to extract the value of E° from the formal potential, E°′, which is the potential *measured* under non-standard conditions). At bare platinum or glassy carbon electrodes, however, the potential at which NADH is oxidized is ~1 V positive to E°′ (~0.4 V vs. SCE at pH 7.5) (*29,30*). Minimizing this 1 V overpotential is crucial if NAD^+ dependent catalysts are to be used in biofuel cells. The goal is to find a system (i.e. a combination of mediators and catalysts) that rapidly oxidizes NADH at its formal potential for a given pH. All approaches explored thus far have exhibited substantial voltage losses. Alternatively, enzymes that oxidize methanol in the absence of NAD^+/NADH have been tried in fuel cells. This approach however, has only been demonstrated using unstable dye mediators with high formal potentials (see the section on enzymatic biofuel cells).

Enzymes can be efficient catalysts, although the hallmark of most enzymatic catalysts in biological systems is in their *selectivity* in converting substrate to product, and their susceptibility to metabolic control, rather than their catalytic rate acceleration. There are enzymes (e.g. triose phosphate isomerase, superoxide dismutase) however, that operate at or near mass-transport limited rates (*31*). To estimate a qualitative upper limit for the current that might plausibly be generated in a biofuel cell using an enzyme as catalyst, one can assume a specific activity of 10^3 U mg^{-1} of enzyme (1 U ≡ 1 μmol of product produced per min). For a reaction producing one electron per molecule of substrate, 1 mg of enzyme, operating at concentrations of substrates at which the active site is saturated, would produce 1 mmol of electrons per min, or 96.5 coulombs per min. (Faraday constant = 96,487 coulombs per mol of electrons; 1 ampere = 1 coulomb sec^{-1}; 1 watt = 1 volt ampere).

Since 1 ampere = 1 coulomb sec^{-1}, 1 mg of enzyme with an activity of 10^3 U mg^{-1} would be capable of catalyzing sufficient reaction to generate a current of ~1.6 amps, provided there are no other losses in the system.

In fact, most enzymes have much lower specific activities (values of 0.1 to 10 U mg^{-1} are common), and enzyme-catalyzed reactions usually operate at less than their theoretical rates. Thus, in practice, the catalytic activity of an enzymatic system might plausibly be such that it would generate 1 μA to 1 mA of current per mg of enzyme. By contrast, the current generated by a platinum catalyst in a fuel cell operating with dihydrogen is in the order of ~5-15 mA mg^{-1} of platinum. (This value is based on typical current densities of 100-300 mA cm^{-2} for a dihydrogen/dioxygen fuel cell (see reference 2) using Pt-gauze electrodes; 2.12 mg Pt cm^{-2} gauze, 100 mesh woven from 0.762 mm diameter platinum wire).

The attractions of biofuel cells are several, despite their possible kinetic disadvantages relative to a dihydrogen/dioxygen fuel cell using platinum anodic and cathodic catalysts. First, they may be able to use fuels such as glucose, wastes containing paper or other biologically derived materials that cannot be handled by any existing transition metal-based fuel cell. Neither non-biological nor biological fuel cells have been operated successfully with unfunctionalized hydrocarbons but there are microorganisms that can use such hydrocarbons as a carbon source; the enzymes involved in these processes are potential candidates for use in fuel cells. Second, biofuel cells are intrinsically non-polluting and their enzymes operate at low temperatures in water close to pH 7. Third, enzymes can, in principle, be produced inexpensively by fermentation using genetic engineering. Fourth, enzymes may be able to contribute to the solution of problems such as the overpotential at the dioxygen cathode that have resisted solution by conventional fuel cell technology: oxidative metabolism using dioxygen is highly optimized in biological systems. Fifth, if it were possible to find a system using enzymes that operated on fuels with specific activities close to 10^3 U mg^{-1}, these enzymes probably *would* offer advantages in rate over platinum. Sixth, biofuel cells may be applicable to certain speculative but important applications such as a fuel cell that could be implanted in humans and provide power for other implanted prosthetic devices.

Overview of the Literature of Biofuel Cells

Several comprehensive reviews on the development of biofuel cells appeared prior to 1985 (*10,11,13,32-34*). The literature subsequent to 1985 will be the focus of this section, with key studies prior to 1985 included only for historical perspective. The literature described in this section is divided into three categories of subject: (1) microbial biofuel cells, (2) photomicrobial biofuel cells, and (3) enzymatic biofuel cells. The experimental conditions used with, and the characteristics of, the biofuel cells in each category are tabulated in Tables I-III, respectively. Throughout this section, we refer to chemical mediators by their acronyms. The chemical names or structures of these mediators can be found at the end of this chapter under the heading "Legend of Symbols and Selected Chemical Structures".

In the 1960's, the United States space program became interested in developing biofuel cells as an energy-saving waste disposal system for extended space flights. Initially, biofuel cells were operated using microorganisms and were classified according to how the microorganism was utilized; either as a manufacturer of an electroactive product (i.e. dihydrogen) that was fed directly into a conventional fuel cell, or in conjunction with a redox mediator that served to transfer electrons to the electrode from an intercepted metabolic pathway of the microorganism. The advantages of microbial biofuel cells that generate dihydrogen (or, in principle, other conventional fuels) are: (1) a variety of complex organic substances such as carbohydrates, fats, and proteins can be utilized as fuels, and (2) known non-

Table I. Summary of Microbial Biofuel Cells

Reference	*Experimental Conditions and Results*
(35)	(glucose oxidase), *E. Coli*, *Nocardia* oxidizing glucose and using MB as mediator: Pt-sheet anode and cathode (60 in^2 each), GOD: 0.28 V at open-circuit; no current given, *E. Coli*: 0.625 V at open-circuit; no current given, *Nocardia*: 0.3 V at open-circuit; 2 mA at 1 kΩ.
(36)	*Pseudomonas methanica* producing H_2 (from methane) using NSIDCP as mediator: platinized-Pt anode and cathode (12.6 cm^2 each), 0.5-0.6 V at open-circuit; 2.8 $\mu A/cm^2$ at 0.35 V.
(38)	agar gel-immobilized *C. butyricum* producing H_2 (from molasses): 2 Pt-blackened Ni anodes (100 mesh, 10.4 cm diameter), 2 Pd-blackened Ni cathodes (250 mesh, 10.4 cm diameter), 0.55 V (fuel cell I) and 0.66 V (fuel cell II) at 1 Ω, 0.95 V at 100 Ω for fuel cells I and II connected in series, 40 mA/cm^2 at 1 Ω; 550 mA at 2 Ω and 1.3 V for 7 days.
(39)	*C. butyricum* producing H_2 (from waste water): 5 Pt-blackened Ni anodes (100 mesh, 14.5 cm diam.), 5 Pd-blackened Ni cathodes (100 mesh, 14.5 cm diam.), 0.62 V at 1 Ω; 0.8 A at 2.2 V for 7 days, 5 fuel cells connected in series.
(40)	*C. butyricum* producing H_2 (from molasses): 2 carbon fiber (impregnated with Pt-blackened carbon pellets) anodes and cathodes, 200 °C, 0.9-0.93 V at open-circuit (two fuel cells connected in parallel); 10-14 A.
(41)	*Proteus vulgaris*, *E. coli*, *Bacillus subtilis*, or *Alcaligenes eutrophus* oxidizing glucose or succinate using ABB, BCB, BV^{2+}, DCPIP, DMST-2, GCN, NMB, PES, PTZ, RS, SF, TB, or TH^+ as mediator: RVC anode (35 x 50 x 7 mm, 800 cm^2 surface area), Pt-foil cathode poised at 0.43 V vs. SCE (10 x 40 mm), 0.2 M $K_3[Fe(CN)_6]$ catholyte, *Proteus vulgaris*: 0.64 V at open-circuit; 0.8 mA at 560 Ω, *E. coli*: open-circuit voltage not given; ~0.8 mA at 560 Ω, *Bacillus subtilis*: open-circuit voltage not given; ~0.8 mA at 560 Ω, *Alcaligenes eutrophus*: 0.55 V at open-circuit; current not given.
(43)	*Proteus vulgaris* oxidizing glucose using TH^+ as mediator: RVC anode (35 x 50 x 7 mm, 800 cm^2 surface area), Pt-foil cathode (10 x 40 mm), 0.2 M $K_3[Fe(CN)_6]$ catholyte, open-circuit voltage not given; 3.5 mA at 100 Ω (calc. from V/I).
(44)	*Proteus vulgaris* oxidizing sucrose using TH^+ as mediator: carbon anode and cathode, 0.1 M $K_3[Fe(CN)_6]$ catholyte, open-circuit voltage not given; 3.5 mA at 100 Ω and 0.35 V (calc.).
(45)	*E. coli* oxidizing glucose using TH^+, DST-1, DST-2, TST-1, or FeCyDTA as mediator: materials and dimensions of anode and cathode not given, open-circuit voltage not given; < 0.7 mA at 560 Ω.

Continued on next page

Table I. *Continued*

Reference	*Experimental Conditions and Results*
(*46*)	*Lactobacillus plantarum, Streptococcus lactis,* or *Erwinia dissolvens* oxidizing glucose using FeEDTA, FeCyDTA, FeDTPA, FeTTHA, or FeEDADPA as mediator: materials and dimensions of anode and cathode not given, *L. plantarum and S. lactis*: 0.2 V at open-circuit, 0.05 V at 560 Ω, *E. dissolvens*: 0.5 V at open-circuit, 0.4 V at 560 Ω.
(*47*)	*Bacillus W1* oxidizing glucose using MV^{2+}, BCB, HNQ, or FeEDTA as mediator: materials and dimensions of anode and cathode not given, open-circuit voltage not given; 10 mA/g dry wt. at 500 Ω.
(*48*)	*Enterobacter aerogenes* producing H_2 (from glucose): Pt-blackened stainless steel net anode (25 cm^2), Pt cathode, 0.01 M $K_3[Fe(CN)_6]$ catholyte, 1.04 V at open-circuit, 0.79 V at 500 Ω; 60 $\mu A/cm^2$.
(*49,50*)	*E. coli* oxidizing glucose using HNQ as mediator: GC anode (12.5 cm^2) or anode comprised of packed bed of graphite particles (surface of packed bed was ~200 cm^2), O_2 gas-diffusion cathode, 0.53 V at 10 kΩ; GC anode: 0.18 mA/cm^2, packed bed anode: 1.3 mA/cm^2 cross-sectional area of the bed.
(*51*)	*Desulfovibrio desulfuricans* oxidizing sulfide: 3 porous graphite anodes impregnated with $Co(OH)_2$ (100 cm^2), 3 porous graphite cathodes impregnated with Fe(phthal) (100 cm^2), 2.8 V at open-circuit for 3 fuel cells connected in series; 2 days, < 1 A after 1 day.

Table II. Summary of Photomicrobial Biofuel Cells

Reference	*Experimental Conditions and Results*
(*52*)	*Anabaena variabilis* oxidizing glycogen and using HNQ as mediator: materials and dimensions of anode and cathode not given, 0.1 M $K_3[Fe(CN)_6]$ catholyte, 0.4 V at 400 Ω; 1-2 mA at 200 Ω.
(*53*)	*Anabaena variabilis* oxidizing glycogen and using HNQ as mediator: materials and dimensions of anode and cathode not given, 0.1 M $K_3[Fe(CN)_6]$ catholyte, open-circuit voltage not given; 400 μA at 200 Ω for 19 h.
(*54*)	immobilized *Rhodospirillum rubrum* producing H_2: carbon-fiber (impregnated with Pt-blackened carbon pellets) anodes and cathodes (2 each) operated at 200 °C, dimensions not given, 0.2-0.35 V at open-circuit for 2 fuel cells connected in parallel; 0.5-0.6 A for 6 h.

Table III. Summary of Enzymatic Biofuel Cells

Reference	Experimental Conditions and Results
(55)	*methanol dehydrogenase* oxidizing methanol and formaldehyde using PMS or PES as mediator: Pt-gauze anode (50 mesh, 7 cm^2), Pt-gauze cathode (80 mesh, 4.4 cm^2), 0.3 V at open-circuit; 3.7 mA at 10 Ω.
(56)	PQQ-linked *alcohol dehydrogenase* oxidizing methanol using TMPD as mediator: Pt-gauze anode (50 mesh, 7 cm^2), Pt-gauze cathode (80 mesh, 4.4 cm^2), 0.16 V at 133 kΩ; 20 μA/cm^2 at 10 Ω and 20 °C.
(57)	*methanol* and *formate dehydrogenase* oxidizing methanol using PMS or PES as mediator: carbon fabric anode and cathode (40 cm^2 each), open-circuit voltage not given; 0.9 mA at 10 Ω and 0.065 V.
(58)	*glucose oxidase* oxidizing β-D-glucopyranose in the presence of elemental Fe: Pt-sheet anode and cathode (64 cm^2 each), at open circuit: 0.35 V without Fe, 0.75 V with Fe; 0.21 μA/cm^2 without Fe, 17 μA/cm^2 with Fe.
(59)	immobilized *D-glucose dehydrogenase* oxidizing glucose using MB as mediator: graphite disk anode (16 cm^2, 740 rev/min), Pt-gauze cathode poised at 0.56 V vs. SCE, open-circuit voltage not given; 0.2 mA cm^{-2} at ~0.8 V.
(60)	*glucose oxidase* oxidizing glucose in anode using DCPIP as mediator; *chloroperoxidase* chlorinating barbituric acid in cathode: Pt-gauze anode (50 mesh, 7.5 cm^2), Au-sheet cathode (22.8 cm^2), 0.6 V at open-circuit; 350 μA.

biological fuel cell technology can be utilized. Unfortunately, microbial biofuel cells have uniformly generated current densities too low to be of practical value as power sources.

The availability of isolated enzymes has made it possible to eliminate the microorganism from the biofuel cell and to substitute with individual enzymes, the result being a substantial gain in volumetric catalytic activity. Individual enzymes offer the advantage of much higher catalytic capacity (in terms of catalytic activity per unit weight or volume), but at substantially higher cost than enzymes present in microorganisms. Genetic engineering has not yet been applied to the area of biofuel cells, either to lower the cost or to improve the characteristics of the enzymes used. Enzymatic biofuel cells are typically more efficient than microbial biofuel cells in converting fuel to electricity for two reasons: the cellular membrane that interferes with electron transfer is absent and fuel-consuming microbial growth is eliminated. For practical reasons however, enzymatic biofuel cells are limited to fuels that are oxidized to CO_2 with a minimal number of enzymatic transformations. Moreover, unsupported enzymes are often more vulnerable to denaturation than enzymes encapsulated within a microbial membrane, and enzymatic biofuel cells are typically less stable than those using living organisms.

Microbial Biofuel Cells. One of the earliest examples of electrical energy derived from both microorganisms and enzymes was demonstrated by Davis and Yarbrough in 1962 (*35*). The microorganisms used in the fuel cell were *Nocardia* and *Escherichia coli* and the enzyme used was glucose oxidase (GOD). The authors began their study using a fuel cell configured with GOD, presumably containing bound flavin adenine dinucleotide (FAD), glucose, buffer, and a Pt-sheet electrode in the dinitrogen-purged anodic compartment. The cathodic compartment, containing a Pt-sheet electrode and buffer, was saturated with dioxygen. GOD catalyzes the oxidation of β-D-glucose to δ-D-gluconolactone with the concomitant reduction of FAD to $FADH_2$ (equation 1). Under biological conditions, $FADH_2$ is reoxidized by dioxygen (equation 2), but because the anodic chamber had been purged to remove dioxygen, this reaction did not occur. Methylene blue (MB) was added to the anode compartment both to reoxidize $FADH_2$ and to shuttle the reducing equivalents made available by the enzymatic reaction to the anode (equations 3 and 4).

$$\beta\text{-D-glucose} + \text{GOD/FAD} \longrightarrow \delta\text{-D-gluconolactone} + \text{GOD/FADH}_2 \quad (1)$$
$$\text{GOD/FADH}_2 + O_2 \longrightarrow \text{GOD/FAD} + H_2O_2 \quad (2)$$
$$\text{GOD/FADH}_2 + \text{MB} \longrightarrow \text{GOD/FAD} + \text{MBH} \quad (3)$$
$$\text{MBH} \xrightarrow{\text{anode}} \text{MB} \quad (4)$$

In the absence of MB, the fuel cell yielded no electricity. V_{cell} at open-circuit was 0.28 V and voltages between 0.075 to 0.09 V could be maintained under a load of 1 kΩ. Because the glucose-GOD-MB system under a dinitrogen atmosphere generated reduced MB that produced electricity, the authors reasoned that a facultative anaerobe would also produce measurable current. *Escherichia coli* was added to the anode compartment in place of GOD, and a V_{cell} of 0.625 V at open-circuit was obtained. Under a 1 kΩ load, V_{cell} could be maintained at 0.52 V for over 1 h. The use of ethane as fuel was attempted using an ethane-oxidizing *Nocardia* as catalyst. No current was produced even in the presence of MB. When glucose was substituted for ethane, metabolic activity increased. An open-circuit voltage of 0.3 V and a maximum current of 2 mA under a 1 kΩ load was reported. This value of maximum current seems unlikely, given the upper limit of 0.3 mA imposed by Ohms law (V = IR).

A unique microbial biofuel cell that utilized methane as fuel was demonstrated by van Hees (*36*). To our knowledge, no other methane-fed biofuel cell has been reported in the literature. The anodic compartment, containing *Pseudomonas methanica*, 1-naphthol-2-sulfonate indo-2,6-dichlorophenol (NSIDCP), and a platinum anode, was purged of dioxygen and supplied with methane. The cathode compartment housed a platinum cathode in a sterile medium and was supplied with dioxygen in the form of air. Inclusion of NSIDCP in the anode compartment had no effect on V_{cell} at open-circuit, suggesting that this dye did not function as an electrochemical mediator. This conclusion was further supported by enclosing the microorganisms in dye-permeable dialysis tubing and observing that the value for V_{cell} at open-circuit was established only very slowly even in the presence of NSIDCP; this observation suggests that microbial contact with the anode was necessary for electron transfer. The power density was low (0.98 $\mu W\ cm^{-2}$) and prolonged open-circuit conditions or high current drain were observed to be detrimental to the obligate aerobe used in this fuel cell. The explanation given for these observations was elaborate and the reader is referred to the original text. In summary, the authors concluded that biofuel cells utilizing obligate aerobes are best suited to functioning under moderate current drain.

Karube, Suzuki *et al* studied several microorganisms for their production of dihydrogen from glucose (*37-40*). Of the organisms studied, *Clostridium butyricum* IFO3847 produced dihydrogen at the highest rate and was therefore used to supply dihydrogen to the anode compartment of a conventional fuel cell. Immobilization of the microorganism in a gel of agar-acetylcellulose both stabilized the dihydrogen-producing system and increased the rate of dihydrogen production. For every mol of glucose consumed, 0.6 mol of dihydrogen and 0.2 mol of formic acid were produced. Two kg of wet gel containing ~200 g of wet microorganism produced 40 ml (1.7 mmol) of dihydrogen per min, equivalent to ~5 A.

Four different fuel cell configurations were studied; the two most recent are described here. One configuration, a fuel cell stack of five fuel cells using Pt-blackened Ni-gauze anodes and Pd-blackened Ni-gauze cathodes, was operated at room temperature. Dihydrogen was passed from a 5 L reservoir containing ~2 kg of immobilized whole cells to the anode compartments of the fuel cell stack at a rate of 40 ml min^{-1} while dioxygen was passed through the cathode compartments. The fuel cell system (dihydrogen producing microbial reservoir and five fuel cells) was operated for 7 days and maintained a current of 0.8 A at 2.2 V. The poor current efficiency was not explained; a high temperature phosphoric acid fuel cell stack was, however, used in the subsequent report.

A second configuration used ~3 kg of immobilized microorganisms in a 200 L fermentation vessel containing 150 L of molasses. The biogas produced, containing both dihydrogen and carbon dioxide, was passed into the anode compartment of a phosphoric acid fuel cell stack (two fuel cells in parallel operated at 200 °C) at a rate of 400-800 ml min^{-1} while dioxygen was passed into the cathode compartment at a rate of 1-1.5 L min^{-1}. The open-circuit voltage was between 0.9-0.93 V and a stable current of 10-14 A was maintained for four hours with a power output of 9-13 W. Current density, power density, and electrode size were not given. The authors analyzed the electrical energy balance for the microbial/phosphoric acid fuel cell configuration and found that only 1% of the energy consumed to operate the system (pumps and heaters) was generated by the microbial fuel cell.

Delaney, *et al* ., studied fuel cell performance using different microorganism-mediator-substrate combinations (*41,42*). V_{cell} at open-circuit was predicted to range from 0.9-1.2 V based on the difference between $E^{\circ\prime}$ of the different mediators used in the anode compartment and $E^{\circ\prime}$ of the potassium ferricyanide used in the cathode compartment. V_{cell}, however, was found to be 0.2-0.4 V lower than predicted, indicating activation overpotentials at the electrodes. The fuel cell that gave the best performance contained *Proteus vulgaris*, thionine (TH^+), and glucose in the anodic compartment. Typical current-voltage curves were obtained by discharging the fuel cell through a range of loads. The coulombic yield of the fuel cell was determined by integration of the area under the current-time curve. (Coulombic yield is the total number of coulombs passed through the fuel cell, as determined by integration of the current-time curve in units of amp•s = coulombs, relative to the theoretical number that might be obtained by complete conversion of the reactants to products). Coulombic yields of 32-62% were obtained based on the theoretical maximum coulombic equivalent for complete oxidation of glucose in accordance with equation 5. (The theoretical maximum coulombic equivalent of glucose is 24 mol electron mol^{-1} glucose x 96,487 C mol^{-1} electron = 2.3 x 10^6 C mol^{-1} glucose):

$$C_6H_{12}O_6 + 6\ H_2O \rightarrow 6\ CO_2 + 12\ H^+ + 24\ e^- \quad (5)$$

Thurston *et al* argue that if fuels such as carbohydrates are to be used, microorganisms will be more economical catalysts than individual enzymes because the multi-step oxidations involved with complex fuels would require multiple, expensive enzyme purifications (*43*). *Proteus vulgaris* was chosen as the biocatalyst for glucose oxidation and ^{14}C-labeled glucose was used as fuel to establish the distribution of products. The amount of CO_2 evolved from the oxidation of glucose was determined by scintillation methods and a sodium hydroxide trap. Conversion of glucose to CO_2 was 40-50%, and coulombic yields indicated near unit current efficiency (based on 50% conversion of glucose). The most significant loss in energy was from the conversion of 30% of glucose to acetate; it is possible that this side reaction might be avoided by metabolic or genetic modification of the microorganism.

Bennetto *et al* used sucrose as fuel in a TH^+ mediated microbial biofuel cell containing *Proteus vulgaris* as biocatalyst (*44*). Equation 6 gives the stoichiometry of complete oxidation of sucrose.

$$C_{12}H_{22}O_{11} + 13\ H_2O \rightarrow 12\ CO_2 + 48\ H^+ + 48\ e^- \quad (6)$$

The biofuel cell was discharged through a resistance of 100 Ω after the addition of various amounts of sucrose. The amount of CO_2 evolved from the anode compartment confirmed near complete oxidation of sucrose. The cell current peaked at 3.5 mA and the production of electricity from the oxidative conversion of sucrose was >95%, determined from both coulombic yield and CO_2 evolution. At higher concentrations of sucrose, the buffering capacity of the anodic medium was exceeded due to acid formation. Preliminary tests using molasses as fuel suggested that >70% of the carbohydrate was oxidized. Neither the open-circuit voltage nor electrode size were given.

Since TH^+ has frequently been used as a mediator in biofuel cells, Lithgow *et al* prepared sulfonated derivatives of thionine to determine the effect of ring substitution on mediation of electron transfer from *E. Coli* to the anode (*45*). Changing from TH^+ to 2-sulfonated thionine (DST-1) and 2,6-disulfonated thionine (DST-2) resulted in an increase in the efficiency of mediated electron transfer. This increase was reflected by changes in cell current under a 560 Ω load: 0.35 mA for TH^+, 0.45 mA for DST-1, and 0.6 mA for DST-2. The low efficiencies of the fuel cells operating with TH^+ and DST-1 were attributed to interference with electron transfer by adsorption of mediator to the microbial membrane, and a resulting decrease of the rate of oxidation of glucose at that membrane.

The authors also compared the thionine derivatives to ferric cyclohexane-1,2-diamine-N,N,N′,N′-tetraacetic acid (FeCyDTA) and found that lower concentrations of both TH^+ and the sulfonated derivatives gave better mediation than the iron chelate. The significant inference from this report was that the power output of microbial biofuel cells could be improved by introducing hydrophilic groups into several mediators.

Vega and Fernandez studied mediation using several ferric chelates -- FeCyDTA, ferric diethylenetriamine pentaacetic acid (FeDTPA), ferric ethylenediamine diacetic acid dipropionic acid (FeEDADPA), ferric ethylenediamine tetraacetic acid (FeEDTA), and ferric triethylenetetramine hexaacetic acid (FeTTHA) -- with microbial biofuel cells containing *Lactobacillus plantarum*, *Streptococcus lactis*, and *Erwinia dissolvens* (*46*). *Lactobacillus plantarum* and *Streptococcus lactis* grow on dairy products and *Erwinia dissolvens* grows on coffee wastes. The objective of this study was to determine which microorganism/mediator combination would be the most efficient in a microbial biofuel cell. Fuel cells using *Erwinia dissolvens* mediated with FeCyDTA ran for five consecutive days, with coulombic yields between 80 and 90%. *Lactobacillus plantarum* and *Streptococcus lactis* did not produce any significant electrical output with any of the ferric chelates. The main results from this study were: 1) different microorganisms, using the same mediators and the same amount of fuel, gave different coulombic yields, 2) biofuel cells using *E. dissolvens* as the biocatalyst gave higher coulombic yields than those using *E. Coli*. The interaction of the microbial membrane with the mediator was suggested to be the source of these effects.

Akiba *et al* studied a microbial biofuel cell that operated at basic pH (*47*). Operating a fuel cell at basic pH circumvents problems associated with fuel cells operated at neutral pH, such as low ionic conductivity and inefficient operation of the O_2 cathode. Problems associated with CO_2 rejection are significant however, in fuel cells operating at basic pH. Methyl viologen (MV^{2+}), brilliant cresyl blue (BCB), and 2-hydroxy-1,4-naphthoquinone (HNQ) were employed as mediators, either alone or in combination with FeEDTA. The anode compartment was loaded with 35-50 mg of *Bacillus W1*, an unstated amount of mediator, 10 μmol of glucose, and 0.1 M KH_2PO_4 buffer (pH ranging between 10.0-11.0). The fuel cell was discharged through resistors of 300 or 500 Ω at 35 °C. The coulombic yields were highest at pH 10.5 and were > 93.5% with the different mediators. The rate of mediator reduction was however, found to be an order of magnitude slower than that at neutral pH. Because the reduced mediators are less stable at higher operating temperatures, the cells were not examined at such temperatures. No open-circuit voltages were given.

Clostridia are known for their production of dihydrogen, and they have frequently been used in microbial fuel cells. Because *Clostridia* are obligate anaerobes, however, both dihydrogen production and microbial growth are oxygen sensitive. Tanisho *et al* isolated a strain of the facultative anaerobe, *Enterobacter aerogenes*, and found it to produce dihydrogen from glucose (*48*). This microorganism, grown under aerobic conditions, consumes dioxygen to oxidize organic substrates; under anaerobic conditions, it degrades substrate into dihydrogen.

During operation of the fuel cell, the anolyte was circulated between the microbial reservoir and the anodic chamber of the fuel cell with a peristaltic pump. The reservoir contained microorganisms, phosphate buffer, ammonium sulfate, sodium citrate, magnesium sulfate, and glucose. Peptone, typically included in culture broth, was not included as it often causes microorganisms to adhere to the electrode. The fuel cell was operated with circulating microorganisms for 35 h; nutrients were added after 25 h. Whenever the anode compartment was opened to the atmosphere, the operating voltage would temporarily decrease, but would recover within ten minutes. The microorganisms were reported to produce 11 mol of dihydrogen/L culture/h (equivalent to 269 L dihydrogen/h at 1 atm and 25°C). The authors assert that if all dihydrogen produced by the 330 ml of culture contained in the reservoir were utilized by the fuel cell (we calculate this to be 3.63 mol dihydrogen/h based on 11 mol of dihydrogen/L culture/h), then 330 mA would be generated. In fact, complete oxidation of 3.63 mol of dihydrogen/h would generate a current of 195 A; some elements of the report are thus internally incompatible. Experimentally, an average current of only 1.4 mA was observed. The authors suggested switching to a molten carbonate fuel cell (MCFC) working on a gas mixture of dihydrogen and carbon dioxide to obtain higher current densities. Operating a MCFC at 650°C has, of course, a substantial cost in energy relative to cells operating close to room temperature.

Sell *et al* employed HNQ as a redox mediator in a fuel cell using *E. coli* as the biocatalyst for the oxidation of glucose (*49,50*). The fuel cell was operated for 10 h under a 250 Ω load, during which time V_{cell} decayed from 250 mV to 75 mV, presumably due to glucose consumption. The mechanism for HNQ reduction was not identified, but it was determined that HNQ is not reduced by isolated ADH. The significant finding in this report was that higher voltage and current density could be obtained by switching from a glassy carbon (GC) disc anode (0.18 mA/cm^2) to a packed bed of graphite particles (1.3 mA/cm^2 of cross-sectional area of bed); thus the rate of electron transfer at the electrode seemed to limit current.

Habermann and Pommer demonstrated a microbial fuel cell utilizing the sulfate-reducing microorganism, *Desulfovibrio desulfuricans* (*51*). Sulfide, produced from the metabolic reduction of sulfate, was oxidized at the anode using air as the terminal oxidant according to equations 7-9:

$$SO_4^{2-} + 8\,H^+ + 8\,e^- \xrightarrow{\text{bacterium and nutrients}} S^{2-} + 4\,H_2O \quad (7)$$

$$S^{2-} + 4\,H_2O \xrightarrow{\text{anode}} SO_4^{2-} + 8\,H^+ + 8\,e^- \quad (8)$$

$$2\,O_2 + 8\,H^+ + 8\,e^- \xrightarrow{\text{cathode}} 4\,H_2O \quad (9)$$

The authors listed glucose, fructose, cane sugar, starch, and hydrocarbons as the substances selected to provide the reducing equivalents required to convert SO_4^{2-} to S^{2-}, and to meet the nutrient requirements of the microorganism.

The fuel cell consisted of an anode fabricated from porous graphite impregnated with cobalt hydroxide. In the presence of sulfate-reducing microorganisms, cobalt hydroxide was converted into a cobalt oxide/cobalt sulfide mixture. As a result, reducing equivalents in the form of sulfide were concentrated within the porous network of the graphite anode during conditions of open-circuit. The cathode was fabricated from graphite activated with iron-phthalocyanine [Fe(phthal)]. The electrolyte used was a solution containing between 0.1 and 5% by weight sodium sulfate, 0.1% by weight urea and dextrose, and trace elements. The highest percentage of sulfate reduction by *Desulfovibrio desulfuricans* occurred when using 0.5 % by weight sodium sulfate (pH 8.5) electrolyte. Three fuel cells were connected in series and gave an open-circuit voltage of 2.8 V. The current level after one day of operation was < 1 mA.

Photomicrobial Biofuel Cells. Electric current is generated by microbial fuel cells when electrons from the oxidation of endogenous or exogenous substrates are transferred to an electrode. The use of photosynthetic microorganisms in fuel cells permits incident light to be converted into current. Photosynthetic microorganisms are envisaged to function in a fuel cell in two ways. First, energy can be stored as products of the photosynthetic apparatus during illumination and subsequently released and processed as in non-photosynthetic microbial biofuel cells. (A photomicrobial biofuel cell operated in this mode is different from a conventional biofuel cell only in that the energy ultimately converted to reduced mediator is derived from light rather than from a chemical fuel added separately). Second, energy might be abstracted directly from the photosynthetic apparatus by a mediator that transports electrons to the anode.

Tanaka *et al* assembled a biofuel cell using the photosynthetic microorganisms, *Anabaena variabilis* (*52,53*). Charge was mediated by the light-stable mediator, HNQ. In the dark, electrons from the oxidation of endogenous glycogen powered this biofuel cell. The fuel cell was discharged under a 400 Ω load and maintained V_{cell} at

0.4 V for >12 h. Illumination of the microorganism-containing anode compartment increased the coulombic yield by 60% over the dark anode due to the additional source of electrons provided by the photosynthetic oxidation of water.

Karube and Suzuki utilized a purple photosynthetic microorganism, *Rhodospirillum rubrum*, in a fuel cell (*54*). Wastewater containing acetate, propionate, and butyrate was used as the medium from which dihydrogen was produced by the immobilized microorganism. The organic acids in the wastewater are claimed to be the hydrogen source for the microorganisms. The authors present the difference in organic acid concentration between the inlet and outlet streams of the bioreactor as evidence to support this claim. Whether or not hydrogen from the organic acids was incorporated into the dihydrogen produced by the microorganism cannot be determined from the results of this study. The conditions employed to optimize microbial production of dihydrogen were pH 7.0, 30°C, and a light intensity of 5000 lux. Dihydrogen, produced by two reactors (20 L each) containing immobilized microorganisms, was transferred to the anode compartment of a phosphoric acid fuel cell at 19-31 ml min^{-1} while dioxygen was fed to the cathode compartment at 100 ml min^{-1}. The fuel cell operated between 0.2 and 0.35 V and delivered 0.5-0.6 A for 6 h.

Enzymatic Biofuel Cells

C_1 Compounds as Anodic Fuel. Plotkin,*et al* studied the oxidation of methanol to formate using a NAD^+-independent methanol dehydrogenase (MDH) (*55*). The redox mediators investigated were phenazine methosulfate (PMS) and phenazine ethosulfate (PES). At pH 9.5, the optimal pH for enzymatic activity, PES was more stable than PMS. The authors claim that a current less than 3 mA would represent 90% of the maximum current possible based on the specific activity of the enzyme (0.42 μmol of methanol oxidized/min/mg MDH). Our calculations indicate however, that 4.6 mg of MDH with the stated specific activity should produce 12.4 mA of current. Consequently, < 25% of the current available from the enzymatic reaction was observed. Aside from poor current efficiency, the main drawback to this fuel cell was the accumulation of formate in the anode compartment.

Davis *et al* studied the oxidation of methanol in a biofuel cell using a quinoprotein-linked ADH as catalyst (*56*). This ADH catalyzes the oxidization of methanol, other primary alcohols and formaldehyde. The authors evaluated several mediators (BCB; PMS; PES; safranine-O; phenolsafranine; brilliant blue-B; malachite green; 1-methoxyphenazine methosulfate; methylviolet-6B; N,N,N′,N′-tetramethyl-4-phenylenediamine (TMPD)) and found TMPD to be the best. The biofuel cell, using TMPD as mediator, operated at 20°C and pH 10.5 with a current fluctuation of less than 10% over a 24 h period. Maximum output of power (12 μW) was obtained at 0.2 mA and 60 mV.

To eliminate the accumulation of formate in the anode compartment, Yue and Lowther used both MDH and FDH in their biofuel cell (*57*). Both PMS and PES were used as mediators for electron transfer. Because PMS and PES are unstable, current efficiencies were poor (36%) and decreased with time (27% in one hour). Immobilization of the enzymes onto the anode increased the average current by a factor of two. The power density was calculated to be 1.5 $\mu W/cm^2$ at 10 Ω based on the surface area of the anode; no open-circuit voltage was given.

Glucose as anodic fuel. A biofuel cell that could be implanted in the human body as a power source for implanted electronic devices (i.e. pacemakers and cochlear implants) would be very useful. To be practical, a fuel cell for this use should utilize endogenous substances, such as O_2 and glucose.

Yahiro *et al* (*58*) provided one of the earliest demonstrations of a glucose/O_2 fuel cell, using GOD as catalyst. V_{cell} for the GOD system was 300-500 mV. The introduction of elemental iron to the anode compartment increased both V_{cell} and current density to 750 mV and 17 $\mu A/cm^2$ at 300 mV, respectively. The probable mechanism by which elemental iron influenced the performance of the fuel cell is as a stoichiometric reducing agent (the authors proposed a different, less plausible, mechanism). It is not obvious from the reported information whether elemental iron functions as a catalyst or fuel or both.

Persson *et al* used N,N-dimethyl-7-amino-1,2-benzophenoxazinium ion (MB^+) as the mediator in a GOD catalyzed glucose/O_2 fuel cell (*59*). MB^+ was chosen over other mediators because of its higher stability and irreversible adsorption to graphite. The reduction potential of MB^+ is -0.175 V vs. SCE; it therefore, spontaneously oxidizes NADH. An indirect fuel cell configuration was used where the immobilized enzyme was separated from the anode. During operation of the fuel cell, NADH resulting from the oxidation of glucose by the immobilized GOD was pumped to the anode compartment, where NADH reduced the adsorbed MB^+ to MBH. The MBH was then electrochemically oxidized, and NAD^+ was recirculated. An "O_2 simulated" cathode was used in this study by poising an electrode at 0.56 V vs. SCE with an external potentiostat. The fuel cell delivered a current density of 0.2 mA cm^{-2} at ~ 0.8 V for more than 8 h using the poised cathode. When immobilized enzymes were used, the fuel cell was stable to storage for months. The mediator, however, needed to be replaced after several hours of use.

Laane *et al* demonstrated the applicability of a biofuel cell for the production of biochemicals (*60*). GOD was used as the catalyst in the anode compartment, and chloroperoxidase (CPO) was used as the catalyst in the cathode compartment. The mediator for electron transfer in the anode compartment was 2,6-dichlorophenol indophenol (DCPIP). The chemistry that occurred in the fuel cell is summarized in equations 10-13:

$$\beta\text{-D-glucose} + \text{DCPIP} \xrightarrow{\text{GOD}} \text{D-gluconic acid} + \text{DCPIPH}_2 \quad (10)$$
$$\text{DCPIPH}_2 \xrightarrow{\text{Pt anode}} \text{DCPIP} + 2\,H^+ + 2\,e^- \quad (11)$$

$$O_2 + 2H^+ + 2e^- \xrightarrow{\text{Au cathode}} H_2O_2 \quad (12)$$
$$H_2O_2 + \text{barbituric acid} + \text{HCl} \xrightarrow{\text{CPO}} \text{5-chlorobarbituric acid} + H_2O \quad (13)$$

After operating a small fuel cell for three days, 10 mg of gluconic acid and 8 mg of 5-chlorobarbituric acid were produced. The current level was no greater than 1 mA and turnover numbers of 1 x 10^4 and 1 x 10^7 were obtained for glucose oxidase and chloroperoxidase, respectively.

Assessments and New Approaches

The motivation for examining fuel cells is their potential in three areas: first, to convert the free energy of a redox reaction (typically the complete oxidation of fuel by dioxygen) to electrochemical work in a way that circumvents the thermodynamic limitations of heat engines; second, to build power-generating systems that operate without the moving parts and high internal temperatures and pressures of internal combustion and turbine engines, and that avoid the pollutants (CO, NO_x, hydrocarbons) produced by these engines; third, to make use of fuels that cannot be used by other power sources. Underlying these general motivations is the idea that biology offers a range of catalysts (enzymes) that might be exploited for use in power generation, and that fuel cells offer one type of system in which enzymes might be used.

The research conducted so far does not establish whether biofuel cells are, in reality, contenders for practical use. Although certain enzymes do have high catalytic efficiency, most are only moderately active: the uniqueness of enzymes lies in their selectivity and amenability to metabolic control, rather than in their ability to accelerate reactions. The competition for enzymes as catalysts in fuel cells is platinum and related metals. In terms of activity and stability as an anode operating with dihydrogen as a fuel, these metals are difficult to surpass.

There have been many small-scale demonstrations of power produced from biofuel cells (although none have demonstrated net power production: i.e., demonstrated that power produced was in excess of power consumed to produce the enzymes, fuels, reagents and cell materials, and to operate the cell). The principle question is: "Is the activity available in biological catalysts sufficient, and in the right form, to provide the basis for a practical fuel cell?" Future work should focus on this question.

In general, there seem to be two approaches to a practical biofuel cell. The first is to use microorganisms to convert fuels that cannot presently be used in conventional fuel cells into appropriate form: the conversion of glucose to dihydrogen represents an obvious example. The second is to use isolated enzymes to generate reducing equivalents that can be used in new types of fuel cells: the conversion of methanol to CO_2 with production of NADH or FADH represents an example.

The fundamental difficulty with approaches to fuel cells based on microorganisms is their low volumetric catalytic activity. If containment vessels are inexpensive enough, however, the capacity for self-renewal of the catalysts by microbial growth, and their ability to accept biologically-derived substrates, might make them interesting. The report of Karube, Suzuki *et al*, in which a 200 L vessel generated sufficient dihydrogen to produce 10 W of power, but in which the power generated was only 1% of the energy budget (pumps and heaters) of the cell as it was operated, illustrates the challenge. Unfortunately, the most efficient bioconversions do not generate dihydrogen, but rather species such as acetic acid, a fuel that cannot be used by currently available fuel cells.

Approaches based on isolated enzymes have the capability, in principle, to equal transition metal-based systems in volumetric catalytic activity, but a careful study of the systems tradeoffs must be performed before the real potential of these systems can be defined. Important questions in this area are whether satisfactory redox mediators can be found or whether the fuel cell must be run in a direct configuration, the nature of membrane separators and electrodes that will resist fouling by components of biological systems, the potential of enzymes to decrease the overvoltage at the dioxygen cathode, and the potential for enzymes from thermophiles to permit operation of the fuel cell at higher temperatures.

To increase the value of research in this area, more emphasis should be placed on reporting the electrochemical details of the systems being examined. Typically, power density (W cm^{-2} of electrode) is the parameter most often used to compare the performance of different fuel cells. Unfortunately, power density is rarely reported in the biofuel cell literature (although it may not be the ideal parameter to use when one or both electrodes are inexpensive, high surface area systems such as carbon felts or packed carbon beds). If V_{cell} and current output at a given load are reported and the dimensions of both electrodes included, power density can be calculated. Examination of Tables I-III reveals that it is also rare to find all three of these parameters (V, A, cm^2) reported. Future reports on biofuel cells should include V_{cell} at open-circuit, current and voltage at a given load, and the geometrical area (or other relevant descriptions) of both anode and cathode. Of the literature reviewed, the highest power density that could be calculated from the values reported for microbial biofuel cells was 533 μW cm^{-2} (based on the calculated surface area of both sides of the anodes and cathodes) (*38*). The enzymatic biofuel cell with the highest power density (160 μW cm^{-2} based on the current density reported) was reported by Persson *et al* (*59*). The corresponding value for a phosphoric acid cell using dihydrogen as fuel is ~100 mW cm^{-2} (*2*).

Table IV. Applications for Biofuel Cells and Their Desirable Features

Application	*Fuel/Type of Catalyst; Desirable Features*
Implantable	glucose/O_2/using human derived enzymes; biocompatible.
Developing Countries	carbohydrates or agricultural and municipal wastes/using microorganisms; robust, low-cost components.
Portable	methanol or ethanol/using thermophilic enzymes; high activity (to achieve high power density).
Space	human waste/microbial; simplicity in regeneration of catalysts, dual use in environmental management and power generation.
Waste Control	waste/microbial; able to handle a large range of feeds.
Cogeneration	detoxify waste in anode and produce synthetic compound in cathode/microbial in anode, enzymatic in cathode; detoxify industrial waste while simultaneously producing a valuable chemical product in the cathode, generate power.

The open-circuit voltage indicates the loss in voltage from the theoretical maximum cell voltage. A comparison of this parameter for different fuel cells provides information on the relative rates of electron transfer of fuels or different mediators at different electrode surfaces. The highest open-circuit voltage reported for a microbial biofuel cell was 1.04 V where dihydrogen (produced by *Enterobacter aerogenes* from glucose) was oxidized at the anode and 0.01 M $K_3[Fe(CN)_6]$ was reduced at the cathode (*48*). The highest open-circuit voltage (0.6 V) reported for an enzymatic biofuel cell used DCPIP as the mediator in the anode compartment and dioxygen in the cathode compartment (*60*).

What constitutes a practical biofuel cell? The answer to this question depends upon the application. For example, space applications suggest systems in which the renewability of the catalyst by fermentation is a valuable characteristic, while applications above room temperature will require thermophilic catalysts. Table IV summarizes possible applications for biofuel cells and the projected important parameters (*61,62*).

Conclusions

The generation of power, albeit in small quantities, using biological catalysts in a fuel cell has been exhaustively demonstrated. The poor current and power densities reported in the literature reveal that key obstacles to developing a practical biofuel cell are increasing the catalytic rates of the system of microorganisms or enzymes used, and maximizing the rates of heterogeneous electron transfer, without losing cell voltage. In the next phase of biofuel cell research, robust and highly active enzymes appropriate for use in biofuel cells must be identified and characterized. If careful analysis indicates that these enzymes have the theoretical capacity to form the basis for practical biofuel cells, their costs must be lowered (and perhaps their properties improved) through genetic engineering. Systems that do not require NAD^+ should be explored as well as mediators whose $E^{\circ\prime}$ is equal to $E^{\circ\prime}$ of the fuel or the common cofactor systems (NAD^+/NADH and FAD^+/FADH). The engineering aspects of the cells -- component lifetimes, membrane performance, energy budget -- require much closer attention than they have received.

The challenge to developing a practical biofuel cell is straightforward: the external circuit and the products of enzymatic catalysis must be linked in a manner that maximizes cell voltage, charge flow and systems life, and minimizes costs.

Legend of Symbols and Selected Chemical Structures

ADH = alcohol dehydrogenase
AldDH = aldehyde dehydrogenase
FAD^+/FADH = oxidized/reduced form of flavin adenine dinucleotide
FDH = formate dehydrogenase
FeCyDTA = Fe(III)-cyclohexane-1,2-diamine-N,N,N′,N′-tetraacetic acid
FeDTPA = Fe(III)-diethylenetriamine pentaacetic acid
FeEDADPA = Fe(III)-ethylenediamine diacetic acid dipropionic acid
FeEDTA = Fe(III)-ethylenediamine tetraacetic acid
FeTTHA = Fe(III)-triethylenetetramine hexaacetic acid
GC = glassy carbon
HNQ = 2-hydroxy-1,4-naphthoquinone
NAD^+/NADH = oxidized/reduced form of nicotinamide adenine dinucleotide
NSIDCP = 1-naphthol-2-sulfonate indo-2,6-dichlorophenol
RVC = reticulated vitreous carbon

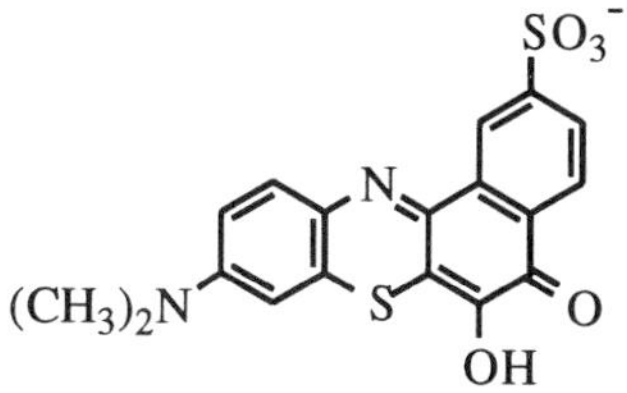

ABB = Alizarin Brilliant Blue

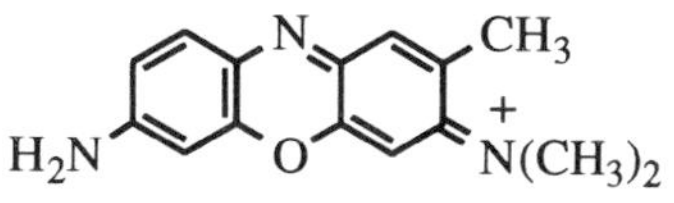

BCB = Brilliant Cresyl Blue

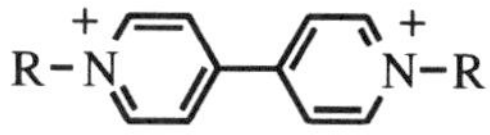

R = benzyl: BV^{2+} = Benzyl Viologen

R = Me: MV^{2+} = Methyl Viologen

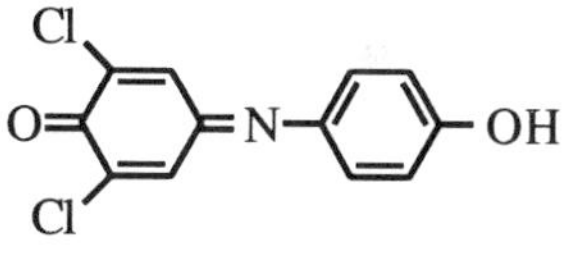

DCPIP = 2,6-dichlorophenolindophenol

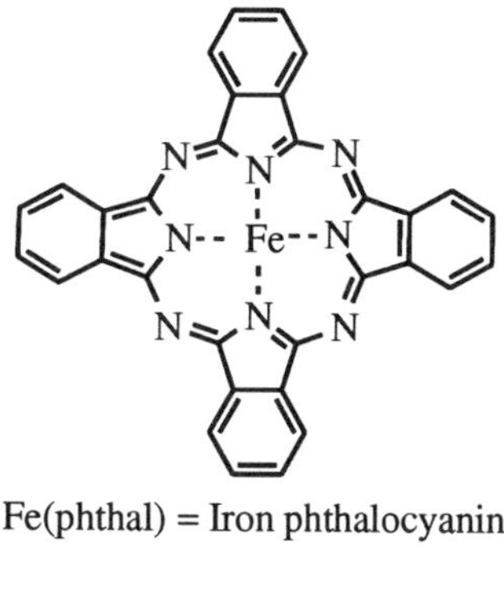

Fe(phthal) = Iron phthalocyanine

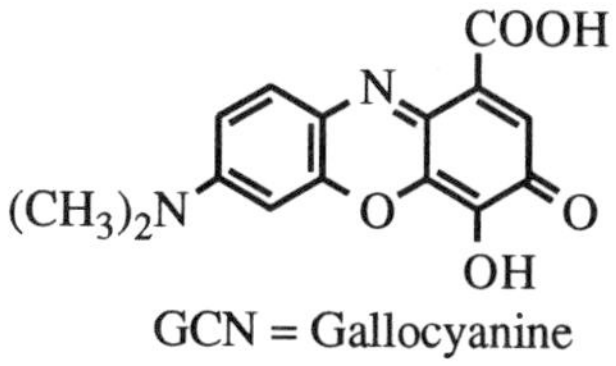

GCN = Gallocyanine

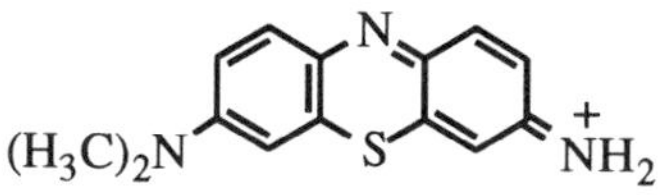

MB = Methylene Blue

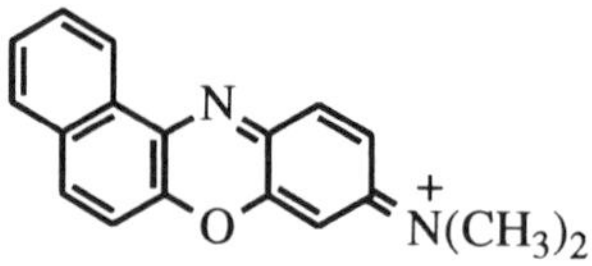

MB^+ = Meldola's Blue

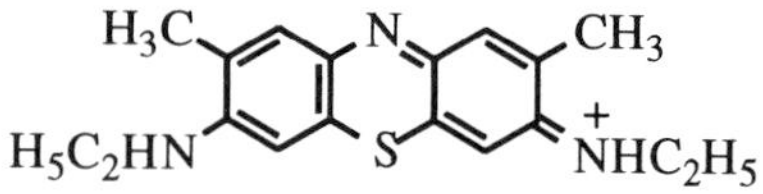

NMB = New Methylene Blue

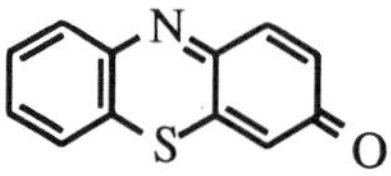

PTZ = Phenothiazinone

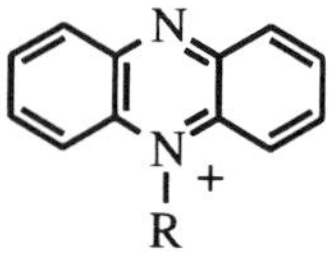

R = Me: PMS = Phenazine methosulphate
R = Et: PES = Phenazine ethosulphate

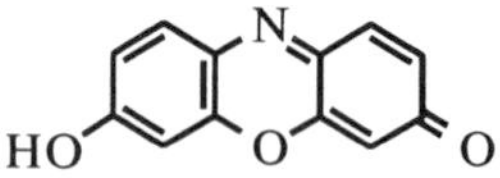

RS = Resorufin

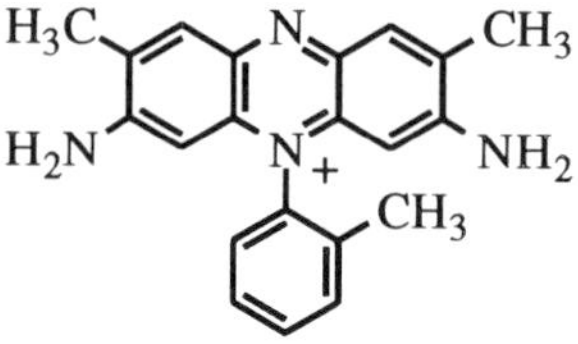

SF = Safranine-O

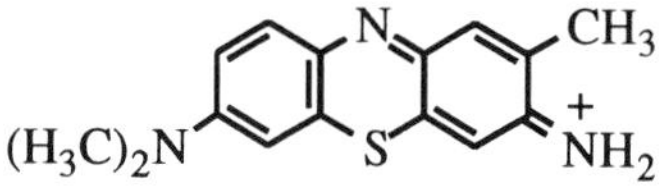

TB = Toluidine Blue-O

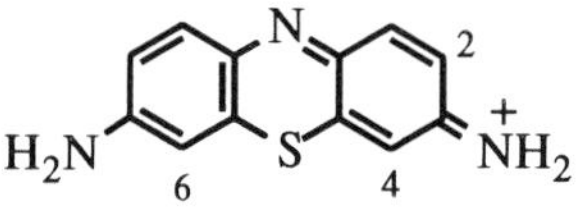

TH^+ = Thionine

DMST-2 = N,N-dimethyl-disulfonated thionine
DST-2 = 2,6-disulfonated thionine
TST-1 = 2,4,6-trisulfonated thionine

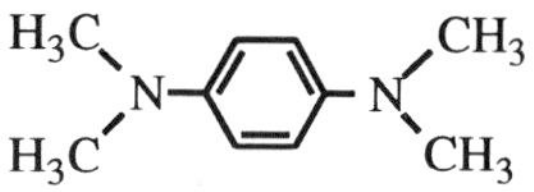

TMPD = N,N,N',N'-tetramethyl-4-phenylene diamine

Acknowledgments

This work was supported in part by a postdoctoral fellowship awarded to G.T.R.P. from the National Science Foundation and in part by support to G.M.W. from the Advanced Research Projects Agency. We wish to thank Dr. Hugo Bertschy and Professor Steven H. Bergens for helpful contributions to the manuscript.

Literature Cited

1. Vijh, S. K. *J. Chem. Ed.* **1970**, *47*, 680.
2. Kleywegt, G. J.; Driessen, W. L. *Chem. Br.*, **1988**, *24*, 447.
3. Prentice, G. *Chem. Tech.* **1984**, *14*, 684.
4. Oniciu, L. In *Fuel Cells*; Hammel, J., Ed.; Abacus Press; Tunbridge Wells, UK, 1976.

5. de Bethune, A. J. *J. Electrochem. Soc.* **1960**, *107*, 937.
6. Liebhafsky, H. A. *J. Electrochem. Soc.* **1959**, *106*, 1068.
7. Liebhafsky, H. A. *J. Electrochem. Soc.* **1961**, *108*, 601.
8. Kordesch, K. *Ber. Bunsenges. Phys. Chem.* **1990**, *94*, 902.
9. *Assessment of Research Needs for Advanced Fuel Cells*; Penner, S. S., Ed.; Energy; Pergamon Press: NY, 1986.
10. Van Dijk, C.; Laane, C.; Veeger, C. *Recl. Trav. Chim., Pays-Bas* **1985**, *104*, 245.
11. Aston, W. J.; Turner, A. P. F. *Biotech. and Gen. Eng. Rev.* **1984**, *1*, 89.
12. Disalvo, E. A. *Bioelectrochem. Bioenerg.* **1980**, *7*, 807.
13. Wingard, Jr., L. B.; Shaw, C. H.; Castner, J. F. *Enzyme Microb. Technol.* **1982**, *4*, 137.
14. Parsons, R.; VanderNoot, T. *J. Electroanal. Chem.* **1988**, *257*, 9.
15. McNicol, B. D. *J. Electroanal. Chem.* **1981**, *118*, 71.
16. Wong, C-H.; Whitesides, G. M. *J. Org. Chem.* **1982**, *47*, 2816.
17. Lee, L. G.; Whitesides, G. M. *J. Am. Chem. Soc.* **1985**, *107*, 6999.
18. Laval, J-M.; Bourdillon, C.; Moiroux, J. *J. Am. Chem. Soc.* **1984**, *106*, 4701.
19. Blankespoor, R. L.; Miller, L. L. *J. Electroanal. Chem.* **1984**, *171*, 231.
20. Gorton, L. *J. Chem. Soc., Faraday Trans. 1* **1986**, *82*, 1245.
21. Chenault, H. K.; Whitesides, G. M. *Appl. Biochem. Biotech.* **1987**, *14*, 147.
22. Chenault, H. K.; Simon, E. S.; Whitesides, G. M. *Biotech. Gen. Eng. Rev.* **1988**, *6*, 221.
23. Bonnefoy, J.; Moiroux, J.; Laval, J-M.; Bourdillon, C. *J. Chem. Soc., Faraday Trans. 1* **1988**, *84*, 941.
24. Persson, B. *J. Electroanal. Chem.* **1990**, *287*, 61.
25. Persson, B.; Gorton, L. *J. Electroanal. Chem.* **1990**, *292*, 115.
26. Biade, A-E.; Bourdillon, C.; Laval, J-M.; Mairesse, G.; Moiroux, J. *J. Am. Chem. Soc.* **1992**, *114*, 893.
27. Miyawaki, O.; Yano, T. *Enzyme Microb. Technol.* **1992**, *14*, 474.
28. Rodkey, F. L. *J. Biol. Chem.* **1955**, *213*, 777.
29. Moiroux, J.; Elving, P. J. *Anal. Chem.* **1978**, *50*, 1056.
30. Moiroux, J.; Elving, P. J. *J. Electroanal. Chem.* **1979**, *102*, 93.
31. Albery, W. J.; Knowles, J. R. *Biochemistry* **1976**, *15*, 5631.
32. Sisler, F. D. *Prog. Ind. Microbiol.* **1971**, *9*, 1.
33. Turner, A. P. F.; Aston, W. J.; Higgins, I. J.; Davis, G.; Hill, H. A. O. *Biotechnol. Bioeng. Symp.* **1982**, *12*, 401.
34. Suzuki, S.; Karube, I. *Appl. Biochem. Bioeng.* **1983**, *4*, 281.
35. Davis, J. B.; Yarbrough, Jr., H. F.; *Science* **1962**, *137*, 615.
36. van Hees, W. *J. Electrochem. Soc.* **1965**, *112*, 258.
37. Karube, I.; Matsunaga, T.; Tsuru, S.; Suzuki, S. *Biotechnol. Bioeng.* **1977**, *19*, 1727.
38. Suzuki, S.; Karube, I.; Matsunaga, T.; Kuriyama, S.; Suzuki, N.; Shirogami, T.; Takamura, T. *Biochimie* **1980**, *62*, 353.
39. Karube, I.; Suzuki, S.; Matsunaga, T.; Kuriyama, S. *Ann. N. Y. Acad. Sci.* **1981**, 369, 91.
40. Suzuki, S.; Karube, I.; Matsuoka, H.; Ueyama, S.; Kawakubo, H.; Isoda, S.; Murahashi, T. *Ann. N. Y. Acad. Sci.* **1983**, *413*, 133.
41. Delaney, G. M.; Bennetto, H. P.; Mason, J. R.; Roller, S. D.; Stirling, J. L.; Thurston, C. F. *J. Chem. Technol. Biotechnol.* **1984**, *34B*, 13.
42. Roller, S. D.; Bennetto, H. P.; Delaney, G. M.; Mason, J. R.; Stirling, S. L.; Thurston, C. F. *J. Chem. Technol. Biotechnol.* **1984**, *34B*, 3.
43. Thurston, C. F.; Bennetto, H. P.; Delaney, G. M.; Mason, J. R.; Roller, S. D.; Stirling, J. L. *J. Gen. Microbiol.* **1985**, *131*, 1393.

44. Bennetto, H. P.; DeLaney, G. M.; Mason, Roller, S. D.; Stirling, J. L.; Thurston, C. F. *Biotech. Lett.* **1985**, *7*, 699.
45. Lithgow, A. M.; Romero, L.; Sanchez, I. C.; Souto, F. A.; Vega, C. A. *J. Chem. Res., Synop.* **1986**, *5*, 178.
46. Vega, C. A.; Fernandez, I. *Bioelectrochem. Bioenerg.* **1987**, *17*, 217.
47. Akiba, T.; Bennetto, H. P.; Stirling, J. L.; Tanaka, K. *Biotechnol. Lett.* **1987**, *9*, 611.
48. Tanisho, S.; Kamiya, N.; Wakao, N. *Bioelectrochem. Bioenerg.* **1989**, *21*, 25.
49. Sell, D.; Kraemer, P.; Kreysa, G. *Appl. Microbiol. Biotechnol.* **1989**, *31*, 211.
50. Kreysa, G.; Sell, D.; Kraemer, P. *Ber. Bunsen-Ges. Phys. Chem.* **1990**, *94*, 1042.
51. Habermann, W.; Pommer, E.-H. *Appl. Microbiol. Biotechnol.* **1991**, *35*, 128.
52. Tanaka, K.; Tamamushi, R.; Ogawa, T. *J. Chem. Technol. Biotechnol.* **1985**, *35B*, 191.
53. Tanaka, K.; Kashiwagi, N.; Ogawa, T. J. *Chem. Technol. Biotechnol.* **1988**, *42*, 235.
54. Karube, I.; Suzuki, S. *Methods Enzymol.* **1988**, *137*, 668.
55. Plotkin, E. V.; Higgins, I. J.; Hill, H. A. O. *Biotech. Lett.* **1981**, *3*, 187.
56. Davis, G.; Hill, H. A. O.; Aston, W. J.; Higgins, I. J.; Turner, A. P. F. *Enzyme Microb. Technol.* **1983**, *5*, 383.
57. Yue, P. L.; Lowther, K. *Chem. Eng. J.* **1986**, *33B*, 69.
58. Yahiro, A. T.; Lee, S. M.; Kimble, D. O. *Biochim. Biophys. Acta* **1964**, *88*, 375.
59. Persson, B.; Gorton, Lo.; Johansson, G.; Torstensson, A. *Enzyme Microb. Technol.* **1985**, *7*, 549.
60. Laane, C.; Pronk, W.; Granssen, M.; Veeger, C. *Enzyme Microb. Technol.* **1984**, *6*, 165.
61. Srinivasa, S. *J. Electroanal. Chem.* **1981**, *118*, 51.
62. Glazebrook, R. W. *Electric Vehicle Developments*, **1982**, *4*, 18.

RECEIVED April 22, 1994

Pretreatment and Fermentation of Biomass

Chapter 15

Pretreatment of Lignocellulosic Biomass

James D. McMillan

Bioprocessing Branch, Alternative Fuels Division, National Renewable Energy Laboratory, 1617 Cole Boulevard, Golden, CO 80401–3393

Pretreatments for lignocellulosic materials include mechanical comminution, alkali swelling, acid hydrolysis, steam and other fiber explosion techniques, and exposure to supercritical fluids. These processes act by a variety of mechanisms to render the carbohydrate components of lignocellulosic materials more susceptible to enzymatic hydrolysis and microbial conversion. A variety of methods are effective on representative biomass feedstocks such as agricultural residues, herbaceous crops, and hardwoods. This chapter reviews pretreatment techniques, focussing on the importance of biomass structure and composition in determining pretreatment efficacy and the mechanisms by which different pretreatments act. The chapter concludes by recommending approaches for achieving further improvements in pretreatment technologies.

Production of fuels and chemicals from renewable lignocellulosic materials is accomplished by hydrolyzing polysaccharide components to soluble sugars which can be fermented to desired end products. Some type of pretreatment is generally required to render the carbohydrate fraction susceptible to enzymatic and microbial action because such materials are only partially digestible in their native form. Comprehensive reviews of pretreatment are provided by Millett et al. *(1)*, Chang et al. (*2*), Lin et al. (*3*), Fan et al. (*4*), and Dale (*5*).

The specific objectives of pretreatment are dictated by the overall objectives of a biomass conversion process. First, pretreatment must be energetically and chemically efficient, i.e economical, for a biomass conversion process to be profitable. Second, pretreatment must promote effective conversion of available carbohydrate to fermentable sugars so that high product yield can be achieved; pretreatment must maximize the formation of sugars or the ability to subsequently form sugars by

0097–6156/94/0566–0292$10.34/0

enzymatic hydrolysis. Hence, degradation or loss of carbohydrate must be avoided. Because it is also desirable to maximize the rate of enzymatic conversion, pretreatment must yield a highly digestible material that is not inhibitory to cell metabolism or extracellular enzyme function. Therefore, it is preferable to avoid the formation of inhibitory products and the need for detoxification or washing; high sugar losses occur if pretreated material is washed prior to enzymatic hydrolysis. Finally, pretreated materials are most efficiently hydrolyzed using low enzyme loadings, so the potential for nonspecific binding of enzymes to lignin and other fractions of pretreated biomass must be minimized.

Historically, pretreatments were investigated as a means to improve the digestibility of lignocellulosic biomass to increase its value as an inexpensive feed or feed supplement for ruminants. The efficacy of using pretreatment to improve lignocellulosic digestibility has been recognized at least since 1919, when Beckmann patented an alkali pretreatment based on soaking in sodium hydroxide (NaOH) for improving the in vitro digestibility of straws by ruminants (*1*). Numerous pretreatment processes have been investigated since then, most based on a combination of mechanical, physical, and chemical processing steps. Pretreatments that have been used or proposed for use in the context of renewable fuels and chemicals production from lignocellulosic materials include alkali swelling, acid hydrolysis, steam and other fiber explosion techniques, as well as exposure to supercritical fluids. This chapter reviews these pretreatment techniques, focussing on their mechanisms of action.

Biomass Composition and Structure

There are many different types of lignocellulosic biomass, including agricultural residues, herbaceous crops, deciduous (hardwood) and coniferous (softwood) trees, and municipal solid wastes (MSW). These biomass types exhibit a wide range of susceptibilities to pretreatment and saccharification because of structural and compositional differences. Thus, before discussing specific pretreatment processes, some background on the chemical composition and physical structure of lignocellulosic materials will be provided. Discussion will focus on woody and herbaceous materials.

Woody and herbaceous biomass species are composed mostly of cellulose, hemicellulose, and lignin, but also contain ash and other so-called extraneous materials. Typical compositions of representative lignocellulosic materials are shown in Figure 1. Cellulose is the main component, followed by hemicellulose and lignin; the paper fraction of MSW is comprised mostly of cellulose. Hardwoods are composed of about 50% cellulose (dry basis), 23% hemicellulose, and 22% lignin. Herbaceous materials and agricultural residues contain a somewhat higher proportion of hemicellulose (30-33%) relative to cellulose (38-45%), and have lower levels of lignin (10-17%). The composition and amount of extraneous components vary widely among the different biomass types. Extraneous components include extractives, fats, oils, protein, and ash. Extractives are compounds that are soluble in water or organic solvents. Extractive components in woody biomass include terpenes (isoprene alcohols and ketones), resins (fats, fatty acids, alcohols, resin acids, and phytosterols), and phenols (primarily tannins) (*4*). Nonextractives are mainly inorganic components

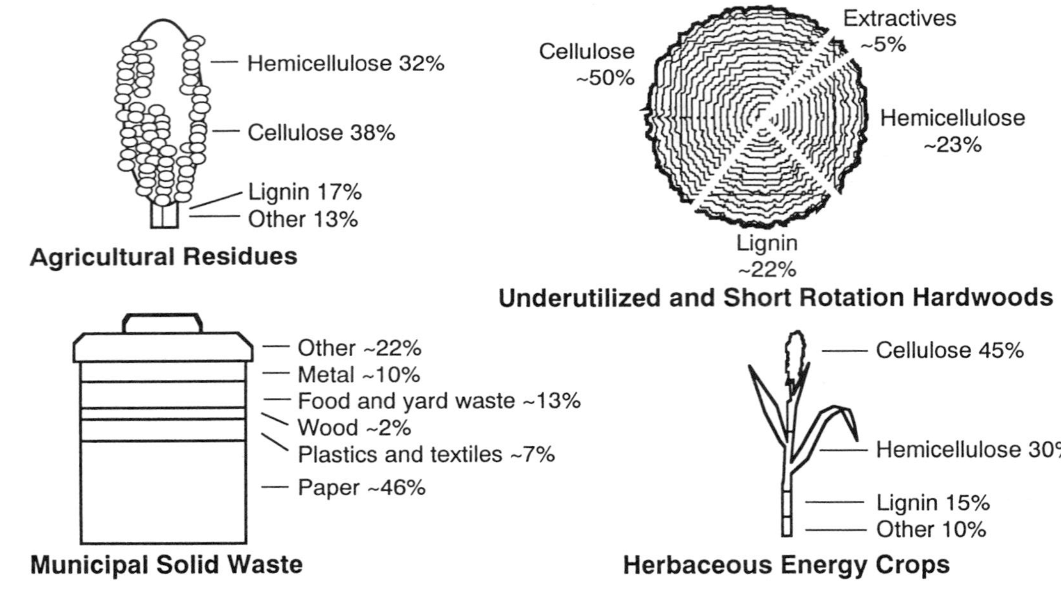

Figure 1. Composition of cellulosic biomass types. Cellulosic biomass consists of cellulose, hemicellulose and lignin, and some extractives, as shown here for representative examples of agricultural residues (corn cobs), hardwoods, municipal solid wastes, and herbaceous plants.

such as alkali earth carbonates and oxalates, as well as some non-cell-wall materials like starches, pectins, and proteins. Nonextractive silica crystals are abundant in herbaceous materials, but not in woody species.

Cellulose and hemicelluloses, collectively referred to as holocellulose, are high molecular weight polysaccharides. Cellulose is a linear polymer of anhydro D-glucose units connected by β-1,4-glycosidic bonds. Native cellulose exists as a fibrous crystal with the so-called Cellulose-I lattice structure. However, Cellulose-I is converted to Cellulose-II by physicochemical treatments such as intercrystalline swelling; mercerized or regenerated cellulose has the Cellulose-II crystalline lattice structure (*2*). Cellulose crystals (Cellulose-I or Cellulose-II) are organized into compacted crystallite microfibrils measuring 35 x 40 Å in width and about 500 Å in length, having an average degree of polymerization (DP) of about 1000. A widely accepted model of cellulose crystal structure is the folding chain model in which segments of linear cellulose polymer are folded back and forth along the major axis of the fibrillar crystallite to make up individual microfibrils.

Hemicelluloses are shorter chain, amorphous polysaccharides of cellulans and polyuronides (*6*). Cellulans are heteropolymers made up of hexosans (mannan, galactan, and glucan) and pentosans (xylan and arabinan). Polyuronides are similar to cellulans, but contain appreciable quantities of hexuronic acids as well as some methoxyl, acetyl, and free carboxylic groups. Xylan and glucomannan are the dominant carbohydrate components of hemicellulose. Hardwood hemicelluloses contain mostly xylan, whereas softwood hemicelluloses contain mostly glucomannan.

Lignin is composed of polymerized phenylpropanoic acids in a complex three-dimensional structure. Individual monomers are held together by ether and carbon-carbon bonds (*4*). Lignin forms by a free radical polymerization mechanism and has a random structure. Lignin and hemicellulose are believed to form an effective sheath around cellulose fibers which adds structural strength to the biomass matrix. Covalent bonds may exist between hemicellulose and lignin, but this is uncertain. The amount of lignin in softwoods (25%-35%) is appreciably greater than in hardwoods (18%-25%) and herbaceous species and agricultural residues (10%-20%) (*4,7-11*). Figure 2 illustrates the compositional differences between trembling aspen hardwood and white spruce softwood.

The distribution of cellulose, hemicellulose, and lignin within a typical wood fiber cell wall is shown in Figure 3. This diagram demonstrates that lignin is the dominant component in the outer portion of the compound middle lamellae, such that a lignin seal effectively exists at the outer periphery of individual wood fibers. The percentage of lignin in the lignocellulosic matrix decreases with increasing distance into the fiber cell wall, with the percentages of lignin in the primary wall and in the S1 layer of the secondary wall much higher than in the S2 and S3 sections of the secondary wall.

Another important structural feature is the asymmetry of fibrous and woody biomass materials. Heat and mass transfer studies by Brownell et al. (*12*), Tillman et al. (*13*), and others demonstrate that the fibrous structure of lignocellulosic materials strongly favors transport in the axial rather than radial direction. Tillman et al. (*13*), for example, determined that acid diffusivities are two-to threefold higher in the direction of the fiber axis than in radial or tangential directions.

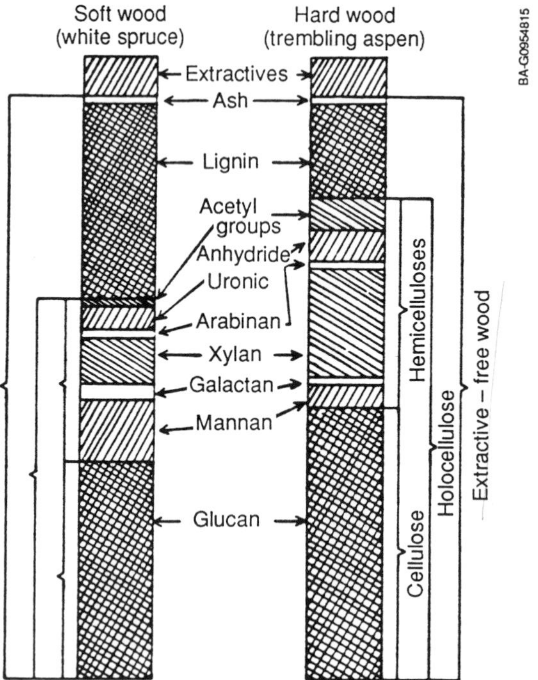

Figure 2. Comparison of compositions of hardwood and softwood. Reprinted with permission from ref. *4*. (Copyright 1982; Springer-Verlag: New York, NY.)

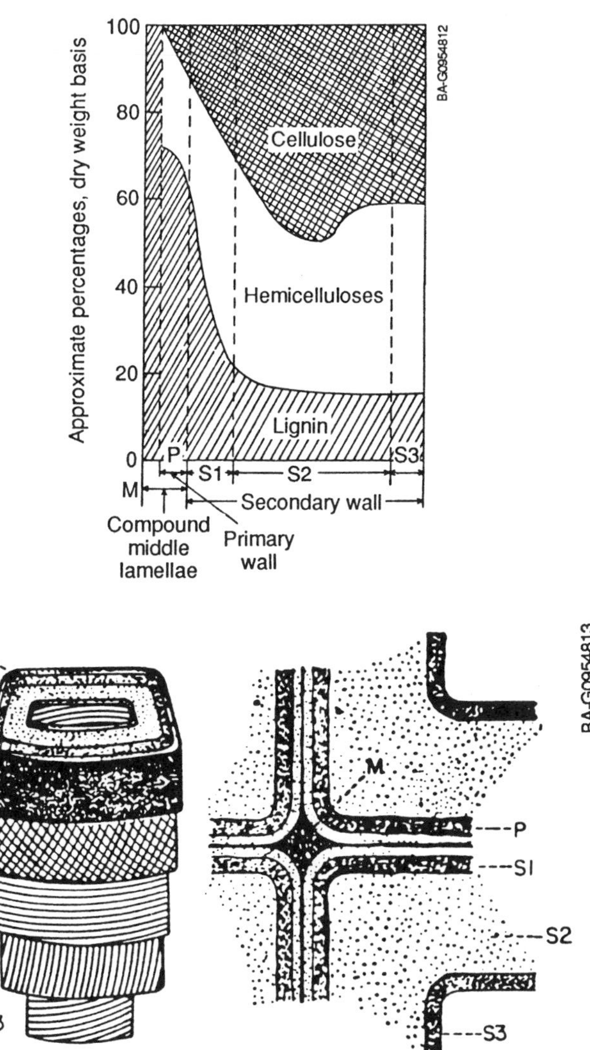

Figure 3. Distribution of cellulose, hemicellulose, and lignin in the wood fiber cell wall. Reprinted with permission from ref. *4*.

Despite gross similarities, however, significant structural differences exist between different biomass types. For example, softwoods and hardwoods have disparate structural features (*7,14*). Coniferous (soft) woods are composed primarily of tracheid fiber cells, which are elongated, tubelike structures with closed ends that are square in cross section. Earlywood fibers in softwoods have a pronounced lumen and are thin-walled, whereas latewood fibers have thicker walls and smaller lumens. Deciduous (hard) woods, on the other hand, are composed of vessels and sclerenchyma or librilform fiber cells. Vessels are long, continuous tubes that serve as a conduit for water and have a much larger diameter than the individual fiber cells.

Differences in the structure of coniferous and deciduous woods contribute to the dramatic differences in susceptibility to various pretreatment techniques that exist between hardwood and softwood species. The presence of vessels in the hardwoods permits greater penetration of heat, chemicals, and enzymes into the biomass matrix, making hardwoods easier to pretreat.

Factors Affecting Enzymatic Digestibility

Major factors that have been identified to influence the digestibility or reactivity of lignocellulosic materials are porosity, cellulose fiber crystallinity, lignin content, and hemicellulose content. Prevailing hypotheses regarding these factors are outlined below.

Porosity (Accessible Surface Area). It is generally believed that the initial rate of cellulose hydrolysis is a function of the specific accessible surface area of cellulose, with the rate increasing with increasing coverage of cellulase enzyme (*15*). Quantifying the rate of enzymatic hydrolysis as a function of accessible surface area has proven to be difficult. Many techniques have been used to estimate surface area, including nitrogen adsorption, mercury porosimetry, dye adsorption (*16,17*) and solute exclusion (*15,18,19*). Results indicate that accessible surface area increases following pretreatment, although there is disagreement about how much. An important but still unresolved issue is how lignin affects accessible surface area (see below).

Crystallinity. Many researchers believe that the rate of cellulose hydrolysis is determined not by the amount of adsorbed cellulase enzyme but rather by the DP of cellulose or the extent of cellulose crystallinity (also referred to as the ratio of amorphous to crystalline cellulose) (*4,20*). This hypothesis is supported by the fact that size reduction, which has little effect on accessible surface area because of the high length to width ratio of lignocellulosic fibers, dramatically improves digestibility. This view is disputed by other researchers like Schurz (*21*) who observe no change in crystallinity during hydrolysis and attribute differences in rate to differences in the type or quality of surface adsorption.

Lignin Content. The reactivity of lignocellulosic materials to cellulase enzyme action generally varies in negative proportion to lignin content. Figure 4 shows data from Stone et al. (*15*) on the digestibility of softwoods as a function of the amount of lignin removed by sulfite pulping. A large increase in 24-h saccharification yields occurs with increasing extent of delignification. The lower digestibility of the dried material

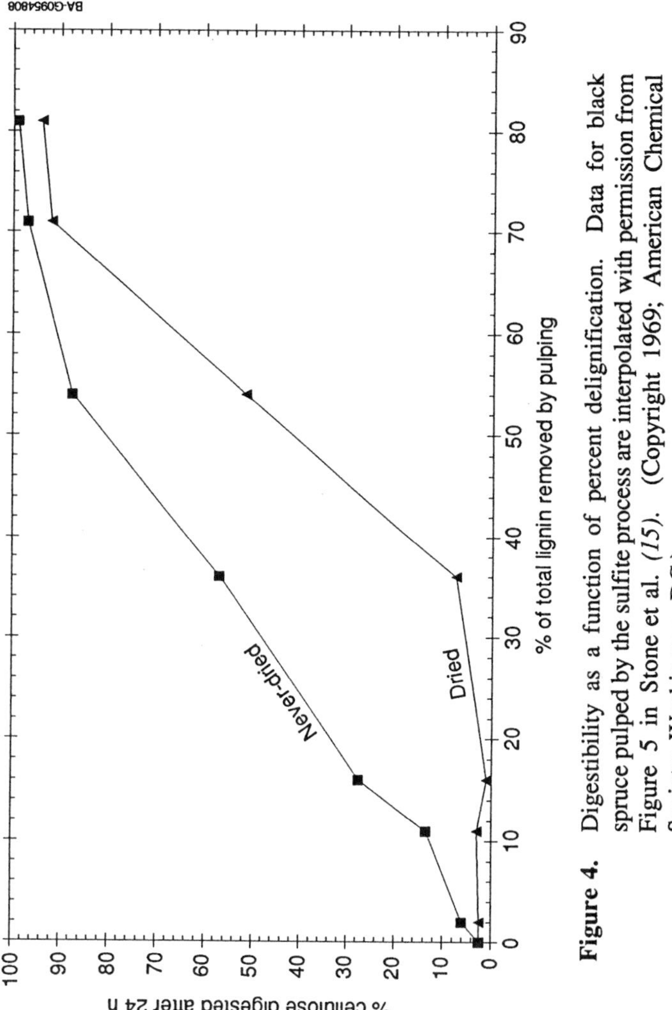

Figure 4. Digestibility as a function of percent delignification. Data for black spruce pulped by the sulfite process are interpolated with permission from Figure 5 in Stone et al. *(15)*. (Copyright 1969; American Chemical Society: Washington, DC.)

is attributed to reduced pore volume caused by drying. Figure 5 demonstrates that the in vitro digestibility of alkali-pretreated hardwoods increases substantially as lignin content decreases.

Lignin is hypothesized to interfere with hydrolysis by blocking access to cellulose fibers and irreversibly binding hydrolytic enzymes (*2*). This is supported by research showing that cellulase enzyme binds to lignaceous materials (*22*). The location and nature of lignin-carbohydrate bonding may also affect enzymatic digestibility, since materials with similar lignin content such as jute and hay can exhibit markedly different digestibilities (*23*). It is unclear, however, to what extent the presence of lignin physically restricts or otherwise blocks enzyme binding to lignocellulosic surfaces.

Hemicellulose Content. Removal of hemicellulose dramatically improves the digestibility of hardwood species (*24*) by increasing porosity and thereby the specific surface area accessible to hydrolytic enzymes (*18,19*). Figure 6 shows enzymatic digestibility as a function of the amount of xylan removed by dilute acid pretreatment of aspen wood. Digestibility increases in an approximately linear fashion with increasing extent of xylan removed.

There is still debate about which factors most influence enzymatic hydrolysis of pretreated lignocellulosic materials. In addition to the factors discussed above, the composition and amount of extraneous components, which vary widely between different biomass types, strongly influence resistance to enzymatic attack and/or susceptibility to thermo-chemi-mechanical pretreatments. Often a small fraction of a pretreated material remains refractory to enzymatic conversion. Many researchers, including Fan et al. (*4*) and Lynd (*25*), believe that external surface area and reactivity determine the rate and extent of enzymatic hydrolysis. This theory ascribes sub-theoretical saccharification yields to highly crystalline regions of cellulose that remain impervious to enzymatic attack, even following pretreatment. Other researchers hypothesize that structural features control the extent to which a lignocellulosic material is accessible to enzymatic attack. Stone et al. (*15*), Grethlein (*26*), and Grohmann (*27*), for example, believe that the porosity of the lignin-hemicellulose sheath controls enzyme accessibility and hydrolysis rates. This view attributes sub-theoretical yields to regions of cellulose that remain inaccessible to enzymatic attack.

It is difficult to determine which of these hypotheses is correct, especially since the relative importance of enzyme accessibility and cellulose crystallinity vary depending on the lignocellulosic material used and the pretreatment method employed. Examples undoubtedly exist where either accessible surface area or reactivity control the rate and extent of hydrolysis. However, susceptibility to enzymatic hydrolysis is generally affected by both factors.

Pretreatment Techniques

Methods of pretreating lignocellulosic biomass that will be described in some detail in the following sections include mechanical comminution, alkali swelling, acid hydrolysis, steam and other fiber explosion techniques, and exposure to supercritical fluids. There are also a variety of pulping techniques that are potential pretreatments, but it is doubtful that pulping processes can be economically used to pretreat

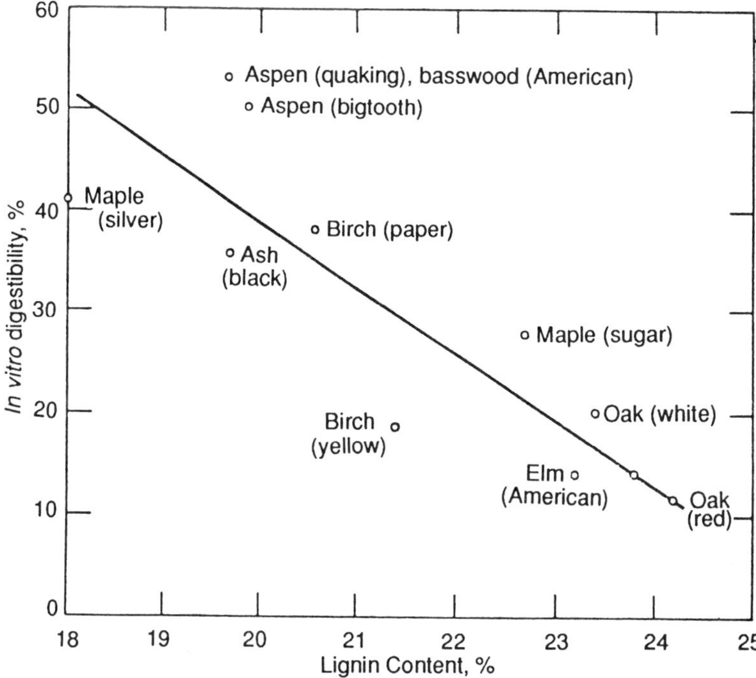

Figure 5. Effect of lignin content on the *in vitro* digestibility of NaOH-pretreated hardwoods. Adjacent to individual data points are the names of the hardwood species to which the point refers. Adapted with permission from ref. *1*. (Copyright 1976; John Wiley & Sons: New York, NY.)

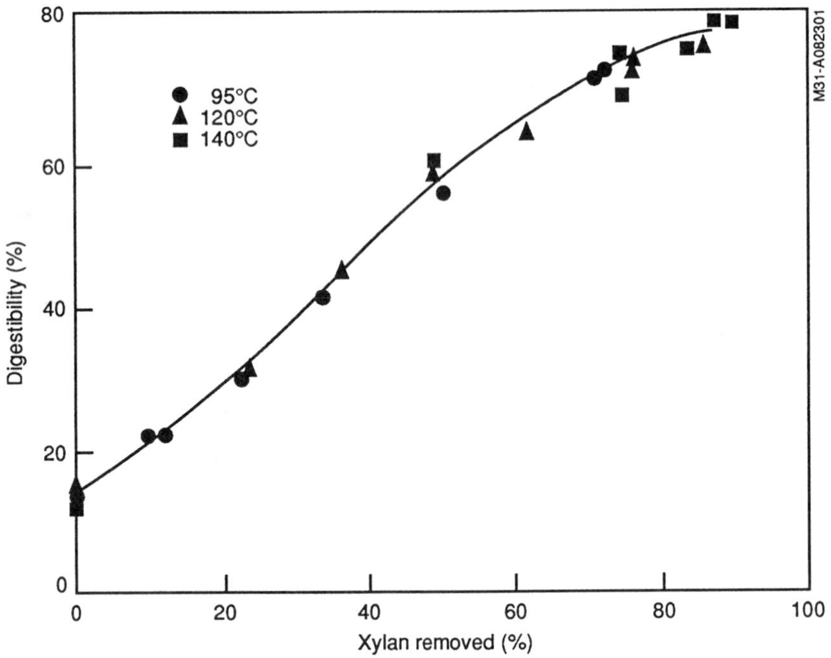

Figure 6. Digestibility as a function of percent xylan removed. Reprinted with permission ref. *24*. (Copyright 1985; John Wiley & Sons: New York, NY.)

lignocellulosics in the context of renewable fuels and chemicals production. Pulping processes, which are generally applied for relatively high-valued products such as paper, generally exhibit relatively low overall carbohydrate yields and are chemical intensive. Large capital investments are typically required to install integrated chemical recovery systems. For these reasons, potential pretreatments based on pulping techniques are not considered here.

Mechanical Comminution. All pretreatment processes involve an initial mechanical step in which the biomass is comminuted by a combination of chipping, grinding, and milling. Steam explosion processes use explosive decompression to significantly reduce the particle size of coarsely chipped biomass, whereas other pretreatment processes typically employ a secondary grinding or milling step to further reduce the particle size of chipped biomass. Chipped biomass has a characteristic dimension of 1 cm to 3 cm, in comparison to 0.2 mm to 2.0 mm for milled or ground material.

Power requirements for comminution are strongly influenced by particle size and biomass type. Power requirements increase rapidly with decreasing particle size, as shown in Figure 7 for ball milling of municipal solid waste. More than 25% of the total energy of the substrate is required to mill this material to particle sizes below 150 μm. Energy requirements for milling coarse biomass chips into fine particles are also strongly influenced by substrate. For example, the specific milling energy required to produce 0.64-cm (0.25-in) or smaller particles is about two-and-a-half-fold higher for aspen wood chips than for corn cobs; for particles 1.27-cm (0.50-in.) or smaller, the difference is greater than eightfold (*28*).

Much of the pretreatment research reported in the literature is carried out using 60 mesh (250 μm = 0.25 mm) wood flour as a substrate (e.g., the extensive work of Grethlein et al.). The energy costs for milling biomass particles to such small sizes are likely to be prohibitively high for low-profit-margin, high-volume, industrial-scale pretreatment processes (*1*).

Some researchers have concluded that milling processes, especially vibratory ball milling, increase the reactivity of cellulose, in addition to increasing the external surface area. Millett et al. (*1*) report that vibratory ball milling breaks down the crystallinity of cellulose to an extent that complete digestion of milled biomass is possible. Fan et al. (*4*) also conclude that ball milling causes a reduction in cellulose crystallinity, although they observe that the efficacy of this approach is strongly dependent on the type of material being milled. They cite the additional advantage that ball-milled material is often more dense than the native material and can therefore be slurried to a higher solids loading.

Alkali Swelling. Soaking in alkaline solutions, most notably dilute NaOH, has been used to pretreat lignocellulosic materials. The efficacy of alkali pretreatment is dependent upon lignin content. As reference to Figure 5 shows, for hardwoods soaking in NaOH shows increasing efficacy as lignin content decreases from 24% to 18%. No effect of dilute NaOH pretreatment is observed for softwoods in which the lignin content is 26%-35% (*1*). The efficacy of dilute NaOH pretreatment is somewhat higher for straws than for hardwoods owing, in part, to the lower lignin content of straws.

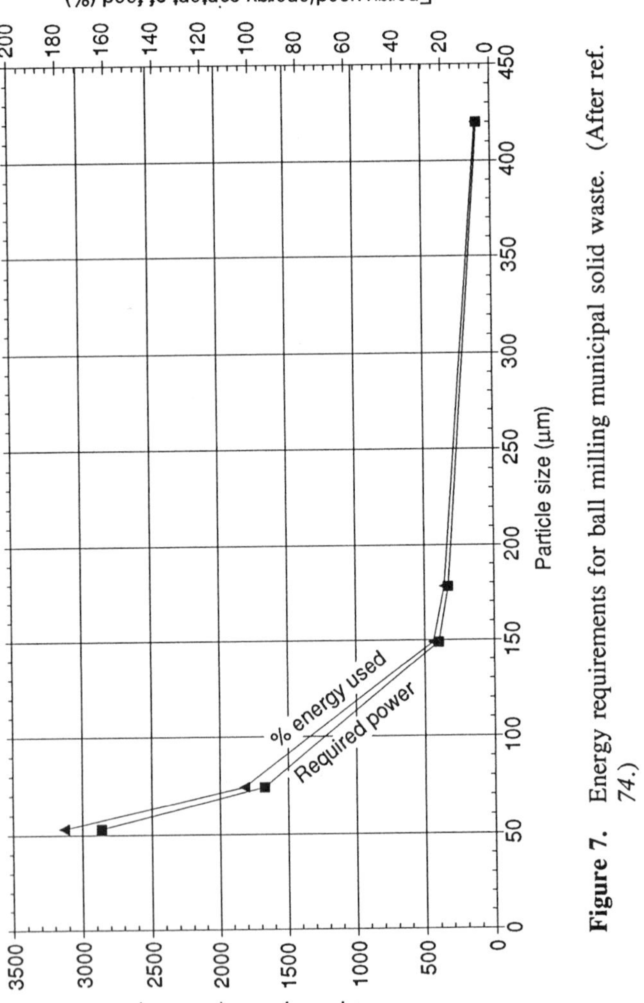

Figure 7. Energy requirements for ball milling municipal solid waste. (After ref. 74.)

Treatment with alkali causes lignocellulosic materials to swell, and increased swelling leads to higher digestibility. Certain swelling agents such as NaOH, amines, and anhydrous NH_3 cause limited or intercrystalline swelling. NaOH treatment followed by water washing yields a lignocellulosic residue with a polyionic character, since sodium ions remain to act as counter charges to carboxylate ions. This polyionic character promotes swelling relative to NaOH treatment followed by acid washing in which the sodium ions are displaced by protons.

Other chemicals, such as concentrated sulfuric acid (H_2SO_4) or hydrochloric acid (HCl), or high concentrations of cellulose solvents like cupran, cuen, or cadoxen, cause intracrystalline swelling; these chemicals penetrate into the cellulose crystal structure and totally dissolve (or hydrolyze) holocellulose (*3,29*). Although they are powerful cellulose solubilization agents, concentrated acids and metal chelate cellulose solvents are toxic, corrosive, and hazardous in nature. For these reasons, in addition to high recovery costs, cellulose solvents may not be appropriate for large-scale pretreatment processes (*1*).

The mechanism of alkali pretreatment is postulated to be saponification of intermolecular ester bonds crosslinking xylan hemicelluloses and other polymeric materials, such as lignin or other hemicelluloses (*23*). Saponification of the uronic ester linkages in 4-O-methyl-D-glucuronic acids pendant along the xylan chain readily occurs in the presence of alkali. (So does saponification of acetyl groups pendant along the xylan backbone, which has an autocatalytic effect on hemicellulose hydrolysis when it occurs--see below.) The removal of crosslinks apparently permits swelling beyond normal water-swollen dimensions. Pore volume measurements indicate that intraparticle porosity and channel size increase following dilute alkali treatment (*23*). There is also a phase change in the cellulose crystal structure (*1-4*).

Gaseous and aqueous ammonia (NH_3) cause swelling by a similar mechanism (*30*). At ambient conditions, longer reaction times are generally required; for comparable reaction times, digestibility is lower following NH_3 treatment than it is following NaOH treatment. In NH_3 alkali treatment, rather than saponification of ester bonds to yield ionized carboxyl groups, ammonolysis of ester bonds occurs to form amides. The decreased swelling characteristics observed following NH_3 pretreatment indicate reduced polyionic character relative to water-washed NaOH-treated materials. However, chemically combined nitrogen (via amide formation) or residual ammonia/ammonium increases overall nitrogen content, and thus may improve the value of NH_3-treated materials as feeds for ruminants.

Dilute Acid Hydrolysis. Exposure to concentrated acid and then later to dilute acid was originally used to directly saccharify lignocellulosic materials (*29*). Above moderate temperatures, however, direct saccharification suffered from low yields because of sugar decomposition. Thus, exposure to dilute acid at high temperature has been developed as a pretreatment prior to enzymatic saccharification to improve overall saccharification rates and yields. Whereas older acid-based saccharification processes largely destroyed the predominantly xylan hemicellulosic fraction, more recently developed processes use less severe conditions and achieve high xylan to xylose conversion yields. Achieving high xylan to xylose conversion yields is necessary to achieve favorable overall process economics because xylan accounts for up to a third of total carbohydrate in many lignocellulosic materials (*31*). NREL currently favors dilute acid hydrolysis as the pretreatment process of choice for a commercial biomass-to-ethanol process (*32,33*).

In NREL's dilute acid process, chipped and/or milled biomass particles of nominal 1-mm size are impregnated with approximately 1% (w/w) H_2SO_4 (liquid basis) and then incubated at 140°-160°C for a period ranging from several minutes to an hour. NREL researchers have characterized the susceptibility of a variety of short rotation woody and herbaceous crops and agricultural residues to this dilute acid pretreatment process (*9-11,24,34-36*). High-temperature dilute acid treatment causes hemicellulose to hydrolyze. Hemicellulose removal increases porosity and improves enzymatic digestibility, as shown in Figure 6. Hemicellulose hydrolysis rates vary with temperature, but for most short rotation woody species and herbaceous crops, complete hemicellulose hydrolysis occurs in 5-10 min at 160°C, or in 30-60 min at 140°C. Maximum enzymatic digestibility of the cellulosic fraction usually coincides with complete hemicellulose removal.

Dilute acid hydrolysis forms the basis of many pretreatment processes. For example, steam explosion and autohydrolysis pretreatments are also based on high-temperature dilute acid-catalyzed hydrolysis of biomass. The dominant factors influencing hydrolysis yields for a variety of pretreatment processes can be understood by examining the kinetics of dilute acid hydrolysis.

The overall goal of dilute acid pretreatment is to achieve high yield while minimizing the breakdown of sugars into decomposition products. Cellulose hydrolysis, hemicellulose hydrolysis, and sugar degradation reactions can be considered as pseudo first-order processes, as shown by Equation (1). High selectivity pretreatment is achieved by maximizing the ratio of the rate constants, k_{i1}/k_{i2}, where the subscript i refers to rate constants for either cellulose (c) or hemicellulose (h) hydrolysis. This is accomplished by using acid-catalyzed, high-temperature, short-residence-time processes.

$$
\begin{array}{ccccc}
 & & \textit{glucose} & & \\
\textit{cellulose} & \xrightarrow{k_{c1}} & \textit{cellobiose} & \xrightarrow{k_{c2}} & \textit{degradation products} \\
 & & \textit{oligosaccharides} & & \\
 & & & & \\
 & & \textit{pentoses} & & \\
\textit{hemicellulose} & \xrightarrow{k_{h1}} & \textit{hexoses} & \xrightarrow{k_{h2}} & \textit{degradation products} \\
 & & \textit{oligosaccharides} & &
\end{array}
\qquad (1)
$$

Literature and data on the kinetics of dilute acid pretreatment primarily exist for two types of processes, low solids loading (5%-10% [w/w]), high-temperature (T > 160°C), continuous-flow processes (*37-42*) and higher solids loading (10%-40% [w/w]), lower temperature (T < 160°C), batch processes (*24,43-48*). The kinetics of high-temperature wood saccharification catalyzed by dilute acid were first extensively investigated by Saeman (*43*), who demonstrated that cellulose hydrolysis and monomer sugar decomposition follow first-order kinetics. Although Root (*44*) and Kwarteng (*39*) subsequently showed that decomposition kinetics become more complex as degradation products accumulate, the approximation is usually made that kinetics remain first order. A generalized relationship for sequential hydrolysis and decomposition of both cellulose and hemicellulose is formulated by representing reactants and products on an equivalent sugar monomer basis. Equation (2) shows the general case in which biomass carbohydrate species A (cellulose or xylan) hydrolyzes to B (glucose/cellobiose or xylose), which then decomposes to C (hydroxymethylfurfural from glucose and furfural from xylose).

$$A \xrightarrow{k_1} B \xrightarrow{k_2} C \tag{2}$$

Expressions for the net rate of formation of components A, B, and C can be integrated to determine the concentrations of A and B as a function of time and initial conditions. Equations (3) and (4) give the concentrations of A and B, respectively, for the case of batch reaction with no heat or mass transfer limitations (*43,49*).

$$C_A = C_{A_o} e^{-k_1 t} \tag{3}$$

$$C_B = C_{A_o} \frac{k_1}{k_1 - k_2} \left[\left(1 + \frac{C_{B_o}}{C_{A_o}} \left(\frac{k_1 - k_2}{k_1} \right) \right) e^{-k_2 t} - e^{-k_1 t} \right] \tag{4}$$

In these equations C_i represents the concentration of component i at time t and C_{i_o} denotes the initial concentration.

Grethlein (*49*) showed that the concentration of product B is maximized at time t_{opt}.

$$t_{opt} = \frac{1}{k_2(K-1)} \ln\left(\frac{K}{Q}\right) \tag{5}$$

In Equation 4, K is the selectivity ratio defined by

$$K = \frac{k_1}{k_2} \tag{6}$$

and Q is the lumped initialization and selectivity parameter given by

$$Q = 1 + \frac{C_{B_o}}{C_{A_o}} \left(\frac{K-1}{K} \right). \tag{7}$$

Substitution of Equation (5) into Equation (4) yields for following expression for the maximum concentration of product B. (Note: the expression published in Grethlein [*49*] is incorrect.)

$$C_B(t_{opt}) = C_{A_o}\left(\frac{Q^K}{K}\right)^{\frac{1}{K-1}} \tag{8}$$

Figure 8 shows maximum yield as a function of selectivity ratio, K, calculated using Equation (8) assuming that initially no B is present. Conversion yields of 95% require K values above 100.

Equations (3) and (4) are useful for describing xylan hydrolysis and decomposition at high temperatures (T > 160°C) where hemicellulose (xylan) hydrolysis occurs rapidly following first-order kinetics (*43,44*). However, at lower temperatures (T < 160°C) hemicellulose hydrolysis is not homogeneous, with a portion hydrolyzing rapidly while the remainder hydrolyzes more slowly. Grohmann et al. (*24*) concluded that in the initial stages of hydrolysis both fast and slow fractions of xylose are hydrolyzed to xylose by parallel first-order reactions. The fraction of slow-reacting xylan is estimated to be 0.20-0.32 (*24,47,48*). Equation (9) depicts the modified reaction scheme for lower temperature (T < 160°C) hemicellulose hydrolysis, where A_f and A_s represent the fast and slow reacting portions of xylan, respectively, and k_{1f} and k_{1s} are the corresponding rate constants.

$$\begin{matrix} A_f \xrightarrow{k_{1f}} \\ \\ A_s \xrightarrow{k_{1s}} \end{matrix} \rightarrow B \xrightarrow{k_2} C \tag{9}$$

The amount of each fraction can be calculated as the product of the total xylan concentration expressed on an equivalent monomer basis, A_o, and the respective weight fraction, with Φ denoting the weight fraction of total xylan that hydrolyzes slowly.

$$\begin{aligned} A_f &= (1-\Phi)A_o \\ A_s &= \Phi A_o \end{aligned} \tag{10}$$

Summing the contributions of xylose formed from fast and slow reacting hemicellulose portions, using Equation (4) to describe the net rate of formation of xylose from a single hemicellulose fraction, gives an expression for the total

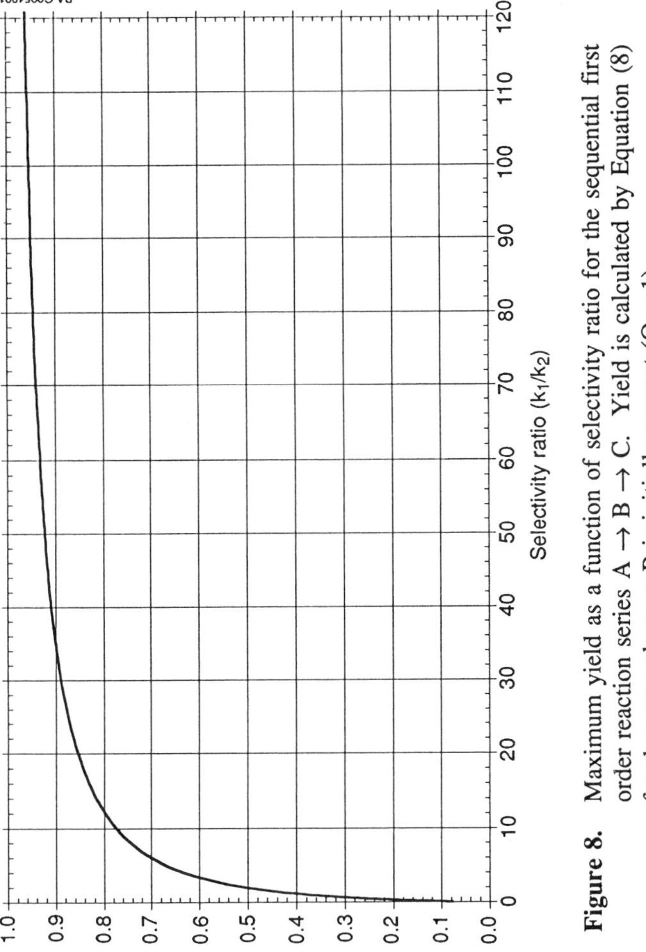

Figure 8. Maximum yield as a function of selectivity ratio for the sequential first order reaction series A → B → C. Yield is calculated by Equation (8) for the case when no B is initially present (Q = 1).

concentration of xylose as a function of time. Equation (11) describes the total xylose concentration for the case when the initial concentration of xylose is zero. It is difficult to directly compare reported kinetic rate constants for dilute acid hydrolysis because studies generally use different substrates and different sample preparation methods (extent of grinding or drying prior to dilute acid hydrolysis, etc.).

$$C_B = C_{A_0}\left(\frac{k_{1s}}{k_2 - k_{1s}}\Phi\left(e^{-k_{1s}t} - e^{-k_2 t}\right) + \frac{k_{1f}}{k_2 - k_{1f}}(1-\Phi)\left(e^{-k_{1f}t} - e^{-k_2 t}\right)\right) \tag{11}$$

Studies also use a variety of batch and continuous reactor types, and employ different methods to initiate and quench reaction. In addition, studies account for the dependence on acid loading in different ways. Some studies do not consider acid concentration dependence (*24*). In many others, the effect of acid concentration is not correlated in terms of hydrogen ion molarity but rather in terms of weight percent acid on a dry substrate basis. In most cases, the neutralizing ability of biomass on the effective acid concentration is overlooked (*39,46*). Biomass species differ significantly in their ability to neutralize acids, so determination of the effective acid concentration is important.

The dependence on acid loading is typically incorporated into an Arrhenius-type relationship, as shown in Equation (12), where k_i represents the kinetic rate constant for reaction i, k_{i_0} the preexponential factor, C_{H^+} the effective acid concentration, n the exponent on acid concentration, E_i the activation energy, R the universal gas constant, and T the temperature.

$$k_i = k_{i_0}\, C_{H^+}^{n}\, e^{\frac{-E_i}{RT}} \tag{12}$$

Maloney et al. (*47*) carried out an analysis of the effect of acid concentration on hydrolysis rates that considered both the neutralization capacity of biomass and the reduced dissociation of H_2SO_4 at high temperature. It was assumed that only 1 mol of hydrogen ions was obtained from each mol of H_2SO_4, since at the high temperatures (100°-170°C) used in their hydrolysis experiments more than 97% of the sulfate existed as bisulfate. The neutralizing capacity of the biomass was calculated assuming that 1 mol of hydrogen ions was neutralized by each mol of inorganic cations that was dissolved during hydrolysis. Atomic absorption spectroscopy was used to quantify the extent of inorganic cation (Ca, Mg, and Mn) dissolution during hydrolysis. Correcting for both factors yielded a neutralization capacity for white birch of 3.5 g H_2SO_4/kg biomass. The exponents on acid concentration determined using corrected acid concentration values were essentially 1 for both fast and slow reacting xylan fractions, indicating that dependence on effective acid concentration is roughly linear. Exponents on acid concentration reported in the literature range from about 0.5 to 1.5, probably because researchers have not adequately considered high-temperature acid dissociation and/or the neutralizing capacity of biomass.

Kim and Lee (*48*) review the kinetics of acid-catalyzed hydrolysis of hemicellulose and provide a useful tabulation of previously published kinetic rate constants. There is wide variation in reported rate constant values, reflecting differences in the substrate and pretreatment conditions used, the model employed, and the method by which acid concentration was correlated. Unfortunately, only a few researchers have performed studies in which both hydrolysis and decomposition rates were determined. Kim and Lee (*48*) carried out the only study including both fast and slow reacting xylan fractions as well as xylose decomposition.

Figure 9 shows simulations of xylan hydrolysis yield profiles, calculated using Equation (11), for an acid loading of 1.0% by weight and a slow reacting xylan fraction of 0.30. Values for the individual rate constants are those reported for Kim and Lee (*48*) for hydrolysis of 0.15- to 0.425-µm screened sawdust particles of southern red oak hardwood using a solid to liquids ratio of 1:1.6 and an H_2SO_4 loading of 1-5 wt %. Kim and Lee (*48*) corrected the effective acid concentration used in their experiments by accounting for the buffering capacity of southern red oak, which they had previously determined to be 7.5-9.0 g H_2SO_4/kg biomass. This correction reduced the effective weight percent acid used in their experiments by about 0.5 wt %. Note that although simulations are shown for temperatures from 120°C up to 220°C, the simulations at temperatures of 160°C and above are extrapolations, since Kim and Lee (*48*) only determined rate constants at temperatures from 120°C to 140°C. Simulation results indicate a maximum xylose yield of about 80% which increases slightly with increasing pretreatment temperature.

Figure 10 shows characteristic normalized profiles (simulated) for total potential xylose (xylan plus xylose), free xylose (hydrolyzed xylan), and xylose decomposition products as a function of hydrolysis time at 160°C. (Other conditions are identical to those used to generate Figure 9). Only after free xylose has accumulated in solution does xylose degradation become significant.

Figure 11 shows selectivity ratios based on the rate constants determined by Kim and Lee (*48*). The selectivity ratio for fast xylan hydrolysis relative to xylose degradation, k_{1f}/k_2, increases with temperature, from a value of 42 at 120°C to a value of 67 at 220°C. A smaller increase is observed in the selectivity ratio for slow xylan hydrolysis, k_{1s}/k_2, with the value rising over the same temperature range from about 6 to 8. The temperature sensitivity of the selectivity ratios reflects a lower activation energy for xylose degradation (112.6 kJ/g mol [26.9 kcal/g mol]) relative to fast (120.1 kJ/g mol [28.7 kcal/g mol]) and slow (118.0 kJ/g mol [28.2 kcal/g mol]) xylan hydrolysis.

Steam Explosion. Steam explosion is a variation of the high-temperature dilute acid hydrolysis technique in which chipped biomass is treated with saturated steam in a pressure vessel, which is then flashed, causing explosive disruption of the biomass by liberated steam. Steam explosion pretreatments are typically carried out using saturated steam at temperatures of 160°-260°C, which corresponds to pressures of 0.69-4.83 MPa (100-700 psia) (*50*). Residence times are typically tens of seconds to several minutes. Steam explosion is the most commercialized of the pretreatment techniques, and numerous reviews of steam explosion are available (*12,51,52*). In general, both the rate and extent of enzymatic hydrolysis achievable following steam explosion pretreatment increase with increasing duration of treatment. This, however, refers only to the washed solids remaining after pretreatment; because of the formation of degradation products that are inhibitory to microbial growth, steam-

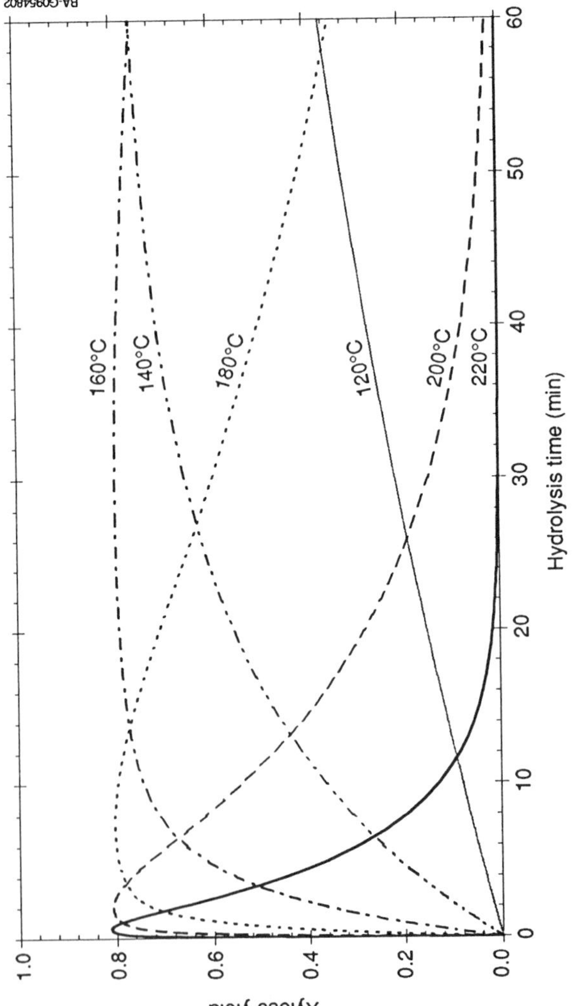

Figure 9. Xylose yield as a function of time. Yield is calculated by Equation (11) using the rate constants of Kim and Lee (*48*) assuming an acid loading of 1% by weight and a slow reacting xylan fraction of 0.30.

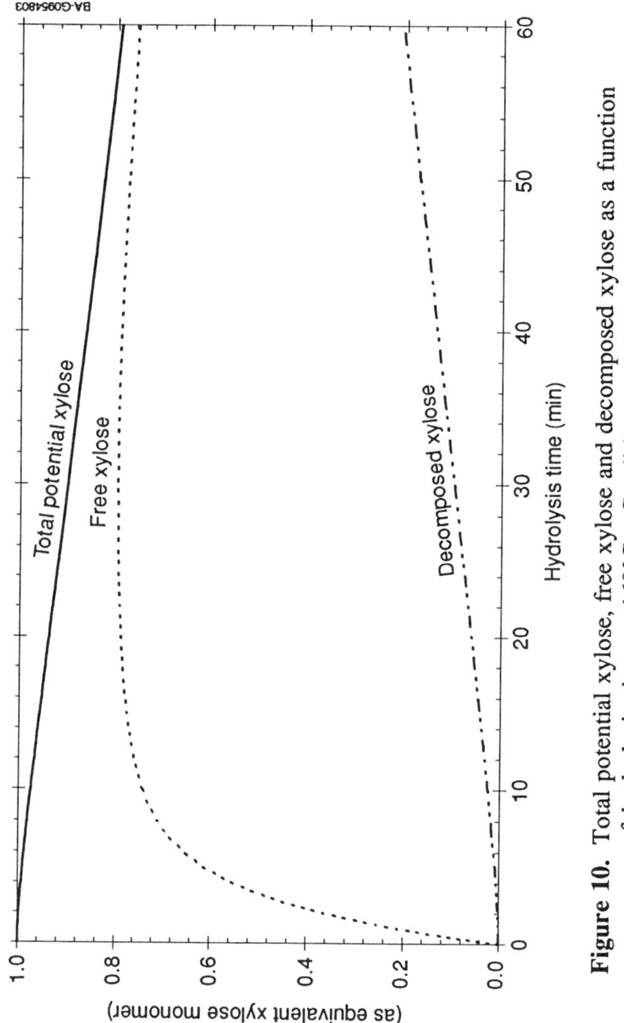

Figure 10. Total potential xylose, free xylose and decomposed xylose as a function of hydrolysis time at 160°C. Conditions are the same as specified in Figure 9.

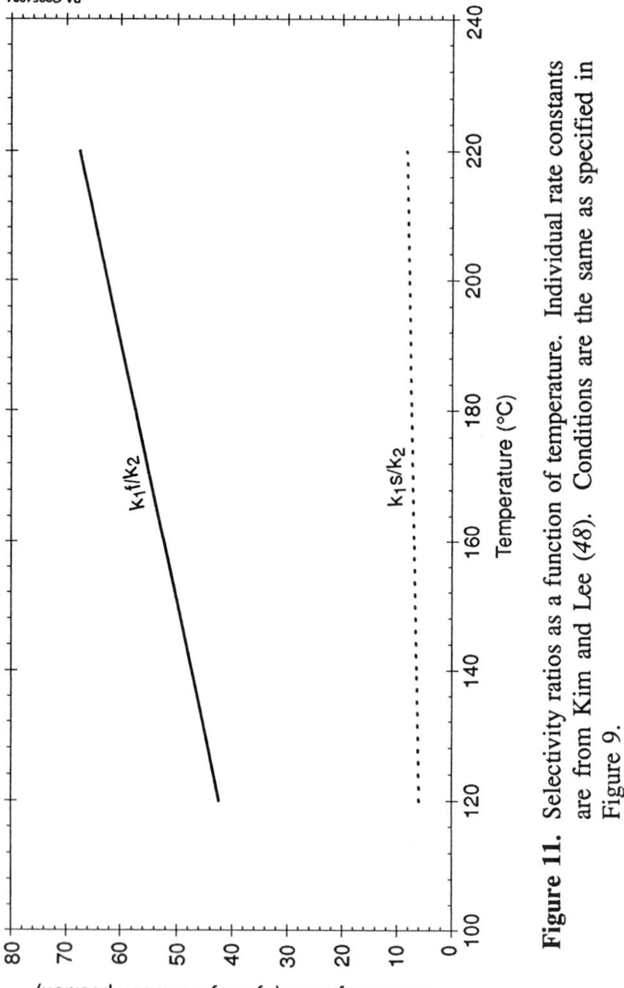

Figure 11. Selectivity ratios as a function of temperature. Individual rate constants are from Kim and Lee (*48*). Conditions are the same as specified in Figure 9.

exploded biomass often must be washed prior to enzymatic hydrolysis and subsequent fermentation. Thus, although the extent and rate of hydrolysis of the solids remaining following pretreatment and washing increase with the severity of treatment, overall saccharification yields can fall because washing removes solubilized sugars, such as those generated by hydrolysis of hemicellulose.

Steam explosion pretreatment can cause extensive hemicellulose degradation and lignin modification, particularly when carried out using high steam temperatures (T > 220°C) or long reaction times (*51,53*). Using 3.2-mm aspen chips, Brownell and Saddler (*51*) observed full solubilization of the pentosan fraction in 1 min using 250°C steam, with the pentosan fraction more than 60% destroyed after 2 min of treatment. Impregnation with H_2SO_4 or sulfur dioxide (SO_2) has been shown to improve the selectivity of the steam explosion process to favor hydrolysis over pyrolysis and degradation (*51,54,55*). In addition to permitting more complete hemicellulose solubilization and reducing sugar degradation, washed material performs better as a substrate for enzymatic hydrolysis when an acid catalyst is used. The use of an acid catalyst also enables lower steam temperatures to be used (*12*).

Steam requirements are dominated by the need to heat the moisture content of biomass, and can be reduced by using dried or partially dried wood (*51*). Decreasing the chip size to less than 6 mm and/or increasing incubation time and reducing steam temperature reduces heat transfer heterogeneity, which can cause overcooking (degradation) to occur at the outside of chips or undercooking to occur at the center (*51*). Temperature studies carried out by Brownell et al. (*12*) demonstrate the importance of steam (heat) penetration along the fiber axis. Convective heat transfer occurs in air-dried chips because steam can enter the end grain and uniformly penetrate into the biomass by moving through open vessels. In contrast, for wet or green chips in which pores and vessels are filled with water (or steam condensate), or when vessels are blocked by tyloses (*14*), rapid convection cannot occur and heat transfer occurs by conduction alone.

There are two potential mechanisms of action in steam explosion pretreatment. First, rapid solubilization of hemicellulose opens up the pore structure of biomass, similar to what occurs in dilute acid hydrolysis. This has been amply demonstrated to occur following high-temperature and acid-catalyzed treatments. Second, explosive decompression exerts a mechanical shear on the biomass, which may increase the specific surface area of the material by defibrating individual cellulose microfibrils or by otherwise expanding the lignocellulose matrix. Although there is little doubt about the importance of the first mechanism, controversy exists regarding the importance of explosive decompression. The study by Brownell et al. (*12*), for example, concluded that explosion was unnecessary, since hydrolysis yields were similar using pressure drops ranging from 0.34 MPa (50 psia) to 3.24 MPa (470 psia).

Clark and Mackie (*52*) showed that SO_2 impregnation prior to steam explosion increases the susceptibility of softwood to enzymatic hydrolysis, with digestibilities of greater than 85% reported for several of the conditions examined. The effect of SO_2 diminished above loadings of 3% (w/w), but was pronounced at loadings between 0.5% and 3% (w/w). Digestibility correlated strongly with the carbohydrate content of the water-insoluble fraction, with a very sharp increase in digestibility (0%-100%) occurring as carbohydrate content decreased from 67% to 50%. However, digestibility studies were performed only on the water-insoluble fraction that remained after washing. Water-insoluble mass yields following steam explosion and washing were less than 60% in all cases, indicating significant solubilization and/or destruction of hemicellulose components. These results demonstrate that in some instances more sugar is solubilized by steam explosion than is subsequently obtained by enzymatic hydrolysis of the water-insoluble fraction.

Using the solute exclusion technique, Wong et al. (*56*) showed that increased porosity (accessible surface area) is the primary mechanism by which SO_2 impregnation-based steam explosion increases softwood digestibility. They postulated that lignin redistribution during or following steam explosion pretreatment might also be a factor. Reduced digestibility was observed when steam-exploded material was washed with caustic solution rather than water, presumably because of partial pore collapse.

Ammonia Explosion. Ammonia explosion, or ammonia fiber explosion (AFEX) as it is more commonly known, is a process in which ground (1-2 mm characteristic dimension) prewetted lignocellulosic material at a moisture content of 0.15-0.30 kg water/kg dry biomass is first placed in a pressure vessel with liquid NH_3 at a loading of about 1 kg NH_3/kg dry biomass. The vapor pressure of NH_3 is high (P_{vap} = 1.06 MPa [154 psia] at 300 K), and the pressure becomes quite high in the closed pressure vessel; the vessel pressure reaches about 1.24 MPa (180 psia) when the process is carried out at ambient temperature (*57*). The mixture is then incubated at the reaction pressure for sufficient time, typically on the order of tens of minutes to an hour, to enable NH_3 to penetrate into the lignocellulosic matrix. Finally, a valve is opened to flash the reaction mixture to a lower but not necessarily ambient pressure. Pretreated samples are air-dried to remove residual NH_3.

AFEX pretreatment has been demonstrated to markedly improve the saccharification rates of numerous herbaceous crops and grasses. Materials pretreated using the AFEX process include alfalfa, corn stover, rice stover, wheat straw, barley straw, bermuda grass, bagasse, and kenaf core (*58-60*). Following AFEX pretreatment, overall hydrolysis yields on these materials are 80% to 90% of theoretical (*58*); overall hydrolysis yield refers to the combined cellulose and hemicellulose conversion yield based on total reducing sugars measured by the dinitrosalicylic acid method (*61*). AFEX pretreatment has not proven as effective on hardwoods and softwoods, and results of a study on AFEX pretreatment of bermuda grass, bagasse, and newspaper suggest decreasing AFEX effectiveness with increasing lignin content. AFEX treatment of Bermuda grass (~5% lignin) and bagasse (~20% lignin) resulted in hydrolysis yields of over 90% of theoretical, whereas the hydrolysis yield on AFEX-treated newspaper (~30% lignin) was only about 40% (*59*). Similarly, hydrolysis yields were below 50% when aspen chips (~25% lignin) were subjected to AFEX pretreatment (*62*).

There appear to be two mechanisms by which AFEX pretreatment acts. First, the reactivity of cellulose is increased by exposure to liquid NH_3. As discussed previously, exposure to liquid NH_3 causes swelling and may also cause partial decrystallization of crystalline cellulose. Penetration of liquid NH_3 into the lignocellulose matrix is relatively rapid because of the tenfold lower viscosity and surface tension of liquid NH_3 relative to liquid water (*57*). Second, accessible surface area increases following AFEX treatment, probably as a combined effect of hemicellulose hydrolysis and explosive decompression. Base-catalyzed hemicellulose hydrolysis is likely, since in the presence of water liquid NH_3 forms ammonium hydroxide. The explosive decompression phenomenon is similar to what occurs in steam explosion pretreatment. When NH_3 that has penetrated into the lignocellulose matrix during the high-pressure AFEX incubation period is suddenly flashed, there is a rapid expansion within the lignocellulose matrix, which significantly increases the specific surface area of the biomass. Electron micrographs of untreated and AFEX-treated alfalfa qualitatively demonstrate that specific surface area increases following AFEX treatment (*57*). This is supported by the fact that the density of alfalfa falls from 290 kg/m^3 to 180 kg/m^3 following AFEX treatment, and the water holding capacity of the treated material increases by 50% (*57*).

AFEX pretreatment improves somewhat when the process is carried out at higher than ambient temperatures, and remains effective when blowdown pressures are increased from atmospheric to around 0.30 to 0.46 MPa (44 to 66 psia) (*59,60*). Compression costs represent a major fraction of AFEX process operating costs, so this latter finding may significantly reduce the cost of AFEX treatment, provided NH_3 can be recovered efficiently. Another cited benefit is that AFEX pretreatment does not appear to require small particle sizes for efficacy (*60*). However, hydrolysis yields are increased by further grinding following AFEX treatment.

Carbon Dioxide Explosion. Dale and Moreira (*57*) briefly examined the potential of using carbon dioxide (CO_2) explosion rather than ammonia- or steam-based explosion to pretreat alfalfa. Alfalfa was incubated (for an unspecified length of time at presumably ambient temperature) at a loading of 4 kg CO_2/kg fiber, corresponding to a pressure of 5.62 MPa (815 psia). Following incubation, the pressure was rapidly released. Seventy-five percent of theoretical glucose was released during a 24-h enzymatic hydrolysis of the exploded material. Higher yields (90%) were achieved using AFEX treatment at markedly lower pressures (see above), so further investigation of CO_2 pretreatment was abandoned.

Puri and Mamers (*63*) investigated incubating biomass with steam and CO_2 at high pressure, and then subjecting the material to explosive decompression. It was hypothesized that in the presence of steam or wet biomass, CO_2 would form carbonic acid that would increase the autohydrolysis rate. Substrates examined were air-dried wheat straw hammer milled to pass through a 1-mm screen, bagasse (used as received), and green eucalyptus chips screened to pass through a 25-mm mesh. Some improvement in the extent of enzymatic hydrolysis was observed with increasing pressure at low solids loadings, with yields (expressed as percent loss of organic matter on an ash-free, oven-dry basis) increasing from about 60% to 75%. However, no effect of pressure was seen at higher solids loadings, even though hydrolysis yields remained similar.

Supercritical Fluid Extraction. A variety of pretreatments based on supercritical fluid (SCF) extraction of lignins, resins, and waxy materials from lignocellulosic materials have been investigated. Three types of SCF solvents have been examined: pure SCF solvents that are liquids at room temperature and pressure (*64-67*) pure SCF fluids that are gases at ambient conditions (*68-70*) and binary and ternary mixtures of such fluids (*66,67,70,71*). A review of SCF-based lignocellulosic treatments is provided by Kiran (*72*).

SCF pretreatment using solvents that are gases at ambient conditions is attractive because of the greater ease with which gases can be recovered for reuse relative to liquids. Gases that have been investigated for use in SCF pretreatments include CO_2, ethylene, nitrous oxide, and butane. The nonflammable, noncorrosive, nontoxic, and inexpensive nature of CO_2 makes SCF extraction using CO_2 particularly appealing. However, efforts to use pure SCF CO_2 to pretreat lignocellulosic materials have so far proven unsuccessful (*69*), presumably because interactions between SCF CO_2 and lignocellulosic materials are nonreactive (*70*). Somewhat better results have been obtained using binary mixtures of CO_2-SO_2 and CO_2-water. For example, SCF CO_2-SO_2 treatment of aspen wood chips for 4 h at 130°C improved rates and yields of enzymatic hydrolysis. This treatment removed 61% of the lignin while maintaining 83% of the total carbohydrate (*67*). Treatment selectively removed arabinogalactan and glucomannan fractions of the hemicellulose, retaining most of the xylan. SCF CO_2-water mixtures appear to be less reactive with lignin components than SCF CO_2-SO_2, but interact reactively with carbohydrate components (*70*).

SCF NH_3 has been used to pretreat a variety of lignocellulosic materials (*68*). Enzymatic hydrolysis yields of greater than 90% of theoretical were achieved following SCF NH_3 pretreatment of agricultural residues (bagasse and corn stalks) and hardwoods (aspen and red oak). Hemicellulose hydrolysis yields showed greater variability, with a low of 65% of theoretical for bagasse and a high of 90% for corn stalks. SCF NH_3 pretreatment showed poor efficacy on softwoods, however. NH_3 treatment was most effective at temperatures and pressures near the critical point, with a temperature optimum of 150°C noted; hydrolysis yields decreased at conditions away from the critical point. No influence of moisture content was observed for water loadings ranging from 10% to 55% on a dry biomass basis. SCF NH_3 pretreatment was effective on coarsely chipped biomass, although fine grinding prior to enzymatic hydrolysis was necessary to achieve high enzymatic hydrolysis yields.

SCF NH_3 treatment extracted very few components, despite the strong solvent power of SCF NH_3, leading Chou (*68*) to hypothesize that components solubilized during treatment recondensed or precipitated when the reactor cooled. Microscopic examination of treated materials showed swollen fiber bundles and shrunken vacuoles, consistent with an alkali swelling mechanism. In conjunction with this study, Weimer et al. (*73*) examined changes in the physical and chemical structure of hardwood and softwood chips following SCF NH_3 treatment. Pretreated materials had an increased nitrogen content as a result of amidation reactions. Solute exclusion studies showed that the total pore volume of the samples increased following pretreatment, but without a corresponding increase in pore width. Increased porosity (pore volume) occurred without a change in cellulose, hemicellulose, or lignin content.

Weimer et al. (*73*) hypothesized that SCF NH_3 pretreatment caused ammonolysis of ester linkages between hemicellulose and lignin but did not solubilize lignocellulosic components. Reduced hemicellulose-lignin bonding presumably increased swelling and thus total pore volume, whereas retention of hemicellulose and lignin components prevented the maximum pore width from increasing. The maximum pore width was 50 Å and 12-38 Å for NH_3-treated hardwood (red oak) and softwood (Douglas fir), respectively. The higher digestibility of pretreated hardwoods relative to pretreated softwoods was attributed to the inability of the treatment to increase the maximum pore width in softwood to above ~43 Å, the minimum pore width estimated to be required for penetration of cellulase enzyme.

Prospects for Developing Improved Pretreatments

From a mechanistic standpoint, there are notable similarities among many pretreatment methods. With the exception of comminution, for example, pretreatments usually employ catalysts (dilute acid, SO_2, alkali, etc.) because this enables lower temperature operation, which increases yield. Another commonality is that many pretreatments are performed at sufficiently high temperature to hydrolyze hemicellulose. The significance of the resistance of lignocellulosic materials to pretreatment is evidenced by the severe conditions ($T > 140°C$ in the presence of a catalyst) generally required for effective pretreatment. Dilute acid hydrolysis and steam explosion pretreatments, for example, are effective only at temperatures above 140°C. CO_2-based explosive decompression and many SCF treatments also appear to require high-temperature operation for efficacy. The ability to achieve effective pretreatment at lower temperatures using NH_3 is evidently a consequence of the ability of NH_3 to penetrate into lignocellulosic materials more rapidly than most other pretreatment chemicals.

Overall carbohydrate yield is the most important factor in commercial-scale biomass conversion processes. Research to improve pretreatment processes must therefore focus on minimizing, or preferably eliminating, degradation of the carbohydrate fraction of lignocellulosic biomass. Although already developed to a significant extent, processes requiring the use of high temperature, such as dilute acid- and steam explosion-based processes, may be unattractive in the long term because the high-temperature formation of degradation products reduces yields. Acid or base catalysis enables processes to be carried out at lower temperatures (or using shorter residence times), which reduces carbohydrate degradation. However, neutralization of acidic or alkaline pretreatment hydrolyzates may impose additional chemical requirements and process complexity and/or generate byproducts that represent a waste disposal burden. Moreover, acid- and base-catalyzed processes may require the use of expensive materials of construction that can withstand corrosive processing.

Obstacles. The development of more effective and efficient pretreatments is hindered by incomplete knowledge about lignocellulose structure, the nature of interactions between lignocellulose components and pretreatment chemicals, and the factors controlling enzymatic hydrolysis. Further information about any of these topics will hasten the development of improved pretreatment processes. One particular issue that needs to be resolved is how to conceptualize the hydrolysis process, since both accessible surface area and cellulose reactivity are hypothesized to be the controlling factor. Although the literature shows that hydrolysis rates and susceptibilities to pretreatments vary widely among materials, it has not been established to what extent these differences are due to differences in porosity (accessible surface area) and reactivity. Two fundamental questions need to be answered: What determines the extent to which surfaces are accessible to enzymes?; and What determines specific reactivity (reactivity per unit accessible surface area)? Pretreatment processes that are more energetically and chemically efficient can be developed once the effects of structural and compositional characteristics on accessible surface area and specific reactivity are more thoroughly understood.

Besides a general need for fundamental studies, there is a specific need to standardize how experiments are conducted and how results are reported (*5*). Substrate standards for comparing pretreatments are also needed. Utilizing standardized materials at specified particle sizes and moisture contents for representative feedstocks (hardwoods, herbaceous crops, etc.) would provide a benchmark for different researchers to compare their results with one another. A uniform method for reporting enzymatic hydrolysis results also needs to be established; reporting hydrolysis yields in terms of percent of theoretical conversion might be the best approach. Accessible surface area estimation is another area for which a reference benchmark needs to be developed. No comparative study of accessible surface area estimation methods is available, so it is difficult to compare results obtained by different methods.

Opportunities. In the near term, the potential to achieve further improvements in existing processes needs to be explored. The economic feasibility of modifying existing processes to enable more rapid enzymatic hydrolysis of pretreated biomass must be evaluated. Sensitivity analyses should be performed to understand the extent to which additional energy and/or chemical inputs can be used to improve bioconversion rates and yields or to reduce energy requirements. For example, a sensitivity analysis might be used to determine the point at which increases in hydrolysis rate and overall yield no longer compensate for increases in the cost of pretreatment. Sensitivity analyses might also examine the importance of potential waste disposal costs and/or chemical recovery requirements so as to quantify the advantages of avoiding byproduct formation. Results of such sensitivity analyses can be used to draw conclusions regarding the best options for developing improved pretreatment processes.

Modifying existing pretreatments to achieve limited delignification also warrants investigation. Low rates of enzymatic hydrolysis on materials pretreated using dilute acid- or steam explosion-based methods to remove hemicellulose imply that lignin remaining after treatment either directly blocks enzymatic attack, competitively binds enzyme(s), or otherwise inhibits enzyme binding at the cellulose surface. It needs to be determined to what extent each of these mechanisms is operative. The structural organization of individual fiber cells suggests that partial delignification, for example, selective delignification of the outer portion of the compound middle lamellae, might dramatically improve the rate and/or extent of enzymatic hydrolysis of pretreated hardwoods. The ability to achieve selective delignification depends on the path by which pretreatment chemicals penetrate into the wood matrix; i.e., do penetrating chemicals (or enzymes) enter a lignocellulosic matrix from the fiber cell wall lumen side and diffuse into the matrix, or do they enter through openings (pits) in the fiber cell wall and diffuse from the outside inward? The channel system within lignocellulose is undoubtedly an important structural feature (*21*), although the precise pathway by which chemical reagents or enzymes penetrate into a lignocellulosic matrix is unclear. Radially uniform penetration of delignifying agents into fiber cells may not be possible, however, as a result of lignocellulosic structural asymmetry which favors transport in the axial rather than radial direction. Further study is required to determine if selective delignification of the outer (or inner) portion of the compound middle lamellae is feasible.

In the longer term, improved pretreatments which overcome undesirable features of existing processes and offer the potential of achieving high yields with little or no byproduct formation should be pursued. Lower temperature processes are particularly attractive because they eliminate the problem of yield losses caused by high-temperature sugar degradation. At present, comminution and NH_3 pretreatment are the only processes with proven efficacy at temperatures below 140°C; further studies are required to assess the potential of CO_2-based pretreatments at low temperatures. NH_3-based pretreatments will be particularly advantageous if large chips can be used, as some of the studies suggest. However, NH_3-based pretreatments require moderate- to high-pressure operation (1.03-1.38 MPa [150-200 psia] for subcritical AFEX or 10.34-13.79 MPa [1500-2000 psia] for SCF NH_3 treatment) and a feasibility study needs to be performed to examine the influence of compression requirements on the economics of NH_3 treatment.

Acknowledgments

This work was supported by the Ethanol from Biomass Project of the DOE Biofuels Systems Division.

Literature Cited

1. Millett, M.A.; Baker, A.J.; Satter, L.D. *Biotech. Bioeng. Symp.* **1976**, *6*, 125-153.
2. Chang, M.M.; Chou, Y.C.T.; Tsao, G.T. *Adv. Biochem. Eng.* **1981**, *20*, 15-42.
3. Lin, K.W.; Ladisch, M.R.; Schaefer, D.M.; Noller, C.H.; Lechtenberg, V.; Tsao, G.T. *AIChE Symp. Ser. No. 207* **1981**, *77*, 102-106.
4. Fan, L.T.; Lee, Y-H.; Gharpuray, M.M. *Adv. Biochem. Eng.* **1982**, *23*, 157-187.
5. Dale, B.E. *Annual Reports on Fermentation Processes* **1985**, *8*, 299-323.
6. Chum, H.L.; Douglas, L.J.; Feinberg, D.A.; Schroeder, H.A. In *Evaluation of Pretreatments of Biomass for Enzymatic Hydrolysis of Cellulose*; SERI/TP-231-2183; Solar Energy Research Institute: Golden, CO, 1985.
7. Wenzl, H.F.J. In *The Chemical Technology of Wood;* Academic Press: New York, NY, 1970.
8. Pettersen, R.C. In *The Chemistry of Solid Wood*; Rowell, R., Ed.; Advances in Chemistry Series 207; American Chemical Society, Washington, DC, 1984; pp 57-126.
9. Torget, R.; Werdene, P.; Himmel, M.; Grohmann, K. *Appl. Biochem. Biotech.* **1990**, *24/25*, 115-126.
10. Torget, R.; Walter, P.; Himmel, M.; Grohmann, K. *Appl. Biochem. Biotech.* **1991**, *28/29*, 75-86.
11. Torget, R.; Himmel, M.; Grohmann, K. *Appl. Biochem. Biotech.* **1992**, *34/35*, 115-123.
12. Brownell, H.H.; Yu, E.K.C.; Saddler, J.N. *Biotech. Bioeng.* **1986**, *28*, 792-801.
13. Tillman, L.M.; Lee, Y.Y.; Torget, R. *Appl. Biochem. Biotech.* **1990** *24/25*, 103-113.

14. Parham, R.A.; Gray, R.L. In *The Chemistry of Solid Wood*; Rowell, R., Ed.; Advances in Chemistry Series 207; American Chemical Society: Washington, DC, 1984; pp 3-56.
15. Stone, J.E.; Scallan, A.M.; Donefer, E.; Ahlgren, E. *Adv. Chem. Ser.* **1969**, *95*, 219-241.
16. Poots, V.J.P.; McKay, G. *J. Appl. Poly. Sci.* **1979**, *23*, 1117-1129.
17. Hradil, G.; Calo, J.M.; Wunderlich, Jr., T.K. Bimonthly report #3 for Brown University subcontract No. XK-7-07031-8, submitted to the Solar Energy Research Institute, 1988.
18. Burns, D.S.; Ooshima, H.; Converse, A.O. *Appl. Biochem. Biotech.* **1989**, *20/21*, 79-94.
19. Converse, A.O.; Ooshima, H.; Burns, D.S. *Appl. Biochem. Biotech.* **1990**, *24/25*, 67-73.
20. Fan, L.T.; Lee, Y-H.; Beardmore, D.H. *Biotech. Bioeng.* **1980**, *22*, 177-199.
21. Schurz, J. *Holzforschung.* **1986**, *40:4*, 225-232.
22. Ooshima, H.; Burns, D.S.; Converse, A.O. *Biotech. Bioeng.* **1990**, *36*, 446-452.
23. Tarkow, H.; Feist, W.C. *A Mechanism for Improving the Digestibility of Lignocellulosic Materials with Dilute Alkali and Liquid* NH_3; Advance Chemistry Series 95; American Chemical Society: Washington, DC, 1969; pp 197-218.
24. Grohmann, K.; Torget, R.; Himmel, M. *Biotech. Bioeng. Symp.* **1985**, *15*, 59-80.
25. Lynd, L., Dartmouth College, Thayer School of Engineering, personal communication, 1990.
26. Grethlein, H.E. *Bio/Technology* **1985**, *3*, 155-160.
27. Grohmann, K., Solar Energy Research Institute, personal communication, 1990.
28. Himmel, M.; Tucker, M.; Baker, J.; Rivard, C.; Oh, K.; Grohmann, K. *Biotech. Bioeng. Symp.* **1985**, *15*, 39-58.
29. Sherrard, E.C.; Kressman, F.W. *Ind. Eng. Chem.* **1945**, *37*, 5-8.
30. Wang, P.Y.; Bolker, H.I.; Purves, C.B. *Tappi J.* **1967**, *50*, 123-124.
31. Hinman, N.D.; Wright, J.D.; Hoagland, W.; Wyman, C.E. *Appl. Biochem. Biotech.* **1989**, *20/21*, 391-401.
32. Schell, D.J.; Torget, R.; Power, A.; Walter, P.J.; Grohmann, K.; Hinman, N.D. *Appl. Biochem. Biotech.* **1991**, *28/29*, 87-97.
33. Hinman, N.D.; Schell, D.J.; Riley, C.J.; Bergeron, P.W.; Walter, P.J. *Appl. Biochem. Biotech.* **1992**, *34/35*, 639-649.
34. Grohmann, K.; Torget, R.; Himmel, M. *Biotech. Bioeng. Symp.* **1986**, *17*, 135-151.
35. Torget, R.; Himmel, M.; Wright, J.D.; Grohmann, K. *Appl. Biochem. Biotech.* **1988**, *17*, 89-104.
36. Spindler, D.D.; Wyman, C.E.; Grohmann, K.; Torget, R.W. In *Evaluation of Pretreated Woody Crops for the Simultaneous Saccharification and Fermentation Process;* FY 1988 Ethanol from Biomass Annual Report, SERI/SP-231-3520; Solar Energy Research Institute: Golden, CO, 1989.
37. Knappert, D.; Grethlein, H.; Converse, A. *Biotech. Bioeng.* **1980**, *22*, 1449-1463.
38. Knappert, D.; Grethlein, H.; Converse, A. *Biotech. Bioeng. Symp.* **1981**, *11*, 67-77.
39. Kwarteng, I.K. Dartmouth College, Thayer School of Engineering; Hanover, NH, PhD thesis, 1983.

40. Grethlein, H.E.; Allen, D.C.; Converse, A.O. *Biotech. Bioeng.* **1984**, *26*, 1498-1505.
41. Brennan, A.H.; Hoagland, W.; Schell, D.J. *Biotech. Bioeng. Symp.* **1986**, *17*, 53-70.
42. Converse, A.O.; Kwarteng, I.K.; Grethlein, H.E.; Ooshima, H. *Appl. Biochem. Biotech.* **1989**, *20/21*, 63-78.
43. Saeman, J.F. *Ind. Eng. Chem.* **1945**, *37*, 43-52.
44. Root, D.F. Department of Chemical Engineering, University of Wisconsin: Madison, WI, PhD thesis, 1956.
45. Horwath, J.A.; Mutharasan, R.; Grossmann, E.D. *Biotech. Bioeng.* **1983**, *25*, 19-32.
46. Cahela, D.R.; Lee, Y.Y.; Chambers, R.P. *Biotech. Bioeng.* **1983**, *25*, 3-17.
47. Maloney, M.T.; Chapman, T.W.; Baker, A.J. *Biotech. Bioeng.* **1985**, *27*, 355-361.
48. Kim, S.B.; Lee, Y.Y. *Biotech. Bioeng. Symp.* **1987**, *17*, 71-84.
49. Grethlein, H.E. *J. Appl. Chem. Biotechnol.* **1978**, *28*, 296-308.
50. Perry, R.H.; Green, D.W.; Maloney, J.O. *Perry's Chemical Engineers' Handbook*; sixth edition; McGraw-Hill: New York, NY, 1984.
51. Brownell, H.H.; Saddler, J.N. *Biotech. Bioeng. Symp.* **1984**, *14*, 55-68.
52. Clark, T.A.; Mackie, K.L. *J. Wood Chem. Tech.* **1987**, *7:3*, 373-403.
53. Schultz, T.P.; Biermann, C.J.; McGinnis, G.D. *Ind. Eng. Chem. Prod. Res. Dev.* **1983**, *22*, 344-348.
54. Mackie, K.L.; Brownell, H.H.; West, K.L.; Saddler, J.N. *J. Wood Chem. Tech.* **1985**, *5:3*, 405-425.
55. Galbe, M.; Zacchi, G. *Biotech. Bioeng. Symp.* **1986**, *17*, 97-105.
56. Wong, K.K.Y.; Deverell, K.F.; Mackie, K.L.; Clark, T.A. *Biotech. Bioeng.* **1988**, *31*, 447-456.
57. Dale, B.E.; Moreira, M.J. *Biotech. Bioeng. Symp.* **1982**, *12*, 31-43.
58. Dale, B.E.; Henk, L.L.; Shiang, M. *Dev. Ind. Microbiol.* **1985**, *26*, 223-233.
59. Holtzapple, M.T.; Jun, J-H.; Ashok, G.; Patibandla, S.L.; Dale, B.E. *Appl. Biochem. Biotech.* **1990**, *28/29*, 59-74.
60. Holtzapple, M.T.; Jun, J-H.; Patibandla, S.; Dale, B.E. *Ammonia Fiber Explosion (AFEX) Pretreatment of Lignocellulosic Wastes;* Paper presented at the American Institute of Chemical Engineers National Meeting, Chicago, IL, November, 1990.
61. Miller, G.L. *Anal. Chem.* **1959**, *41*, 426-428.
62. Dale, B.; Holtzapple, M. *Technical Summary of Ammonia Freeze Explosion*, 1989, unpublished.
63. Puri, V.P.; Mamers, H. *Biotech. Bioeng.* **1983**, *25*, 3149-3161.
64. McDonald, E.C.; Howard, J.; Bennett, B. *Fluid Phase Equilibria* **1983**, *10*, 337-344.
65. Reyes, T.; Bandyopadhyay, S.S.; McCoy, B.J. *J. Supercritical Fluids* **1989**, *2*, 80-84.
66. Li, L.; Kiran, E. *Tappi J.* **1989**, 183-190.
67. Shah, M.M.; Lee, Y.Y.; Torget, R. *Pretreatment of Aspen Wood Aimed at High Hemicellulose Retention*; Proceedings of the Ethanol from Biomass Annual Review Meeting, National Renewable Energy Laboratory: Golden, CO, 1990.
68. Chou, Y.C.T. *Biotech. Bioeng. Symp.* **1986**, *17*, 18-32.
69. Ritter, D.C.; Campbell, A.G. *Biotech. Bioeng. Symp.* **1986**, *17*, 179-182.

70. Li, L.; Kiran, E. *I&EC Research* **1988**, *27*, 1301-1312.
71. Beer, R.; Peter, S. In *Supercritical Fluid Technology*; Penninger, J.M.L.; Radosz, M.; McHugh, M.A.; Krukonis, V.J., Eds: Elsevier Science: The Netherlands, 1985; pp 385-396.
72. Kiran, E. *Journal of Research at the University of Maine* **1987**, *III:2*, 24-32.
73. Weimer, P.J.; Chou, Y-C. T.; Weston, W.M.; Chase, D.B. *Biotech. Bioeng. Symp.* **1986**, *17*, 5-18.
74. Datta, R. *Proc. Biochem.* **1981**, *16*, 16-19, 42.

RECEIVED March 17, 1994

Chapter 16

Bioconversion of Wood Residues

Mechanisms Involved in Pretreating and Hydrolyzing Lignocellulosic Materials

L. P. Ramos[1] and J. N. Saddler[2]

[1]Department of Chemistry, Federal University of Parana, P.O. Box 19.081, Curitiba, Parana, Brazil 81531
[2]Department of Forest Products Biotechnology, Faculty of Forestry, University of British Columbia, Vancouver, British Columbia V6T 1Z4, Canada

The enhanced susceptibility of acid-catalysed steam-treated residues to enzymatic hydrolysis will be discussed with reference to the effects that steam-explosion has on the structure of lignocellulosic residues. The peroxide-treated fraction derived from SO_2-impregnated steam-treated eucalyptus was used to investigate the mechanism whereby fungal cellulases degrade cellulose. The effect of enzymatic hydrolysis on the crystallinity, degree of polymerisation and fibre length of the substrate is described.

Several pretreatment methods have been shown to open up the structure of cellulose and increase its susceptibility to enzymatic hydrolysis. These structural changes enhance hydrolysis primarily because they result in a considerable increase in the substrate surface area (or pore volume) which is available to cellulolytic enzymes (*1-4*).

Among all of the pretreatment options studied to date, steam explosion appears to be one of the most effective methods for pretreating and fractionating lignocellulosic residues (*5,6*). This method, which combines both physical and chemical elements, causes the rupture of the wood cell wall structure and the subsequent exposure of its inner layers to microbial and/or enzymatic attack. The effectiveness of pretreatment is therefore usually correlated with the development of substrate accessibility to cellulases, although the optimal recovery of both hemicelluloses and lignin is also important from an economic viewpoint (*7,8*).

Several major changes in the structure of the cellulose have been demonstrated after steam explosion of cellulosic residues. Various authors have reported a gradual decrease in the degree of polymerization (DP) of cellulose and a substantial increase in the crystallinity index (CrI) of the substrate (*9,10*). The apparent increase in crystallinity seems to be the result of fusion between cellulose crystallites which were originally separated by a matrix of hemicellulose and lignin. Therefore, the gradual

0097–6156/94/0566–0325$08.00/0

removal of hemicelluloses and lignin appears to trigger the reorientation of the cellulose molecules which, after pretreatment, assume a distinct crystalline form. Steam explosion has also been shown to improve the alkali solubility of several cellulosic fractions (*11,12*). A significant decrease in both crystallinity and degree of polymerization of cellulose was reported when more drastic pretreatment conditions were used. Changes in the crystallinity of the substrate were explained by a gradual reduction of intermolecular hydrogen bonding between the carbon atoms C-3 and C-6 of adjacent anhydroglucose residues.

Previous work has shown that softwood species are far more difficult to pretreat than hardwoods. This recalcitrance which is attributed to differences in their hemicellulose and lignin structure and content, has been partly alleviated by using an acid catalyst prior to pretreatment (*13,14*). The influence of the lignin component on both the hydrolysis rate and enzyme recycling is also very important and needs to be fully characterized. There is some evidence that enzyme recycling is facilitated by lignin removal but it is not clear whether one batch of enzymes can be effectively used to hydrolyse fresh substrate (*15*). It has also been proposed that the alkali extraction of lignin does not enhance the hydrolysis of pretreated hardwoods (*16,17*) and that oxidative post-treatments such as alkaline hydrogen peroxide are required to increase the accessibility of steam-treated softwoods (*14*).

The gradual decrease in the hydrolysis rate of pretreated cellulose has often been associated with increases in substrate recalcitrance during hydrolysis (*18*). This usually involves several factors such as lignin accumulation and redistribution, increased cellulose crystallinity and reduction of the surface area available to the enzyme due to a partial coating of the residual substrates with irreversibly adsorbed enzymes and/or lignin. At the fibre level, the rapid fragmentation of the substrate can be studied by looking at changes in distribution of fibre length and/or substrate particle size during hydrolysis. At the molecular level, the effects of enzymatic hydrolysis on the crystallinity and degree of polymerization of cellulose must be addressed.

In this paper we have tried to discuss how a pretreatment method such as steam explosion can help enhance the enzymatic hydrolysis of cellulose. We have tried to identify the major substrate- and enzyme-related factors which limit hydrolysis and influence the mode of action of the enzymes.

Materials and Methods

Materials. Four shipments of 6 year old logs of *Eucalyptus viminalis* Labill were kindly supplied by the Brazilian Institute for Agricultural Research, Embrapa ("Empresa Brasileira de Apoio à Pesquisa Agropecuária", Colombo, Paraná, Brazil). Enzymatic hydrolyses of cellulosic substrates were performed using Celluclast, a commercial cellulase preparation from *T. reesei*, and Novozym, a β-glucosidase preparation from *A. niger*. Both enzyme preparations were obtained as gifts from Novo Nordisk, Denmark.

Substrate Preparation. Logs of *E. viminalis* were debarked and chipped (to an approximate size of 36 x 7 mm (S.D. 1.5 mm) using a commercial chipper at Forintek Canada (Ottawa, Ontario). SO_2- impregnated (1%, w/w) green chips were steam-treated using a 2 L steam gun at 210°C for 50 s (*19*). Optimal pretreatment conditions were defined as those in which the best substrate for hydrolysis was obtained with the least amount of soluble sugars lost by degradation reactions. The resulting steam-treated substrates were extracted twice with water at 5% (w/v) solids concentration at room temperature. These water insoluble fractions (WI) were subsequently extracted with alkali (NaOH 0.4% (w/v)) to give the WIA fraction.

The peroxide-treated fraction (WIA-H_2O_2) was obtained by treating the water-and-alkali insoluble residue with alkaline hydrogen peroxide as described previously (*17*). Pretreated materials were washed extensively with water and drained until a moisture content of 75-80% (w/w) was obtained. Samples were routinely stored at 4°C.

Substrate Analysis. The Klason lignin (acid-insoluble lignin) content of the cellulosic substrates was determined according to TAPPI Standard Method T222 os-74. Acid-soluble lignin was determined by TAPPI Useful Method 250. The total lignin content of the substrate was calculated as a percentage of the original weight of the test specimen. The carbohydrate content of the substrate was determined from the acid hydrolysate of a Klason lignin determination using HPLC (*20*).

Fibre screening. The peroxide-treated fraction derived from SO_2-impregnated steam-exploded eucalyptus chips (WIA/H_2O_2) was screened using a Bauer-McNett fibre classifier to eliminate most of the fines and fibre bundles. A series of fibre screens with 14, 28, 48, 150 and 200 mesh collected two major fibre populations, one between the 48 and 150 mesh screens (F-150 fraction) and another between the 150 and the 200 mesh screens (F-200 fraction). The fibre length distribution of these fractions, and the partially hydrolysed residues, was determined using a Kajaani FS-200 fibre analyser.

Substrate Hydrolysis. The amount of both filter paper and cellobiase activities of the enzyme preparation used for hydrolysis was always calculated in relation to the cellulose content of the substrate and adjusted to 10 filter paper units (FPU) and 25 cellobiase units (CBU) per gram of cellulose. The substrates were hydrolysed in duplicate at 45°C at pH 4.8. The release of soluble sugars was monitored by HPLC (*21*) and the extent of enzymatic hydrolysis of cellulose was calculated as a fraction of the theoretical amount of glucose originally present in the substrate.

The enzymatic hydrolysis profile of several screened samples was carried out for selected incubation times at 2% (w/v) substrate concentration. Washed residues, which were obtained at the different times during the hydrolysis, were immediately analysed for their fibre length distribution. The remainder of the residue was solvent exchanged with *t*-butanol and freeze-dried for further storage in a desiccator.

All enzymatic activities were determined in the enzyme mixture according to the modified I.U.P.A.C. procedure (*22,23*). When the enzymatic activities of 17 distinct enzyme preparations, ranging from 0.17 to 7.05 FPU mL^{-1}, were analysed by linear regression, a R^2 value of 0.9812 (standard error of 0.2768) was obtained. This comparison of the enzymatic activities was carried out to ensure that different batches and dilutions of enzymes used were comparable.

Determination of Substrate Structural Changes During Hydrolysis. The molecular weight distribution of both the original substrates and the partially hydrolysed residues were obtained by gel permeation chromatography (GPC) of their tricarbanyl derivatives (*24-26*). The degree of polymerization (DP) of cellulose was obtained by dividing the molecular weigth (MW) of the tricarbanylated polymer by the corresponding MW of the tricarbanyl derivative of anhydroglucose (DP = MW/519). Both the number average MW (MW_N) and the weight average MW (MW_W) of cellulose tricarbanylate, from which the DP_N and DP_W of cellulose were derived, were determined as described previously (*27*). The Mark-Houwink coefficients used in the present analysis were those reported by Valtasaari and Saarela (*28*).

The X-ray diffractogram of each set of samples was recorded by a Siemens diffractometer equipped with a D-5000 rotating anode X-ray generator (*26*). A peak resolution program was used to calculate both the crystallinity index of cellulose and the dimensions of the crystallite. This program allowed the resolution of the X-ray diffraction pattern into the contribution of each of the diffraction planes. The crystallinity index was defined as the ratio of the resolved peak area to the total area under the unresolved peak profile (*29*). The crystallinity index of cellulose was also calculated by the empirical method described by Segal *et al.* (*30*).

Results and Discussion

Steam Pretreatment for Enhanced Enzymatic Hydrolysis of Wood Residues. Chips derived from 6-7 year-old stems of *E. viminalis* were shown to be highly amenable to steam pretreatment and subsequent fractionation. Water and alkali extraction removed most of the hemicellulose and lignin originally present in the wood chips, resulting in a cellulosic residue that could be more readily hydrolysed to glucose (Figure 1A, B). The peroxide-treated fraction (WIA/H_2O_2) obtained from steam-treated eucalyptus provided the best substrate for hydrolysis, particularly at high substrate concentrations (Figure 1B) (*17*). The greater accessibility of the WIA/H_2O_2 fraction was thought to be primarily due to its very low lignin content, although the peroxide treatment also increased the degree of fibrillation and/or delamination of the fibres, causing a reduction in the crystallinity and/or degree of polymerization of cellulose and a subsequent increase in the available surface area of the substrate. However, the peroxide treatment of the alkali-washed substrate derived from eucalyptus did not appear to result in as dramatic an increase in accessibility as had been obtained previously with aspen or spruce (*16,14*).

When compared to aspen, all of the steam-treated fractions derived from *E. viminalis* appeared to be more accessible to enzymatic hydrolysis, particularly at high substrate concentrations of 10% (w/v) (*17*). The better hydrolysis yields obtained with pretreated eucalyptus was probably due to the age of the tree. As mentioned earlier, pretreatment of eucalyptus was carried out using chips derived from a 6-7 year old stem, whereas in the earlier work using aspen, pretreatment was performed using chips derived from an approximately 40 year old tree. The stem wood of young eucalyptus trees is almost exclusively composed of sapwood and is relatively free of polyphenolic deposits and extractives. Owing to its morphological characteristics and structural properties, sapwood has a higher porosity than heartwood and is generally more permeable to chemical impregnation and steam. Therefore, young stems of

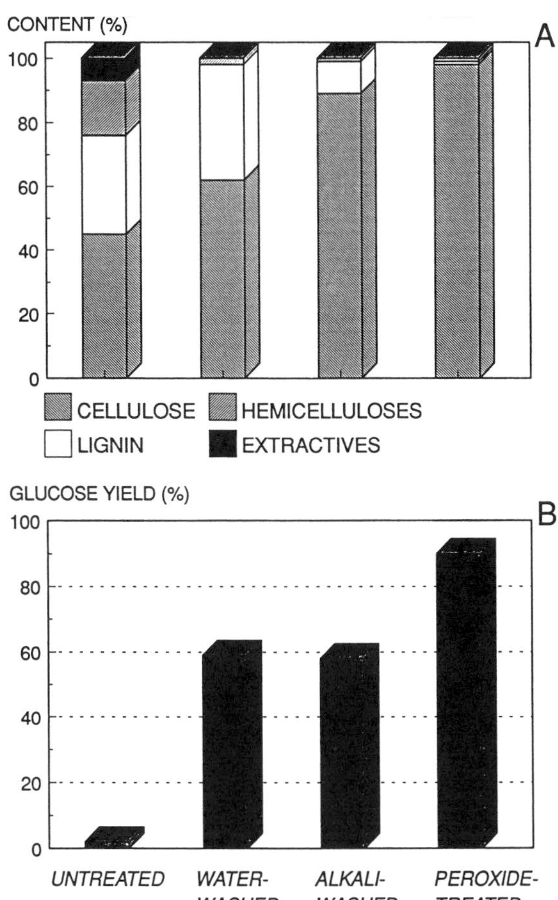

Figure 1. Characterization of the various cellulosic substrates derived from SO_2-impregnated steam-exploded eucalyptus chips with regard to their (A) chemical composition and (B) enzymatic hydrolysis. Hydrolyses were carried out for 48 h at 45°C, pH 4.8 using a 5% (w/v) cellulose concentration and an enzyme loading containing 10 FPU g^{-1} and 25 CBU g^{-1} cellulose.

E. viminalis may be more easily pretreated than adult stems of aspen because of their higher content of sapwood in relation to heartwood tissue.

It has been shown that softwoods, such as spruce (*14*) and radiata pine (*13*), are far more difficult to pretreat than hardwoods (*16*). This greater ease of pretreatment of hardwoods over softwoods has often been associated with differences in the morphology and chemical composition of the various species. To overcome the relative recalcitrance of softwoods, several authors have introduced acid catalysts, such as sulphuric acid and sulphur dioxide (SO_2), prior to high-pressure steaming (*13*). In addition to reducing the steam requirements for optimum pretreatment, acid catalysis also enhances the recovery and survival of pentoses in the water-soluble stream and the susceptibility of the substrate to hydrolysis. The former advantage is mostly associated with hardwoods, whereas the latter advantage applies to both wood types. The higher hydrolysis and survival of hemicellulose sugars has primarily been attributed to the higher stability of pentoses in acid solutions (*31*), whereas the enhanced hydrolyzability of the substrate is a result of a more drastic disruption of the cellulose structure. For instance, the degree of depolymerization (DP) of cellulose was shown to decrease from about 10^4 to 250 during acid-catalysed steam treatment (*10*). Comparable reductions in cellulose DP have been demonstrated during the mineral acid hydrolysis of cotton and other cellulosic materials (*32-34*).

Grethlein (*1*) reported that the initial rate of hydrolysis of cellulosic substrates was linearly correlated with the distribution of micropores which were accessible to the enzymes. The lower susceptibility of softwood substrates to hydrolysis, compared to hardwood substrates, was associated with a smaller increase in the distribution of these micropores during pretreatment. The development of pore volumes accessible to the enzymes has also been demonstrated to be a key factor in determining the degree of enhancement of substrate accessibility for steam-treated aspen (*4*), poplar (*35*) and radiata pine (*2*) chips. Drying was also shown to be detrimental to hydrolysis as it results in a dramatic reduction in the available surface area of the substrate (*35,36*).

The pre-impregnation of eucalyptus chips with 1% (w/w) SO_2 enhanced the recovery, fractionation and hydrolysis of steam-treated substrates (*19*). The best substrate for hydrolysis was produced at 210°C for 50 s. Further increases in the temperature or residence time of pretreatment resulted in lower recovery of water-soluble sugars without any improvement in the enzymatic digestibility of the substrate (Figure 2). It was concluded that pretreatment at high temperatures and short residence times were desirable since lower levels of pentosan degradation were obtained. These results were consistent with studies previously reported on acid catalysed steam pretreatment of aspen and spruce (*14,16,37*). However, higher SO_2 loadings of 1.6 and 2.5% (w/w) were required to enhance the hydrolysis of aspen and spruce, respectively. Previously we had shown that SO_2 gas is partially incorporated within the wood chips as sulphurous acid and then subsequently converted into sulphuric acid during steam pretreatment. Our results suggested that a relatively low SO_2 concentrations of 1% (w/w) is all that is required to generate enough sulphuric acid, provided the wood chips have a relatively high moisture content (green conditions). Lower SO_2 loadings are also desirable from an economic and environmental perspective.

It has been suggested that, during hydrolysis of steam-treated substrates, lignin acts as a physical barrier which hinders the contact between the substrate and the enzymes (*38*). Factors such as the irreversible adsorption and non-specific binding of enzymes onto the lignin-containing residue have often been implicated (*15,39*). Although alkali washing of steam-treated hardwoods usually resulted in a substrate with a lower lignin content, it was generally only as readily hydrolysed as was the water-washed substrate (Figure 1B) (*4,16,19*). It appeared that the beneficial effects of alkali extraction, such as lignin removal and cellulose swelling, were counteracted by other factors which had a detrimental effect on hydrolysis, such as the possible redistribution of the residual lignin and modifications in the crystalline state of the cellulose. In previous work on the hydrolysis of steam-treated *E. regnans* (*40*) a considerable decrease in the accessibility of the steam-treated substrate was reported after alkali washing, despite the removal of a substantial amount of the residual lignin. In this case, the age of the tree appeared to influence the fractionation of hemicelluloses and lignin after steam pretreatment. Despite the relatively poor benefit of lignin removal on cellulose hydrolysis, the extraction of the lignin from the pretreated residues is still desirable as it results in a substrate with a comparatively higher cellulose content from which a higher glucose yield per gram of substrate can be achieved. However, alkali-washing of steam-treated softwoods generally resulted in substrates which were more recalcitrant to enzymatic degradation than the respective water-washed substrate (*2,14*). This was more dramatic in the case of spruce, where alkali extraction seemed to redeposit the lignin in a fashion which greatly reduced the ease of hydrolysis (*14*). Extraction of the residual alkali-insoluble lignin by oxidative agents, such as sodium chlorite (*36*) and hydrogen peroxide (*17*), was shown to increase the susceptibility to enzymatic hydrolysis of pretreated substrates derived from both hardwood and softwood residues. It was suggested that it was the redistribution of this residual alkali-insoluble lignin which limited complete hydrolysis of alkali-washed substrates. After alkali washing, this highly condensed, modified lignin appeared to reprecipitate on the surface of the substrate, causing a reduction of both the available surface area and the ability of the cellulose fibres to swell in water.

The lignin component of softwoods has a greater guaiacyl to syringyl ratio than hardwoods. However, several authors have demonstrated that the lignin component of hardwood vessels is also of a guaiacyl type and that this lignin differs from the lignin found in other cells of the hardwood tissue (*41,42*). High levels of deterioration in hardwoods are usually associated with the degradation of parenchyma cells and libriform fibres, where lignin of the syringyl type is predominant. Therefore, the high chemical and biological resistance of plant vessels can be partially attributed to the occurrence of a more recalcitrant guaiacyl lignin type in their structure.

We have observed that, during enzyme recycling, the presence of small amounts of lignin led to a limited recovery of desorbed enzymes (*43*). A partial loss of β-glucosidase activity was also observed when lignin containing residues were hydrolysed. Although some components of the cellulase system may associate with lignin (*44*), it seemed more probable that residual amounts of cellulose are still present within the unhydrolysed material and this caused the retention of some key components of the cellulase complex. In earlier work on the simultaneous recycling of both the adsorbed enzymes and the partially hydrolysed residue, the cellulases that were originally added still retained most of their activity, even after 24 h of incubation with the substrate (*45*).

Elucidating the mode of action of cellulases adsorbed onto the surface of the substrate is one of the fundamental steps towards gaining a better understanding of the mechanism by which cellulose is hydrolysed. We have shown (*45*) that when low lignin content substrates were used, one "batch" of cellulase could be recycled multiple times and that, rather than waiting for complete hydrolysis of the substrate, the enzyme-containing residue could be recovered and added to a fresh batch of substrate where a similar hydrolysis profile could be obtained. It seemed that all the cellulase components required for the efficient hydrolysis of pretreated eucalyptus remained associated with the substrate and could not be removed by a thorough washing with hydrolysis buffer. As most of the endoglucanases and cellobiohydrolases which have been characterized to date have been shown to contain a cellulose binding domain (CBD) within their protein structure (*46*), it appears that, for at least the enzymes of the *Trichoderma* Celluclast system, this ensures a close association of the main cellulase components with the substrate until the residual substrate is completely hydrolysed.

Changes in the Morphology and Fine Structure of the Substrate During Enzymatic Hydrolysis. The rate of hydrolysis of cellulosic substrates depends on several factors associated with the nature of the substrate and the enzyme system used. Many authors support the hypothesis that it is the increased recalcitrancy of the substrate that gradually compromises the rate by which cellulose is hydrolysed (*18*). This increase in recalcitrancy may be reflected by the structural modification of the substrate, and would include changes in the morphology (particle size and pore volume), crystallinity and the degree of polymerization (DP) of the cellulose.

The initial fast rate by which cellulose is hydrolysed has been associated with the occurrence of more accessible regions at the surface of the substrate where hydrolysis is facilitated. At the fibre level, these more accessible regions are often associated with cracks and defects of the fibre. At the molecular level, they are characterized by a larger pore volume and/or available surface area (*1,2*) and a lower crystallinity index (amorphous cellulose) (*47,48*). The rapid fragmentation of fibers (Figure 3), which occurred regardless of the initial fiber length of the substrate, probably resulted from the random attack of enzymes at eroded walls or defects on the fiber. If the residual cellulose at the site where the fiber was eventually attacked was primarily amorphous, fragmentation would be the result of the localized action of endoglucanases on the structurally disordered areas of the fiber (*49*). This would gradually increase both the surface area of the substrate and the availability of chain ends for the action of exoglucanases.

The residue obtained after 24 h hydrolysis of the WIA/H_2O_2 fraction for 24 h was shown to be further fragmented into a large population of smaller particles. These particles appeared to be narrow rod-like structures which were about 5 - 10 μm in length and probably corresponded to small bundles of cellulose microfibrils. Peters *et al.* (*50*) also observed a rapid accumulation of small fragments during hydrolysis of microcrystalline cellulose by bacterial cellulases. Although our results were consistent with this preliminary study, we measured fibre length distributions rather

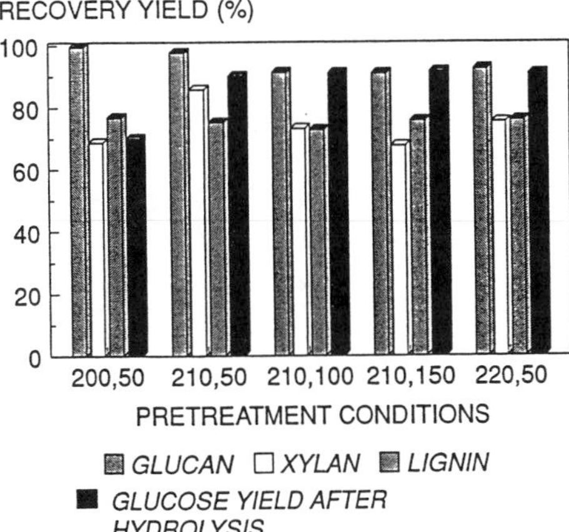

Figure 2. Influence of the temperature and time of steaming on both the overall recovery yield of three major components of the wood chips and the susceptibility of the water-washed cellulosic residue to enzymatic hydrolysis. Hydrolyses were carried out for 48 h using a 2% (w/v) substrate concentration and an enzyme loading containing 10 FPU g^{-1} and 25 CBU g^{-1} cellulose.

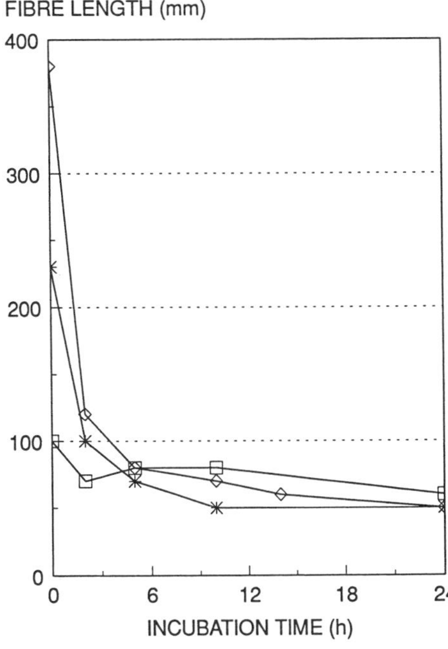

Figure 3. Effect of enzymatic hydrolysis on the average fibre length of the several screened fractions derived from peroxide-treated eucalyptus (WIA/H_2O_2). Hydrolyses were carried out at 2% (w/v) with 10 FPU g^{-1} and 25 CBU g^{-1} cellulose. Open squares, unscreened fraction; asterisks, F-200 fraction; open lozenges, F-150 fraction.

than particle counts. It appeared that fragmentation was an essential step during the enzymatic hydrolysis of cellulose by both fungal (*51*) and bacterial (*50*) cellulases.

Several studies have demonstrated that there is no significant correlation between variations in the degree of polymerization (DP) of cellulose and the development of substrate accessibility to enzymes (*48,52*). Most of these studies have utilized viscosimetric assays to determine the DP of cellulosic fractions (*52*). However, this method is restricted to the determination of the average DP of a population of cellulose molecules. Therefore, an alternative method has to be used if a degradative process such as enzymatic hydrolysis is to be characterized. The GPC method, which consists of analyzing the molecular weight distribution of cellulose by gel permeation chromatography (GPC) of its tricarbanyl derivative, has been successfully used by various groups (*10,25,33,10*). Compared to other methods, GPC was considered to be more representative as: MW distributions rather than MW averages were obtained; the tricarbanyl derivatives were more chemically stable than other cellulose derivatives; less cellulose degradation occurrs during sample preparation, and; better reproducibility was obtained by the GPC method (*53*).

To obtain a better knowledge of the mechanism of cellulose fragmentation, we investigated the probable changes in the molecular weight (MW) distribution of several cellulosic substrates during hydrolysis. A small modification of the GPC method was introduced in order to avoid substantial losses of oligosaccharides with low DP. Instead of precipitating the cellulose derivative in methanol or water in order to purify it (*32*), the reaction products were recovered by simple evaporation of the reaction solvents. Screened fractions derived from peroxide-treated eucalyptus were assayed and found to consist of a narrow distribution of cellulose molecules with a weight average DP (DP_W) of approximately 370 anhydroglucose units. Other authors have reported similar values for the DP_W of acid-depolymerised residues obtained from several cellulosic materials (*25,34*). Acid depolymerization generally resulted in a rapid decline in the DP of cellulose from a couple of thousands to a minimum value of about 200 anhydroglucose units, which is the limit DP of the method. As a result, the low DP_W of the WIA-H_2O_2 fraction was readily attributed to the acidic nature of the pretreatment method used (SO_2-catalysed steam treatment). Interestingly, the depolymerization of native cellulose by *Postia placenta* (*Poria placenta*) also reduced the DP_W of cotton cellulose from 2111 to 238 (*34*). It was suggested that cellulose depolymerization occurred through the initial degradation of the amorphous regions of the substrate, thus forming crystallites.

Enzymatic hydrolysis of the WIA/H_2O_2 fraction resulted in a gradual decrease in the DP of cellulose (Figure 4). Fiber screening had no effect on this behavior as both F-150 and F-200 fractions showed the same trends during hydrolysis. This shift in cellulose DP was not followed by any considerable increase in polydispersity, indicating that there was no accumulation of short oligomers during hydrolysis. In fact, it appeared that, once a cellulose molecule at the surface of the substrate was attacked by the concerted action of endo- and exoglucanases, the macromolecule was broken down to soluble oligomers very quickly. Therefore, it appeared that the *Trichoderma* cellulases present in the Celluclast preparation were able to simultaneously attack both the crystalline and amorphous regions of the substrate, resulting in a progressive erosion followed by a extensive collapse of the cellulose structure.

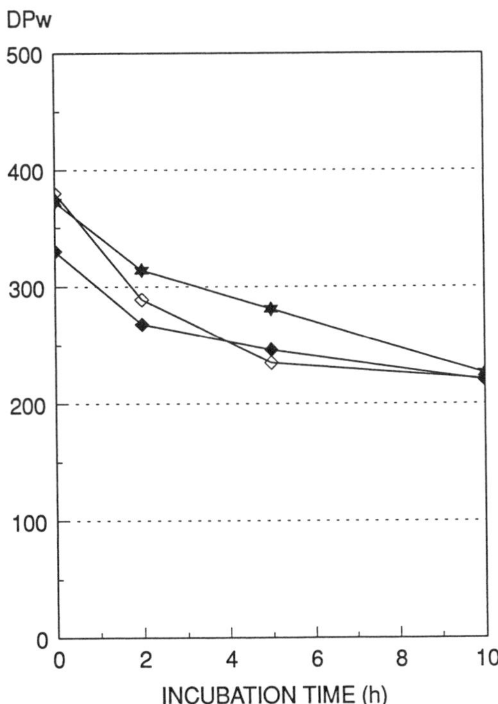

Figure 4. Effect of enzymatic hydrolysis on the weight average degree of polymerization (DP_W) of the several screened fractions derived from peroxide-treated eucalyptus (WIA/H_2O_2). Hydrolyses were carried out at 2% (w/v) with 10 FPU g^{-1} and 25 CBU g^{-1} cellulose. Filled lozenges, unscreened fraction; open lozenges, F-200 fraction; filled asterisks, F-150 fraction.

The gradual depolymerization of cellulose, which was observed during hydrolysis of the WIA/H_2O_2 fraction, can be mostly attributed to exoglucanase catalysis. Although this hypothesis should still recognize the importance of endoglucanase (EG) action, particularly in the early stages of hydrolysis, it appears that a much more drastic shift in cellulose DP would be expected if EG activity predominated. As the population of cellooligomers present in the WIA/H_2O_2 fraction averaged a cellulose DP of only 380, this hypothesis agrees with the general concept that exoglucanases have a higher specificity for substrates that are characterized by having a higher availability of chain ends.

In the earlier work of Reese *et al.* (*54*), it was suggested that the crystalline regions of cellulose are less susceptible to hydrolysis than the amorphous regions. This hypothesis is supported by the complex polymorphism of pretreated cellulose and the different reactivity of each polymorph to cellulolytic enzymes. In fact, a linear correlation of the degree of crystallinity of cellulose and its enzymatic saccharification has been reported for several pretreated substrates with varying degrees of crystallinity D.H. (CrI) (*18,48,52*). However, other authors have suggested that the increased surface area of cellulose available to enzyme molecules may be of a greater importance (*55,56*). It seems that the apparent controversy regarding the effect of cellulose crystallinity on hydrolysis is partly due to technical difficulties in measuring the crystallinity of cellulose (*57*).

It seemed probable that, during hydrolysis of the WIA/H_2O_2 fraction, the gradual decrease in cellulose DP could be accompanied by a structural reorganization of the substrate. In fact, regardless of the method used to determine crystallinity, the CrI of the F-150 fraction decreased during hydrolysis (Figure 5). These observations reinforced our previous hypothesis that the *Trichoderma* Celluclast preparation was capable of simultaneously hydrolysing both the amorphous and crystalline regions of the substrate. An alternative explanation is that there is a relative increase in the amorphous character of the substrate during the early stages of hydrolysis, a process previously referred to as amorphogenesis (*58*). The gradual increase in the amount of enzymes adsorbed onto the residual substrate during the latter stages of hydrolysis could also result in a relative increase in the area of the diffractogram associated with the amorphous character.

There was no clear indication that hydrolysis of both the F-150 and F-200 fractions had triggered any lattice transformation of the substrates (*i.e.*, transformation of one cellulose polymorph to another). The X-ray diffractograms of the partially hydrolysed residues showed the same strong equatorial reflections at the 101, 101 and 002 diffraction planes, which are characteristic of the cellulose I polymorph. There was little change in the resolution between the 101 and 101 reflections during hydrolysis. However, hydrolysis resulted in a progressive broadening of the 002 reflection of the F-150 fraction. This may have resulted from either a gradual disordering in the fine structure of the substrate or a preferential attack of the enzymes at the 002 plane of the crystallites.

Chanzy and Henrissat (*59*) applied transmission electron microscopy (TEM) to study the enzymatic breakdown of cellulose microcrystals from the alga *Valonia macrophysa* by purified *T. reesei* cellulases. It was shown that the concerted action of exo- and endocellulases caused a reduction in both the width and length of the microcrystals. These authors suggested that the enzymatic degradation of bacterial

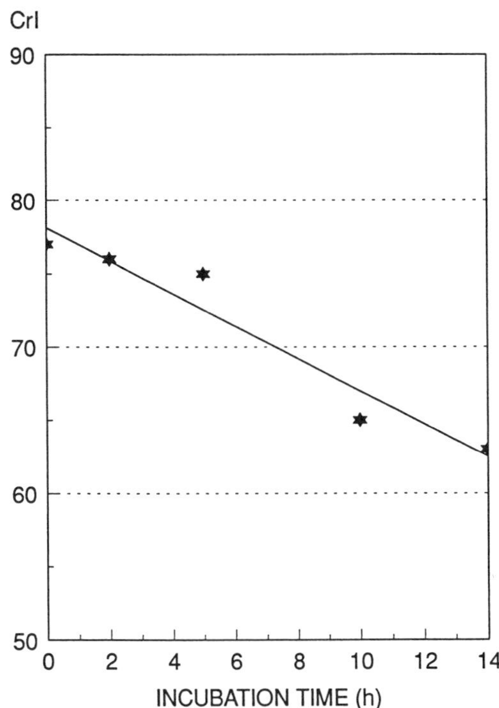

Figure 5. Effect of enzymatic hydrolysis on the crystallinity index (CrI) of the F-150 fraction derived from peroxide-treated eucalyptus (WIA/H_2O_2). Hydrolysis was carried out at 2% (w/v) with 10 FPU g^{-1} and 25 CBU g^{-1} cellulose.

cellulose occurred preferably at one plane of the crystallite and that the *Trichoderma* cellobiohydrolase II (CBH II) was responsible for an unidirectional attack at the end of the crystallite. A reduction in CrI, followed by a gradual decrease in the average thickness of the cellulose crystallites, was also reported during hydrolysis of bleached substrates derived from several cellulosic residues (*60,61*). Although these results resemble some of our observations, they are not strong enough to confirm that both *Penicillium pinophillum* and *T. reesei* cellulases share the same mode of action. In fact, there has been some evidence that the mechanism of cellulose degradation differs from one fungal species to another (*62*).

The results obtained to date suggest that effective hydrolysis of lignocellulosic residues is highly dependent on factors such as hemicellulose removal and increased surface area of the substrate. The characterization of hydrolysis through the determination of changes in cellulose DP and crystallinity was also shown to be very helpful in determining the mode of action of the enzymes on pretreated substrates.

Conclusion

The steam pretreatment of *Eucalyptus viminalis* chips was characterized. Pretreatment parameters such as steam temperature, residence time, addition of SO_2 as an acid catalyst, and initial moisture content of the chips were evaluated in order to optimize recovery, fractionation and enzymatic hydrolysis of pretreated materials.

Wood residues derived from young eucalyptus trees appeared to be more easily and readily fractionated than old stems of aspen and spruce, yielding better substrates for hydrolysis. This, together with the fast growing rates and short rotation times which can be obtained with several *Eucalyptus* sp., suggests a potential application of these species for bioconversion, particularly in the southern hemisphere. Due to the increased utilization of *Eucalyptus* spp. for pulp and paper production, residues from the processing of this feedstock will also be readily available at a relatively low cost.

Post-treatment with alkaline hydrogen peroxide was shown to enhance both the hydrolysis rate and the efficiency of enzyme recycling. A rapid reduction in fiber length (fragmentation), followed by an almost complete saccharification of cellulose, was observed for several fractions derived from peroxide-treated eucalyptus. A gradual decrease in the degree of polymerization (DP) of this fraction also reflected its high susceptibility to hydrolysis. Substrate fragmentation occurred before any substantial weight loss was observed, whereas depolymerization occurred at the same time. As enzymatic attack resulted in a gradual decrease in both cellulose DP and crystallinity, it was apparent that hydrolysis of this substrate was predominantly due to exoglucanase activity. However, it is apparent that the mode of action of *Trichoderma* cellulases (Celluclast) is highly dependent upon the fine structure of the pretreated substrate that is being hydrolysed.

Literature Cited

1. Grethlein, H.E. *Bio/Technology* **1985**, *3*, 155-160.
2. Wong, K.K.Y.; Deverell, K.F.; Mackie, K.L.; Clark, T.A.; Donaldson, L.A. *Biotechnol. Bioeng.* **1988**, *31*, 447-456.

3. Grethlein, H.E.; Converse, A.O. *Biores. Technol.* **1991**, *36*, 77-82.
4. Excoffier, G.; Toussaint, B.; Vignon, M.R. *Biotechnol. Bioeng.* **1991**, *38*, 1308-1317.
5. Brownell, H.H. Progress Report for the Project No. 04-53-12-402, Forintek Canada: Ottawa, Ontario, 1989.
6. Saddler, J.N.; Ramos, L.P.; Breuil, C. In *Bioconversion of Forest and Agricultural Plant Wastes*; Saddler, J.N., Ed.; C.A.B. International: London, UK, 1993.
7. Brownell, H.H.; Saddler, J.N. *Biotechnol. Bioeng.* **1987**, *29*, 228-235.
8. Nguyen, Q.A.; Saddler, J.N. *Biores. Technol.* **1991**, *35*, 275-282.
9. Puls, J.; Poutanen, K.; Korner, H.-U.; Viikari, L. *Appl. Microbiol. Biotechnol.* **1985**, *22*, 416-423.
10. Miller, D.; Sutcliffe, R.; Saddler, J.N. In *TAPPI Proceedings of the International Symposium on Wood and Pulping Chemistry*; Technical Association of the Pulp and Paper Industry: Atlanta, GA, 1989; pp 9-11.
11. Yamashiki, T.; Matsui, T.; Saitoh, M.; Matsuda, Y.; Okajima, K.; Kamide, K. *British Polym. J.* **1990**, *22*, 201-212.
12. Yamashiki, T.; Matsui, T.; Saitoh, M.; Okajima, K.; Kamide, K.; Sawada, T. *British Polym. J.* **1990**, *22*, 121-128.
13. Clark, T.A.; Mackie, K.L. *J. Wood. Chem. Technol.* **1987**, *7*, 373-403.
14. Schwald, W.; Smaridge, T.; Chan, M.; Breuil, C.; Saddler, J.N. In *Enzyme Systems for Lignocellulose Degradation*; Coughlan, M. P., Ed.; Elsevier: New York, NY, 1989b; pp. 231-242.
15. Clesceri, L.S.; Sinitsyn, A.P.; Saunders, A.M.; Bungay, H.R. *Appl. Biochem. Biotechnol.* **1985**, *11*, 433-443.
16. Schwald, W.; Breuil, C.; Brownell, H.H.; Chan, M.; and Saddler, J.N. *Appl. Biochem. Biotechnol.* **1989a**, *20/21*, 29-44.
17. Ramos, L.P.; Breuil, C; Saddler, J.N. *Appl. Biochem. Biotechnol.* **1992b**, *34/35*, 37-47.
18. Fan, L.T.; Gharpuray, M.M.; Lee, Y.-H. *Cellulose hydrolysis*; Springer-Verlag: New York, NY, 1987, pp 5-120.
19. Ramos, L.P.; Breuil, C.; Kushner, D.N.; Saddler, J.N. *Holzforschung* **1992a**, *46*, 149-154.
20. Irick, T.J.; West, K.; Brownell, H.H.; Schwald, W.; Saddler, J.N. *Appl. Biochem. Biotechnol.* **1988**, *17/18*, 137-149.
21. Schwald, W.; Chan, M.; Breuil, C.; Saddler, J.N. *Appl. Microbiol. Biotechnol.* **1988**, *28*, 398-403.
22. Ghose, T.K. *Pure and Appl. Chem.* **1986**, *59*, 257-268.
23. Chan, M.; Breuil, C.; Schwald, W.; Saddler, J.N. *Appl. Microbiol. Biotechnol.* **1989**, *31*, 413-418.
24. Schroeder, L.R.; Haigh, F.C. *TAPPI* **1979**, *62*, 103-105.
25. Miller, D.; Senior, D.; Sutcliffe, R. *J. Wood Chem. Technol.* **1991**, *11*, 23-32.
26. Ramos, L.P.; Nazhad, M.M.; Saddler, J.N. *Enzyme Microb. Technol.* **1993b** (in press).
27. Yau, W.W.; Kirkland, J.J.; Bly, D.D. *Modern Size-exclusion Liquid Chromatography*; John Wiley & Sons: New York, NY, 1979, pp 7-12.
28. Valtasaari, L.; Saarela, K. *Paperi ja Puu* **1975**, *57*, 5-10.

29. Wims, A.M.; Myers Jr., M.E.; Johnson, J.L.; Carter, J.M. *Adv. X-ray Anal.* **1986**, *29*, 281-290.
30. Segal, L.; Creely, J.J.; Martin, A.E.; Conrad, C.M. *Text. Res. J.* **1959**, *29*, 786-793.
31. Springer, E.L.; Harris, J.F. *Svensk. Papperstidn.* **1982**, *85*, 152-154.
32. Wood, B.F.; Conner, A.H.; Hill Jr., C.G. *J. Appl. Polym. Sci.* **1986**, *32*, 3703-3712.
33. Lauriol, J.-M.; Comtat, J.; Froment, P.; Pla, F.; Robert, A. *Holzforschung* **1987**, *41*, 165-169.
34. Kirk, T.K.; Ibach, R.; Mozuch, M.D.; Conner, A.H.; Highley, T.L. *Holzforschung* **1991**, *45*, 239-244.
35. Grous, W.R.; Converse, A.O.; Grethlein, H.E. *Enzyme Microb. Technol.* **1986**, *8*, 274-280.
36. Saddler, J.N.; Brownell, H.H.; Clermont, L.P.; Levitin, N. *Biotechnol. Bioeng.* **1982**, *24*, 1389-1402.
37. Mackie, K.L.; Brownell, H.H.; West, K.L.; Saddler, J.N. *J. Wood. Chem. Technol.* **1985**, *5*, 405-425.
38. Ucar, G.; Fengel, D. *Holzforschung* **1988**, *42*, 141-148.
39. Converse, A.O.; Ooshima, H.; Burns, D.S. *Appl. Biochem. Biotechnol.* **1990**, *24/25*, 67-73.
40. Dekker, R.F.H.; Karageorge, H.; Wallis, A.F.A. *Biocatalysis* **1987**, *1*, 45-59.
41. Musha, Y.; Goring, D.I.A. *Wood Sci. Technol.* **1975**, *9*, 55-58.
42. Saka, S.; Goring, D.A.I. In *Biosynthesis and Biodegradation of Wood Components;* Higuchi, T., Ed.; Academic Press: New York, NY, 1985, pp 469-503.
43. Ramos, L.P.; Breuil, C.; Saddler, J.N. *Enzyme Microb. Technol.* **1993a**, *15*, 19-25.
44. Sutcliffe, R.; Saddler, J.N. *Biotechnol. Bioeng. Symp.* **1986**, *17*, 749-762.
45. Ramos, L.P.; Saddler, J.N. *Appl. Biochem. Biotechnol.* **1993c** (in press).
48. Sinitsyn, A.P.; Gusakov, A.V.; Vlasenko, E.Y. *Appl. Biochem. Biotechnol.* **1991**, *30*, 43-59.
46. Esterbauer, H.; Hayn, M.; Abuja, P.M.; Claeyssens, M. In *Enzymes in Biomass Conversion*; Leatham, G.F.; Himmel, M.E., Eds.; American Chemical Society: Washington, DC, 1991, Vol. 460; pp 301-312.
47. Bertran, M.S.; Dale, B.E. *Biotechnol. Bioeng.* **1985**, *27*, 177-181.
49. Nieves, R. A., Ellis, R. P., Todd, R. J., Johnson, T. J. A., Grohmann, K.; Himmel, M.E. *Appl. Env. Microbiol.* **1991**, *57*, 3163-3170.
50. Peters, L.E.; Walker, L.P.; Wilson, D.B.; Irwin, D.C. *Biores. Technol.* **1991**, *35*, 313-319.
51. Oltus, E.; Mato, J.; Bauer, S.; Farkas, V. *Cellulose Chem. Technol.* **1987**, *21*, 663-672.
52. Puri, V.P. *Biotechnol. Bioeng.* 1984, **26**, 1219-1222.
53. Kossler, I.; Danhelka, J.; Netopilik, M. *Svensk. Papperstidn. Research* **1981**, *18*, R137-R140.
54. Reese, E.T.; Siu, R.G.H.; Levinson, H.S. *J. Bacteriol.* **1950**, *59*, 485-497.
55. Caulfield, D.F.; Moore, W.E. *Wood Sci.* **1974**, *6*, 375-379.
56. Rivers, D.B.; Emert, G.H. *Biotechnol. Lett.* **1987**, *9*, 365-368.

57. Sidiras, D.K.; Koullas, D.P.; Vgenopoulos, A.G.; Koukios, E.G. *Cellulose Chem. Technol.* **1990**, *24*, 309-317.
58. Coughlan, M.P. *Biotechnol. Gen. Eng. Rew.* **1985**, *3*, 39-109.
59. Chanzy, H.; Henrissat, B. *FEBS Lett.* **1985**, *184*, 285.
60. Betrabet, S.M.; Paralikar, K.M. *Cellulose Chem. Technol.* **1977**, *12*, 241-252.
61. Betrabet, S.M.; Paralikar, K.M. *Cellulose Chem. Technol.* **1978**, *12*, 241-252.
62. Kleman-Leyer, K.; Agosin, E.; Conner, A.H.; Kirk, T.K. *Appl. Env. Microbiol.* **1992**, *58*, 1266-1270.

RECEIVED January 19, 1994

Chapter 17

Development of Genetically Engineered Microorganisms for Ethanol Production

Stephen K. Picataggio, Min Zhang, and Mark Finkelstein

Applied Biological Sciences Branch, National Renewable Energy Laboratory, 1617 Cole Boulevard, Golden, CO 80401–3393

Although cellulosic biomass is a favorable feedstock for fuel ethanol production, substantial hurdles remain before a typical hydrolysate can be utilized efficiently as a fermentation substrate. Rapid and efficient conversion of pentose sugars, particularly xylose, remains one of the key economic bottlenecks in a biomass to ethanol process. Despite the development of recombinant strains with improved xylose fermentation performance, high ethanol yields from lignocellulosic hydrolysates, and increased product concentrations and ethanol tolerances are key targets that have yet to be achieved. The genetic modifications that can have the greatest impact on the economic feasibility of these fermentations include amplification and deregulation of rate-limiting reactions; metabolic engineering that redirects the normal carbon flow to ethanol, improves glycolytic efficiency, or reduces futile cycling; introduction of genes that broaden the substrate range; and manipulations that improve ethanol tolerance, osmotolerance, thermotolerance, and resistance to the inhibitory compounds normally present in lignocellulosic hydrolysates.

With the fuel alcohol industry producing a large-volume, low-value product, many of the raw materials used by the potable distiller are too expensive to consider as a feedstock. Currently, only sugarcane and corn are routinely used as feedstocks for the production of fuel ethanol. Brazil produces more than 3 billion gallons of ethanol per year from sugarcane, while most of the fuel ethanol produced in the United States is made from corn. In order to make this process competitive with the prevailing cost of gasoline, the U.S. government grants a credit of approximately $0.50 per gallon to ethanol producers. Our challenge in the 21st century is to develop new, economically viable feedstocks as well as the microorganisms that can efficiently and rapidly convert these feedstocks to fuel ethanol.

0097–6156/94/0566–0342$08.18/0

Cellulosic biomass is a favorable feedstock because it is readily available and, because it has no food value, it is less expensive than corn or sugarcane. However, there remain substantial hurdles that must be overcome before a typical lignocellulosic hydrolysate can be efficiently utilized as a substrate by fermentative microorganisms. The typical feedstock contains approximately 30-60% glucose, 15-30% xylose, 10-20% lignin, and 5-30% of a mixture of arabinose, mannose, galactose, and a variety of other minor pentose and hexose sugars. Many microorganisms are capable of efficiently fermenting glucose, but the conversion of pentose sugars, particularly xylose, to ethanol remains one of the key economic bottlenecks in a biomass to ethanol conversion scheme. In addition, the feedstock will probably be presented to the selected microorganism at elevated temperature, low pH, and high salt concentration. The few organisms that can grow on all the sugars in lignocellulosic hydrolysates typically grow slowly and demonstrate marginal yields and productivities. The rapid and efficient utilization of all of these component sugars is an absolute requirement for a commercial process. In all likelihood, the ideal ethanol-producing microorganism will have the productivity of bacteria, the selectivity of yeast, and a broad substrate utilization range. It will also be a facultative anaerobe and it will tolerate the inhibitory compounds present in dilute-acid prehydrolysates.

Unfortunately, no one microorganism is known to possess all these traits. Recent advances in the application of recombinant DNA technology, while encouraging, have not yet yielded an organism with all of the desired features of an ideal ethanologen. Hence, processes based on their use have not been commercialized. However, the recombinant approach to add the ability to produce ethanol to a microorganism with an already broad substrate utilization range shows substantial promise. This paper reviews the recent advances in the development of recombinant microorganisms for ethanol production and highlights those strain development strategies that have demonstrated some measure of success and which appear to hold the most promise.

Strain Development Strategies

Saccharomyces cerevisiae. The historical importance of *S. cerevisiae* in industrial fermentations cannot be overstated and, as one might expect, there is a long list of advantages to consider for its potential application in biomass fermentations. *S. cerevisiae* ferments glucose through the Embden-Meyerhoff-Parnas pathway (Figure 1) and demonstrates high fermentation selectivity, with ethanol as virtually the sole product (only small amounts of glycerol and acetate are formed in order to maintain intracellular cofactor balance). This yeast is known for its high ethanol tolerance and some strains can continue fermentation even at ethanol concentrations in excess of 30% w/v (*1*)! As a facultative anaerobe, *S. cerevisiae* is capable of anaerobic growth and, because it has a restricted respiratory system capable of a Crabtree effect, there is little yield loss to biomass during the fermentation. This effect is of fundamental importance in carbohydrate fermentations because it normally uncouples substrate utilization from respiration, even in the presence of oxygen, providing that some fermentable sugar is present (*2,3*). Its facilitated diffusion sugar transport system maximizes intracellular energy efficiency by eliminating the need for anaerobically-generated ATP. Also, because fermentation occurs under completely anaerobic conditions, there is no yield loss due to aerobic ethanol reassimilation.

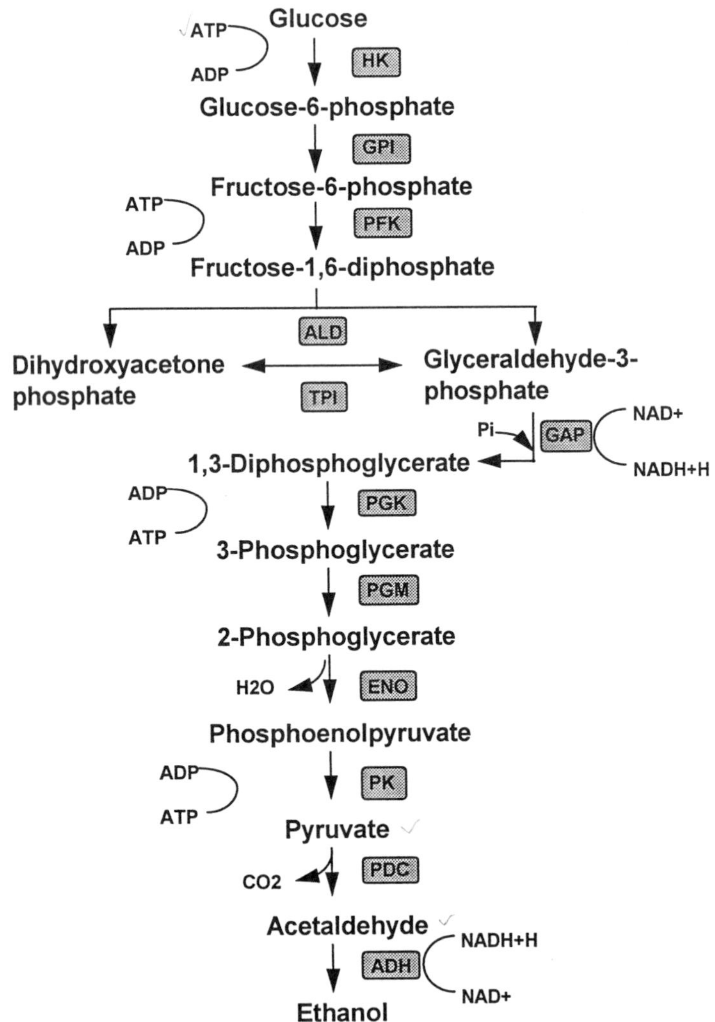

Figure 1. Embden-Meyerhoff-Parnas Pathway. HK, Hexokinase; GPI, Glucose-6-phosphate isomerase; PFK, Phosphofructokinase; ALD, Aldolase; TPI, Triose-phosphate isomerase; GAP, Glyceraldehyde-3-phosphate dehydrogenase; PGK, Phosphoglycerate kinase; PGM, Phosphoglycerate mutase; ENO, Enolase; PK, Pyruvate kinase; PDC, Pyruvate decarboxylase; ADH, Alcohol dehydrogenase.

This yeast's ability to ferment sugars at low pH provides protection against bacterial contamination during prolonged cultivation and precludes the need for the base addition required in bacterial fermentations. Other key advantages are its ability to grow and ferment sugars in the presence of lignocellulosic hydrolysates and its superior resistance to acetate at low pH under anaerobic conditions (*4*). The high indigenous levels of glucose-inducible pyruvate decarboxylase (PDC) and alcohol dehydrogenase (ADH) ensure rapid fermentation rates and high specific productivities and provide resistance to glucose catabolite repression during fermentation of mixed sugar hydrolysates. The availability of flocculent strains permit the development of processes based on cell-recycle and provide the opportunity to significantly reduce the costs associated with inoculum preparation. Certainly, the utility of *S. cerevisiae* in a simultaneous saccharification and fermentation process (SSF) has been well established (*5-12*) and its industrial scale-up is well understood. A well-developed gene transfer system is also available for metabolic engineering.

Clearly, the only real disadvantage with *S. cerevisiae* is its limited substrate range (although strains with higher thermotolerance would be more compatible in SSF processes utilizing cellulases with optimal enzymatic activities around 50°C). Unfortunately, this yeast lacks both a xylose-assimilation pathway and adequate levels of key pentose phosphate pathway enzymes. Xylose uptake by the facilitated transport system is relatively slow and occurs only in the presence of other metabolites, such as ribose (*13*). Like many other yeasts, *S. cerevisiae* ferments xylulose, but more slowly and not as efficiently as glucose (*14-17*). Senac and Hahn-Hagerdal (1990) found a significant accumulation of the intermediate sedoheptulose-7-phosphate in xylulose-grown cells compared to glucose-grown cells, suggesting transaldolase as a rate-limiting enzyme in the pentose phosphate pathway. The transaldolase specific activity, while essentially the same for glucose and xylulose-grown cells, was several orders of magnitude lower than that reported for xylose-assimilating yeasts such as *Candida utilis*. Similar accumulation in the presence of iodoacetate, a specific inhibitor of glyceraldehyde-3-phosphate dehydrogenase (GAP), suggested that slow xylulose fermentation could be the result of competition between transaldolase in the pentose phosphate pathway and GAP in glycolysis for the common intermediate, D-glyceraldehyde-3-phosphate (*18*). However, cell extracts in which up to 100-fold more transaldolase activity was added demonstrated the same sugar conversion rate and a decreased rate of ethanol formation (*19*).

Several attempts to genetically engineer xylose utilization into *S. cerevisiae* by introduction of bacterial xylose isomerase genes have been unsuccessful (*20,21*), apparently due to improper folding of the heterologous protein in the highly-reducing yeast cytoplasm. Greater success has been achieved in cloning and expressing the xylose reductase and xylitol dehydrogenase genes from xylose-assimilating yeasts. *S. cerevisiae* transformed with the *Pichia stipitis* xylose reductase gene was incapable of growth on xylose as the sole carbon source or of ethanol production (*22*), but converted xylose almost exclusively to xylitol (*23*). When transformed with the *P. stipitis* xylose reductase and xylitol dehydrogenase genes, recombinant *S. cerevisiae* fermented xylose as a sole carbon source, though incompletely and at a considerably slower rate than glucose (*24*). The low ethanol yields indicated that, contrary to aerobic glucose metabolism, xylose utilization was almost entirely oxidative. Further analysis of these recombinants indicated that the incomplete xylose conversion to ethanol was the result of both cofactor imbalance and an insufficient capacity for xylulose conversion through the pentose phosphate pathway (*25*). No growth was observed in the absence of respiration and, under such conditions, maximal ethanol yields were 34% of theoretical with xylitol and ethanol as the major fermentation products. Inefficient ATP generation resulting from yield loss to xylitol coupled with slow xylose metabolism are believed to be the cause of growth arrest in the absence of respiration.

In another approach, the level of xylulokinase activity in *S. cerevisiae* was increased up to 230-fold following amplification of its xylulokinase gene on a high-copy number plasmid (*26*). The resulting strain could ferment xylulose up to 130% faster than the parental strain. Efforts to construct a xylose-fermenting *S. cerevisiae* continue and a genetically engineered yeast in which the genes encoding xylose reductase, xylitol dehydrogenase and xylulokinase were coordinately expressed can apparently ferment 5% xylose to ethanol within two days and with very little xylitol formation (N. Ho, personal communication). It is conceivable, however, that this approach will result in the development of a xylose-fermenting *S. cerevisiae* that is limited in the same fashion as the xylose-assimilating yeasts - notably in its oxygen requirement to maintain xylose transport and cofactor balance, its formation of xylitol byproduct and its aerobic ethanol reassimilation. Despite its numerous advantages, this yeast appears to have a fundamental deficiency in pentose fermentation that may preclude its use in commercial xylose fermentations for the time being.

Opportunities to improve this technology involve cloning a novel xylose isomerase into *S. cerevisiae* to provide a xylose-assimilation pathway. One such approach involves the cloning of a xylose isomerase from an acidophilic *Lactobacillus* (*27*). Longer-term opportunities to improve the metabolic efficiency of fermentation involve amplification and deregulation of the key rate-limiting pentose phosphate pathway enzymes, deregulation of the typical yeast diauxic response to the presence of mixed sugars, and elimination of CO_2 loss through futile cycling (*16,19,28-30*). In addition, introduction of a facilitated xylose transport system will probably be an essential element of an efficient recombinant yeast system.

Xylose-Assimilating Yeasts. A survey of over 400 yeasts (*31*) failed to identify any strains capable of fermenting xylose under strictly anaerobic conditions. This was somewhat surprising considering their historical importance in glucose fermentations and their ability to ferment xylulose (albeit not as efficiently as glucose). Upon further investigation it was found that 63% of the 466 species of yeasts examined were able to assimilate xylose under aerobic or microaerophilic conditions (*32*). It is now generally accepted that *Candida shehatae, Pachysolen tannophilus* and *Pichia stipitis* are the three wild-type yeasts best suited for xylose "fermentation". Unlike bacteria, which utilize xylose by direct isomerization to xylulose via xylose isomerase, these yeasts utilize a two-step pathway in which xylose is first reduced by an NAD(P)H-dependent xylose reductase to xylitol, which is then oxidized to xylulose by an NAD-dependent xylitol dehydrogenase (Figure 2). Xylulose is subsequently phosphorylated by xylulokinase to form xylulose-5-phosphate and then metabolized to ethanol through the pentose phosphate and Embden-Meyerhoff-Parnas pathways (*15,33*). The different cofactor specificities of these two enzymes; however, limits the efficiency by which these yeasts convert xylose to xylulose (*25,34*).

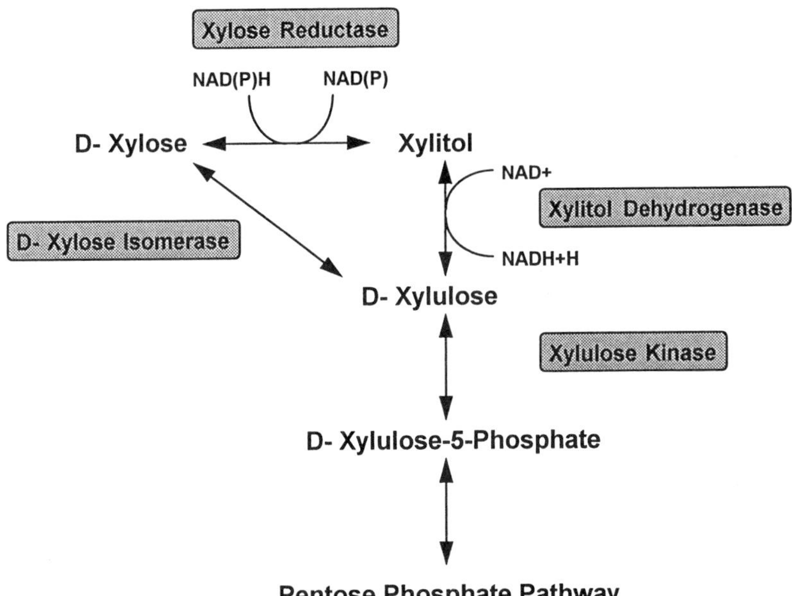

Figure 2. Xylose Assimilation Pathways in Bacteria and Yeast.

Generally, the xylose-assimilating yeasts are capable of achieving ethanol yields from 78-94% of theoretical and final ethanol concentrations of up to 5% (w/v), but demonstrate relatively low ethanol productivities (0.3 - 0.9 g/L/h), especially in the absence of oxygen (0.1 - 0.2 g/L/h). One advantage of their use in xylose fermentations is their low pH optima, ranging from pH 3.5 - 4.5, which precludes the need for the base addition typical in bacterial fermentations. *P. stipitis* demonstrates the best overall performance in terms of complete sugar utilization, minimal coproduct formation and insensitivity to temperature and substrate concentrations (*35*). Coproduct formation is negligible under ideal fermentation conditions and high selectivity is considered one of the key advantages to their use. Because the strains developed thus far are wild-type strains, no specialized containment equipment is required upon scale-up.

While ethanol yields approaching theoretical have been reported for *P. stipitis* under ideal fermentation conditions on pure xylose (*36*), typical yields range from 80-90% of theoretical due to considerable formation and accumulation of xylitol (*35*). The accumulation of this intermediate is believed to result from an inhibition of xylitol dehydrogenase by excess NADH formed in the absence of sufficient respiration (*37*). Furthermore, unlike glucose fermentations with *S. cerevisiae*, there is also significant yield loss to biomass formation because of the absence of the Crabtree effect. Perhaps the greatest disadvantage with the use of these yeasts is their fundamental requirement for low levels of oxygen (2 mmol/L/h) to maintain cell viability, xylose transport and ethanol productivity. Whereas cells lose viability rapidly in the absence of sufficient oxygen, excess oxygen causes them to completely cease ethanol production and respire the substrate to form biomass (*37,38-40*). Clearly, the degree of control necessary to maintain this narrow operational margin on an industrial scale would be difficult and cost-prohibitive. Other disadvantages with these yeasts include: low to moderate ethanol tolerance (maximum ethanol concentrations ranging from 3-5% (w/v)); poor growth and fermentation performance on lignocellulosic hydrolysates (because of their sensitivity to inhibitory components); comparatively low volumetric productivity (0.3-0.9 g ethanol/L/h); low temperature optima (<30°C); low specific growth rates; and aerobic reassimilation of ethanol.

The application of recombinant technology to improve the fermentation performance of these yeasts has been limited to the development of suitable gene transfer systems. Most efforts have been directed to a comparison of their performance characteristics in xylose fermentations. The gene transfer systems developed for *P. stipitis* are based on resistance to the antibiotic genticin (*41*) or on complementation of an *ura3* auxotroph with the native URA3 gene cloned on autonomously replicating or integrative shuttle vectors (*42*). In addition, many of the xylose assimilation genes have been cloned, including the xylose reductase and xylitol dehydrogenase genes from *P. stipitis* (*2-24*) and the xylulokinase gene from *P. tannophilus* (*43*). Unlike most xylose reductases that are linked solely to an NAD(P)H cofactor, the *P. stipitis* enzyme also has an NADH-dependent activity which is thought to allow reoxidation of the NADH generated by the NAD-dependent xylitol dehydrogenase reaction under anaerobic conditions. Attempts are now under way to modify the *P. stipitis* xylose reductase cofactor requirement by site-specific mutagenesis so that it requires only NADH (N. Ho, personal communication). This would eliminate the cofactor imbalance commonly encountered with these yeasts.

Opportunities to improve these yeasts involve metabolic engineering to increase their ethanol tolerance. It has been postulated that their low ethanol tolerance is, in part, due to the aerobic reassimilation of ethanol leading to an intracellular accumulation of toxic levels of acetaldehyde (*44*). Genetic disruption of the ethanol assimilation genes (ADHII analogues) would have the beneficial effects of not only increasing ethanol tolerance but simultaneously preventing product reassimilation. Another opportunity would be to deregulate expression of the *pdc* and *adh* genes to prevent their repression by the small amounts of oxygen necessary for xylose transport and cell viability. A third potential opportunity would be to develop a comparable system around *Candida tropicalis*, a xylose-assimilating yeast with inherently superior hydrolysate resistance and an available gene transfer system (*45*).

Simultaneous Fermentation and Isomerization of Xylose (SFIX). As noted above, the greatest single disadvantage to the use of *S. cerevisiae* in xylose fermentations is its lack of a xylose-assimilation pathway. While this is not a handicap with the xylose-assimilating yeasts, their comparatively low ethanol tolerances, oxygen requirements, volumetric productivities, temperature optima and specific growth rates, combined with their aerobic reassimilation of ethanol, limit commercial application. An alternative approach couples the use of exogenously added xylose isomerase for conversion of xylose to the more readily fermentable xylulose with the simultaneous anaerobic fermentation by ethanol tolerant yeasts, such as *S. cerevisiae*, *Schizosaccharomyces pombe* or *C. tropicalis*. Because the xylulose:xylose ratio is low at equilibrium (1:5) (*46*) and the yeast ferments the xylulose as rapidly as it is formed, the SFIX process allows for complete xylose conversion in a single-step process (*47, 48*).

Providing these yeast with xylulose instead of xylose allows one to exploit many of the benefits associated with yeast glucose fermentations. The main disadvantage with this approach is the high cost for commercial xylose isomerase. This cost burden is compounded by the relatively poor stability of the enzyme (*4*), the incompatible pH optima between isomerization (pH 7.0) and fermentation (pH 4.0) (*14,15*), and the unfavorable equilibrium constant of xylose isomerase (*17,46*). Theoretical yields from pure xylulose of >90% have been reported and rival the efficiency of glucose fermentations (*15*). However, xylulose fermentations are typically up to ten-fold slower than glucose fermentations (*15*). While theoretical ethanol yields from xylose of up to 85% have been reported, there is significant yield loss to xylitol and to CO_2 through the oxidative portion of the pentose phosphate pathway (*4,16,19,28-30*).

The use of recombinant technology to improve the economics of the SFIX process has focussed on reducing the cost of xylose isomerase by overproduction in genetically engineered bacteria or, alternatively, on eliminating its cost altogether by attempting to obtain functional expression of cloned heterologous xylose isomerases in yeast. As an example of the former approach, the gene encoding the *Escherichia coli* xylose isomerase has been cloned under the control of the lambda P_L promoter to obtain enzyme overexpression following induction of a pre-grown culture at 42°C (*49*). The overproduced enzyme, representing over 38% of the total cell protein, was found to be identical to the native enzyme by several biochemical and immunological criteria (*50*). Similar strategies to overexpress xylose isomerase in *E. coli* by cloning the gene under the control of its native (*51*), *lac* (*52*), *tac* (*53*) and *att-nutL-p-att-N* (thermal inverting) (*54*) promoters have also been reported.

While several attempts to genetically engineer xylose utilization into *S. cerevisiae* by introduction of bacterial xylose isomerase genes have failed (*20,21*), similar efforts have apparently met with some success in *S. pombe* transformed with the *E. coli* xylose isomerase gene. Transformants demonstrated both low level xylose isomerase activity and the ability to grow very slowly on xylose as the sole carbon source (*55,56*). The xylose isomerase activity detected in transformants grown on xylose (6.2 nmol/h/mg) was very low compared to levels in induced wild-type *E. coli* (107 nmol/m/mg) (*51*), and only about 4-fold higher than that detected in the untransformed host grown on a mixture of xylose and xylulose. Subsequent studies have demonstrated enzymatic activity in non-denaturing PAGE zymograms, but the migration patterns from recombinant *S. pombe* were distinct from the native *E. coli* xylose isomerase (*57*). Xylose-inducible expression was apparently directed from either the native *E. coli xylA* promoter or from some fortuitous promoter on the shuttle vector. Nevertheless, the transformants demonstrated ethanol yields from xylose that were 80% of theoretical. Ethanol productivity was very low (0.15-0.19 g/L/h) and xylitol formation remained a significant problem. The detection of both xylose reductase and xylitol dehydrogenase activities in these transformants (*57*), coupled with their inability to ferment xylose under anaerobic conditions (*58*), may suggest the presence of a typical, albeit disfunctional, yeast xylose assimilation pathway in *S. pombe*.

Despite its novel approach to reduce or eliminate many of the disadvantages associated with other yeast fermentations, the main limitation with the SFIX process is the high cost and poor performance characteristics of xylose isomerase. Absolute theoretical yields will be required to obtain the credit necessary to offset the cost of xylose isomerase. Potential opportunities to metabolically engineer SFIX yeasts involve eliminating the yield loss to xylitol by genetic disruption of the xylose reductase and xylitol dehydrogenase genes (*48*). Approaches to improve the metabolic efficiency of fermentation include amplification and deregulation of the key rate-limiting enzymes and elimination of CO_2 loss through the oxidative portion of the pentose phosphate pathway. As with *S. cerevisiae*, introduction of a facilitated xylose transport system will be a key element of an efficient recombinant yeast system. Reducing the cost of xylose isomerase by cloning and overproduction, or eliminating its cost altogether by successfully expressing the gene in an SFIX yeast, is an absolute prerequisite for a commercial process.

Recombinant *Escherichia coli*. Historically, the xylose-fermenting bacteria were viewed as being capable of high fermentation rates, but only at the expense of low ethanol yields due to the formation of numerous coproducts (i.e. acetate, succinate, lactate, formate, etc.). The construction of recombinant *E. coli* strains capable of directly fermenting xylose to ethanol represents the most significant recent advance in xylose fermentation research. By introducing the *pdc* and *adh* genes from *Zymomonas mobilis* cloned under the control of an *E. coli lac* promoter (PET operon), near-theoretical ethanol yields could be achieved with a microorganism not formerly known to produce ethanol (*59*). These results demonstrate the feasibility of improving bacterial ethanol production by introducing the highly efficient *Z. mobilis* PET operon into a host capable of utilizing a broad range of substrates. They also illustrate the use of metabolic engineering to redirect intracellular carbon flow by introducing enzymes with kinetic properties that favor ethanol formation by out-competing other enzymes for the key metabolic intermediate pyruvate.

The key advantages to the use of these recombinant strains in xylose fermentation are their high ethanol yields and volumetric productivities, their broad substrate utilization range, and, by virtue of being a facultative anaerobe, their lack of an oxygen requirement to maintain fermentative capacity. *E. coli* ATCC 11303, containing plasmid pLOI297, has been shown to achieve 96% of theoretical ethanol yield from xylose with a maximum productivity of up to 0.7 g ethanol/L/h during anaerobic batch fermentations in a rich medium maintained at near-neutral pH (*60*). Generally, these fermentations last about 2 days and yield maximum final ethanol concentrations of 4.8% (w/v). Higher-than-theoretical ethanol yields have been reported and are believed to result from ethanol formation from medium buffers and components other than xylose (*61*). While the original strains required antibiotic selection for plasmid maintenance, strains with improved stability have since been developed by integration of the PET operon into the *E. coli* chromosome (*62*). The substrate utilization range for these strains has been reported to encompass glucose, lactose, mannose, galactose, arabinose and xylose (*61*). In addition, recombinant *Klebsiella oxytoca* strains have been developed along similar lines which are capable of fermenting cellobiose, xylobiose and xylotriose to ethanol (*63*).

Despite these advances, a number of significant disadvantages remain in the use of these strains for commercial xylose fermentations. Coproduct formation is still a problem despite expression of the kinetically superior *Z. mobilis* PDC and ADH enzymes to levels representing up to 17% of the total cellular protein (*64*). These acidic coproducts not only limit the conversion yield but adversely affect cell viability. Second-generation strains with reduced coproduct formation have since been developed, but exhibit lower productivity than the plasmid-bearing strain (*62*). Another disadvantage with an *E. coli*-based process is that its pH optimum requires the fermentations to be conducted at near-neutral pH, resulting in increased costs for base addition and the potential for contamination during prolonged large scale cultivation. Furthermore, a nutrient-rich and expensive growth medium is presently required to achieve the maximum reported yields and productivities. The low ethanol tolerance of *E. coli* and its sensitivity to the inhibitory compounds contained in real hydrolysates are perhaps the greatest fundamental limitations to commercial application in the conversion of lignocellulosic hydrolysates. *E. coli* is especially sensitive at low pH and under anaerobic conditions to the acetic acid generated during dilute-acid pretreatment of lignocellulosic biomass. In addition to its being an enteric microorganism lacking GRAS status, efforts to commercialize processes based on the use of *E. coli* could be hindered further by its recent implication in contaminated foodstuffs. Like any recombinant strain, scale-up will require specialized containment equipment.

Opportunities to improve these strains involve the use of metabolic engineering to eliminate the acidic coproducts and, consequently, maximize fermentation yield, growth rate and cell density. However, some coproduct formation may be necessary to maintain cofactor balance during fermentation.

Zymomonas mobilis. *Z. mobilis* is a Gram-negative, facultative anaerobe that has been utilized for centuries in the tropical areas of the world as a natural fermentative agent for preparation of alcoholic beverages, such as pulque and palm wines produced from plant saps (*65*). Because of its potential value in industrial ethanol production, much attention has been paid to the genetics and biochemical engineering of this fermentative bacterium in the past 15 years (*65-74*). In comparison to ethanol production by *Saccharomyces carlsbergensis*, *Z. mobilis* has demonstrated several advantages, including two- to three-fold higher specific glucose uptake rates and productivities with ethanol yields of up to 97% of theoretical (*66*). The high specific productivity is the result of reduced yield loss to biomass formation during fermentation. Whereas yeasts produce 2 mol ATP/mol glucose through fermentation via the Embden-Meyerhoff-Parnas pathway (Figure 1), *Zymomonas* ferments glucose anaerobically by the Entner-Douderoff pathway (Figure 3) and produces only 1 mol ATP/mol glucose (*75-76*). In addition, the existence of kinetically superior PDC and ADH enzymes results in high ethanol fermentation selectivity.

Because *Zymomonas* is acid tolerant and grows over a pH range from 3.5-7.5, industrial fermentations based on its use, like yeast fermentations, are resistant to bacterial contamination during prolonged large scale cultivation. *Zymomonas* is naturally resistant to many of the antibiotics (*65*) used for treatment of contaminated batch fermentations. *Zymomonas* also tolerates many of the inhibitors present in industrial feedstocks and demonstrates comparable performance to *S. cerevisiae* in fermentations of steam-pretreated salix (EH) and spent sulfite liquor (SSL) hydrolysates (*27*). While EH contains high concentrations of both glucose and acetic acid and low concentrations of microbial inhibitors, SSL contains low glucose and acetic acid concentrations but high levels of microbial inhibitors. *Zymomonas* naturally tolerates 1% NaCl (*65*), and industrially-useful mutants with improved salt tolerance (*77*), flocculence (*78*), ethanol tolerance (*79, 80*), and thermotolerance (*81*) have also been developed.

A key advantage to the use of *Zymomonas* is its ability to grow at high sugar concentrations (>25% glucose) and to produce and tolerate ethanol at concentrations up to 13% (w/v) (*66*). Unlike *S. cerevisiae*, *Zymomonas* does not require small amounts of oxygen for lipid synthesis (*77*). Although it can grow in the presence of oxygen, aerobic growth does not result in higher cell yields or growth rates compared to anaerobic conditions (*73*). In fact, aeration leads to a diversion of reducing power and a consequent decrease in ethanol production.

Several independent comparative performance trials have suggested that *Zymomonas* may become an important industrial ethanol-producing microorganism because of its 5-10% higher yield, up to 5-fold higher productivity and its lower biomass formation compared to traditional yeast fermentations. However, as is the case with yeast, success at the industrial scale appears to be strain dependent, with some strains being more susceptible to glucose or ethanol inhibition or to lactic acid bacterial contamination (*82*). Industrial practice with *Zymomonas* fermentations has been limited, although the Glucotech process developed at the University of Queensland, Australia, has been demonstrated on dry-milled milo at industrial scales of up to 586,000 L, apparently without contamination problems and with higher yields and productivities than comparable yeast fermentations (*83*). Distillers grain from this fermentation has received FDA approval for use as an animal feed. The *Zymomonas*-

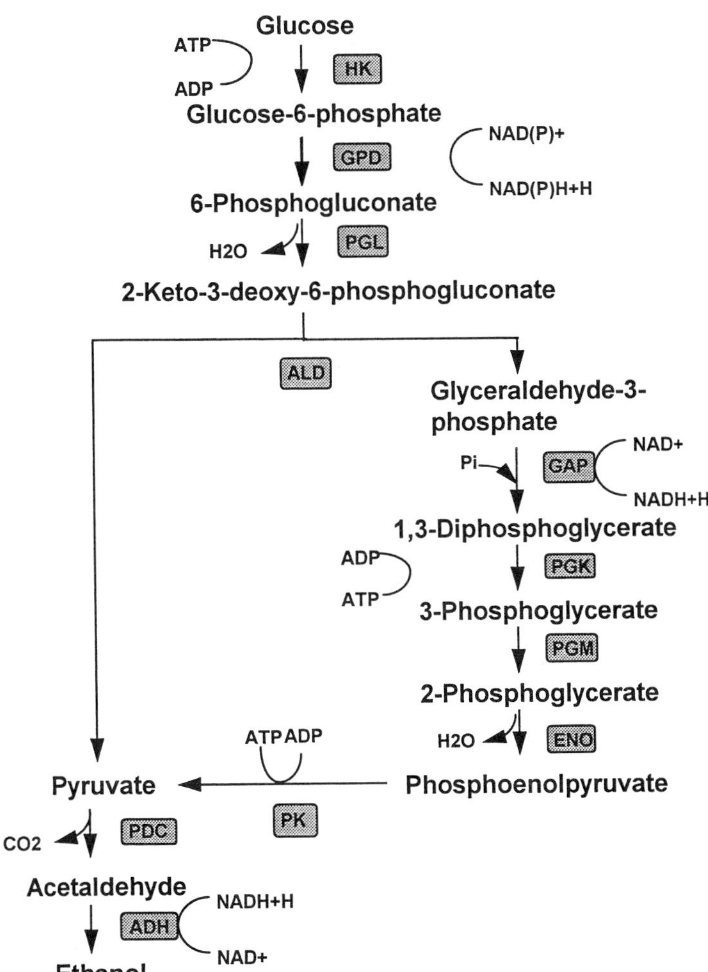

Figure 3. Entner-Douderoff Pathway. HK, Hexokinase; GPD, Glucose-6-phosphate dehydrogenase; 6-Phosphogluconolactonase; ALD, 2-Keto-3-deoxy-6-phosphogluconate aldolase; GAP, Glyceraldehyde-3-phosphate dehydrogenase; PGK, Phosphoglycerate kinase; PGM, Phosphoglycerate mutase; ENO, Enolase; PK, Pyruvate kinase; PDC, Pyruvate decarboxylase; ADH, Alcohol dehydrogenase.

based Bio-Hol process developed at the University of Toronto has also been scaled-up to 3000 L with cell recycle, demonstrating 97% theoretical yield and producing 12% ethanol at 14 g/L/h with a residence time of less than 7 h (*84*). With feedstock accounting for about 70% of all production costs, the 5-10% improvement in conversion yield afforded by *Zymomonas* resulted in an extra 40 L of ethanol per ton of corn. *Z. mobilis* appears to offer a number of advantages for industrial ethanol production, including high ethanol yield and tolerance, high specific productivity, low pH optimum, considerable tolerance to inhibitors in feedstocks and GRAS status. In addition, *Z. mobilis* has been used as a host for heterologous cellulase gene expression (*85*) and consequently could be compatible in an SSF process.

Despite these apparent advantages, *Z. mobilis* has a narrow substrate utilization range, which is limited to only glucose, fructose and sucrose. Thus, opportunities exist to genetically engineer this organism for the fermentation of other hexose and pentose sugars. Gene transfer systems based on conjugation, transformation and electroporation using native or broad-host range plasmids already have been developed (*86-90*). Expression vectors have been constructed to maximally express heterologous genes (*91-93*). Considerable research has been directed towards the development of lactose-fermenting strains which can be used to produce ethanol from whey. Both the lactose operon (*91,94*) and the *lac* transposon, Tn951 (*95,96*), have been introduced into *Z. mobilis* and shown to express beta-galactosidase activity. Although the recombinant strains were unable to grow on lactose, they were able to ferment it to ethanol. The primary reasons for the lack of growth on lactose appear to be insufficient expression of lactose permease and product inhibition by galactose. Introduction of a *gal* operon appears essential to permit effective lactose utilization. Attempts to introduce a xylose catabolic pathway from either *Xanthomonas* or *Klebsiella* into *Zymomonas* have only met with limited success (*97,98*). Although the genes were functionally expressed in *Z. mobilis,* none of the recombinant strains were capable of growth on xylose as the sole carbon source. Recent studies have shown that *Zymomonas* also lacks one of the key enzymes in the pentose-phosphate pathway (transaldolase) (*98*), and linkage between the pentose-phosphate and Entner-Douderoff pathways will be essential for effective pentose utilization. Thus, introduction of both xylose operon and key pentose pathway genes will be necessary to enable *Zymomonas* to effectively ferment xylose.

Lactobacillus. *Lactobacillus* is used commercially in the preparation of a variety of food and feed products (*99*) and provides several potential advantages in biomass fermentations. These include high ethanol tolerance, resistance to the inhibitors present in lignocellulosic hydrolysates, fermentation at low pH, thermotolerance and GRAS status. These gram-positive, non-spore-forming bacteria ferment many of the carbohydrates found in biomass, such as glucose, starch, cellobiose, lactose, xylose, arabinose and ribose, and produce high concentrations of lactate (*100-103*). One of their most desirable properties is the ability of some strains to tolerate ethanol concentrations as high as 20% (*103-105*)! Anaerobic fermentations are typically conducted under conditions which are resistant to contamination during prolonged large scale cultivation (pH 5.4-6.2 and temperatures of 30-45°C). Since the lactobacilli are facultative aerobes, the strict exclusion of oxygen from the fermentor is unnecessary. In addition, lactobacilli show considerable resistance to the inhibitory agents found in lignocellulosic hydrolysates (*27*).

Some lactobacilli ferment glucose to lactate as the sole end product with yields of over 95% (*102-103*). Catabolism occurs via glycolysis, converting 1 mol of hexose to 2 mol of lactic acid. These obligate homofermentative strains are typically thermophilic and resistant to glucose catabolite repression. A xylose-fermenting homofermentative strain would be an ideal host for ethanol production. Unfortunately, these strains are incapable of both pentose fermentation and ethanol production. While it is not yet clear if these strains lack the genes necessary for xylose utilization, their inability to ferment pentose sugars is thought to result from the absence of a phosphoketolase pathway, which metabolizes xylulose-5-phosphate to acetyl-phosphate and the glyceraldehyde-3-phosphate precursor to ethanol. However, since glucose fermentation occurs via the key intermediate pyruvate, successful introduction and expression of the xylose and PET operon genes could result in high yields of ethanol from lignocellulosic hydrolysates. A survey of 31 *Lactobacillus* strains identified several potential hosts for the PET operon that were at once capable of growth at ethanol concentrations up to 16% (v/v) and at least 90% conversion of glucose, cellobiose, lactose or starch to lactic acid, ethanol and acetic acid (*103*). Unfortunately, no strains were found that were able to convert xylose at similar efficiencies. Thus far the xylose operon from *Lactobacillus pentosus* has been cloned and expressed in other heterofermentative lactobacilli (*106*). Since a gene transfer system already has been developed (*107*), similar experiments can now be conducted with obligate homofermentative lactobacilli. Cloning and amplification of genes encoding key pentose-phosphate pathway enzymes may also be necessary to optimize ethanol production in homofermentative strains.

In contrast, pentose sugars are readily fermented by facultative heterofermentative strains to equimolar amounts of lactate and acetate via the phosphoketolase pathway. While glucose and xylose are fermented at similar rates, sugar consumption rates are relatively low compared to *S. cerevisiae* and *Zymomonas* (*66,106*). Another approach to the development of ethanologenic lactobacilli involves the introduction of a PET operon into these strains. Since the K_m of most lactate dehydrogenases (LDH) for pyruvate (0.37 to 10 mM) (*108*) are higher than that of PDC (0.4 mM), the latter has the potential to out-compete LDH for the key intermediate pyruvate. However, targeted inactivation of the genes that encode enzymes which compete for key intermediates may also be necessary to eliminate coproduct formation.

Clostridium. The clostridia have been of commercial interest for the fermentative production of chemicals and fuels for many years. There has been renewed interest in their use for ethanol production from renewable biomass substrates because of the ability of some species to grow at high temperature and to ferment a variety of low-cost substrates to solvents, such as acetone, butanol and ethanol. Despite this interest, however, there remained a number of significant drawbacks associated with the commercial use of these obligate anaerobes dating from the 1950's.

Acetone and butanol production by *Clostridium acetobutylicum* in particular has received considerable attention. Although certain strains have been reported to produce industrial solvents at concentrations of up to 10-12%, ethanol tolerance is poor and final ethanol concentrations are too low (<4%) for economic recovery (*109*). While saccharides are converted via glycolysis to the common intermediate pyruvate further catabolism occurs through an interconnected series of electrochemical transformations leading to the formation of various coproducts, such as H_2, acetate, butyrate, and isopropanol. As a result, product yields are typically quite low. The high pH optimum of these gram-positive bacteria makes prolonged large scale cultivation susceptible to contamination. Low growth rates, cell densities and productivities also limit commercial fermentation with these organisms. Special equipment would be required to maintain anoxic conditions and to compensate for the high gas pressures generated during the fermentation.

The presence of restriction enzymes in these bacteria has slowed the use of genetic engineering to eliminate many of these undesirable properties. Although a breakthrough has recently been made to circumvent this problem in one species (*110*), serious reservations remain concerning the use of any spore-forming recombinants in industrial fermentation processes due to their ability to survive many of the treatments designed for their containment. Despite such limitations, these organisms contain several enzymes with properties suited for the metabolic engineering of other ethanologenic hosts, and thus provide a source of novel genes.

Future Prospects

A recent economic analysis of xylose fermentation has identified high ethanol yields and product concentrations as the most important factors influencing production costs, with volumetric productivity being of secondary importance (*111*). Accordingly, the microbial characteristics that appear to be indispensable or, at the very least, desirable in a commercial biomass-to-ethanol process are presented in Table I. Despite the development of recombinant strains with improved xylose fermentation performance, many important issues remain unresolved concerning their use in industrial processes. For example, high ethanol yields from lignocellulosic hydrolysates, increased ethanol product concentrations and ethanol tolerances are key targets that have yet to be achieved. Aspects such as these drive the need to develop novel recombinant microorganisms capable of rapid, high yield fermentation of xylose and other sugars to ethanol.

A review by Stokes et al. (1983) on recombinant genetic approaches for efficient ethanol production identified several economic bottlenecks in biomass conversion that potentially could be relieved through genetic improvement of microbial strains. Ten years later, while some of these bottlenecks have been eliminated through recombinant technology, many others remain. The types of genetic modifications that can have the greatest impact on the economic feasibility of these fermentations include: amplification and deregulation of rate-limiting enzymatic reactions in fermentative pathways; metabolic engineering that redirects the normal carbon flow to ethanol as the sole fermentation product, improves glycolytic efficiency and reduces futile cycling; introduction of genes encoding pathways that broaden the substrate utilization range of ethanologenic hosts; and genetic manipulations that improve ethanol tolerance, osmotolerance, thermotolerance, and resistance to the inhibitory compounds normally present in lignocellulosic hydrolysates. These latter traits may best be achieved (at least in the near term) by selection for superior mutants or adaptation of production strains to increasingly hostile fermentation conditions. Generally, the strategy that has so far proven to be most fruitful involves the introduction of the

Table I

Microbial Performance Characteristics in a Biomass-to-Ethanol Process

Essential Traits	Desirable Traits
• High conversion yield	• High sugar consumption rate
• High ethanol tolerance	• High specific productivity
• Resistance to hydrolysates	• High specific growth rate
• No oxygen requirement	• High volumetric productivity
• Low fermentation pH	• Hexose/pentose co-fermentation
• High fermentation selectivity	• Minimal nutrient requirements
• Broad substrate utilization range	• High salt tolerance (acetate)
	• Capable of Crabtree effect
	• Entner-Douderoff pathway
	• Facilitated sugar transport
	• GRAS status
	• Non-sporeforming
	• Non-conjugative
	• Amenable to scale-up
	• Availability of "industrial"strains
	• Compatibility with SSF
	• Thermotolerance
	• High shear tolerance
	• Cellulase producer
	• Availability of a gene transfer system

"Ethanol Production Operon" (PET operon) into microbial hosts with an inherently broad substrate utilization range. The complementary approach of introducing sugar assimilation pathways into ethanologenic hosts has also received considerable attention, but has not yet achieved a similar measure of success.

Literature Cited

1. Brown, S.W.; Oliver, S.G.; Harrison, D.E.F.; Righelato, R.C. *Eur. J. Appl. Microbiol. Biotechnol.* **1981**, *11*, 151-155.
2. Wills, C.; Jornvall, H. *Eur. J. Biochem.* **1979**, *99*, 323-331.
3. Wills, C.; Kratofil, P.; Londo, M.; Martin, T. *Arch. Microbiol.* **1981**, *210*, 777-785.
4. Linden, T.; Hahn-Hagerdal, B. *Enz. Microb. Technol.* **1989**, *11*, 583-589.
5. Gauss, W.F.; Suzzuki, S.; Takagi, M. **1976**, U.S. Patent 3,990,994.
6. Takagi, M.; Abe, S.; Suzuki, S.; Emert, G.H.; Yata, N. In *Proceedings Bioconversion Symposium*; Indian Institute of Technology: Delhi, India, 1977; pp 551-557.
7. Blotkamp, P.J.; Takagi, M.; Pemberton, M.S.; Emert, G.H. *Amer. Inst. Chem. Eng.* **1978**, 85-90.
8. Gonde, P.; Glondin, B.; Leclerc, M.; Ratomahenina, R.; Arnaud, R.; Galzy, P. *Appl. Environ. Microbiol.* **1984**, *46*, 430-436.
9. Lastik, S.M.; Spindler, D.D.; Grohmann, K. In *Wood and Agricultural Residues*; Soltes, E.J., Ed.; Academic Press: New York, NY, 1983; pp 239-250.
10. Lastik, S.M.; Spindler, D.D.; Terrel, S.; Grohmann, K. *Biotechnol.* **1984**, *84,* 277-289.
11. Spindler, D.D.; Wyman, C.E.; Mohagheghi, A.; Grohmann, K. *Biotechnol. Bioeng.* **1987**, *18*, 279-284.
12. Szczodrak, J.; Targonski, Z. *Biotechnol. Bioeng.* **1988**, *31*, 300-303.
13. van Zyl, C.; Prior, B.A.; Kilian, S.G.; Kock, J.L.F. *J. Gen. Microbiol.* **1989**, *135*, 2791-2798.
14. Chiang, L-C.; Gong, C-S.; Chen, L.F.; Tsao, G.T. *Appl. Environ. Microbiol.* **1981**, *42*, 284-289.
15. Gong, C-S.; Chen, L.F.; Chiang, L-C.; Flickenger, M.C.; Tsao, G.T. *Appl. Environ. Microbiol.* **1981**, *41*, 430-436.
16. Senac, T. ; Hahn-Hagerdal, B. *Appl. Environ. Microbiol.* **1990**, *56*, 120-126.
17. Wang, P.Y.; Johnson, B.F.; Schneider, H. *Biotechnol. Lett.* **1980**, *2*, 273-278.
18. Ciriacy, M.; Porep, H.J. In *Biomolecular Engineering in the European Community;* Magnien, E., Ed.; Martinus Nijhoff Publishers: Dordrecht, The Netherlands, 1986; pp 677-679.
19. Senac, T.; Hahn-Hagerdal, B. *Appl. Environ. Microbiol.* **1991**, *57*, 1701-1706.
20. Sarthy, A.V.; McConaughy, B.L.; Lobo, Z; Sundstrom, J.A.; Furlong, C.E.; Hall, B.D. *Appl. Environ. Microbiol.* **1987**, *53*, 1996-2000.
21. Amore, R.; Wilhelm, M.; Hollenberg, C.P. *Appl. Microbiol. Biotechnol.* **1989**, *30*, 351-357.
22. Takuma, S.; Nakashima, N.; Tantirungkij, M.; Kinoshita, S.; Okada, H.; Seki, T.; Yoshida, T. *Appl. Biochem. Biotechnol.* **1991**, *28/29*, 327-340.

23. Hallborn, J.M.; Walfridsson, U.; Airaksinen, H.; Ojama, H.; Hahn-Hagerdal, B.; Penttila, M.; Keranen, S. *Bio/Technology.* **1991**, *9*, 1090-1095.
24. Kotter, P.; Amore, R.; Hollenberg, C.P.; Ciriacy, M. *Curr. Genet.* **1990**, *18*, 493-500.
25. Kotter, P.; Ciriacy, M. *Appl. Microbiol. Biotechnol.* **1993**, *38*, 776-783.
26. Deng, X.X.; Ho, N. *Appl. Biochem. Biotechnol.* **1990**, *24/25*, 193-199.
27. Olsson, L.; Linden, T.; Hahn-Hagerdal, B. *Appl. Biochem. Biotechnol.* **1992**, *34/35*, 359-368.
28. Lohmeier-Vogel, E.; Hahn-Hagerdal, B. *Appl. Microbiol. Biotechnol.* **1985**, *21*, 167-172.
29. Hahn-Hagerdal, B.; Berner, S.; Skoog, K. *Appl. Microbiol. Biotechnol.* **1986**, *24*, 167-172.
30. Lohmeier-Vogel, E.; Skoog, K.; Vogel, H.; Hahn-Hagerdal, B. *Appl. Environ. Microbiol.* **1989**, *55*, 1974-1980.
31. Barnett, J.A. *Adv. Carbohydr. Chem. Biochem.* **1976**, *32*, 125-234.
32. Jeffries, T.W. In *Yeast Biotechnology and Biocatalysis*; Verachtert, H. and De Mot, R., Eds.; Marcel Deckker: New York, NY, 1990; pp 349-394.
33. Gong, T.S.; Clatpool, T.A.; McCreacken, L.D.; Mauw, C.M.; Ueng, P.P.; Tsao, G.T. *Biotechnol. Bioeng.* **1983**, *15*, 85-102.
34. Bruinenberg, P.M.; van Dijken, J.P.; Scheffers, W.A. *J. Gen. Microbiol.* **1983**, *129*, 965-971.
35. McMillan, J.D. *Xylose Fermentation to Ethanol: A Review*; TP-421-4944; National Renewable Energy Laboratory: Golden, CO, 1993.
36. Skoog, K.; Hahn-Hagerdal, B. *Appl. Environ. Microbiol.* **1990**, *56*, 3389-3394.
37. Bruinenberg, P.M.; de Bot, P.H.M.; van Dijken, J.P.; Scheffers, W.A. *Appl. Microbiol. Biotechnol.* **1984**, *19*, 256-260.
38. Dellweg, H.; Debus, D.; Methner, H.; Schulze, D.; Saschewag, I. In *Symposium Technische Mikrobiologie, Berlin*; Dellweg, H., Ed.; Inst. Garungsgewerbe u Biotechnol.: Berlin, GDR, 1982, Vol. 5; pp 200-207.
39. Jeffries, T.W. *Biotechnol. Lett.*, **1981**, *3*, 213-218.
40. Alexander, M.A.; Chapman, T.W.; Jeffries, T.W. *Appl. Microbiol. Biotechnol.* **1988**, *28*, 478-486.
41. Ho, N.W.Y.; Petros, D.; Deng, X.X. *Appl. Biochem. Biotechnol.* **1991**, *28/29*, 369-375.
42. Jeffries, T.W. *Proceedings of the Eighth International Symposium on Yeasts*; Atlanta, GA, August, 23-28, 1992.
43. Stevis, P.E.; Huang, J.J.; Ho, N. *Appl. Environ. Microbiol.*, **1987**, *53*, 2975-2977.
44. Skoog, K.; Hahn-Hagerdal,B.; Degn, H.; Jacobsen, J.P.; Jacobsen, H.S. *Appl. Environ. Microbiol.* **1992**, *58*, 2552-2558.
45. Haas, L.O.C.; Cregg, J.M.; Gleeson, M.A.G. *J. Bacteriol.* **1990**, *172*, 4571-4577.
46. Tewari, Y.B.; Steckler, D.K.; Goldberg, R.N. *Biophys. Chem.* **1985**, *22*, 181-185.
47. Lastick, S.M.; Mohagheghi, A.; Tucker, M.P.; Grohmann, K. *Appl. Biochem. Biotechnol.* **1990**, *24/25*, 431-439.
48. Lastick, S.M.; Tucker, M.Y.; Beyette, J.R.; Noll, G.R.; Grohmann, K. *Appl. Microbiol. Biotechnol.* **1989**, *30*, 574-579.
49. Lastick, S.M.; Tucker, M.Y.; Macedonski, V.; Grohmann, K. *Biotechnol. Lett.* **1986**, *8*, 1-6.

50. Tucker, M.Y.; Tucker, M.P.; Himmel, M.E.; Grohmann, K.; Lastick, S.M. *Biotechnol. Lett.* **1988**, *10*, 79-84.
51. Wovca, M.G.; Steuerwald, D.L.; Brooks, K.E. *Appl. Environ. Microbiol.* **1983**, *45*, 1402-1404.
52. Stevis, P.E.; Ho, N. *Enzyme Microb. Technol.* **1985**, *7*, 592-596.
53. Batt, C.A.; Novack, S.R.; O'Neill, E.O.; Ko, J.; Sinskey, A.J. *Biotechnol. Prog.* **1986**, *2*, 140-144.
54. Wong, D.W.S.; Yee, L.N.H.; Batt, C.A. *J. Ind. Microbiol.* **1989**, *4*, 1-5.
55. Ueng, P.P.; Volpp, K.J.; Tucker, J.V.; Gong, C.S.; Chen, L.F. *Biotechnol. Lett.* **1985**, *7*, 153-158.
56. Chan, E-C.; Ueng, P.P.; Chen, L.F. *Appl. Biochem. Biotechnol.* **1989**, *20/21*, 221-232.
57. Chan, E-C.; Ueng, P.P.; Chen, L.F. *Appl. Microbiol. Biotechnol.*, **1989**, *31*, 524-528.
58. Chan, E-C.; Ueng, P.P.; Elder, K.L.; Chen, L.F. *J. Ind. Microbiol.* **1989**, *4*, 409-418.
59. Ingram, L.O.; Conway, T.; Clark, D.P.; Sewell, G.W.; Preston, J.F. *Appl. Environ. Microbiol.* **1987**, *53*, 2420-2425.
60. Ohta, K.; Alterthum, F.; Ingram, L.O. *Appl. Environ. Microbiol.* **1990**, *56*, 463-465.
61. Alterthum, F.; Ingram, L.O. *Appl. Environ. Microbiol.* **1989**, *55*, 1943-1948.
62. Ohta, K.; Beall, D.S.; Mejia, J.P.; Shanmugam, K.T.; Ingram, L.O. *Appl. Environ. Microbiol.* **1991**, *57*, 893-900.
63. Wood, B.E.; Ingram, L.O. *Appl. Environ. Microbiol.* **1992**, *58*, 2103-2110.
64. Ingram, L.O.; Conway, T. *Appl. Environ. Microbiol.* **1988**, *54*, 397-404.
65. Swings, J.; DeLey, J. *Bacteriol. Rev.* **1977**, *41*, 1-46.
66. Rogers, P.L.; Lee, K.L.; Tribe, D.E. *Biotechnol. Lett.* **1979**, *1*, 165-170.
67. Rogers, P.L.; Lee, K.J.; Skotnicki, M.L.; Tribe, D.E. *Adv. Biotechnol.* **1981**, 2, 189-194.
68. Rogers, P.L.; Lee, K.L.; Skotniki, M.L.; Tribe, D.E. *Adv. Biochem. Eng.* **1982**, *23*, 37-84.
69. Rogers, P. L.; Skotnikii, M.L.; Lee, K.J.; Lee, J.H. *CRC Crit. Rev. Biotechnol.* **1984**, *1*, 273-346.
70. Montenecourt, B.S. In *Biology of Industrial Microorganisms*; Demain, A.L.; Solomon, N., Eds.; Benjamin-Cummings Publishing: Menlo Park, CA, 1985; pp 261-289.
71. Bringer-Meyer, S.; Sahm, H. *FEMS Microbiol. Rev.* **1988**, *54*, 131-142.
72. Viikari, L. *CRC Crit. Rev. Biotechnol.* **1988**, *7*, 237-261.
73. Ingram, L.O.; Eddy, C.K.; Mackenzie, K.F.; Conway, T.; Alterthum, F. *Devlop. Indust. Microbiol.* **1989**, *30*, 53-69.
74. Doelle, H.W.; Kirk, L.; Crittenden, R.; Toh, H.; Doelle, M. *Crit. Rev. Biotechnol.* **1993**, *13*, 57-98.
75. Swings, J.; DeLey, J. In *Bergey's Manual of Systematic Bacteriology*; Krieg, N.R.; Holt. J.G., Eds.; Williams & Wilkins: Baltimore, MD, 1984, Vol. 1; pp 576-580.
76. Gottschalk, G. *Bacterial Metabolism*; Second Edition; Springer-Verlag: New York, NY, 1986; pp 114-118.

77. Ingram, L.O.; Carey, V.C.; Dombek, K.M.; Holt, A.S.; Holt, W.A.; Osman, Y. A.; Wlia, S.K. *Biomass* **1984**, *6*, 131-143.
78. Lee, J.H.; Skotnichi, M.L.; Rogers, P.L. *Biotechnol. Lett.* **1982**, *4*, 615-620.
79. Skotnicki, M.L.; Lee, K.J.; Tribe, D.E.; Rogers, P.L. In *Genetic Engineering of Microorganisms for Chemicals*; Hollaender, A.; De Moss, R.D.; Kaplan, S.; Konisky, J.; Savage, D.; Wolfe, R.S., Eds.; Plenum Press: New York, NY, 1982; pp 271-290.
80. Skotnicki, M.L.; Warr, R.G.; Goodman, A.E.; Lee, K.J.; Rogers, P.L. *Biochem. Symp.* **1983**, *48*, 53-86.
81. Berthelin, B.; Zucca, J.; Mescle, J.F. *Can. J. Microbiol.* **1985**, *31*, 934-937.
82. Doelle, M.B.; Millichip, R.J.; Doelle, H.W. *Process Biochem.* **1989**, *24*, 137-140.
83. Millichip, R.J.; Doelle, H.W. *Process Biochem.* **1989**, *24*, 141-145.
84. Lawford, H.G.; Ruggiero, A. In *Proceedings 7th Canadian Bioenergy R&D Seminar*; Hogan, E.N., Ed.; National Research Council: Ottowa, Canada, 1989; pp. 401-410.
85. Misawa, N.; Okamoto, T.; Makamura, K. *J. Biotechnol.* **1988**, *7*, 167-178.
86. Skotnicki, M.L.; Tribe, D. E.; Rogers, P.L *Appl. Environ. Microbiol.* **1980**, *40*, 7-12.
87. Browne, G.M.; Skotnichi, M.L.; Goodman, A.E.; Rogers, P.L. *Plasmids* **1984**, *12*, 211-214.
88. Tonomura, K.; Okamoto, T.; Yasui, M.; Yanase, H. *Agric. Biol. Chem.* **1986**, *50*, 805-808.
89. Okamoto, T.; Nakamura, K. *Biosci. Biotechl. Biochem.* **1992**, *56*, 833.
90. Lam, C.K.; O'Mullan, P.; Eveleigh, D.E. *Appl. Microbiol. Biotechnol.* **1993**, *39*, 305-308.
91. Yanase, H.; Kurii, J.; Tonomura, K. *Agric. Biol. Chem.* **1986**, *50*, 2959-2961
92. Conway, T.; Byun, O.K.; Ingram, L.O. *Appl. Environ. Microbiol.* **1987**, *53*, 235-241.
93. Reynen, M.; Reipen, I.; Sahm, H.; Sprenger, G.A. *Mol. Gen Genet.* **1990**, *223*, 335-341.
94. Byun, M.O.-K.; Kaper, J.B.; Ingram, L.O. *J. Indust. Microbiol.* **1986**, *1*, 9-15.
95. Carey, V.C.; Walia, S.K.; Ingram, L.O. *Appl. Environ. Microbiol.* **1983**, *46*, 1163-1168.
96. Strzeleccki, A.T.; Goodman, A.E.; Rogers, P.L. *J. Biotechnol.* **1986**, *3*, 197-205.
97. Liu, C.-Q.; Goodman, A.E.; Dunn, N.W. *J. Biotechnol.* **1988**, *7*, 61-77.
98. Feldmann, S.D.; Samh, H.; Sprenger, G.A. *Appl. Microbiol. Biotechnol.* **1992**, *38*, 354-361.
99. Chassy, B.M. *Trends in Biotechnol.* **1985**, *3*, 273-275.
100. Sharpe, M.E. In *Lactic Acid Bacteria in Beverage and Food*; Carr, J.G.; Cutting, C.V.; Withing, G.C., Eds.; Academic Press: New York, NY, 1975; pp 1653-1678.
101. Kandler, O. *Antonie van Leeuwenhoek* **1983**, *49*, 209-244.
102. Kandler, O.; Weiss, N.I. In *Bergey's Manual of Systematic Bacteriology*; Sneath, P.H.A.; Mair, N.S.; Sharpe, M.E.; Holt. J.G., Eds.; Williams & Wilkins: Baltimore, MD, 1984, Vol. 2; pp 1208-1234.
103. Gold, R.S.; Meagher, M.M.; Hutkins, R.; Conway, T. *J. Indust. Microbiol.* **1992**, *10*, 45-54.

104. Kitahara, K.; Kaneko, T.; Goto, O. *J. Gen. Appl. Microbiol.* **1957**, *3*, 102-110.
105. Uchida, K.; Mogi, K. *J. Gen. Appl. Microbiol.* **1973**, *19*, 233-249.
106. Posno, M.; Heuvelmans, P.T.H.M.; van Giezen, M.J.F., Lokman, B.C.; Leer, R.J.; Pouwels, P.H. *Appl. Environ. Microbiol.* **1991**, *57*, 2764-2766.
107. Chassy, B.M. *FEMS Microbiol. Rev.* **1987**, *46*, 297-312.
108. Garvie, E.I. *Microbiol. Rev.* **1980**, *44*, 106-139.
109. Zeikus, J.G. In *Biology of Industrial Microorganisms*, Demain, A.L.; Solomon, N., Eds.; Benjamin-Cummings Publishing: Menlo Park, CA, 1985; pp 79-114.
110. Mermelstein, L.D.; Papoutsakis, E.T. *Appl. Environ. Microbiol.* **1993**, *59*, 1077-1081.
111. Hinman, N.D.; Wright, J.D.; Hoagland, W.; Wyman, C.E. *Appl. Biochem. Biotechnol.* **1989**, *20/21*, 391-410.
112. Stokes, H.W.; Picataggio, S.K.; Eveleigh, D.E. In *Advances in Solar Energy*; American Solar Energy Society: New York, NY, 1983; pp 113-132.

RECEIVED January 19, 1994

Chapter 18

Kinetic Consequences of High Ratios of Substrate to Enzyme Saccharification Systems Based on *Trichoderma* Cellulase

Michael H. Penner and Ean-Tun Liaw

Department of Food Science and Technology, Oregon State University, Corvallis, OR 97331–6602

Trichoderma derived cellulase systems are known to exhibit an apparent substrate inhibition. This phenomena has been evaluated using a microcrystalline cellulose substrate and a complete *T. reesei* enzyme preparation. Rates of saccharification were defined as total glucose equivalents solubilized per unit time. The extent of substrate inhibition, at 50°C, pH 5.0, was shown to be dependent on the ratio of total enzyme to total substrate. The properties associated with this type of inhibition are consistent with the general concept that there is a kinetic advantage when independent enzymes occupy/act at the same substrate loci.

Complete cellulase preparations from *T. reesei* and *T. viride* have been shown to exhibit an "apparent substrate inhibition". "Apparent substrate inhibition" in this case refers to any decrease in the rate of cellulose saccharification which accompanies an increase in substrate concentration. This behavior is noteworthy with respect to both basic and applied aspects of cellulose saccharification. In an applied sense, such behavior may prove to be an important factor in terms of saccharification reactor design. From a more basic perspective, it is of interest to determine the mechanistic basis for this behavior. Substrate inhibition may be viewed as a fundamental kinetic property and, hence, reflect a mechanism that is germane to the enzyme system as a whole. The inhibition of *Trichoderma* cellulase catalyzed cellulose saccharification at relatively high substrate concentrations has been observed in several laboratories. Substrate inhibition of *T. viride* derived cellulase systems have been observed with ball-milled (*1*), microcrystalline (*2*) and unspecified cellulose substrates (*3*). The enzyme system from *T. reesei* has similarly been observed to exhibit substrate inhibition in apparent initial velocity studies (*4,5*) and in time dependent assays (*6*).

Apparent substrate inhibition is not uncommon when enzymes are acting at relatively high substrate concentrations. In classical soluble enzyme/soluble substrate systems this behavior is generally attributed to the formation of dead-end or abortive

0097–6156/94/0566–0363$08.00/0

complexes in which two substrates bind per active site (*7*). However, these types of complexes may be less important for heterogeneous systems involving insoluble substrates. It has been postulated that the observed substrate inhibition of cellulose saccharification is the result of decreasing the bulk aqueous phase of reaction mixtures as substrate concentrations increase (*4*). This mechanism is most probable for reaction mixtures in which there are very large decreases in reaction mixture "free water" due to the association of reaction mixture water with the additional substrate. Although this explanation appears plausible under some conditions, it does not appear to be a likely mechanism for all the systems in which substrate inhibition has been observed. For example, an apparent substrate inhibition has been observed in reaction mixtures containing a microcrystalline cellulose substrate at concentrations in which greater than 80% of the reaction mixture aqueous phase is apparently still "free water" (*2,5*). Similarly, the observation that the optimum substrate concentration for some reaction mixtures is enzyme dependent (*2,5*) and that a purified component of the *T. reesei* enzyme system does not show substrate inhibition under similar conditions (*5*) also indicate that the phenomena is not solely a function of the reaction mixture bulk aqueous phase. The complexity of the observed substrate inhibition is further illustrated by the observation that, under apparently equivalent experimental conditions, not all cellulose substrates generate substrate-activity profiles indicative of substrate inhibition (*2*).

Methods

Enzyme and substrate. Complete cellulase was produced by *Trichoderma reesei* QM9414 in our laboratory using shake-flask cultures as described by Mandels *et al.* (*8*). The stock *Trichoderma* culture used for enzyme production was graciously provided by M. Mandels (U.S. Army Natick Research and Development Command, Natick, MA). Powdered cellulose, Solka-Floc SW40 (James River Inc., Berlin, NH) was used as the primary energy source for cellulase induction. Enzyme was separated from mycelia by filtration after 7 days of incubation. The pH of the enzyme solution was adjusted to 4.8 and the solution concentrated approximately 20-fold using a Millipore PM7178 membrane. The enzyme was then precipitated by the addition of 2 volumes of acetone at 4°C, separated by centrifugation, washed twice with cold acetone, and dried under vacuum. The resulting powder constituted the complete cellulase preparation, which had a specific activity of 1.27 FPU per mg solids. The microcrystalline cellulose starting material, Avicel PH105, was commercially obtained from FMC Corp. (Philadelphia, PA).

Enzymatic saccharification assays. The standard enzymatic saccharification assay was performed in 50 mM sodium acetate buffer, pH 5.0, at 50°C. Cellulose substrate was added to the reaction flask (25 ml, Erlenmeyer) containing buffer and equilibrated to 50°C. The reaction was then initiated by the addition of buffered enzyme solution. The final volume for each reaction mixture was 10 ml. Reaction mixtures were agitated at 160 rpm in a constant-temperature, orbital shaking water bath. Saccharification reactions were terminated at 2 hours by filtering the reaction mixture through a 0.22 μm pore size membrane filter followed by immediate assay for total glucose equivalents using a coupled glucose oxidase/peroxidase assay following treatment of the filtrate with *A. niger* cellobiase.

Results and Discussion

Substrate-activity profiles for *T. reesei* cellulases. Representative substrate-activity profiles which reflect the apparent substrate inhibition of the complete *T. reesei* cellulase system acting on a microcrystalline cellulose substrate are provided in Figure 1. Each curve of Figure 1 represents a series of reaction mixtures containing equivalent amounts of enzyme, but differing with respect to substrate concentration. The data clearly illustrate that this enzyme/substrate system does not obey classical saturation kinetics. Instead, the rate of saccharification increases to a maximum with increasing substrate concentrations, after which further increases in substrate concentration result in corresponding decreases in rates of saccharification. In the context of this paper, the rate of saccharification refers to the total amount of saccharide solubilized per unit time. The substrate concentration corresponding to the fastest rate of saccharification may, for many practical purposes, be considered the "optimum" substrate concentration. The optimum substrate concentration for a given reaction mixture is likely to be dependent of the enzyme source, substrate, and reaction conditions. Optimum substrate concentrations for the conditions employed in this study ranged from 0.25 to 1.0%. The curves of Figure 1 indicate that the reaction rate decreases asymptotically with increasing substrate concentrations above the optimum, the extent of substrate inhibition reaching an apparent maximum of approximately 75%. The "extent of substrate inhibition" referred to in this work was calculated as follows:

$$ESI_i = [(MORS - RS_i) \div MORS]\ 100$$

where: ESI_i = extent of substrate inhibition at substrate concentration "i"
MORS = maximum observed rate of saccharification
RS_i = rate of saccharification at substrate concentration "i"

Clearly, this working definition is only applicable for substrate concentrations greater than that which corresponds to the maximum observed rate of saccharification. Figure 1 also suggests that increasing the enzyme concentration of a given reaction mixture results in an apparent shift in the optimum substrate concentration, the optimum substrate concentration being higher for those reaction mixtures containing more enzyme, as reported previously (*2,5*).

The relationship between enzyme concentration and optimum substrate concentration is more easily visualized when the rate of saccharification is presented with respect to the ratio of total enzyme (FPU) to total substrate (g) in the reaction mixture. A plot of this type is presented in Figure 2. The Figure illustrates that the maximum rate of saccharification for a reaction mixture containing a constant amount of enzyme is dependent on the ratio of total enzyme to total substrate. This behavior is in contrast to well-behaved Michaelis-Menten systems in which substrate concentrations above that required for enzyme saturation have little or no effect on the reaction rate. The optimum enzyme to substrate ratio is likely to be dependent

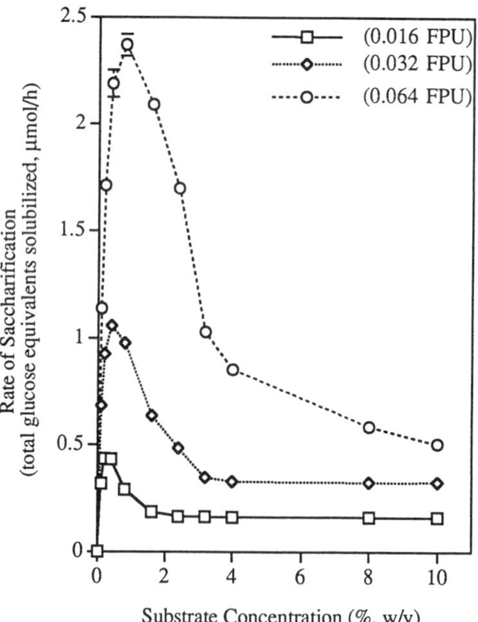

Figure 1. Substrate-activity profile for the *Trichoderma reesei* cellulase system acting on a microcrystalline cellulose substrate. Reaction conditions were 50 mM sodium acetate, pH 5.0 at 50°C with a total reaction volume of 10 ml. Total enzyme loads were 0.016, 0.032 or 0.064 FPU reaction mixture. Substrate concentrations were as indicated along the abscissa. "Rate of saccharification," as indicated along the ordinate, is measured as μmol glucose equivalents solubilized per hr. per 10 ml reaction mixture.

on the particular substrate and enzyme used in any given study. In the system presented here, the optimum enzyme to substrate ratio was approximately 0.8 FPU per gram substrate.

Enzyme concentration and the rate of saccharification. The rate of saccharification of microcrystalline cellulose is expected to be a function of the amount of enzyme adsorbed on the substrate surface. A relatively simple scenario would be that the rate of saccharification is directly proportional to the amount of bound enzyme and that the relationship between bound enzyme and the enzyme concentration of the reaction mixture is described by a Langmuir-type sorption isotherm. In this case, a plot of initial saccharification rate versus enzyme concentration for a given substrate load would yield a linear relationship between enzyme concentration and reaction rate at relatively low enzyme concentrations. However, as enzyme concentrations increase this relationship will break down, such that further increases in enzyme concentration will result in progressively smaller increases in saccharification rates. Eventually, it is expected that the substrate surface will be saturated and, hence, further increases in enzyme concentration will have relatively little effect on the rate of saccharification. Graphically, this behavior would take the form of a right rectangular hyperbola. Experimentally, this simple relationship is not observed. The experimental data reflects a sigmoidal relationship between enzyme concentration and rates of cellulose saccharification (Figure 3). The sigmoidal curve is a consequence of the same behavior which was termed "substrate inhibition" when referring to Figure 1. A data set consisting of the four lowest enzyme concentrations included in Figure 3 is plotted in Figure 4a. This plot demonstrates the linear relationship between enzyme concentration and rate of reaction at these low enzyme concentrations. The change in saccharification rate per unit change in enzyme concentration is approximately 9.5 under these conditions. Each data point in Figure 4a corresponds to a total enzyme (FPU) to total substrate (g) ratio of less than 0.05 (for reference, see Figure 2). A second data set, consisting of the next four higher enzyme concentrations from Figure 3, corresponding to total enzyme (FPU) to total substrate (g) ratios ranging from 0.1 to 1.4, are plotted in Figure 4b. The change in saccharification rate per unit change in enzyme concentration at these enzyme to substrate ratios was 34.4, approximately 3.6-fold greater than that observed for the data of Figure 4a. The higher slope for the data set of Figure 4b relative to that of Figure 4a can be rationalized with reference to Figure 2, which indicates that the extent of "substrate inhibition" progressively declines as reaction mixture enzyme to substrate ratios are increased from approximately 0.08 to approximately 0.6.

Mechanistic basis of apparent substrate inhibition. Sigmoidal curves, such as that of Figure 3, are commonly observed for biological systems demonstrating positive cooperativity. A form of positive cooperativity that exists for cellulolytic enzyme systems is synergism. Synergism between cellulolytic enzymes is widely recognized (*9,10,11*) and has been the focus of several studies with the *Trichoderma* cellulases (*12,13,14*). A mechanism which is consistent with the data presented above is one which embraces the concept of synergism between component enzymes. In its simplest form, the mechanism is based on the assumption that each enzyme component has an equal affinity for all loci on the substrate surface, and that the

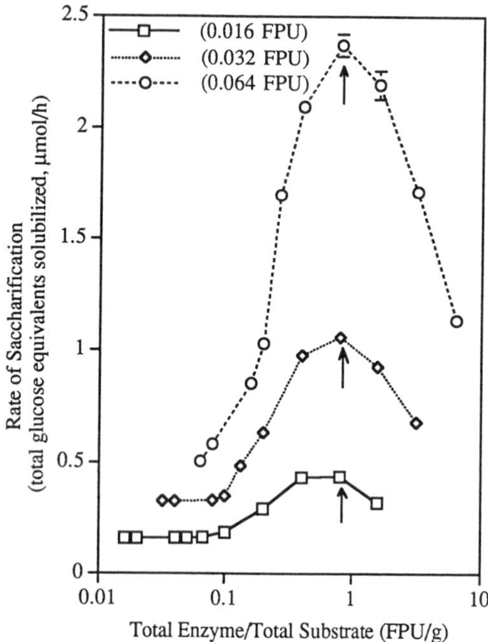

Figure 2. Effect of enzyme to substrate ratio on rate of saccharification of microcrystalline cellulose. Reaction conditions, enzyme loads and substrate concentrations were as defined in Figure 1.

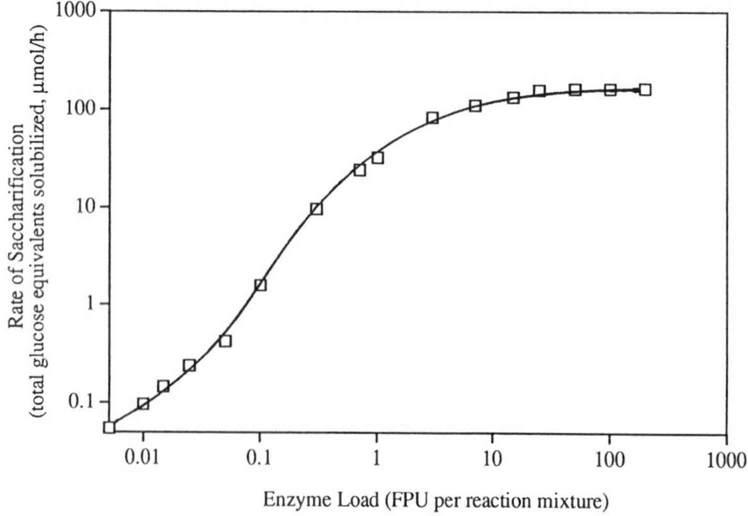

Figure 3. Effect of total enzyme load on rate of saccharification of microcrystalline cellulose. Reaction conditions as defined in Figure 1 with a substrate concentration of 5% (w/v). Enzyme loads (total enzyme, FPU, per reaction mixture) varied as indicated along the abscissa.

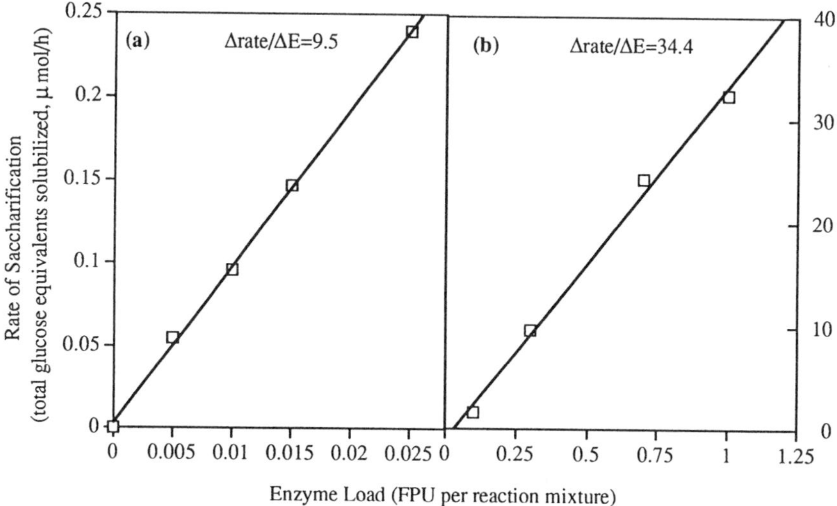

Figure 4. Correlations between enzyme load and rate of saccharification. Reaction conditions as defined in Figure 1 with a substrate concentration of 5% (w/v). Enzyme loads (total enzyme, FPU, per reaction mixture) ranged from (a) 0.005 to 0.025 and (b) 0.1 to 1.0.

maximum kinetic efficiency for a pair of potentially synergistic enzymes occurs only when the two enzymes occupy the same locus. The mechanism does not require that the pair of synergistic enzymes be at the same locus simultaneously. If there is a kinetic advantage for two enzymes to act sequentially at the same locus, then a form of substrate inhibition may also be exhibited.

A mechanism which provides a kinetic advantage for enzymes acting at a "common locus" suggests that the observed substrate inhibition is the result of competitive binding of the pertinent enzyme components between common loci, where they display a relatively higher activity due to a synergistic effect, and physically distinct loci, where they display a relatively lower activity when the enzymes are acting independently. In this mechanism, additional substrate above that required for enzyme saturation would promote the formation of physically distinct single enzyme-substrate complexes, which for this system are kinetically inferior. The mechanism suggests that the substrate inhibition characteristic of the complete enzyme system would not be observed for the purified enzyme components acting independently, which has been demonstrated for the major component of the *T. reesei* system (*5*). The mechanism also suggests an asymptotic decrease in activity with increasing substrate concentration, as depicted in the data of Figure 1. The activity observed at relatively high substrate concentrations and, thus, the maximum extent of substrate inhibition, would be dictated by the activity of each of the enzymes acting independently.

The mechanism suggested in this paper is presented in its most general form. A better understanding of the basic enzymology which governs this system, including information on possible enzyme-enzyme interactions at common loci, the relative importance of sequential versus simultaneous action of multiple enzymes at common loci and the values of applicable kinetic and thermodynamic constants for enzymes acting at various loci, will obviously be helpful in terms of developing a comprehensive mechanistic model which can account for the kinetic behavior reported here.

Literature Cited

1. Van Dyke, B.H.,Jr. Ph.D. Thesis, Massachusetts Institute of Technology, Sept. 1972.
2. Liaw, E-T. and Penner, M.H. *Appl. Environ. Microbiol.* **1990**, *56*, 2311-2318.
3. Okazaki, M. and Moo-Young, M. *Biotech. Bioeng.* **1978**, *20*, 637-663.
4. Lee, Y-H and Fan, L.T. *Biotech. Bioeng.* **1982**, *24*, 2383-2406.
5. Huang, X. and Penner, M.H. *J. Agric. Food Chem.* **1991**, *39*, 2096-2100.
6. Ryu, D.D.Y. and Lee, S.B. *Chem. Eng. Commun.* **1986**, *45*, 119-134.
7. Fromm, H.J. *Initial Rate Enzyme Kinetics*; Springer-Verlag: NY, 1975; pp. 121-160.
8. Mandels, M.; Medeiros, J.E.; Andreotti, R.; Bissett, F. H. *Biotechnol. Bioeng.* **1981**, *23*, 2009-2026.
9. Wood, T.M. and McCrae, S.E. In *Hydrolysis of Cellulose: Mechanisms of Enzymatic and Acid Catalysis*, Brown, R.D.Jr. and Jurasek, L. Eds.; Advances in Chemistry Series 181; American Chemical Society, Washington, DC, 1978; pp 181-209.

10. Wood, T.M., McCrae, S.I., Wilson, C.A., Bhat, K. M. and Gow, L.A. In *Biochemistry and Genetics of Cellulose Degradation*; Aubert, J.P., Beguin, P. and Millet, H. Eds.; FEMS Symposium No. 43; Academic Press, London, 1988, pp. 31-52
11. Irwin, D.C., Spezio, M., Walker, L.P., and Wilson, D.B. *Biotechnol. & Bioeng.*, **1993**, *42*, 1002-1013.
12. Henrissat, B., Driguez, H., Viet, C. and Pettersson, L.G. *Bio/Technology*. **1985**, *3*(8), 722-726.
13. Woodward, J., Hayes, M.K., and Lee, N.E. *Bio/Technology*. **1988**, *6*(3), 301-304.
14. Heitz, H.-J., Witte, K., and Wartenberg, A. *Acta Biotechnologica*, **1991**, *11*(3), 279-286.

RECEIVED April 29, 1994

Chapter 19

Pectin-Rich Residues Generated by Processing of Citrus Fruits, Apples, and Sugar Beets

Enzymatic Hydrolysis and Biological Conversion to Value-Added Products

K. Grohmann[1] and R. J. Bothast[2]

[1]U.S. Citrus and Subtropical Products Laboratory, 600 Avenue S, Northwest, P.O. Box 1909, Winter Haven, FL 33883–1909
[2]National Center for Agricultural Utilization Research, U.S. Department of Agriculture, 1815 North University Street, Peoria, IL 61604

Processing of citrus, apple and beet crops to juice and crystalline sugar annually generates several million tons of residues which are sold as a cattle feed or cause disposal problems. These residues are very rich in carbohydrates and are attractive potential feedstock for microbial conversions to value added liquid fuels and other products. The residues are rich in pectin and in the case of apple pomace and citrus processing residues they also contain large amounts of soluble sugars. All polysaccharides in these residues are easily hydrolysed to monomeric sugars by mixtures of cellulolytic and pectinolytic enzymes. Microbial conversions of sugar rich hydrolysates from these residues will require identification and development of microorganisms that can utilize galacturonic acid and five carbon sugars.

It has been estimated that the annual production of cellulosic biomass could supply 10 times our energy needs and 100 times our food needs on a global scale (*1*). The term cellulosic biomass is usually applied to woody and other lignified tissues of plants which are currently under utilized and undervalued. While there is no doubt that woody residues and residues from harvesting of grain represent the largest reservoir of the cellulosic biomass necessary to supply huge fuel markets (*2,3*), these tissues are heavily lignified and present serious obstacles to efficient enzymatic hydrolysis of cell wall polysaccharides (*3-7*). Lignin hemicellulose complexes shield the cellulose fibers from enzymes and prevent significant hydrolysis of polysaccharides in these walls. Very harsh mechanical or chemomechanical pretreatments of these tissues are necessary for efficient enzymatic hydrolysis to occur, and even then, the crystalline structure of the major polymer, cellulose, creates difficulties in terms of yield, enzyme loading and rates of enzymatic depolymerization (*4,6-9*).

0097–6156/94/0566–0372$08.00/0

These serious difficulties with enzymatic hydrolysis of lignified plant cell walls may create an erroneous impression that all plant tissues are very resistant to enzymatic hydrolysis and will require chemical pretreatments.

We would like to present in this short review three examples of specialized cellulosic tissues in fruits and tubers in which lignin is apparently absent, cells are cemented by pectin and the tissues are relatively easy to hydrolyze by mixtures of pectinolytic, hemicellulolytic and cellulolytic enzymes. The three examples are processing residues from the production of citrus, apple and sugar beet juices. These residues have been chosen because they are produced in relatively large amounts by mature industries in the U.S. and other countries. All three residues are available at the processing plants, which potentially decreases collection and transportation problems. Moreover, the carbohydrate composition of cell walls of these residues strongly resembles the composition of cell walls of many other fruits, vegetables and tubers used for human consumption (*10-16*), therefore the issues discussed for these selected residues may have broader applications.

Summary of Processing

Since all three crops are grown for processing to sweet juice, the key processing step is the expression of the juice.

Citrus Juice Products. In the case of citrus fruit two types of extractors are used (*17-22*) which either core a fruit or ream fruit halves and express the juice and pulp with minimal contamination by peel juice and peel oil. Peel juice is astringent and unpalatable and citrus peel oil decreases quality and storability of the juice. Other pressing equipment for whole fruit has been tried (*17,20*) and abandoned because of peel oil contamination problems. The juice is then screened to remove the coarse pulp using finishers (*17,19,21,22*) and either pasteurized by heat treatment or concentrated in multiple effect evaporators before cold storage and packaging. The heat treatment of the juice is extremely important because it not only kills spoilage microorganisms but also inactivates enzymes in the juice which destabilize the particulate cloud. One of these enzymes, pectinmethylesterase, has been implicated in the modification of the citrus juice cloud which leads to its rapid coagulation and other problems. Since cloudiness is one of the very important attributes customers expect in citrus juice products, its maintenance is of great importance to the industry.

The other half of the plant is devoted to the processing of residues (*17,18*). Residues, mainly peel, cores, segment membranes and small amounts of seed amount to approximately one half of the weight of fruit (*17,18*) and due to high sugar content are prone to rapid spoilage. These residues are hammermilled to coarse particles and blended with small amounts of calcium oxide. Calcium oxide reacts with pectin in the cell walls and increases the yield of peel juice. The limed, ground residues are partially dewatered in a screw press. The peel juice is concentrated in multiple effect evaporators using waste heat from the peel dryer. Resulting peel molasses are sometimes sold, but often are blended back with dewatered residue and dried in the peel dryer to a cattle feed. Small amounts of a valuable terpene, limonene, are recovered from the vapors of the evaporator. The pressing of the peel juice and evaporation to molasses are driven more by the economy of heat utilization and recovery of limonene than by the value of molasses (*23*).

The citrus juice production is dominated by the processing of several varieties of oranges (*24*) with a much smaller amount contributed by grapefruit. One unique feature of orange crop and orange juice production is a very long harvesting season which in Florida lasts approximately from October until May of the next year. The long harvesting season assures an equally long processing season and supply of residues.

Apple Juice Products. Apple fruit processing does not pose a peel oil contamination problem so important to the citrus processing industry. The whole fruit can be disintegrated and pressed by a variety of pressing machines to produce juice and pressed residue called pomace (*25-28*). Diffusion extractors, adapted from the sugar beet industry, are also used (*25*). The juice yield in the pressing step usually decreases with storage of apples, so the apple tissue is often treated with pectinolytic enzymes before the pressing step to help to disintegrate it. One recent trend (*25,28*) involves liquefaction of apple tissues with a mixture of pectinolytic, cellulolytic and hemicellulolytic enzymes which dramatically increases the juice yield. The authenticity of such a product contaminated with sugars from apple cell walls is of course highly questionable (*28*). Since consumers usually expect a clear apple juice product much of the processing is devoted to clarification of the juice using filtration, precipitation (fining) and pectinolytic enzyme treatments. The clarified juice is either pasteurized or evaporated before cold storage and packaging (*25*). The important enzymes that need to be inactivated or inhibited in apple juice products are polyphenol oxidases (*25,28*) which cause darkening of the juice and formation of precipitates during storage. The processing season can vary from 3-4 months to almost a year depending on local conditions. Introduction of extended storage methods for apples extended the length of supply and the processing season to a whole year (*25*). The disposal of the apple pomace depends on local conditions (*25*). It is often used as a supplemental cattle feed or soil conditioner. A few plants use apple pomace for the production of pectin.

Sugar Beet Processing. Production of juice from sugar beets uses a leaching process, rather than the pressing used with fruit juices (*29,30*). Washed beets are sliced into thin pieces called cosettes, which are then leached with hot water, usually in continuous countercurrent diffusers. The beet tissue is heated above 70°C to modify the cell walls and increase their permeability to sucrose (*29,30*). The leached juice is then clarified by liming, carbonation with carbon dioxide, filtration and treatment with sulfur dioxide. The thin, clarified juice is then converted to crystalline sugar and molasses by an additional sequence of processing steps (*29,30*) which include concentration by evaporation, deionization, additional decolorization, and crystallization or precipitation of sucrose with lime. The leached beet cosettes are blended with spent molasses, dried and sold as a cattle feed supplement in the United States (*30*), but in other areas traditional ensiling for cattle feed or spreading on soil may still be used (*31-33*). Drying is the method choice if long term storage and long distance transportation of all three residues is contemplated. The high water content of wet residues makes long distance transportation unattractive.

Availability and Composition of Residues

Citrus juice processing residues are available in the U.S. mainly in the state of Florida. The smaller California-Arizona crop is usually sold on the fresh fruit market. Approximately 600,000 to one million dry tons of citrus processing residues was produced annually in the 1980's (*34*). Additional large amounts of citrus processing residues are available in other major processing countries such as Brazil (a world leader), Mexico, Spain, China and Israel.

Apples are grown in 35 states of the U.S. (*25*) with Washington state by far the largest producer, followed by New York, Michigan, California and Pennsylvania. More than half of the annual crop is sold as fresh fruit and approximately 44% is processed to juice, canned sauce and other products. The processing of apples in the U.S. generates approximately 1.3 million tons of wet, (i.e., approximately 250-400,000 dry) pomace a year (*35*). Many other countries in the world produce and process apples (*25,28*) so additional amounts of pomace are produced in temperate zones.

The annual production of sugar beets in the United States varied between 20 and 30 million wet tons during the last two decades (*36*) from which approximately 1.6-2.5 million dry tons of residue (marc) have been produced. The northern tier of states from Ohio to the Pacific Northwest are the leading producers in the U.S. Even larger quantities of sugar beets are grown and processed in northern Europe and the European part of the former Soviet Union (*29*).

The production statistics indicate that several million tons of these residues accumulate annually on the worldwide basis. These amounts are nowhere near the large quantities of lignocellulosic residues obtainable from the production of grain crops (2) but are sizable enough to be a resource for the chemical industry.

All three processing residues are characterized by a high content of water (Table I). The two fruit processing residues are also characterized by a high content of soluble sugars (Table I). The soluble sugars in apple juice and pomace are fructose, glucose and sucrose with fructose being the dominant component (*25*). Sorbitol is also present in small amounts. Citrus processing residues also contain glucose, fructose and sucrose, but fructose and glucose are present in nearly equimolar amounts (*17*) since they are produced by hydrolysis of sucrose transported by the sap to the fruit. Immature apples contain small amounts of starch (*25*) which disappears with maturity of the fruit. No starch has been detected in citrus fruit and sugar beets. Sucrose, the dominant sugar in sugar beets, is leached during processing so only traces remain in the residue. There is a significant variation in the soluble sugar content of both apple and citrus processing residues, because soluble sugar content increases during maturation of the fruit on the tree and the cell wall material decreases at the same time (*17,25,41*). Other soluble components in these tissues are protein, minerals and organic acids, namely citric in citrus, malic in apples and oxalic in sugar beets.

Despite the crude nature of analytical methods used to obtain results in Table I, the sum of reducing sugar content and alcohol-insoluble solids indicates that all three processing residues are very rich in carbohydrates. The low crude fiber content also indicates that the processing residues have a low content of cellulose and lignin. These conclusions are supported by detailed analyses of solid residues insoluble in aqueous alcohol (usually 80% v/v) or water and aqueous alcohol (Table II). These

Table I.
Approximate Composition of Processing Residues[1]

Component	Residue Weight Percent		
	Apple Pomace[2]	Citrus Peel[2]	Sugar Beet Marc[2]
Water	66-78	70-82	90-93
Total solids	22-34	18-30	7-10[3]
Soluble sugar	6.6-12.2 (30-36)	5.4-12.0 (30-40)	0.2-0.3 (3-4)
Alcohol Insoluble Solids (AIS)	11.0-20.4 (50-60)	8.1-21.0 (45-70)	6.3-9 (90)
Crude fiber	3.3-8.8 (15-26)	2.2-3.6 (12)	1.4-2.0 (20)
Pectin	5.1-7.8 (23)	3.2-6.9 (18-23)	1.4-2.0 (20)
Crude protein	1.0-1.5 (4.4)	1.2-2.2 (6.6-7.3)	0.6-0.9 (8.7)
Fat (ether extract)	1.5-2.3 (6.8)	0.7-1.5 (4-5)	<0.1 (<1)
Ash	0.3-0.5 (1.4)	0.4-1.5 (2.2-5.0)	0.2-0.4 (3-4)

[1]The results are tabulated in weight percent of wet material. The numbers in parenthesis are tabulated in weight percent of dry total solids.
[2]The results for apple pomace; citrus peel; and sugar beet marc are adapted from references *35, 37-40*; *18,41-44*; *11,16,32,33,45,46*; respectively.
[3]Total solids content can be increased to 15-22% by pressing (*31,32*).

Table II. Composition of Cell Wall Fraction[1,2]

Component[2]	Residue Weight Percent		
	Apple Pomace	Orange Peel	Sugar Beet Marc
Cellulose	20.9	17.5-21.4	22-24
Non-cellulosic glucose	4.3[3]	N.D.[4]	1.8-2.5
Total glucose	25.2-33.3	23.7	21.6-26.5
Galactose	3.0-7.0	8.2	4.2-4.9
Mannose	1.0-3.4	-	0.3-1.5
Arabinose	5.1-14.3	14.2	16.3-20.1
Xylose	5.8-6.6	<5	1.4-1.6
Fucose	0.6-1.2	N.D.	N.D.
Rhamnose	0.3-1.5	<2	1.0-2.25
Total neutral sugars	41.0-67.3	46.1-53.1	44.8-56.8
Galacturonic acid	18.7-28.2	26.0	18.4-23.0
Total sugars	59.7-95.5	72.1-79.1	63.7-79.8
Lignin	N.D.	3.0	1.0-2.0
Protein	9-11	5.8-6.7	3.6-8.0
Ash	1.5-2.0	3.9-4.1	4.4-12.0

[1]Cell wall fraction refers to solid residue insoluble in aqueous alcohol or water and aqueous alcohol. The results for apple pomace; orange peel; and sugar beet marc are adapted from references *10,25,47-49; 44,49; 11,16,45,51;* respectively.
[2]Content of individual sugars is expressed on polymeric (anhydrous) basis.
[3]Non-cellulosic anhydroglucose corresponds to starch content in apple pomace (*25*).
[4]N.D. = not determined.

residues, usually called alcohol-insoluble solids or residues (AIS or AIR) are highly enriched in cell wall components, but are contaminated with small amounts of protein and minerals (Table II).

The cell walls and connecting middle lamella in these tissues are composed of carbohydrates. Lignin, so prevalent in many other plant tissues (e.g. wood) appears to be absent. Pectin, a polymer of galacturonic acid methyl ester is the major component of the middle lamella (*52*) but is also present in cell walls of fruits and tubers. The content or composition of the other polysaccharides are also markedly different from lignified plant tissues. The cellulose content (17-24%) is very low when compared to wood or mature grass tissues (*4*) in which it is approximately 40-50% of the dry weight. The xylan content is also very low. The major hemicelluloses contain arabinose and galactose as their monomeric units. Branched 1,5-α-L-arabinan has been isolated from sugar beet cell walls (*53*) and ß-1,4-D-galactan has been isolated from citrus pectin (*54*), but generally very little is known about composition and structure of major polysaccharides in cell walls of these tissues. This lack of knowledge will complicate the development of efficient systems for enzymatic hydrolysis as discussed below. Discussion of the structure of primary cell walls of flowering plants (*55*) may be applicable to specialized tissues discussed here, but in the absence of solid structural analyses the extrapolation (*56*) from one tissue to another may be misleading. Even the structure of pectins which has been studied for decades has not been unambiguously determined (*57-59*). Despite the lack of detailed knowledge about the structure of cell walls in fruits and tuberous roots, their susceptibility to microbial decomposition and high digestibility in the rumen of cattle have implied for a long time that these tissues are not very resistant to enzymatic hydrolysis. However, the more detailed studies of enzymatic hydrolysis summarized below were not performed until relatively recent times.

Enzymatic Hydrolysis

The first set of polymers that need to be hydrolyzed in fruit and tuber tissues are pectins. Pectins are polymers of α-1,4 linked galacturonic acid and appear to be composed of linear segments of polygalacturonic acid and "hairy" regions containing rhamnose units and side chains of neutral polysaccharides containing arabinose and galactose (*57,58,60-63*). Part of the pectic substances is soluble in water or aqueous solutions of chelating agents. The other part, termed protopectin (*64*) is crosslinked with other polymers and remains insoluble in water. Carboxylic groups in pectic substances are highly esterified with methoxyl groups and pectins from sugar beets, citrus, apple and some other fruit are partially acetylated (*59,65,66*) probably at the C-2 and/or C-3 positions of anhydrogalacturonyl units. Small amounts of ferulic acid esters have also been detected in sugar beet pectins (*11,59,67*).

The hydrolysis of pectic substances requires interaction of several enzymes (*68-70*). Enzymatic cleavage of the polygalacturonic acid backbone is unique among polysaccharides, because it can proceed by two different mechanisms, hydrolysis or ß-elimination. The enzymes that cleave the glycosidic bond of polygalacturonic acid by ß-(trans) elimination are termed polymethylgalacturonate lyases when they cleave esterified (methoxylated) pectin chains or polygalacturonate lyases when they cleave de-esterified polygalacturonic acid (*68*). The combined action of endo- and exo-polygalacturonate lyases produces short oligomers of galacturonic acid terminated on the non-reducing end by δ-4,5-D-galacturonate residues. The combined action of endo- and exo-polymethylgalacturonate lyases produces similar methoxylated oligomers. Monomeric products can be obtained by action of oligogalacturonate lyases (*68,71*). All lyases have relatively high pH optima (pH = 5-9) for activity and many have either an absolute requirement for, or are stimulated by Ca^{2+} ion. The classic hydrolysis of α-1,4-glycosidic linkages in pectin is catalyzed by a combination of pectinmethylesterase and endo-or exopolygalacturonases (*68,70,71*). The existence of polymethylgalacturonases which could hydrolyze polymethylgalacturonate is still under dispute (*68*). A small family of endopolygalacturonases that are highly active in hydrolysis of protopectin have been termed protopectinases (*72*). The pectinmethylesterases (*69*) are carboxylic acid esterases. The products of their action are de-esterified pectin containing increased amounts of carboxylic acid groups, methanol and protons from ionization of carboxylic groups. These enzymes are produced by many microorganisms and plant tissues (*69*). The presence of sufficient amounts of pectinmethylesterase appears to be very important for efficient solubilization and saccharification of pectin rich tissues (*68,70*). The pectinmethylesterase prepares the demethoxylated substrate for polygalacturonases and polygalacturonate lyases (*70*). These enzymes in turn remove end product inhibition of pectinmethylesterase by hydrolyzing demethoxylated pectin to D-galacturonic acid or very short oligogalacturonides which are no longer inhibitory (*68-70*). Very strong synergism between pectinmethylesterase and polygalacturonase or polygalacturonate lyase has been observed (*69,70*) both for pectin and apple cell walls as substrates. The enzymatic hydrolysis of apple cell walls with simple mixtures of purified pectinolytic enzymes also solubilizes considerable amounts of polymers enriched in arabinose, galactose, xylose and rhamnose (*60-63*) and disintegrates the tissue.

The next issue in structure and hydrolysis of fruit and tuber processing residues is the structure and attachment of hemicelluloses rich in arabinose, galactose and perhaps xylose in middle lamella and primary and secondary cell walls of these tissues. Side chains rich in arabinose and galactose are thought to be present in "hairy" regions of pectin, but hemicelluloses rich in these neutral sugars appear to exist as well. Isolation of arabinose from sugar beet and galactan from citrus peel has been mentioned already. The structure and composition of these polymers is relatively unclear at the present time, mainly due to difficulties with extraction and purification (*57,58*). Lack of well defined substrates also hinders isolation and characterization of appropriate hydrolytic enzymes (*73,74*). The observation pertinent to the structure and enzymatic depolymerization of cell walls in processing residues discussed here is the lack of pectin release when all three tissues are treated with purified endo-galactanase and endo-arabinanase enzymes (*61-63,75*). These results indicate that a large part of polymers containing arabinose and galactose may be covalently attached to pectin or enmeshed in it, but the same part of these polymers does not appear to crosslink the pectin to other cell wall components by covalent bonds. The selective enzymatic hydrolysis of cell walls of apple and citrus fruit which could release pectin would provide an important technical alternative to partial hydrolysis of these tissues with dilute mineral acids which is commercially used for the extraction of this valuable polysaccharide. The importance of xylanase activities in enzymatic hydrolysis of these tissues is unclear at the present time due to uncertain structure of xylose containing polymers and xylanolytic activities of some purified cellulase enzymes (*62,63,76*).

Cellulose is the second most abundant polymer in cell walls of apple, orange and sugar beet tissues (Table II) but its structure in these tissues has hardly been investigated (*77*). The enzymatic hydrolysis of cellulose fibers has been investigated rather extensively during the last 25 years and the results have been summarized in recent reviews (*7,8,76,78,79*). The depolymerization of cellulose fibers requires cooperative hydrolysis by endo- and exo-ß-1,4 glucanase enzymes. Due to insoluble and partially crystalline nature of this substrate, the reaction is confined to the surface and proceeds rather slowly. These problems are compounded by encasing of cellulose fibers in the matrix of hemicelluloses and pectin which have to be depolymerized and dissolved before cellulase enzymes can access and depolymerize cellulose. Therefore investigations pertinent to our discussion (*48,50,60-63,70,80-88*) indicate that there is a significant synergism between cellulolytic and pectinolytic enzymes during hydrolysis of cell walls in citrus, apple and sugar beet tissues. The synergism of cellulolytic and pectinolytic enzymes has been reported for all three tissues and both purified and crude enzyme preparations. The crude pectinase preparations appear to be more effective in maceration and solubilization of these tissues than commercial cellulase preparations, and at least one commercial pectinase enzyme (*50,89,90*) contains sufficient amounts of cellulolytic and other hydrolytic activities that it can hydrolyze and solubilize fruit tissues without added cellulase. The addition of ß-glucosidase (*45,81*) can decrease the strong end-product inhibition of currently available cellulases, allow efficient hydrolysis at higher concentrations of substrate solids, and therefore allow production of concentrated sugar solutions. There are many other hydrolytic activities, notably esterases, glycosidases, and debranching enzymes, which are necessary for complete hydrolysis of pectin-rich plant tissues. These enzymes are not covered in this review.

It may appear that the sheer number of different enzymes necessary for complete hydrolysis of pectin-rich plant cell walls would prevent commercial development of enzymatic mixtures for maceration and solubilization of these tissues. Fortunately the key polymers (cellulose, hemicellulose and pectin) have been available to microorganisms mainly in the form of plant tissues, so microorganisms that decompose cellulosic materials (both pectin- and lignin-rich) have to secrete a complex mixture of enzymes to derive significant energy from hydrolysis and assimilation of these tissues. The crude exocellular enzyme preparations used in industry therefore usually contain numerous additional hydrolytic activities besides the ones used for their marketing (*50,70,89-91*). The presence of other hydrolytic activities makes commercial pectinase and cellulase enzymes very valuable for applications described here, because the development of efficient hydrolytic mixtures from individual enzymes would be a complicated and expensive task.

There are two other approaches that have been or can be used to augment the activity of exogenous enzymes in hydrolysis of pectin rich tissues reviewed here. One approach involves exploitation of endogenous hydrolytic enzymes produced in fruits and tubers. The second one involves chemical pretreatments which can solubilize, modify or depolymerize polysaccharides in cell walls and turn them into more active or simpler substrates.

Endogenous Hydrolytic Enzymes. The pectinmethylesterases appear to be ubiquitous in tissues of pectin rich fruits and tubers (*69,70*) where they are probably involved in growth, enlargement and eventual senescence. Acetylesterase active on acetylated pectin is commercially produced from orange peel, as is pectinmethylesterase (*92*). Endo- and exo-polygalacturonases present in many other fruit tissues appear to be absent from orange peel. Other hydrolytic enzymes may be present as well, but either have not been assayed or escaped detection (*93*).

Apple fruit provides a greater variety of endogenous hydrolytic enzymes (*94*). Besides pectinmethylesterase, exo-polygalacturonase (*70*), ß-1,4 glucanase (*94*) and numerous glycosidases (*95*) have been detected in apple fruit. Some of these may be active in hydrolysis of cell wall polymers, because histochemical investigations (*80*) indicate that middle lamella is depolymerized and dissolved during ripening of apples. The endogenous hydrolytic enzymes will be of lesser importance in enzymatic hydrolysis of leached beet cosettes, because they are probably inactivated by hot water during the leaching step.

Interaction of endogenous and exogenous hydrolytic enzymes can provide synergistic effects in hydrolysis of pectin rich fruit and tuber tissues, provided that pH and temperature optima of these enzymes are properly matched.

Another avenue for increased hydrolysis and solubilization of pectin-rich plant cell walls by enzymes is provided by chemical pretreatments.

Chemical Pretreatments. Many polysaccharides in pectin-rich plant cell walls are susceptible to acid-catalyzed hydrolysis. The extraction of pectin from apple and citrus processing residues by hot, dilute mineral acids is the basis of commercial pectin production (*96*). Treatment with hot acid also breaks the crosslinks in pectin-rich cell walls and both depolymerizes and solubilizes hemicelluloses containing arabinose, galactose, xylose and glucose. Only cellulose fibers are quite resistant to the action of dilute mineral acids. The treatment with dilute sulfuric acid has been used for partial solubilization of orange peel (*83,87,97*) either in applications where solubilized carbohydrates have been used for single cell protein production (*97*) or as a pretreatment to obtain enhanced enzymatic saccharification of carbohydrates (*83,87, 97*). Hydrolysis of citrus peel catalyzed by carbon dioxide or indigenous organic acids at elevated temperatures and pressures has been patented (*98*). Similar treatments with dilute mineral acids have been tested for solubilization and pretreatment of apple pomace (*47,56,99,100*) and sugar beet pulp (*46,101,102*). The pretreatment with dilute sodium hydroxide has also been tested for all three residues (*45,83,85,102,103*). The treatment with sodium hydroxide de-esterifies pectin. It also extracts pectin and hemicelluloses by the base catalyzed cleavage of glycosidic bonds between galacturonic acid units and solubilizing action on hemicelluloses.

The chemical pretreatments of pectin rich tissues will have to be carefully evaluated in the future because with newer enzyme preparations they are not necessary, the results are highly influenced by the composition of enzyme preparations and chemical pretreatments add to the complexity and cost of the saccharification. It is probable, however, that chemical pretreatments can increase rates of hydrolysis and decrease consumption of hydrolytic enzymes which are both important economic considerations.

On the technical side, the enzymatic hydrolysis of pectin rich processing residues appears to be much easier than the hydrolysis of lignified cellulosic substrates. Pectin and hemicelluloses provide some barrier to cellulolytic enzymes, but these polymers are easily hydrolysed and solubilized by pectinolytic and hemicellulolytic enzymes. Severe chemomechanical pretreatments that must be used before polysaccharides in lignified plant tissues become susceptible to hydrolysis by enzymes do not appear to be necessary with pectin rich tissues. Inhibition of enzymatic hydrolysis of apple tissues by soluble phenolic compounds has been observed, but this inhibition can be removed by aeration and holding of the pulp which leads to polymerization and precipitation of these compounds (*25,70*).

The low content of insoluble cellulose and lignin leads to rapid liquefaction of these tissues during enzymatic hydrolysis even at high (>8%) concentrations of solids. Our work with orange peel (*50,81*) also indicates that very low amounts of enzymes are needed for relatively rapid (<12 hrs) saccharification of the peel. There is not enough data with thoroughly characterized enzyme preparations to compare rates of enzymatic hydrolysis of cellulose in lignified and pectin rich plant tissues at the present time, but recent industrial and academic interest should remedy the situation in the near future. Treatments with pectinolytic enzymes are used in the fruit juice industry (*25,28,70*) and the GRAS (i.e. generally recognized as safe) status of many of these enzymes can only lead to the expansion of their utilization.

Another important difference between the enzymatic hydrolysis of lignocellulosic tissues and sugar rich residues discussed here is a small contribution of glucose from hydrolysis of cellulose to the overall yield of soluble sugars. Inspection of Tables I and II shows that cellulose is not the most important contributor to soluble sugar production as is the case with lignocellulosic substrates. The release of soluble sugars retained in tissues of citrus and apple processing residues is of utmost importance, followed by the hydrolysis of pectin and hemicelluloses. Complete hydrolysis of cellulose adds only a small percentage to the overall sugar yield and may not be as important as it is with lignocellulosic substrates. Similar considerations apply to enzymatic saccharification of unleached sugar beets, where soluble sugar (sucrose) is by far the most abundant component of total dry solids and cellulose amounts to only a few percent. Therefore, the emphasis in enzymatic hydrolysis of tissues discussed here is actually on maceration and disintegration of tissues and cell walls by pectinolytic and hemicellulolytic enzymes to allow release of entrained soluble sugars. Hydrolysis of cellulosic fibers is of lesser importance and possibly may be omitted. The enzymatic hydrolysis of pectin is more efficient than similar hydrolysis of cellulose and requires very small amounts of enzyme (*104*), which is a very important economic consideration. The formation of galacturonic acid and much smaller amounts of acetic acid by the action of esterases lead to a significant increase in hydrogen ion concentration during enzymatic hydrolysis. The pH of the reaction mixtures decreases to the 3.2-3.5 range, unless significant amounts of base are added. Therefore the preferable enzymes for efficient hydrolysis of pectin rich tissues should be acidophilic and retain activity at pH of 3-4. However, the release of galacturonic acid and five-carbon sugars from hemicelluloses complicates subsequent utilization of enzymatic hydrolysates from juice processing residues. The hydrolyzates may be utilized as feed supplements for ruminants and chicken (105), but there may be problems with other farm animals such as pigs (*105*) which cannot digest five carbon sugars and galacturonic acid. Microorganisms are adapted to utilization of a broad spectrum of sugars, consequently microbial conversions to fermentation products offer a greater variety of options. Although individual microorganisms may have difficulties with utilization of some sugars as is discussed below or display a sequential (diauxic) pattern of sugar utilization they offer many options for the conversion of mixed sugars to value added products. The conversion options that have received significant attention are summarized below.

Microbial Conversions

Background on the fermentability of enzymatic hydrolyzates is provided by extensive research and industrial experience with fermentations of expressed or leached juice, which in the case of apples and sugar beets has been used for industrial production of ethanol. Fermented (hard) apple cider is a common product in apple producing countries (*25*) and productivity of sugar beets is so high that they have been considered from time to time a resource for the production of the fuel ethanol (*106-109*). No serious problems have been encountered with these fermentations and the high content of mineral nutrients in apple and beet juice often makes addition of nutritional supplements unnecessary.

The fermentation of citrus peel and peel juice has been hindered by the presence of antimicrobial compounds, mainly limonene, in peel oil (*110-117*). Levels of limonene as low as 50-100 ppm are strongly inhibitory to yeast (*110*) and anaerobic bacteria (*112-117*). The limonene in peel juice is not inhibitory to aerobic cultures because it is stripped by aeration. It can also be removed by steam stripping which occurs during the concentration of peel juice to molasses in multiple effect evaporators (*23*). A portion of peel molasses is used for industrial ethanol production in Florida and Brazil (*23,111*). We also observed (*81*) that limonene can be removed from peel hydrolyzates by simple filtration.

Besides fermentations of apple juice, several studies dealt with fermentation of apple pomace to ethanol (*38,118,119*) with (*119*) or without (*38,118*) enzymatic hydrolysis. Production of citric acid (*39,120*), biogas (*35*) and single cell protein (*121*) from apple pomace and apple distillery slops (*121*) has also been investigated. Similar studies of ethanol (*46,107,122,123*), biogas (*124*) and single cell protein (*101,125,126*) production from sugar beet pulp have been conducted. Conversion of citrus processing wastes also received considerable attention. Highest interest has been in single cell (fungal) protein production (*127-140*) including cultivation of mushrooms (*141-145*), while anaerobic digestion (*113,115,146,147*) and ethanol production (*81,148*) have been investigated much less. The coupling of enzymatic hydrolysis with ethanolic fermentation by yeasts has not been a very productive choice for all three residues. Yeasts ferment efficiently only six carbon neutral sugars (*149,150*). They generally do not ferment galacturonic acid and five carbon sugars, such as arabinose, and xylose (*151*), although a few strains which ferment concentrated xylose solutions under microaerophilic conditions have been identified (*152*). The increase in the yield of six carbon neutral sugars (glucose and galactose) fermentable by yeasts is only modest after enzymatic hydrolysis of citrus and apple residues (see Tables I and II) or unleached sugar beet pulp. This small increase in the yield of fermentable sugars accounted for the modest (0-30%) increase in observed ethanol yields when enzymatic hydrolysis and yeast fermentation were combined (*46,81,119,122,148*). The situation can improve with application of genetically engineered bacteria which have been constructed for fermentation of five carbon sugars to ethanol (*153*) or with genetic development of yeasts for fermentation of these sugars. Homofermentative production of lactic acid as a value added product suffers from similar limitations in the pattern of sugar utilization as ethanolic fermentation (*154*), even though xylose can be fermented to a mixture of lactic and acetic acid with smaller amounts of ethanol (*155*). Acetone-butanol fermentations can utilize all sugars in enzymatic hydrolysates discussed here, but a large increase in acetate formation has been observed during acetone-butanol fermentation of galacturonic acid (*156,157*). The development of fermentations for other products may require testing on a strain basis (*156*).

Increased attention needs to be paid to isolation and development of microorganisms utilizing arabinose and galacturonic acid. Arabinose is present in hydrolysates of both xylan- and pectin-rich plant tissues, and galacturonic acid is obviously a major sugar in hydrolysates of pectin rich tissues, some of which have been discussed here.

Conclusions

Pectin rich residues from the production of fruit juices and beet sugar offer a unique opportunity for the production of sugars by enzymatic hydrolysis of cellulosic substrates. The absence of lignin and low content of cellulose make these residues highly susceptible to depolymerization and solubilization by mixtures of pectinolytic and cellulolytic enzymes. These enzymes usually contain additional hemicellulolytic activities needed for depolymerization of arabinose, galactose and xylose rich polysaccharides which are also present in cell walls of these residues. The processing residues from apple and citrus juice production contain significant amounts of soluble sugars which are also effectively released by enzymatic treatment. High yields and relatively high rates of enzymatic depolymerization of carbohydrates in these tissues have been observed without any chemical pretreatment. High specific activity and low end product inhibition of commercial pectinolytic enzymes allow production of relatively concentrated sugar solutions at low enzyme loadings. These enzymatic hydrolysates appear to be suitable substrates for further microbial conversion to value added products. The main challenge in the development of microbial conversions of these hydrolysates is the identification and development of microorganisms utilizing galacturonic acid and arabinose which are abundant components of these hydrolysates and quite abundant in many other plant tissues.

Literature Cited

1. Hall, D.O. *Solar Energy* **1979**, *22*, 307.
2. Miller, D.L.; Eisenhauer, R.A. In *Handbook of Processing and Utilization in Agriculture*; Wolff, I.A., Ed.; CRC Press: Boca Raton, FL; 1982, Vol. 2, Part 1; pp 691-709.

3. Grohmann, K.; Himmel, M.E. In *Enzymes in Biomass Conversion*; Leatham, G.F.; Himmel, M.E.; Eds.; ACS Symposium Series # 460; American Chemical Society : Washington, D.C., 1991; pp 2-11.
4. Hayn, M.; Steiner, W.; Klinger, R.; Steinmüller, H.; Sinner, M.; Estebauer, H. In *Bioconversion of Forest and Agricultural Plant Residues*; Saddler, J.N., Ed.; C.A.B. International: Wallingford, U.K., 1993; pp 33-72.
5. Saddler, J.N.; Ramos, L.P.; Breuil, C. In *Bioconversion of Forest and Agricultural Plant Residues*; Saddler, J.N., Ed.; C.A.B. International: Wallingford, U.K., 1993; pp 73-91.
6. Converse, A.O. In *Bioconversion of Forest and Agricultural Plant Residues*; Saddler, J.N., Ed.; C.A.B. International: Wallingford, U.K., 1993; pp 93-106.
7. Marsden, W.L.; Gray, P.P. In *Critical Reviews in Biotechnology*; Stewart, G.G.; Russell, I., Eds.; CRC Press: Boca Raton, FL, 1986, Vol. 3; pp 235-276.
8. Klyosov, A.A. *Biochem.* **1990**, *29*, 10577-10589.
9. Thompson, D.N.; Chen, H.C.; Grethlein, H.E. In *Biotechnology in Pulp and Paper Manufacture: Application and Fundamental Investigations*; Kirk, T.K.; Chang, H., Eds.; Buttersworth-Heinemann: New York, NY, 1990; pp 329-338.
10. Bittner, A.S.; Burritt, E.A.; Moser, J.; Street, J.C. *J. Food Sci.* **1982**, *47*, 1469-1471, 1477.
11. Bertin, C.; Rouau, X.; Thibault, J.F. *J. Sci. Food Agric.* **1988**, *44*, 15-29.
12. Jona, R.; Foha, E. *Sci. Horticult.* **1979**, *10*, 141-147.
13. Testolini, G.; Rossi, E.; Vercesi, P; Porrini, M.; Simonetti, P; Ciappellano, S. *Nutr. Rep. Int.* **1982**, *25*, 859-865.
14. Kochar, G.K.; Sharma, K.K. *J. Food Sci. Technol.* **1992**, *29*, 187-188.
15. Anderson, J.W; Bridges, S.R. *Am. J. Clin. Nutr.* **1988**, *47*, 440-447.
16. Michel, F; Thibault, J.F.; Barry, J.L. *J. Sci. Food Agric.* **1988**, *42*, 77-85.
17. Agricultural Research Service, *Chemistry and Technology of Citrus, Citrus Products and Byproducts*, Agricultural Handbook No. 98, United States Department of Agriculture: Washington, D.C., 1962.
18. Kesterson, J.W.; Braddock, R.J. *By-Products and Specialty Products of Florida Citrus;* Bull. #784, University of Florida: Gainesville, FL, 1976.
19. Rebeck, H.M. In *Production and Packaging of Non-carbonated Fruit Juices and Fruit Beverages*; Hicks, D., Ed.; Van Nostrand Reinhold: New York, NY, 1990, pp 1-32.
20. Hendrix, C.M.; Redd, J.B. In *Production and Packaging of Non-carbonated Fruit Juices and Fruit Beverages*; Hicks, D., Ed.; Van Nostrand Reinhold: New York, NY, 1990, pp 33-67.
21. Chen, C.S; Shaw, P.E.; Parish, M.E. In *Fruit Juice Processing Technology*; Nagy, S.; Chen, C.S.; Shaw, P.E., Eds.; Agscience Inc.: Auburndale, FL, 1993, pp 110-165.
22. Shaw, P.E.; Nagy, S. In *Fruit Juice Processing Technology*; Nagy, S.; Chen, C.S.; Shaw, P.E., Eds.; Agscience Inc. : Auburndale, FL, 1993, pp 166-214.
23. Braddock, R.J.; Kesterson, J.W.; Miller, W.M. *Proc. Int. Soc. Citriculture* **1977**, *3*, 737-738.
24. Brown, M.G.; Kilmer, R.L.; Bedigian, K. In *Fruit Juice Processing Technology*; Nagy, S.; Chen, C.S.; Shaw, P.E., Eds.; Agscience Inc.: Auburndale, FL, 1993, pp 1-22.

25. Downing, D.L. *Processed Apple Products*; Van Nostrand Reinhold: New York, NY, 1989; p 448.
26. Downes, J.W. In *Production and Packaging of Non-carbonated Fruit Juices and Fruit Beverages*; Hicks, D., Ed.; Van Nostrand Reinhold: New York, NY, 1990; pp 158-181.
27. Lea, A.G.H. In *Production and Packaging of Non-carbonated Fruit Juices and Fruit Beverages*; Hicks, D., Ed.; Van Nostrand Reinhold: New York, NY, 1990; pp 182-223.
28. Binning, R.; Possmann, P. In *Fruit Juice Processing Technology*; Nagy, S.; Chen, C.S.; Shaw, P.E., Eds.; Agscience Inc.: Auburndale, FL, 1993; pp 271-317.
29. Fisher, C.H.; Chen, J.C.P. In *Handbook of Processing and Utilization in Agriculture*; Wolff, I.A., Ed.; CRC Press: Boca Raton, FL, 1982, Vol. 2, Part 1; pp 579-610.
30. *Beet Sugar Technology*; McGinnis, R.A., Ed.; Beet Sugar Development Foundation: Denver, CO, 1982; p 855.
31. Demaux, M. *Sucr. Fr.* **1979**, *120*, 357-384.
32. Thier, E. *Zuckerind.* (Berlin) **1982**, *107*, 223-230.
33. Chomyszyn, M. *Int. Z. Landwirtsch.* **1977**, *4*, 357-362.
34. *Florida Citrus Processors' Association, Statistical Summary 1989-90 Season,* Florida Citrus Processors' Association: Winter Haven, Fl, 1991.
35. Jewell, W.J.; Cummings, R.J. *J. Food Sci.* **1984**, 407-410.
36. *The World Almanac and Book of Facts* 1994; Famighetti, R., Ed.; Funk & Wagnalls Corp., Publisher: Mahwah, NJ, 1994; p 126.
37. Bomben, J.L.; Guadagni, D.C.; Harris, J.G. *Food Technol.* **1971**, *25*, 1108-1109, 1112, 1117.
38. Hang, Y.D.; Lee, C.Y.; Woodams, E.E. *J. Food Sci.* **1982**, *47*, 1851-1852.
39. Hang, Y.D.; Woodams, E.E. *MIRCEN J.* **1986**, *2*, 283-287.
40. Walter, R.H.; Sherman, R.M. *J. Agric. Food Chem.* **1976**, *24*, 1244-1245.
41. Ting, S.V.; Deszyck, E.J. *J. Food Sci.* **1961**, *26*, 146-152.
42. Braddock, R.J. In *Citrus Nutrition and Quality*; Nagy, S.N.; Attaway, J.A., Eds.; ACS Symp. Ser. #143; American Chemical Society: Washington, D.C., 1980; pp 273-288.
43. Braddock, R.J.; Graumlich, T.R. *Lebensm.-Wiss. u. Technol.* **1981**, *14*, 229-231.
44. Eaks, I.L.; Sinclair, W.B. *J. Food Sci.* **1980**, *45*, 985-988.
45. Conteras, I.; Gonzales, R.; Ronco A.; Ansenjo, J.A. *Biotechnol. Lett.* **1982**, *4*, 51-56.
46. Schaffeld, G.; Illanes, A.; Mejias, G.; Pinochet, A.M. *J. Chem Tech. Biotechnol.* **1987**, *39*, 161-171.
47. Neukom, H.; Amado, R.; Pfister, M. *Lebensm.-Wiss. u. Technol.* **1980**, *13*, 1-6.
48. Voragen, A.G.J.; Pilnik, W. *Fluess. Obst.* **1981**, 261-264.
49. Gormley, R. *J. Sci. Food Agric.* **1981**, *32*, 392-398.
50. Grohmann, K.; Baldwin, E.A. *Biotechnol. Lett.* **1992**, *14*, 1169-1174.
51. Guillon, F.; Barry, J.L.; Tribault, J.F. *J. Sci. Food Agric.* **1992**, *60*, 69-79.
52. McCready, R.M.; Reeve, R.M. *Agr. Food Chem.* **1955**, *3*, 260-262.

53. McCleary, B.W. In *Gums and Stabilisers for the Food Industry*; Phillips, G.O.; Williams, P.A.; Wedlock, D.J., Eds.; Oxford University Press: Oxford, Great Britain, 1990, Vol. 5; pp 291-298.
54. Labavitch, J.M.; Freeman, L.E.; Albersheim, P. *J. Biol. Chem.* **1976**, *251*, 5904-5910.
55. Darvill, A.; McNeil, M.; Albersheim, P.; Delmer, D. In *The Biochemistry of Plants*; Tolbert, N.E. Ed.; Academic Press: New York, NY, 1980, Vol. 1; pp 91-162.
56. Meurens, M., *Rev. Ferm. Ind. Aliment;* **1978**, *33*, 67-75.
57. Wilkie, K.C.B. In *Biochemistry of Plant Cell Walls*; Brett, C.T.; Hillman, J.R.,Eds.; Cambridge University Press: Cambridge, G.B., 1985; pp 1-37.
58. Selvendran, R.R.; Stevens, B.J.H.; O'Neill, M.A. In *Biochemistry of Plant Cell Walls*; Brett, C.T.; Hillman, J.R., Eds.; Cambridge University Press: Cambridge, G.B., 1985; pp 39-78.
59. *Chemistry and Function of Pectins;* Fishman, M.L.; Jen, J.J., Eds.; ACS Symposium Series #310; American Chemical Society: Washington D.C., 1986.
60. Voragen, A.G.J.; Heutink, R.; Pilnik, W. *J. Appl. Biochem.* **1980**, *2*, 452-468.
61. Knee, M.; Fielding, A.H.; Archer, S.A.; Laborda, F. *Phytochem.* **1975**, *14*, 2213-2222.
62. Renard, C.M.G.C., Searle van Leeuwen, M.J.M.; Voragen, A.G.J.; Thibault, J.F.; Pilnik, W. *Carbohydr. Polym.* **1991**, *14,* 295-314.
63. Renard, C.M.G.C., Schols, H.A.; Voragen, A.G.J.; Thibault, J.F.; Pilnik, W. *Carbohydr. Polym.* **1991**, *15,* 13-32.
64. Kertesz, Z.I.; Baker, G.L.; Joseph, G.H.; Mottern, H.H.; Olsen, A.G. *Chem. Eng. News* **1944**, *22*, 105-106.
65. McComb, E.A.; McCready, R.M. *Anal. Chem.* **1957**, *29,* 819-821.
66. Kravtchenko, T.P.; Voragen, A.G.J.; Pilnik, W. *Carbohydr. Polym.* **1992**, *18*, 17-25.
67. Rombouts, F.M.; Thibault, J.F. *Carbohydr. Res.* **1986**, *154*, 177-188.
68. Stutzenberger, F. In *Encyclopedia of Microbiology*; Lederberg, J., Ed.; Academic Press Inc.: San, Diego, CA, 1992, Vol. 3; pp 327-337.
69. Sajjaanantakul, T.; Pitifer, L.A. In *Chemical Technology of Pectin*; Walter, R.H., Ed.; Academic Press Inc.: San Diego, CA, 1991; pp 135-164.
70. Pilnik, W.; Voragen, A.G.J. In *Food Enzymology*; Fox, P.F., Ed.; Elsevier: London, Great Britain, 1991, Vol. 1; pp 303-336.
71. Rexova-Benkova, L.; Markovic, O. *Adv. Carbohydr. Chem. Biochem.* **1976**, *33*, 323-385.
72. Sakai, T. In *Methods in Enzymology*; Wood, W.A.; Kellogg, S.T., Eds.; Academic Press, Inc.: San Diego, CA, 1988, Vol. 161; pp 335-350.
73. Dekker, R.F.H.; Richards, G.N. *Adv. Carbohydr. Chem. Biochem.* **1976**, *32*, 277-351.
74. McCleary, B.V. In *Enzymes in Biomass Conversion*; Leatham. G.F.; Himmel, M.E., Eds.; ACS Symp. Ser. #460, American Chemical Society: Washington, D.C., 1991; pp 437-449.
75. Thibault, J.F.; DeDreu, R. *Carbohydr. Polym.* **1988**, *9,* 119-131.
76. Gilkes, N.R.; Henrissat, B.; Kilburn, D.G.; Miller, R.C.; Warren, R.A.J. *Microbiol. Rev.* **1991**, *55,* 303-315.

77. Paton, D. *Can. Inst. Food Sci. Technol. J.* **1974**, *7*, 61-64.
78. Wood, T.M. In *Biosynthesis and Biodegradation of Cellulose*; Haigler, C.H.; Weimer, P.J., Eds.; Marcel Dekker: New York, NY, 1991; pp 491-533.
79. Coughlan, M.P. *Bioresource Technol.* **1992**, *39*, 107-115.
80. Ben-Arie, R.; Kislev, N.; Frenkel, C. *Plant Physiol.* **1979**, *64*, 197-202.
81. Grohmann, K.; Baldwin, E.A.; Buslig, B.S. *Appl. Biochem. Biotechnol.* **1994** (in press).
82. Coughlan, M.P.; Mehra, R.K.; Considine, P.J.; O'Rorke, A.; Puls, J. *Biotechnol. Bioeng. Symp. No. 15*, **1985**, 447-458.
83. Ben-Shalom, N. *J. Food Sci.* **1986**, *51*, 720-721, 731.
84. Durr, P.; Schobinger, U.; Zellweger, M. *Lebensm.-Wiss. u.-Technol.* **1981**, *14*, 268-272.
85. Moloney, A.P.; O'Rorke, A.; Considine, P.J.; Coughlan, M.P. *Biotechnol. Bioeng.* **1984**, *26*, 714-718.
86. Marshall, M.R.; Graumlich, T.R.; Braddock, R.J.; Messersmith, M. *J. Food Sci.* **1985**, 1211-1212.
87. Nishio, N.; Oha, Y.; Kawamura, D.; Nagai, S. *Hakko Kogaku Kaishi* **1979**, *57*, 354-359.
88. Echeveria, E.; Burns, J.K.; Wicker, L. *Proc. Fla. State Hort. Soc.* **1988**, *101*, 150-154.
89. Sreenath, H.K.; Santhanam, K.J. *Ferm. Bioeng.* **1992**, *73*, 241-243.
90. Okai, A.E.; Giersner, K. *Z. Lebensm. Unters. Forsch.* **1991**, *192*, 244-248.
91. Buckenhüskes, H.; Omran, H.; Zhang, C.; Gierschner, K. *Food Biotechnol.* **1990**, *4*, 291-299.
92. Williamson, G.; Faulds, C.B.; Matthew, J.A.; Archer, D.B.; Morris, V.J.; Browsey, G.J.; Ridout, M.J. *Carbohydr. Polym.* **1990**, *13*, 387-397.
93. Baldwin, E.A. In *Biochemistry of Fruit Ripening*; Seymour, G.B.; Taylor, J.E.; Tucker, G.A., Eds.; Chapman and Hall: London, G.B., 1993; pp 107-149.
94. Knee, M. In *Biochemistry of Fruit Ripening*; Seymour, G.B.; Taylor, J.E.; Tucker, G.A., Eds.; Chapman and Hall: London, G.B., 1993; pp 325-346.
95. Dick, A.J.; Opoku-Gyamfua, A.; DeMarco A.C. *Physiol. Plant.* **1990**, *80*, 250-256.
96. Pilnik, W. In *Gums and Stabilizers for the Food Industry*; Phillips, G.O.; Williams, P.A.; Wedlock, D.J., Eds.; Oxford University Press: Oxford, Great Britain, 1990, Vol. 5; pp 223-232.
97. Vaccarino, C.; LoCurto, R.; Tripodo, M.M.; Patané, R.; Lagana, G.; Ragno, A. *Biol. Wastes* **1989**, *29*, 1-10.
98. Pavilon, S.J. U.S. Patent 4,952,504, August 28, 1990, 10.
99. Scheluchina, H.P.; Tyrdakova, I.I.; Aimychaemova, M.B. U.S.S.R. Patent 583,137, December 5, 1977, 2.
100. Renard, C.M.G.C.; Voragen, A.G.J.; Thibault, J.F.; Pilnik, W. *Carbohydr. Polym.* **1990**, *12*, 9-25.
101. Illanes, A.; Schaffeld, G. *Biotechnol. Lett.* **1983**, *5*, 305-310.
102. Sidi Ali, L.; Cochet, N.; Ghose, T.K.; Lebeault, J.M. *Biotechnol. Lett.* **1984**, *6*, 723-728.
103. Mamaiaschvili, N.A.; Sicharulidze, H.S.; Kvesitadre, G.I. *Izv. Akad. Nauk. Gruz. SSR, Ser. Biol.* **1986**, *12*, 390-395.

104. Endre, B.; Omran, H.; Gierschner, K. *Lebensm.-Wiss. u.-Technol.* **1991**, *24*, 80-85.
105. Longstaff, M.A.; Knox, A.; McNab, J.M. *Brit. poultry Sci.* **1988**, *29*, 379-393.
106. Lipinski, E.S.; Nathan, R.A.; Sheppard, W.J.; McClure, T.A.; Lawhon, W.T.; Otis, J.L. *Systems Study of Fuels from Sugarcane, Sweet Sorghum and Sugar Beets*; Energy Research and Development Administration, Washington, D.C., 1977; Vol. 1.
107. Kirby, K.D.; Mardon, C.J. In *Proc. 4th Int. Symp. Alcohol Fuels Technol;* Inst. Pesgui. Tecnol. Estado Sao Paulo: Sao Paulo, Brazil, 1981; Vol. 1, pp 13-19.
108. Tourliere, S. *Sucr. Fr.* **1979**, *120*, 337-338.
109. Sodeck, G. *Zuckerind.* **1980**, *105*, 836-841.
110. von Loesecke, H.W. *Citrus Ind.* **1934**, *15*, 8,9,20,21.
111. Graumlich, T.R. *Food Technol.* **1983**, *37*, 94-97.
112. McNary, R. Wolford, R.W.; Patton, V.D. *Food Technol.* **1951**, *5*, 319-323.
113. Lane, A.G. *J. Chem. Tech. Biotechnol.* **1980**, *30*, 345-350.
114. Lane, A.G. *Environ. Technol. Lett.* **1983**, *4*, 65-72.
115. Mizuki, E.; Akao, T.; Saruwatari, T. *Biol. Wastes* **1990**, *33*, 161-168.
116. Murdock, D.I.; Allen, W.E. *Food Technol.* **1960**, *14*, 441-445.
117. Subba, M.S.; Soumithri, T.C.; Suryanarayana, R. *J. Food Sci.* **1967**, *32*, 225-227.
118. Hang, Y.D.; Woodams, E.E. *MIRCEN J.* **1985**, *1*, 103-106.
119. Miller, J.E.; Weathers, P.J.; McConville, F.X.; Goldberg, M. *Biotechnol. Bioeng. Symp. No. 12* **1982**, 183-191.
120. Hang, Y.D.; Woodams, E.E. *Biotechnol Lett.* **1984**, *6*, 763-764.
121. Friedrich, J.; Cimerman, A.; Perdih, A. *Appl. Microbiol. Biotechnol.* **1986**, *24*, 432-434.
122. Beldman, G.; Voragen, A.G.J.; Rombouts, F.M.; Pilnik, W. *Energy From Biomass*; Proc. E.C. Contractors' Meeting, Capri, Italy, 1983, 4, 344-351.
123. Larsen, D.H.; Doney, D.L.; Orien, H.A. *Dev. Ind. Microbiol.* **1981**, *22*, 719-724.
124. Lescure, J.P.; Bourlet, P. *Sucr. Fr.* **1980**, *121*, 85-91.
125. Gupta, R.P.; Singh, A.; Kalra, M.S.; Sharma, V.K.; Gupta, S.K. *Ind. J. Nutr. Dietet.* **1977**, *14*, 302-307.
126. Mathot, P.; Brakel, J. *Med. Fac. Landbouww. Rijskuniv. Gent.* **1990**, *55*, 1539-1545.
127. Lequerica, J.L.; Lafuente, B. *Rev. Agroguim. Tecnol. Aliment.* **1977**, *17*, 71-78.
128. Nishio, N.; Tai, K.; Nagai, S. *Eur. J. Appl. Microbiol. Biotechnol.* **1979**, *8*, 263-270.
129. Karapinar, M.; Okuyan, M. *J. Chem. Tech. Biotechnol.* **1982**, *32*, 1055-1058.
130. Clementi, F.; Moresi, M.; Rossi, J. *Appl. Microbiol. Biotechnol.* **1985**, *22*, 26-31.
131. Rodrigues, J.A.; Bechstedt, W.; Echevarria, J.; Sierra, N.; Delgado, G.; Daniel, A.; Martinez, O. *Acta. Biotechnol.* **1986**, *6*, 253-258.
132. Suha Sukan, S.; Yasin, B.F.A. *Process Biochem.* **1986**, 50-53.
133. Maldonado, M.C.; Navarro, A.; Callieri, D.A.S. *Biotechnol. Lett.* **1986**, *8*, 501-504.
134. Nwabueze, T.U.; Ogantimein, G.B. *Biol. Wastes* **1987**, *20*, 71-75.
135. Moresi, M.; Clementi, F.; Rossi, J. *J. Chem. Tech. Biotechnol.* **1987**, *40*, 223-249.

136. Moresi, M.; Clementi, F.; Rossi, J.; Medici, R.; Vinti, G.L. *Appl. Microbiol. Biotechnol.* **1987**, *27*, 37-45.
137. Vaccarino, C.; Curto, R.L.; Tripodo, M.M.; Patane, R.; Lagana, G.; Schachter, S. *Biol. Wastes* **1989**, *29*, 279-287.
138. Barreto de Menezes, T.J.; de J.G. Salva, T.; Baldini, V.L.; Papini, R.S.; Sales, A.M. *Process Biochem.* **1989**, *24*, 167-171.
139. Kalra, K.L.; Grewal, H.S.; Kahlon, S.S. *MIRCEN J.* **1989**, *5*, 321-326.
140. El-Rafai, A.H.; Ghanem, K.M.; El-Sabaeny, A.H. *Microbios.* **1990**, *63*, 7-16.
141. Black, S.S.; Stearns, T.W.; Stephens, R.L.; McCandless, R.F.J. *Agr. Food Chem.* **1953**, *1*, 890-893.
142. Labaneiah, M.E.O.; Abou-Donia, S.A.; Mohamed, M.S.; El-Zalaki, E.M. *J. Food Technol.* **1979**, *14*, 95-100.
143. Nicolini, L.; von Hunolstein, C.; Carilli, A. *Appl. Microbiol. Biotechnol.* **1987**, *26*, 95-98.
144. Elisaschvili, V.I.; Glonti, N.M.; Katschlischvili, E.T.; Kiknadze, M.O.; Tusischvili, C.A. *Prikl. Biokhim. Mikrobiol.* **1992**, *28*, 362-366.
145. Nicolini, L.; Volpe, C.; Pezrotti, A.; Carilli, A. *Biores. Technol.* **1993**, *45*, 17-20.
146. Lane, A.G. *Food Technol. Austral.* **1979**, *31*, 201-207.
147. Nishio, N.; Kitaura, S.; Nagai, S. *J. Ferment. Technol.* **1982**, *60*, 423-429.
148. Hara, T.; Fujio, Y.; Ueda, S. *Nippon Shokuhin Kogyo Gakkaishi* **1985**, *32*, 241-246.
149. Barnett, J.A.; Payne, R.W.; Yarrow, D. *Yeasts: Characteristics and Identification*; 2nd Ed., Cambridge Univ. Press: Cambridge, Great Britain, 1990.
150. Barnett, J.A. *Adv. Carbohydr. Chem. Biochem.* **1976**, *32*, 125-234..
151. Toivola, A.; Yarrow, D.; van den Bosch, E.; van Dijken, J.P.; Scheffers, W.A. *Appl. Environm. Microbiol.* **1984**, *47*, 1221-1223.
152. Slininger, P.T.; Bothast, R.J.; Ohes, M.R.; Ladisch, M.R. *Biotechnol. Lett.* **1985**, *7*, 431-436.
153. Ingram, L.O.; Altherthum, F.; Ohta, K.; Beall, D.S. *Dev. Ind. Microbiol.* **1990**, *31*, 21-30.
154. Rogosa, M. In *Bergey's Manual of Determinative Bacteriology*; 8th Ed., Buchanan, R.C.; Gibbons, N.E., Eds.; The Williams and Wilkins Co.: Baltimore, MD, 1974; pp *576-593*.
155. Tyree, R.W.; Clausen, E.C.; Gaddy, J.L. *Biotechnol. Lett.* **1990**, *12*, 51-56.
156. Forsberg, C.W.; Donaldson, L.; Gibbins, L.N. *Can. J. Microbiol.* **1987**, *33*, 21-26.
157. Simon, E. *Arch. Biochem.* **1947**, *14*, 39-51.

RECEIVED April 28, 1994

Chapter 20

Silage Processing of Forage Biomass to Alcohol Fuel

Linda L. Henk and James C. Linden

Departments of Microbiology and of Agricultural and Chemical Engineering, Colorado State University, Fort Collins, CO 80523

Sweet sorghum is an attractive fermentation feedstock because forty percent of the dry weight consists of readily fermented sugars. Cellulose and hemicellulose comprise another fifty percent. If such material is to be used on a year-round basis as feedstock for ethanol production, a suitable method of storage is needed. Modified versions of ensiling were investigated. Cellulase and hemicellulase were added to green-chop harvested sweet sorghum prior to the production of silage. Cellulase-amended silage exhibited in situ hydrolysis of cellulose and hemicellulose. The resulting cellulase-amended biomass showed a decreased reactivity to subsequent enzymatic hydrolysis. However, more ethanol was produced during the SSF of cellulase-amended sorghum compared to the control silage receiving no enzyme amendments. The relationships between available carbohydrates and the potential for ethanol production are also presented.

Sweet sorghum has been extensively studied as a substrate for fuel alcohol production for several years. Italians first evaluated sweet sorghum from 1935 to 1942 (*1*). Sweet sorghum is one of the most promising crops that can be grown for biomass in temperate climates (*2*). It is a genetically diversified and drought tolerant plant that produces high yields of fermentable sugars, cellulose and hemicellulose (*3-5*). Sweet sorghum can be treated enzymatically to hydrolyze cellulose and hemicellulose to fermentable sugars (*19,20,43*).

Storage of Sweet Sorghum

Seasonal availability and storability of the raw material are important factors in the use of renewable biomass. Storability and sugar extraction with currently available technology are two serious problems that have limited the use of sweet sorghum as

0097–6156/94/0566–0391$08.00/0

an ethanol feedstock (*6*). Because sweet sorghum is harvested seasonally at a high moisture level (70-80%), rapid deterioration of readily available carbohydrates can occur unless the material is stabilized by some process that allows storage of the material on a long-term basis.

Sugar recovery from (*3,7-11*) and the direct fermentability of (*12-15*) sweet sorghum have been investigated. Previous studies have focused on harvesting and processing sweet sorghum stalks to produce juice for fermentation to ethanol (*15,17*). Sweet sorghum stalks are separated into two fractions: (1) the pith fraction containing most of the juice and sugars, and (2) the rind-leaf fraction containing the hemicellulose and cellulose fractions. Separation of the two fractions increases the press capacity and sugar extraction efficiency. Sugar concentrations in the juice obtained from this process range from 15° to 18° brix (mass of sugar as a percentage of juice mass). Worley and Cundiff suggest that the rind-leaf fraction and the pith presscake be ensiled and fed to cattle or, alternatively, used as a feedstock for ethanol conversion provided that the technology for cellulose conversion to ethanol becomes economical (*17*).

Solid-state fermentation of sweet sorghum to ethanol has been investigated (*12,16*). Kargi, *et al.*, removed the rind and outside fiber from the sweet sorghum stalk prior to a solid-state ethanol fermentation (*16*). The performance of the solid-state system was evaluated using presqueezed (pressed) sorghum with juice added back to the presscake and unsqueezed sorghum. The authors determined that presqueezing did not improve ethanol yield and therefore, was not necessary. At an optimal moisture level of 70%, a final ethanol concentration of 6.7% (w/v) was achieved. Storability of sweet sorghum was not addressed by Kargi, *et al.*, nor was the utilization of the lignocellulosic portion of sweet sorghum.

Gibbons, *et al.*, used a semicontinuous solid-phase fermentation system to convert dried, rehydrated sweet sorghum to fuel ethanol (*12*). Sweet sorghum stalks were cut, dried to 15% moisture, and shredded to pass through a 2.54 cm screen. The shredded material was stored in plastic bags under ambient temperatures prior to the fermentations. The reducing sugar content of the fresh sorghum stalks (70% moisture) was 10.4% (wt/wt). Ethanol yields of 6% (vol/vol) or 85% of theoretical were obtained based on the concentration of reducing sugars. An ethanol concentration of 6% corresponds to an ethanol yield of 179 L/10^3 kg. Cellulose and hemicellulose were not converted to ethanol through enzymatic hydrolysis and fermentation.

In order for a biofuels system to be economical, the conversion of cellulose, hemicellulose and sugars to ethanol must approach theoretical yields (*44*). Cellulose and hemicellulose are to be utilized as ethanol feedstocks in order to keep feedstock costs down. In addition, economical, long-term storage methods must be developed in order to use herbaceous crops as an ethanol feedstock on a year-round basis (45). If sweet sorghum is to be used as an ethanol feedstock on a year-round basis, some means of storing all of the potential carbohydrates including cellulose and hemicellulose must be developed. A modified version of the traditional ensiling process accomplishes this objective (*19,20*).

Ensiling as a Method of Preservation

Ensiling is a traditional agricultural practice that serves as a forage crop preservation method. Complete discussions of the ensiling process can be found in Barnett (*21*) and in MacDonald (*22*). Basically, silage is the product of a lactic acid bacterial fermentation of readily available sugars present in the chopped forage. The ensiling fermentation quickly converts some of these sugars to lactic and acetic acids. The Embden-Meyerhof pathway converts 1 mol of hexoses to 2 mol of lactic acid (homolactic fermentation). The 6-phosphogluconate pathway results in one mol of CO_2, one mol ethanol or acetic acid and one mol of lactic acid (heterolactic). A combination of the resulting anaerobic and acidic conditions retards the growth of spoilage organisms (*22*). Microbial populations in the silage are stabilized thus diminishing consumption of the remaining soluble sugars.

Our early research proved that ensiling is a process whereby sweet sorghum is preserved. The lignocellulosic fraction is rendered reactive to conventional hydrolysis (*19*) and to *in situ* hydrolysis with cellulases and hemicellulases to yield additional fermentable carbohydrates (*20*). The pH and temperature profiles within the silo provide compatible conditions for fungal cellulase enzymes to hydrolyze hemicellulose and cellulose. Linden, *et al.*, showed that between 81% to 85% of the total potential fermentable carbohydrates originally in sweet sorghum were maintained over 155 days of ensiling (*19*). In a subsequent study, total carbohydrate levels were maintained at 95% over a 66 day period (*20*). Total potential fermentable carbohydrates were calculated as the sum of the neutral detergent fibers plus the soluble reducing sugars.

In addition, the structural polysaccharides of sweet sorghum became more reactive to enzymatic hydrolysis after 32 days of ensiling (*19*). Reducing sugar yields due to the enzymatic hydrolysis of cellulose and hemicellulose increased from 137 g/kg dry matter at the start of ensiling to 232 g/kg dry matter after 32 days of ensiling for pressed sweet sorghum. The acidic environment of the silo served as a pretreatment for enhancing the enzymatic hydrolysis of sweet sorghum lignocellulose. A preliminary process design has been presented elsewhere (*46*).

In situ Acid Hydrolysis of Plant Cell Wall Components

In order for sorghum lignocellulose to serve as a substrate for fuel production, accessibility to enzymatic digestion of cellulose must be improved through pretreatment that hydrolyzes hemicellulose and enables removal of lignin. The hemicellulose in sorghum is xylan, a polymer consisting of a β(1-4) xylose backbone which is substituted with acetyl, arabinosyl and glucuronyl residues. These side-chain substituents of hemicellulose are covalently cross-linked by ether and ester linkages to phenolic acids, primarily ferulic and cinnamic acids, that are components of lignin. Such cross-linking anchors lignin to xylan, to maintain the integrity of the plant cell wall (*25*) by imparting resistance to fungal enzyme degradation.

Cleavage of acid-labile glycosidic arabinosyl linkages with xylose residues may be promoted in the acidic environment of ensilage (*20*). The investigations of Morrison (*26*) of changes in the cell wall components during ensiling show that hydrolysis of arabinofuranose side chains occurs preferentially to that of the D-xylopyranose residues on the xylan backbone. The five carbon L-arabinose is present in these polysaccharides as the five-member furanose ring. The glycosidic bond of the furanose ring is more labile under acid conditions than is that of the six-membered pyranose ring (D-xylose). The extent to which side-chains are hydrolyzed has an effect on the rate of subsequent degradation of hemicellulose by xylanase enzymes (*27*).

Morrison has also studied lignin-carbohydrate complexes isolated from ryegrass (*28*). Bonds in lignin-carbohydrate complexes isolated from jute fiber and aspen are predominantly ester linkages between lignin and 4-O-methylglucuronyl sidechains of hemicellulose (*29,30*). These ester linkages are approximately as labile as the L-arabinosyl furanoside side chains of hemicellulose when subjected to acid-catalyzed hydrolysis (*31*). Together, these acid catalyzed hydrolyses could cause sufficient separation of lignin from the hemicellulose main chain, creating pores in the microfibril and allowing access to cellulase enzymes.

The emphasis of this paper is the improved lignocellulosic hydrolysis of sweet sorghum through the combined action of the acidic conditions in the silo and enzymes. The hypothesis guiding our work is the following: The action of cellulase and hemicellulase additives in combination with the acidic environmental conditions of the silo will open the sorghum cell wall and will serve as a dilute acid pretreatment prior to simultaneous saccharification (SSF) to ethanol. As a silage additive, commercial cellulases may improve the digestibility of the cellulose fibers during the ensiling process. Thus, the combination of dilute acid and enzymatic hydrolysis of the sweet sorghum fiber components may produce a more reactive substrate for SSF to ethanol.

Experimental Design

In order to test our hypothesis, sweet sorghum was ensiled with and without cellulase enzymes at an activity level (*33*) of 5 IU/g dry sorghum. Sweet sorghum was harvested by field chopping near Ames, Iowa in September, 1991, immediately frozen and shipped to Colorado State University in the frozen state. Freezing was not meant to be a storage option in this study. It was meant to be used as a short-term preservation method due to the logistics of the study.

Laboratory silos were set up with partially thawed batches of sweet sorghum (26.3 kg) which was mixed with 0.1% Pioneer Silabac, a silage inoculum containing *Lactobacillus plantarum* and *Streptococcus faecium*.

Enzyme amendments were defined as follows: Series A represented the control with no enzyme addition and Series B contained 5 IU of cellulase activity/g dry weight.

Silos were set up in replicates of eighteen, each containing approximately 200 g of sweet sorghum on a wet weight basis (23% dry matter) packed tightly into 0.5 pint canning jars. The silos were stored at 20°C for a total of 60 days. Triplicate silos were harvested at intermediate intervals of time through a sixty day period and kept at -20°C until analyses were conducted. Compositional changes occurring during ensiling were monitored by analyses including pH, glucose (*20*), reducing sugars (*34*), other monosaccharides and organic acids (*19,20*). In addition, fiber reactivity to enzymatic hydrolysis was examined (*19*). Finally, the possible benefits of ensiling toward increased ethanol productivity were measured using adapted SSF procedures (*35*). Because the coarse, fibrous nature of the sweet sorghum used in this study posed severe mixing problems, static fermentation vessels were used as opposed to an agitated system that was standardized for dried and milled materials. The addition of ergosterol and oleic acid and antibiotics was also not included in our study.

The term, unensiled, appears several times in the discussion. Within the context of this paper, unensiled means samples taken at day zero. These samples were taken immediately after Pioneer Silabac and cellulase, if appropriate, were added to the partially thawed sweet sorghum. Unensiled samples were frozen immediately to represent material that was not fermented by lactic acid bacteria.

Compositional Changes Occurring During Ensiling

pH and Organic Acid Content. The silage fermented rapidly, as seen from the drop in pH shown in Figure 1. The pH dropped within seven days from pH 5.8 in both treatment series A and B to pH 4.2. The rapid decline in pH resulted from the fermentation of readily available sugars (sucrose, glucose and fructose) to lactic and acetic acids by lactic acid bacteria. Continued metabolic action reduced the pH to between 3.6 and 4.1 over a 60 day period. HPLC analysis of sorghum ensiled for twenty-eight days showed a lactic acid concentration of 77.1 and 75.1 g/kg dry matter for series A and B, respectively. Acetic acid was present in a nominal amount of 2.5 g/kg.

McDonald characterizes lactic acid silages as those having a pH value between 3.7 and 4.2 with a lactic acid concentration of between 80 to 120 g/kg dry matter or approximately 1 molar (*22*). An acidic environment improves the stability of the silage because acidic conditions retard microbial growth. Therefore, potential fermentable carbohydrates can be stored successfully for long periods of time under acidic ensiling conditions (*19,20,22*).

Extractable Sugars in Sweet Sorghum Silage. The data in Figure 2 represents levels of extractable glucose on a dry weight basis in each silage treatment series throughout the sixty days of storage. Approximately 37 g glucose per kg dry matter was utilized by lactic acid bacteria during the first five days of ensiling. An increase in glucose beyond the initial concentration in treatment B is due to cellulose hydrolysis, *in situ* (*20*). In treatment series B, glucose was maintained at close to the same level at day 60 as at day 0; the average day 60 glucose levels were 165 +/- 5 g/kg dry matter, compared to 159 +/- 12 g/kg dry matter at day 0. This is a definitive indication that cellulase action occurred in the mini-silos. The control treatment series A, without enzymes, had a glucose level of 106 g/kg dry matter, approximately two-thirds of the average for the cellulase-amended silage. In spite of continued metabolic action of the silage microbial population, as indicated by a slight drop in pH after 5 days (Figure 1), enzymes allowed maintenance of glucose at or above levels originally in the sweet sorghum (Figure 2).

The change in reducing sugars over a twenty-eight day period is depicted in Figure 3. Total reducing sugars includes glucose, fructose, xylose, arabinose and, in the case of series B, cellobiose. Cellobiose results from the enzymatic hydrolysis of cellulose *in situ*. As was seen with glucose in Figure 2, an initial decline in available reducing sugar concentration occurred in the first five days, particularly in the control silage. After five days, the reducing sugar content of both series parallel each other. Series B contained twenty-five percent more reducing sugar than series A. Reducing sugar analysis was not conducted on day 60 material.

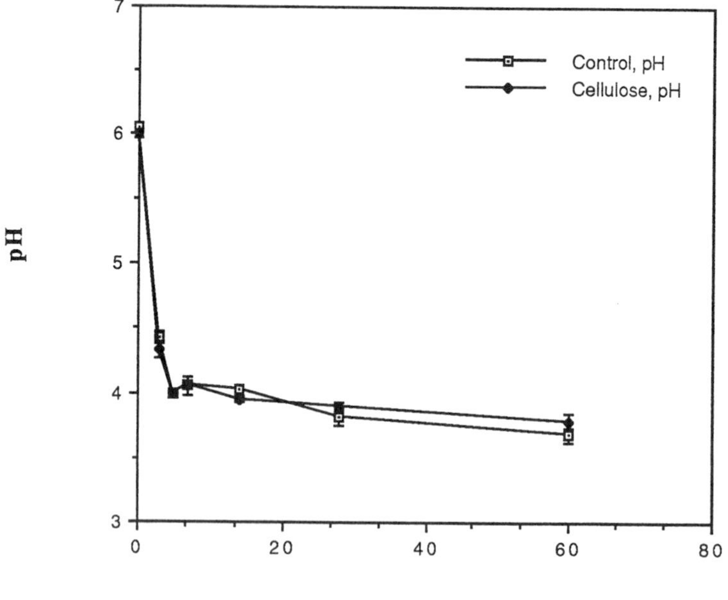

Figure 1. Changes in pH with time.

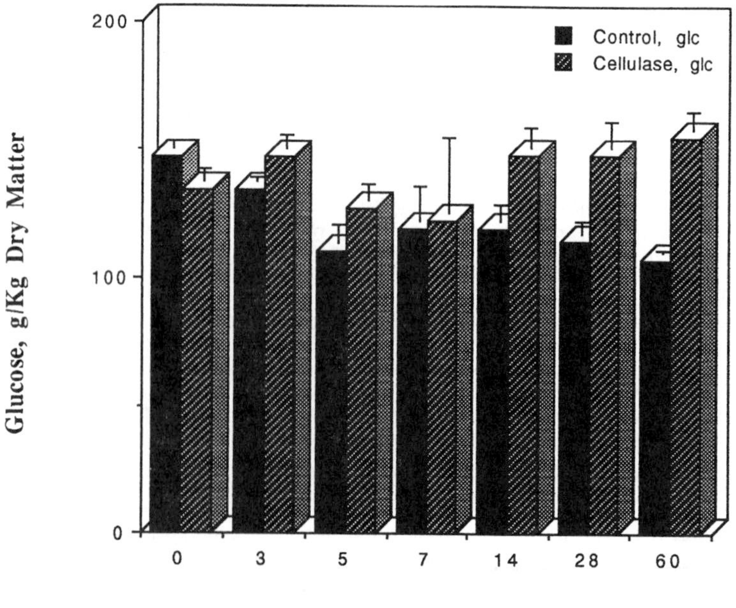

Figure 2. Changes in glucose concentration in sweet sorghum silage ± cellulase amendments.

It is interesting to compare the levels of glucose in series B (Figure 2) to the levels of reducing sugars in the same series (Figure 3). Glucose maintains a constant level between day 14 and day 28 while reducing sugars decline during the same period. Perhaps the microbial population is utilizing sugars other than glucose during this time period. HPLC analysis of these extracts is not complete; however, the dynamic changes of the various sugars should be determined when these analyses are complete. In any case, this clearly shows that glucose is maintained *in situ* in sorghum amended with cellulase over a sixty day time period.

These results have economic implications. If enzymatic hydrolysis of cellulose and hemicellulose can occur within the confines of a silo, hydrolysis costs will be significantly reduced compared to a conventional hydrolysis (*19,46*). Previous studies (*20*) have demonstrated silage stability for at least 120 days, which is important for the long-term production of ethanol from sorghum stored as silage. Stability of silage means that the silo conditions stays acidic and anaerobic thus preserving the silage from spoilage microorganisms such as clostridia and fungus. If spoilage would occur, the carbohydrates would be degraded, rendering the silage useless as an ethanol feedstock.

HPLC analysis of the initial starting material also indicates that fructose is a major component, being present at an average concentration of 146.3 g/kg dry matter compared to 159 g/kg for glucose. Fructose is readily fermented by yeast to ethanol; thus, it is an important component of sweet sorghum as it pertains to ethanol production. However, this sugar is also readily utilized by heterolactic silage microbes with mannitol as a by-product. By day 7, fructose was not detected by HPLC.

The utilization of fructose along with approximately 50 g glucose/kg dry matter is responsible for the production of lactic and acetic acids and carbon dioxide during the silage fermentation. This expenditure of carbohydrate is required in order to produce the anaerobic and acidic conditions necessary to store and maintain a high quality lactic acid silage. In addition to the maintenance requirements, these acid conditions serve as an economical means of dilute acid pretreatment of the lignocellulosic fibers.

Weight Loss and Dry Matter Changes that Occur During Ensiling. Sweet sorghum receiving cellulase during ensiling, series B, exhibited a higher percent weight loss than did the control, series A (Figure 4). Weight loss is presumably due to the production of carbon dioxide and liquid effluent during the ensiling fermentation. Increased dry matter losses and increased effluent with the use of enzymes as additives has also been detected by others (*36,37*). In our study, approximately eight percent of the original material treated with cellulase was lost over a sixty day period as opposed to three percent with the control receiving no cellulase additive. This parameter may have economic implications when the final process is evaluated.

Dry matter losses also occur during the ensiling process as shown in Figure 5. Because oven dry weight was used to determine dry matter, this graph may also reflect the loss of volatile compounds produced by the ensiling fermentation such as volatile organic acids or ethanol.

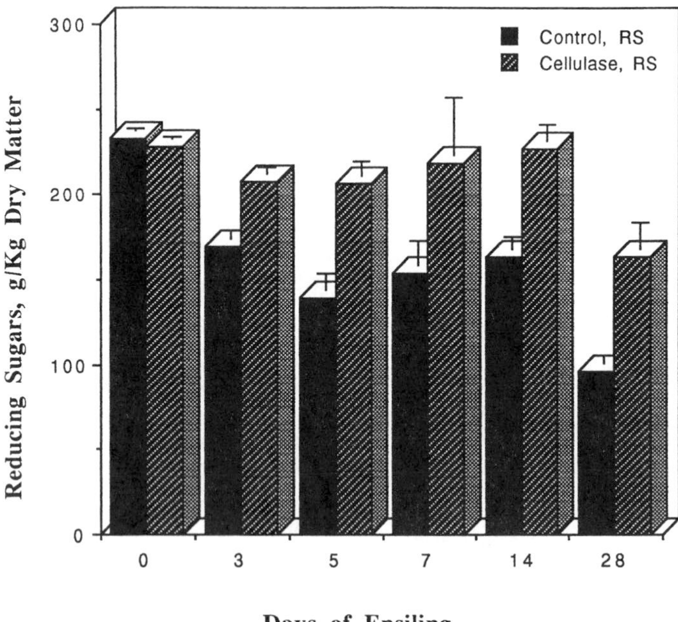

Figure 3. Changes in reducing sugar concentration in sweet sorghum silage ± cellulase amendments.

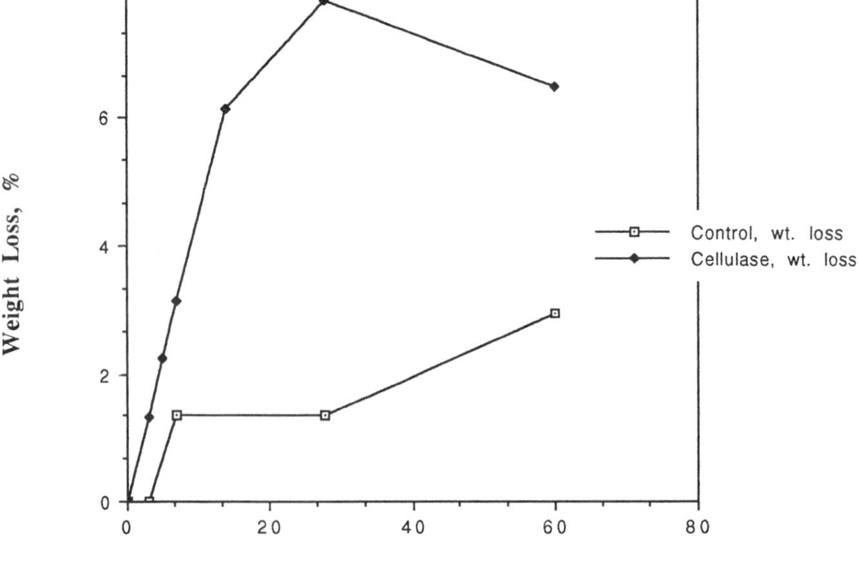

Figure 4. Weight loss occurring during the ensiling period in sweet sorghum silage ± cellulase amendments.

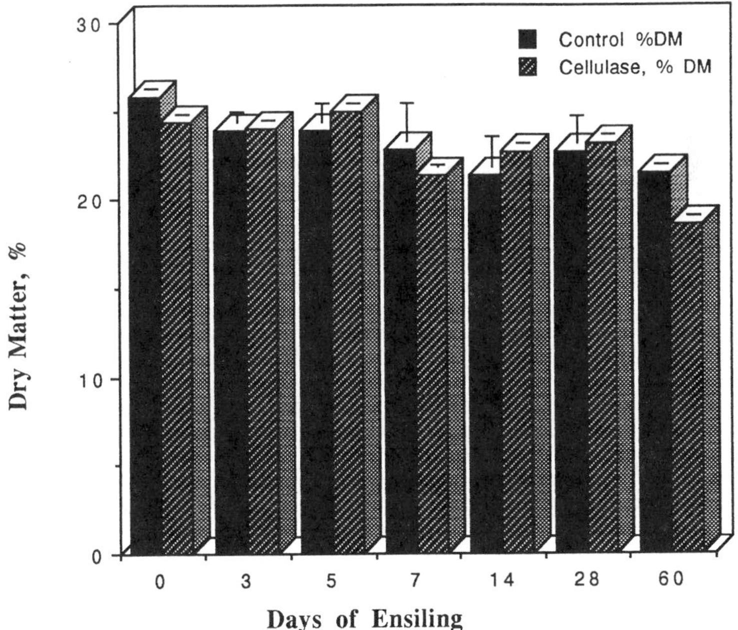

Figure 5. Dry matter changes occurring during the ensiling period in sweet sorghum silage ± cellulase amendments.

Enzymatic Hydrolysis of Ensiled Sweet Sorghum. Two different experiments were conducted to determine the reactive nature of both series of sweet sorghum silage. For the first experiment, conventional hydrolysis was used and for the second, hydrolysis was performed under operating conditions whereby silage material was either extracted or not extracted with dilute sodium hydroxide.

Conventional Enzymatic Hydrolysis. In the first experiment, conventional conditions were used: 40 FPU of cellulase activity per g dry weight, an optimal temperature of 45°C and an agitation rate of 150 rpm. The full potential of cellulose hydrolysis is determined under these conditions. Control silage ensiled for sixty days yields 165 g of glucose/kg dry matter under hydrolysis conditions. This value compares to an initial glucose value of 87 g/kg for the control before hydrolysis for a net hydrolysis glucose value of 77 g/kg. Sorghum ensiled with cellulase contained 144 g of glucose prior to being subjected to conventional hydrolysis; after hydrolysis, the glucose level was 189.3 g/kg for a net increase of 45.3 g/kg.

While it is clear that, overall, there is more glucose produced in the series B silage, the lignocellulosic fibers remaining in the silo may not be as reactive to subsequent enzymatic hydrolysis as the control fibers, as shown by the net hydrolysis for both series. Series B produces an additional 45 g/kg while series A produces 77 g glucose/kg dry matter. Conventional hydrolysis of the two series shows that series A is capable of producing eighty-seven percent (87%) of the total amount of glucose produced from the hydrolysis of cellulase-amended silage.

Jacobs, *et al.* (*36*) show that silage receiving cellulase as an additive exhibited lower rumen digestibilities than silage receiving no additives. The reduction in the organic matter digestibility may be a result of a loss of cellulose during ensiling and that the enzyme additives may degrade the more digestible fraction leaving a less digestible residue. Similarly, Narasimhalu, *et al.* (*38*) show that enzyme treatments lower the acid detergent fiber content of silage and also lower fiber digestibilities in sheep. In contrast, Vanhatalo, *et al.* (*39*) demonstrated improved organic matter and cell wall digestibilities *in vivo* with cattle using enzyme-treated cocksfoot timothy grass. Tengerdy, *et al.* (*23*) detected an increase in *in situ* digestibility in silage prepared from fresh alfalfa but not from wilted alfalfa.

Enzymatic Hydrolysis under Operating Conditions. The second experiment determined the extent of hydrolysis that can occur in the SSF vessel without yeast. Conditions were as follows: an enzyme loading of 25 FPU/g of cellulose, which corresponds to 6.25 FPU/g on a dry weight basis. The hydrolysis was conducted at 38°C, the operating temperature for SSF, under static conditions, with a cellulose loading of 1% (1 g cellulose/100 mL hydrolysate) or a solid-to-liquid ratio of approximately 4 g dry silage/100 mL. Samples were taken at 0, 1, 3, 6, 12, and 24 hours of hydrolysis. HPLC analysis was conducted on 0, 6 and 24 hour samples while total reducing sugars and glucose were determined on all time points.

Results from the hydrolysis of day zero sweet sorghum silage are shown in Table I and of sorghum ensiled for 7 days in Table II. Sucrose, glucose, xylose and arabinose were quantified by HPLC using a Biorad HPX-87P column with a corresponding prefilter (*16*). Cellobiose coelutes with sucrose. Standard curve calibration in the analysis is based on sucrose; therefore, cellobiose is evaluated as

Table I. Sugar Release Upon Enzymatic Hydrolysis of Unensiled Sweet Sorghum Silage

Sample	*Hours*	*Disaccharide*[a]	*Glucose*[a]	*Xylose*[a]	*Arabinose*[a]	*Total Sugars*	*Net Sugars*[c]	*% Recovery*[d] *(Fiber)*	*Corrected Net Sugars*[c]
Control, NX[b]	0	44.4	158.1	3.6	0.0	206.1	0.0	100.0	0.0
	6	58.5	158.2	2.2	0.0	218.9	12.8		12.8
	24	73.2	174.2	7.3	24.5	279.3	73.2		73.2
Control, EX	0	23.9	68.1	44.4	36.9	173.3	0.0	42.8	0.0
	6	153.4	84.1	81.3	54.1	372.9	199.6		85.4
	24	194.2	116.9	148.1	64.4	523.8	350.5		150.0
Cellulase, NX	0	65.7	172.6	0.0	34.2	272.5	0.0	100.0	0.0
	6	71.3	176.3	3.5	43.0	294.1	21.6		21.6
	24	70.6	178.8	6.4	44.3	300.4	27.9		27.9
Cellulase, EX	0	22.1	62.1	53.1	37.8	175.1	0.0	43.9	0.0
	6	100.2	74.9	77.2	44.3	296.6	121.5		53.3
	24	145.0	118.5	129.0	39.1	431.6	256.5		112.6

[a]Units = g sugar/kg of unextracted or extracted fiber; HPLC analysis by BioRad HPX-87P;
[b]NX = not extracted; EX = extracted.
[c]Net = $T_{24} - T_0$. This value represents the amount of sugars obtained strictly due to enzymatic hydrolysis of sweet sorghum. Corrected net based on original dry matter (equal to Extracted, net × % Recovery). This correction bases the values obtained for the hydrolysis of extracted material on original unextracted material so that accurate comparisons can be made between the two materials.
[d]% Recovery is the percentage of fibrous silage material that is recovered after extraction with 0.1N NaOH for one hour under reflux conditions and subsequently washed with 0.1 N NaOH and boiling water.

Table II. Sugar Recovery Upon Enzymatic Hydrolysis of Day 7 Sweet Sorghum Silage

Sample	*Hours*	*Disaccharide*[a]	*Glucose*[a]	*Xylose*[a]	*Arabinose*[a]	*Total Sugars*	*Net Sugars*[c]	*% Recovery*[d] *(Fiber)*	*Corrected Net Sugars*[c]
Control, NX[b]	0	14.8	87.6	0.0	0.0	102.4	0.0	100.0	0.0
	6	41.0	101.7	0.0	6.0	148.7	46.2		46.2
	24	50.0	144.7	4.3	6.5	205.5	103.2		103.2
Control, EX	0	20.5	39.0	37.5	37.5	134.5	0.0	49.4	0.0
	6	131.8	64.5	48.0	40.0	284.3	149.8		74.0
	24	159.0	100.5	139.0	24.0	422.5	288.0		142.3
Cellulase, NX	0	33.2	123.6	4.4	6.1	167.3	0.0	100.0	0.0
	6	33.8	117.6	4.4	5.8	161.6	0.0		0.0
	24	51.5	152.8	6.3	7.2	217.8	50.5		50.5
Cellulase, EX	0	23.3	37.7	32.9	39.4	133.3	0.0	49.9	0.0
	6	107.6	58.8	68.9	50.3	285.6	152.3		76.0
	24	136.8	89.6	105.8	51.0	383.2	249.9		124.7

[a]Units = g sugar/kg of unextracted <u>or</u> extracted fiber; HPLC analysis by BioRad HPX-87P;
[b]NX = not extracted; EX = extracted.
[c]Net = $T_{24} - T_0$. This value represents the amount of sugars obtained strictly due to enzymatic hydrolysis of sweet sorghum. Corrected net based on original dry matter (equal to Extracted, net × % Recovery). This correction bases the values obtained for the hydrolysis of extracted material on original unextracted material so that accurate comparisons can be made between the two materials.
[d]% Recovery is the percentage of fibrous silage material that is recovered after extraction with 0.1N NaOH for one hour under reflux conditions and subsequently washed with 0.1 N NaOH and boiling water.

sucrose. It can be assumed that the disaccharide present at day zero of ensiling is sucrose. Sucrose is rapidly inverted under the acidic conditions to glucose and fructose (*19*). Any increase should be thought of as cellobiose. Because the disaccharide values increase in all samples, cellobiose is being produced but it is not all converted to glucose.

Reasons for cellobiose production during the enzymatic hydrolysis may be two-fold: the cellulase enzyme system may be deficient in β-glucosidase or product inhibition of the enzyme system may be occurring. In the latter case, the hydrolysis is not going to completion to produce glucose as an end-product. The latter may result from the hydrolysis being conducted under static conditions. Because of inadequate mixing, glucose and cellobiose may not be able to diffuse away from the cellulose fibers. The sugars may reach concentrations at which inhibition of β-glucosidase and cellobiohydrolase limits glucose and cellobiose production, respectively. In Tables I and II, glucose present in high concentrations results in limited production of cellobiose, particularly in non-extracted sorghum.

Enzymatic Hydrolysis of Extracted Material. In addition, silages were extracted with 0.1 N sodium hydroxide (*19*) and washed prior to further enzymatic hydrolysis under operating conditions. Linden *et al.* (*19*) showed that extraction with 0.1 N sodium hydroxide improved enzymatic hydrolysis of silage. In addition, the fibers showed increased reactivity to enzymatic hydrolysis as a function of ensiling time. This material had not been treated with cellulolytic enzymes prior to ensiling. Furusaki, *et al.* (*47*) treated sweet sorghum bagasse with 1% NaOH for one hour at 100C. They determined that 78% of the cellulose was hydrolyzed with a glucose yield of 61%.

The purpose of this experiment was to analyze the reactivity of the untreated and cellulase-amended silages as a function of ensiling time. Extraction with alkali removes factors that seem to inhibit cellulose hydrolysis (*19*). The reactivity of cellulose can then be assessed without inhibition occurring during the hydrolysis.

The benefits of ensiling as an acid pretreatment can be seen by the seventh day of the silage fermentation as shown in Figures 6 and 7. Net sugar yield signifies the amount of sugars that are obtained strictly due to the enzymatic hydrolysis of sweet sorghum. This value is obtained by subtracting the sugar values obtained initially (0 hours of hydrolysis) from the final concentration of sugars obtained after 24 hours of hydrolysis. The effects of ensiling are seen in Figure 6 where the net sugar yield is shown for unextracted sweet sorghum silage. The control silage (Series A) is more reactive to enzymatic hydrolysis than the cellulase amended material. In addition, sorghum ensiled for seven days is more reactive than the day zero samples that had not been ensiled.

Corrected net sugar yields for extracted silage are depicted in Figure 7. Use of "corrected net sugar yields" bases the values obtained for the hydrolysis of extracted material on original, unextracted silage so that accurate comparisons can be made. These values are determined by multiplying the sugar values obtained from the hydrolysis of extracted sweet sorghum by the percent recovery of extracted fiber. Extraction minimizes the noticeable effects of ensiling. Control silage after extraction appears to be as reactive at day 0 as compared to day 7. A slight difference due to ensiling is seen with sorghum receiving cellulase. Day 7 cellulase-amended sorghum

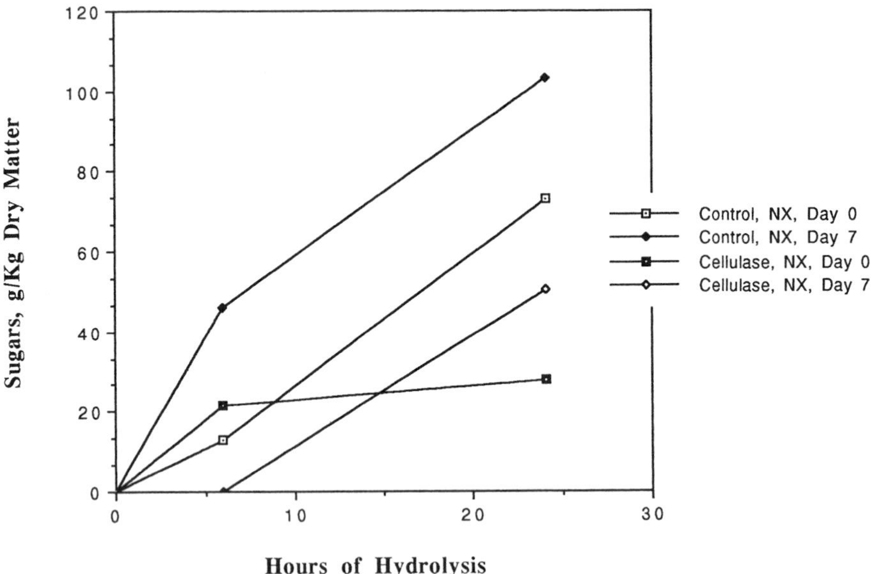

Figure 6. Comparison of net sugar yield for unextracted sweet sorghum silage.

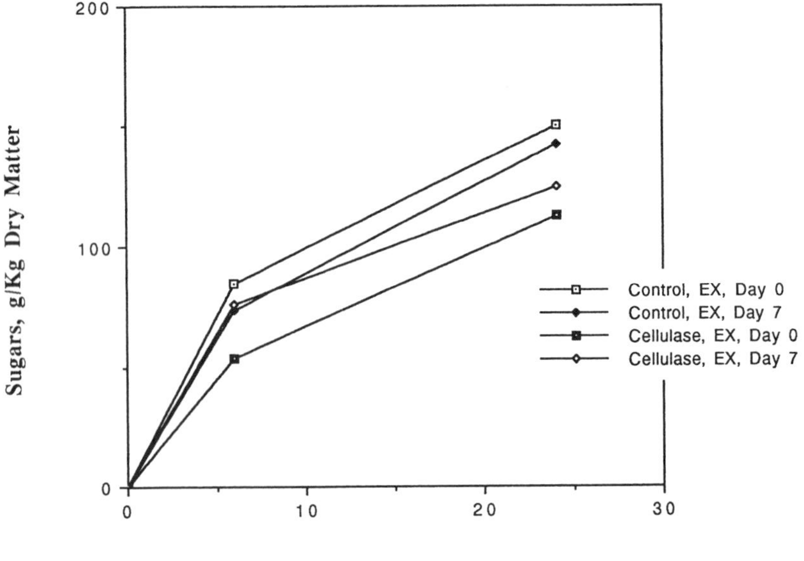

Figure 7. Comparison of corrected net sugar yield for alkaline extracted sweet sorghum silage.

appears to be more reactive than the day zero material. The control still exhibited greater reactivity than cellulase treated silage.

No clear benefits are obvious from the addition of cellulase to the sweet sorghum prior to ensiling as it pertains to the resulting biomass. Extracted fibers from cellulase-amended silage show decreased reactivity, indicating that conditions in the silo are conducive to increased reactivity but that cellulase addition lends very little of producing a more reactive fiber. The expense of adding cellulase to sweet sorghum prior to ensiling may not be warranted when considering the reactivity of the resulting fibers. However, the unextracted cellulase-amended silages contained more soluble carbohydrates prior to enzymatic hydrolysis (Tables I and II, Time zero of hydrolysis). These soluble sugars were easily fermented to ethanol, as will be seen in the discussion about ethanol production from these silages.

Extraction of the silages with 0.1 N NaOH has dramatic effects on the product distribution following enzymatic saccharification under operating conditions using 25 FPU per g cellulose at 38°C for 24 hours. One clear benefit to extraction is the increased hydrolysis of both xylan and cellulose after extraction with sodium hydroxide as seen by the increase in the disaccharide, cellobiose, glucose and xylose (Tables I and II).

What phenomena do the combination of ensiling and extraction catalyze? Extraction clearly made more of the fibers accessible to enzymatic hydrolysis. One condition is mildly acidic (pH 4 for seven days) and the other is alkaline with refluxing for one hour. According to our hypothesis stated above, the xylan backbone is debranched by the acidic condition of the silage, even at ambient temperatures. Xylan is deacetylated by saponification of esters during alkaline extraction. Deacetylated xylans are soluble in alkaline solutions, insoluble in water and are easily hydrolyzed under acidic conditions (*21*). When plant cell walls are treated with alkali, the amount of phenolic acids released correlates positively with increased rumen digestibilities (*40*). These consecutive sets of conditions produced a substrate that was readily hydrolyzed to the component sugars, cellobiose, glucose and xylose. By optimizing the reactor and reducing limitations imposed by the apparent product inhibition, the complete hydrolysis to glucose should be feasible.

Ethanol Production by SSF

After designated periods of storage, the silage was subjected to SSF; the data for ensiled sweet sorghum at day zero and for sorghum ensiled for 60 days are given in Figures 8 and 9, respectively. Day zero sweet sorghum silage produced more ethanol than did material that had been ensiled for sixty days, presumably because easily fermentable sugars are so abundant in the initial material. The effects of cellulase addition are immediately seen as the enzymes digested easily hydrolyzed fibers to glucose which, in turn, was converted to a higher percentage of ethanol when compared to the control sorghum, Figure 8.

For material ensiled for 60 days, cellulase-amended sorghum, series B, did boost ethanol yields by 25% compared to the control, series A, presumably because more sugars were present initially for the yeast (Figure 9). An initial lag period exists for approximately five hours which is followed by a sharp rise in ethanol production for the next four hours. After eight hours, ethanol production has ceased in the control

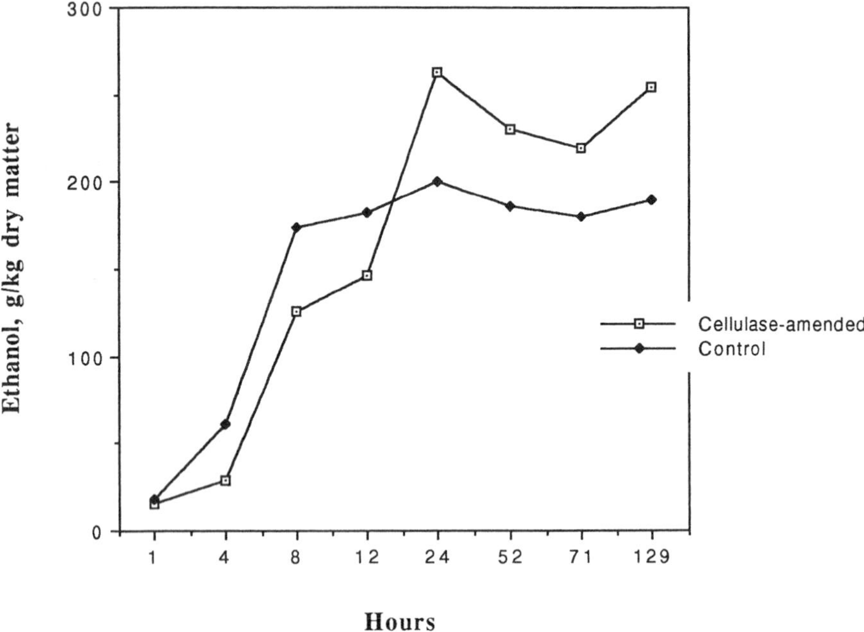

Figure 8. Ethanol production during SSF of unensiled sweet sorghum, day 0.

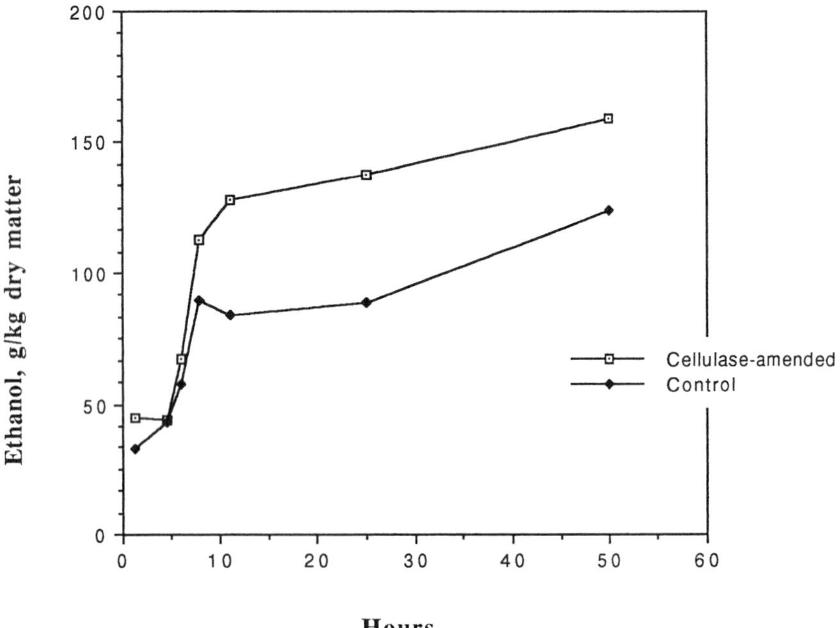

Figure 9. Ethanol production during SSF of ensiled sweet sorghum, day 60.

and is proceeding at a slower, steadier rate in the cellulase-amended silage. Ethanol production in the control increases again after 24 hours and actually proceeds at a faster rate from 24 to 48 hours than does the cellulase-silage. This difference in the rate of ethanol production may be reflective of the more reactive nature of the control fiber compared to the more recalcitrant fibers resulting from the cellulase-amended material. Remember, cellulase enzymes are present in the SSF broth and the information obtained from performing an operational enzymatic hydrolysis should provide some insight regarding the reactivity of the two silages during SSF.

Enzyme inhibition may be occurring during the SSF to ethanol, particularly in the early stages of the fermentations. As was discussed in the previous sections, cellobiose production may be due to an enzyme system deficient in β-glucosidase or due to product inhibition. From examining the rates of ethanol production from both day zero and day 60 silages, readily available sugars are fermented rapidly to ethanol. However, the lignocelluloic fractions are not being adequately hydrolyzed during SSF to continue the production of ethanol. Clearly more work is needed to improve both the hydrolsis yields of fermentable sugars and to improve ethanol production from ensiled sweet sorghum.

Conclusions

Our research intended to study storage effects on sweet sorghum ensiled with or without the addition of cellulase. Our objectives for this study were to

- evaluate the reactivity of the lignocellulosic portion of silage as it relates to the length of ensiling,
- monitor compositional changes in soluble sugar yields in both control and cellulase-amended sweet sorghum,
- assess the use of ensiling as a means of storage on a long-term basis,
- and evaluate sweet sorghum as an ethanol feedstock using SSF technology.

With these goals in mind, the following conclusions can be made:

Ensiling improves the reactivity of the lignocellulosic fibers to enzymatic hydrolysis. These results confirm our previous study on preserving potential fermentables by ensiling (*19*). However, the addition of cellulase may not act in a synergistic manner along with the acid conditions to improve subsequent hydrolysis of the ensiled material.

A rapid decline in pH is necessary in order to preserve the carbohydrates in sweet sorghum. Ensiling sweet sorghum provides an adequate, inexpensive means of storing the potential available carbohydrates found in sweet sorghum for long periods of time.

Benefits of cellulase as a silage additive are seen in the SSF process. Cellulase added to sweet sorghum before ensiling hydrolyzes easily accessible fibers, producing and maintaining glucose levels close to those found initially in sweet sorghum. Economics will be the deciding factor as to whether or not cellulase is beneficial as a silage additive. Will it be more cost effective to add cellulase to the silo so that SSF to ethanol can proceed with a 24-hour feed cycle to the fermentation vessel? If a twenty-four hour SSF is possible using cellulase-amended silage, will it be necessary to add additional cellulase to the SSF? Further study is required before the decision is made regarding the benefits of adding cellulase to sweet sorghum before ensiling.

Process design changes may improve the amount of sugar produced by the enzymatic hydrolysis. These changes include the alkaline extraction of silage stored for various amounts of time with the subsequent fermentation of cellobiose, glucose and xylose. A mixed culture of yeast capable of using all three sugars may be incorporated in future studies. Improved process design includes the following:

- Extracting soluble sugars with a nominal amount of water prior to the alkaline extraction of the lignocellulose. Extracts containing soluble sugars produced *in situ* can be returned to the SSF vessel prior to the ethanolic fermentation. It was noted that soluble sugars were degraded during the alkaline extraction of sweet sorghum silage prior to enzymatic hydrolysis (data not shown). This step will prevent the degradation of soluble sugars during alkaline extraction.

- Performing alkaline extractions of the lignocellulosic portion of the silage in as mild of conditions as possible to still be effective in deacetylating xylan. Mild conditions should prevent further degradation of xylose and other sugars obtained in the extracts.

- Fermenting the xylose obtained through alkaline extraction to ethanol. A mixed culture of yeast such as *Brettanomyces custersii* (*42*) plus *Pichia stipitis* (*43*) that are capable of utilizing cellobiose, glucose and xylose may then be used instead of *Saccharomyces cerevisiae*.

- Improving the reactor design to include some form of agitation. A plug-flow reactor may accomodate the fibrous nature of the sweet sorghum silage and still allow agitation within the vessel. Agitation will diminish product inhibition of the enzyme complex and allow increased mass transport throughout the system.

Other process design changes may include pressing or extraction of juice from freshly chopped sweet sorghum prior to ensiling. Worley and Cundiff (*17*) propose a similar process. After the juice is harvested, the resulting pith could be ensiled and stored on a long-term basis for use as an ethanol substrate. Residual sugars left in the presscake may be sufficient to produce enough lactic acid to store and maintain the material as silage until it could be converted to ethanol. If not, a nominal addition of cellulase and hemicellulase should liberate enough sugars from the presscake to sustain an lactic acid fermentation. McDonald characterizes lactic acid silages as those with a lactic acid concentration of between 80 to 120 g/kg dry matter (*22*). Therefore, it would require between 90 to 180 g of hexose per kg dry matter to produce a stable silage from either the presscake or from fresh sweet sorghum that was not pressed.

Preservation of sweet sorghum as silage remains an attractive method of storage. Economic analysis will determine if the addition of cellulase to sweet sorghum prior to ensiling is worthwhile. The loss in total carbohydrate to the lactic acid fermentation is less than 20%. This loss may be made up in lower equipment costs, particularly for dilute acid pretreatment. Lactic acid may become a valuable by-product in this process thus improving process economics.

Acknowledgments

This work was sponsored by the National Renewable Energy Laboratory Subcontract No. XC-1-11142-1 under the title of Evaluation of Feedstock Pretreatability and Saccharification/ Fermentation Suitability as a Function of Harvesting and Storage. Partial support also comes from the Colorado Institute for Research in Biotechnology graduate fellowship and the Colorado State University Agricultural Experiment Station Project No. 622.

Literature Cited

1. Belletti, A.; Petrini, C.; Minguzzi, A.; Landini, V.; Piazza, C.; Salamini, F. *Maydica.* **1991**, *36*, pp 283-291.
2. Smith, G.A.; Bagby, M.O.; Lewellam, R.T.; Doney, D.L.; Moore, P. *Crop Sci.* **1987**, *27*, 788-793.
3. Bryan, W.L.; Monroe, G.E.; Nichols, R.L.; Gascho, G.J. *American Society of Agricultural Engineers Paper No. 81-3571,* **1981.**
4. Jackson, D.R.; Lawhon, W.T. *Gasohol U.S.A.* **1981**, pp 10-12.
5. Kresovich, S. *Handbook of Biosolar Resources*, CRC Press Inc.: Boca Raton, FL, **1981**, pp 147-155.
6. Coble, C.G.; Shimulevich, I.; Egg, R.P. *Amer. Soc. of Agri. Engineers Paper No. 83-3563*, **1983**.
7. Eiland, B.R.; Clayton, J.E.; Bryan, W.L. *Trans. of the Amer. Soc. of Agri. Engineers* **1983**, *26*, 1596-1600.
8. Toledo, R.T. *Amer. Soc. of Agri. Engineers Paper No. 84-3585*, **1984**.
9. Lamb, M.E.; Von Bargen, K.; Bashford, L.L. *Amer. Soc. of Agri. Engineers Paper No. 82-3110,* **1982.**
10. Monroe, G.E.; Bryan, W.L.; Sumner, H.R.; Nichols, R.L. *Amer. Soc. of Agri. Engineers Paper No. 81-3541.*
11. Reidenbach, V.G.; Coble, C.G. *Trans. of the Amer. Soc. of Agri. Engineers*, **1985**, *28*, 571-575.
12. Gibbons, W.R.; Westby, C.A.; Dobbs, T.L. *Appl. and Environ. Microbiol.* **1986**, *51*, 115-122.
13. Rolz, C.; de Cabreara, S.; Garcia, R. *Biotechnol. Bioeng.* **1979**, *21*, 2347-2349.
14. Cundiff, J.S.; Vaughan, D.H.; Parris, D.J. *Proceedings of the Fifth Annual Solar and Biomass Energy Workshop.* **1985**, pp 133-136.
15. Worley, J.W.; Vaughan, D.H.; Cundiff, J.S. *Bioresource Technol.* **1992**, *40*, 263-273.
16. Kargi, F.; Curme, J.A.; Sheehan, J.J. *Biotechnol Bioeng.* **1985**, *27*, 34-40.
17. Worley, J.W.; Cundiff, J.S. *Trans. ASAE.* **1991**, *34*, 539-547.
18. Lueschen, W.E.; Putnam, D.H.; Kanne, B.K.; Hoverstad, T.R. *J. Prod. Agric.* **1991**, *4(4)*, 619-625.

19. Linden, J.C.; Henk, L.L.; Murphy, V.G.; Smith, D.H.; Gabrielson, B.C.; Tengerdy, R.P.; Czako, L. *Biotechnol. Bioeng.* **1987**, *30*, 860-867.
20. Henk, L.L.; Linden, J.C. *Enz. Microb. Technol.* **1992**, *14*, 923-930.
21. Barnett, A.J.G. *Silage Fermentation;* Butterworths Publications Ltd.: London, England, 1954.
22. McDonald, P. *The Biochemistry of Silage*; John Wiley & Sons, Ltd.: Chichester, England, 1981.
23. Tengerdy, R.P.; Weinberg, Z.G.; Szakacs, G.; Wu, M.; Linden, J.C.; Henk, L.; Johnson, D.E. *J. Sci. Food Agric.* **1991**, *55*, 215-228.
24. Linden, J.C.; Moreira, A.R.; Smith, D.H.; Hedrick, W.S.; Villet, R.H. *Biotechnol. Bioeng. Symp.* **1980**, *10*, 199-212.
25. Biely, P. *Trends Biotechnol.* **1985**, *3*, 286-290.
26. Morrison, I.M. *J. Agric. Sci. Camb.* **1979**, *93*, 581-586.
27. Brice, R.E.; Morrison, I.M. *Carbohyd. Res.* **1982**, *101*, 93-100.
28. Morrison, I.M. *Biochem. J.* **1974**, *193*, 197-204.
29. Das, N.N.; Das, S.C.; Duff, A.S.; Roy, A. *Carbohyd. Res.* **1981**, *94*, 73-82.
30. Joselean, J.P.; Kesraoui, R. *Holzforschung.* **1986**, *40*, 163-168.
31. Fry, S.C. *The Growing Plant Cell Wall: Chemical and Metabolic Analysis*; John Wiley and Sons, Inc.: New York, NY, 1988, pp. 103-119.
32. Knappert, D.; Grethlein, H.; Converse, A. *Biotechnol. Bioeng. Symp.* **1981**, *11*, 67-77.
33. Ghose, T.K. *Pure & Appl. Chem.* **1987**, *59(2)*, 257-268.
34. Miller, G.L. *Anal. Chem.* **1959**, *31*, 426-428.
35. Spindler, D.D.; Wyman, C.E.; Grohman, K. *Biotechnol. Bioengin.* **1989**, *34*, 185-189.
36. Jacobs, J.L.; McAllan, A.B. *Grass and Forage Sci.* **1991**, *46*, 63-73.
37. Leatherwood, J.M,; Mochrie, R.D.; Stone, E.J.; Thomas, W.E. *J. Dairy Sci.* **1963**, *46*, 124-127.
38. Narasimhalu, P.; Halliday, L.J.; Sanderson, J.B.; Kunelius, H.T.; Winter, K. A. *Can. J. Anim. Sci.* **1992**, *72*, 431-434.
39. Vanhatalo, A.; Varvikko, T.; Aronen, I. *Agric. Sci. Finl.* **1992**, *1*, 163-175.
40. VanSoest, P.J.; Mascaranhas-Ferreina, A.; Hartley, R.D. *Animal Feed Science and Tech.* **1984**, *10*, 155-164.
41. Spindler, D.D.; Wyman, C.E.; Grohmann, K.; Philippidis, G.P. *Biotechnol. Lett.* **1992**, *14(5)*, 403-407.
42. Bruinenberg, P.M.; de Bot, P.H.M.; van Dijken, J.P.; Scheffers, W.A. *Appl. Microbiol. Biotechnol.* **1984**, *19*, 256-260.
43. Furasaki, S.; Asai, N.; Hoshikawa, K. *J. Ferment. Technol.* **1985**, *63*, 523-528.
44. Hinman, N.D.; Schell, D.J.; Riley, C.J.; Bergeron, P.W.; Walter, P.J. *Appl. Biochem. Biotechnol.* **1992**, *34/35*, 639-649.
45. Moreira, A.R; Linden, J.C.; Smith, D.H.; Villet, R.H. In *Fuels from Biomass and Wastes,* Klan, D.L. and Emert, G.H., Eds.; Ann Arbor, MI, **1981**, pp 357-374.
46. Linden, J.C.; Murphy, V.G.; Henk, L.L.; Lange, K.D. *Amer. Soc. of Agri. Eng.* **1986**, Paper No. 86-6581.

RECEIVED March 17, 1994

Chapter 21

Conversion of Hemicellulose Hydrolyzates to Ethanol

James D. McMillan

Alternative Fuels Division, National Renewable Energy Laboratory, 1617 Cole Boulevard, Golden, CO 80401–3393

This chapter provides an overview of research directed at conversion of biomass-derived hemicellulose hydrolyzates generated by dilute acid pretreatment. The composition of hemicellulose hydrolyzates differs markedly from synthetic laboratory media formulations commonly used to assess pentose fermentation performance. In addition to containing sugars other than xylose, hemicellulose hydrolyzates contain a variety of components that are potent inhibitors when present at typical levels. As a result, hydrolyzates must be substantially detoxified prior to fermentation to achieve performance objectives. Following a brief review of the subject of pentose fermentation, performance data for conversion of highly detoxified hydrolyzates are tabulated for two organisms currently being evaluated for use in large scale pentose fermentation processes, *P. stipitis*, a wildtype xylose-fermenting yeast, and recombinant *E. coli*, an ethanologenic bacterium. Suspected inhibitory components present in biomass hydrolyzates are then identified, and their relative toxicity is qualitatively compared. Finally, methods commonly used to detoxify hydrolyzates prior to fermentation are summarized. Discussion focuses on relevant process considerations for selecting a biocatalyst for the pentose fermentation step of an integrated ethanol-from-biomass process.

Processes capable of efficiently converting hemicellulose hydrolyzates to ethanol are necessary to achieve high overall biomass-to-ethanol process yield and therefore favorable process economics (*1*). D-xylose, a pentose sugar, is the major carbohydrate component present in hemicellulose hydrolyzates generated by dilute acid pretreatment of a wide variety of lignocellulosic biomass species being considered as ethanol

0097–6156/94/0566–0411$09.26/0

feedstocks, including agricultural residues, herbaceous species, and short rotation hardwoods (*2-4*). Unfortunately, many microorganisms are unable to ferment xylose to ethanol. Notable examples include the yeast *Saccharomyces cerevisiae* and the bacterium *Zymomonas mobilis* (*5,6*), both of which are ethanologens traditionally favored for ethanol production from glucose (*7*). As a consequence, early research on xylose conversion investigated processes based on enzymatic isomerization of xylose to xylulose, which is a pentose sugar that can be fermented to ethanol by many glucose-fermenting organisms, including *S. cerevisiae*.

Since the early 1980s, however, a variety of wildtype yeast and fungi have been recognized that are capable of fermenting xylose to ethanol under oxygen-limited conditions. This has led to research to develop processes based on direct fermentation of xylose, although efforts continue to develop an isomerization-based process (*8-12*). The focus of this chapter is on the direct fermentation of hemicellulose hydrolyzates.

Performance on Rich Synthetic Media

Substantial progress has been made since the early 1980s towards achieving direct high-yield xylose fermentation in synthetic laboratory medium supplemented with pure xylose (*7,12-18*). A number of prospective wildtype yeast strains that perform well on laboratory medium formulations have been identified. The yeast *Pichia stipitis* is arguably the most promising species investigated, owing to its high ethanol selectivity and low xylitol coproduct formation in comparison with other wildtype yeast species. More recently, recombinant bacteria -- most notably strains of *Escherichia coli* and *Klebsiella oxytoca* -- have been genetically engineered to ferment xylose rapidly and at high yield (*6,19,20*). Table I summarizes representative high-yield performance values for wildtype xylose-fermenting yeasts and recombinant bacteria, the two classes of microorganisms now considered to be the best choices for achieving high-yield xylose fermentation. This table shows that, in rich synthetic medium, the best performing wildtype yeast and recombinant bacteria ferment xylose at yields greater than 0.40 g/g, achieve final ethanol concentrations of 4% (w/v) or higher, and exhibit volumetric productivities approaching (yeast) or above 1.0 g/L-h (bacteria). Included in this table are performance goals established at NREL for a pentose fermentation process for converting hemicellulose hydrolyzate. The goals are to ferment pentose sugars to ethanol at a yield of 0.46 g/g (90% of theoretical) and achieve a final ethanol concentration of 2.5% (w/v) or higher at an average volumetric productivity of 0.52 g/L-h. The economic analysis of Hinman et al. (*1*) is important because it shows that yield and final ethanol concentration rather than production rate are the keys to achieving cost-effective pentose fermentation. A comparison of the performance ranges provided in Table I indicates that in rich synthetic media under optimal operating conditions, the best performing microorganisms can achieve targeted performance objectives for yield, final ethanol concentration, and volumetric productivity.

Performance on Hemicellulose Hydrolyzates

Reviews encompassing conversion of hemicellulose hydrolyzates are provided by Schneider (*16*), Prior et al. (*21*), and Lawford and Rousseau (*22,23*). Hemicellulose

Table I. Representative High-Yield Xylose Fermentation Performance Results

Type of Microorganism	Yield (g/g)	Max. Ethanol %(w/v)	Productivity (g/L-h)	Productivity (g/g-h)
wildtype yeast (oxygen-limited)	0.40-0.48	3-5	0.30-0.90	0.20-0.40
recombinant bacteria (anaerobic)	0.43-0.49	2-4	0.30-1.30	0.30-1.30
NREL Performance Goals	0.46	2.5	0.52	

hydrolyzates are significantly different from synthetic laboratory media containing pure xylose. Hydrolyzates generated by dilute acid pretreatment, for example, typically contain a variety of sugar substrates and components that are inhibitory to microbial growth and fermentation. In addition to monomeric sugars other than D-xylose, D-glucose, D-mannose, D-galactose, and L-arabinose (*24*), hydrolyzates sometimes contain appreciable levels of oligosaccharides as a result of incomplete hydrolysis of hemicellulose polysaccharides. Often, a secondary dilute acid hydrolysis step is used after primary pretreatment to hydrolyze oligomeric sugars into monomeric sugars prior to fermentation (*25-28*). In addition to mixed sugars and oligosaccharides, inhibitory components such as acetic acid, furfural, and numerous aromatic compounds are also usually present in pretreated materials.

The presence of inhibitory components significantly reduces performance on untreated hydrolyzates. Consequently, to achieve rapid high-yield fermentation, hydrolyzates are usually detoxified prior to fermentation. Numerous reports of poor conversion performance on insufficiently detoxified hydrolyzates are available (*25,26,29-38*). Consistent with these reports, unpublished studies carried out at NREL show relatively poor bioconversion performance for a variety of xylose-fermenting microorganisms on neutralized poplar hardwood hemicellulose hydrolyzates produced by dilute acid pretreatment. Poor performance is attributed to inhibition of the microorganisms by the toxic components present in pretreated materials. A variety of methods can be used to reduce the concentration of inhibitory components to non-inhibitory levels, thus producing highly fermentable hydrolyzates. Common methods for detoxifying hydrolyzates include overliming and steam stripping, and a variety of other methods have been investigated as well (see below).

Table II shows a summary of favorable performance data on detoxified hydrolyzates by promising strains of *E. coli* and *P. stipitis*. Table II data suggest that these microorganisms can achieve NREL's performance objectives for yield, final ethanol concentration, and volumetric productivity if hydrolyzates are sufficiently detoxified prior to fermentation. As the table shows, both *E. coli* and *P. stipitis* strains have been reported to achieve yields and maximum ethanol concentrations exceeding performance goals. However, the reported yields are generally based on consumed sugar rather than total fermentable sugar and do not consider yield losses resulting from inoculum growth. Although data are not available, it is likely that process yields (g ethanol produced per g total fermentable sugars) would be below 0.46 g/g if consumption of substrate for inoculum cell mass production were included, particularly in cases where high inoculum levels have been used. For volumetric productivity, recombinant *E. coli* strains appear superior. *P. stipitis* only achieved productivities above 0.5 g/L-h when the fermentations were carried out at high cell concentrations.

It should be emphasized that many of the studies cited in Table II were carried out using hydrolyzates that contained significant levels of glucose (e.g., softwood-derived hydrolyzates). High glucose levels, when present, are likely to have a favorable impact on process performance, because of the higher energetic yield of, and the more metabolically efficient pathways for, glucose fermentation relative to pentose fermentation (*18,24*). In addition, in most of the studies cited in Table II the detoxified hydrolyzates were supplemented with rich nutrients (e.g., corn steep liquor, tryptone, and yeast extract), which improves performance relative to that in minimally

Table II. Performance of Candidate Process Microorganisms on Detoxified Hydrolyzates

Organism	Substrate	Pretreatment	Yield (g/g)	Ethanol Conc. (g/L)	Time (h)	Volumetric Productivity[a] (g/L-h)	Reference
Bacteria							
E. coli ATCC 11303 (pLOI297)	aspen	SO_2-Wenger extruder overlimed	0.47	16.9	28	0.60	Lawford and Rousseau (*56*)
E. coli KO11	pine	SO_2-water 30 min, 160 °C overlimed + sulfite	0.42[a]	31[b]	48	0.37[c] 1.53	Barbosa et al. (*54*)
E. coli KO11	corn hulls	dilute acid 30 min, 140 °C overlimed	0.50 (0.45)	38.1	48	0.79	Beall et al. (*55*)
E. coli KO11	corn cobs	dilute acid 10 min, 160 °C overlimed	0.51 (0.51)	20	24	0.83	Beall et al. (*55*) (5 g DCW/L inoculum)
E. coli ATCC 11303 (pLOI297)	aspen	SO_2-Wenger extruder overlimed	0.45	17.2	37	0.46	Lawford and Rousseau (*22*)
E. coli ATCC 11303 (pLOI297)	aspen	SO_2-Wenger extruder overlimed	0.45	29.6	48	0.62	Lawford and Rousseau (*23*)

[a] based on total sugars consumed (pentoses and hexoses); [b] inferred from graph; [c] based on pentose sugars; [d] yields in parenthesis are based on total sugars present initially (pentoses and hexoses); [e] maximum value; [f] average value;[g] neglecting initial lag; N.A. not available.

Continued on next page

Table II. *Continued*

Organism	Substrate	Pretreatment	Yield (g/g)	Ethanol Conc. (g/L)	Time (h)	Volumetric Productivity[a] (g/L-h)	Reference
Yeast							
P. stipitis CBS 5776	red oak	acid	0.46	9.9	N.A.	N.A.	Tran and Chambers (*26*)
P. stipitis R	aspen	SO_2-water 20 min, 150°C enzymes	0.47	41	48	0.85	Parekh et al. (*62*) (8.5 g DCW/L inoculum)
P. stipitis R	mixed pine and spruce	steam, SO_2 enzymes	0.43 (0.41)	55	70	0.79	Wayman et al. (*63*) (16 g DCW/L inoculum)
P. stipitis R	aspen	acid, steam stripped	0.44	8.9	24	0.37	Parekh et al. (*60*)
P. stipitis R	pine	acid, steam stripped	0.47	10.3	24	0.43	Parekh et al. (*60*)
P. stipitis	aspen and/or pine	SO_2-Wenger extruder	0.46[f]	24.0	N.A.	N.A.	Gans et al. (*82*)
P. stipitis CBS 5776	aspen	steam, enzymes	(0.47)	10.9	70	0.20[e]	Wilson et al. (*28*)

Continued on next page

Table II. *Continued*

Organism	Substrate	Pretreatment	Yield (g/g)	Ethanol Conc. (g/L)	Time (h)	Volumetric Productivity[a] (g/L-h)	Reference
P. stipitis CBS 5773	sugarcane bagasse	acid	0.35	24	50[g]	0.48[g]	Roberto et al. (*53*)
P. stipitis	sugarcane bagasse	acid	0.37 0.38	15.6 14.5	N.A.	0.56[e] 0.23[e]	van Zyl et al. (*47*)

[a] based on total sugars consumed (pentoses and hexoses); [b] inferred from graph; [c] based on pentose sugars; [d] yields in parenthesis are based on total sugars present initially (pentoses and hexoses); [e] maximum value; [f] average value #;[g] neglecting initial lag; N.A. not available

supplemented media (*22*). Finally, many of the studies achieved high productivities by inoculating at very high levels (> 5 g dry cell weight per liter [g DCW/L]). The use of large inocula will prevent a high process yield from being achieved, unless the cells can be recycled and reused.

Identification of Suspected Inhibitors

In addition to the acetic acid generated upon hydrolysis of acetylated hemicellulose (e.g., xylan), which is often present at inhibitory levels in hydrolyzates (*16*), a variety of other potentially inhibitory compounds are present in lignocellulosic hydrolyzates (*39,40*). Suspected inhibitory components include compounds that are hydrolyzed or solubilized during pretreatment, as well as products of carbohydrate degradation. Based on their source, three general classes of potentially inhibitory compounds exist:

(1). Compounds present in the raw biomass that are released by hydrolysis, such as extractive components, organic acids, and sugar acids esterified to backbones of hemicelluloses (acetic, glucuronic, and galacturonic), and components resulting from the hydrolysis of lignin (phenolic compounds).

(2) Compounds formed by degradation of solubilized oligomeric and monomeric sugars, such as furfural (from xylose) and hydroxymethylfurfural (HMF) (from glucose), as well as compounds resulting from reaction of other solubilized components such as those listed in (1).

(3) Metal ions and/or other compounds released as a result of equipment corrosion under acidic, high-temperature processing conditions.

Several studies have been carried out to identify suspected inhibitory components in hemicellulose hydrolyzates. A summary of compounds identified in biomass extracts and pretreatment hydrolyzates is provided in Table III. Metal ions are not listed, because levels vary widely depending on pretreatment method and pretreatment reactor materials of construction (*33,41,42*). A total of 16 acids, 8 alcohols, 10 aldehydes, and 1 ketone have been identified in a variety of lignocellulosic materials; many other compounds are also present, as indicated by peaks on HPLC chromatograms and mass spectra, but remain unidentified (*26,43*). Researchers have used a variety of analytical techniques as well as different biomass substrates. As a result, the levels and types of identified components vary widely from study to study, so it is difficult to make generalizations regarding compositional characteristics. Nonetheless, Table III provides a good indication of the types of suspected inhibitory materials that will be present in pretreated biomass feedstocks. As the middle column of this table shows, only a small number of the compounds that have been identified have actually been tested for inhibition.

Toxicity of Inhibitory Compounds

Acetic acid has been shown to be an important inhibitory component to both yeast and bacteria by a variety of researchers (*22,37,44-48*). The acetic acid generated upon

Table III. Compounds Identified in Biomass Extracts and Pretreatment Hydrolyzates

component	substrate	toxicity demonstrated?	microorganism	references
acids				
acetic acid	red oak	yes	*P. stipitis*	Tran and Chambers (*26*)
	eucalyptus	yes	*P. stipitis*	Ferrari et al. (*37*)
	wood and ag. residues	no		Kostenko and Nemirovskii (*69*)
caproic acid	red oak	yes	*P. stipitis*	Tran and Chambers (*26*)
caprylic acid	red oak	yes	*P. stipitis*	Tran and Chambers (*26*)
cinnamic acid	poplar	yes	*S. cerevisiae*	Ando et al. (*50*)
coumaric acid	wood and ag. residues	no		Kostenko and Nemirovskii (*69*)
formic acid	wood and ag. residues	no		Kostenko and Nemirovskii (*69*)
glucuronic acid	eucalyptus	no	*P. stipitis*	Ferrari et al. (*37*)
galacturonic acid	eucalyptus	no	*P. stipitis*	Ferrari et al. (*37*)
m-hydroxybenzoic acid	poplar	yes	*S. cerevisiae*	Ando et al. (*50*)
p-hydroxybenzoic acid	poplar	yes	*S. cerevisiae*	Ando et al. (*50*)
	wood and ag. residues	no		Kostenko and Nemirovskii (*69*)
levulinic acid	wood and ag. residues	no		Kostenko and Nemirovskii (*69*)
pelargonic acid	red oak	yes	*P. stipitis*	Tran and Chambers (*26*)
palmitic acid	red oak	yes	*P. stipitis*	Tran and Chambers (*26*)

Continued on next page

Table III. *Continued*

component	substrate	toxicity demonstrated?	microorganism	references
syringic acid	red oak	yes	*P. stipitis*	Tran and Chambers (*26*)
	poplar	no	*S. cerevisiae*	Ando et al. (*50*)
	birch	no		Buchert et al. (*43*)
syringylglycolic acid	birch	no		Buchert et al. (*43*)
vanillic acid	red oak	yes	*P. stipitis*	Tran and Chambers (*26*)
	poplar	no	*S. cerevisiae*	Ando et al. (*50*)
	birch	no		Buchert et al. (*43*)
	wood and ag. residues	no		Kostenko and Nemirovskii (*69*)
alcohols				
coniferyl alcohol	red oak	no		Tran and Chambers 1986
	poplar	no		Ando et al. 1986
	birch	no		Burchert et al. 1990
dihydroconiferyl alcohol	red oak	no		Tran and Chambers 1986
	birch	no		Burchert et al. 1990
dihydrosinapyl alcohol	red oak	no		Tran and Chambers 1986
	birch	no		Burchert et al. 1990
3,5-dimethoxy-4-hydroxycinnamyl alcohol	poplar	no		Ando et al. 1986
β-oxysinapyl alcohol	red oak	no		Tran and Chambers 1986
sinapyl alcohol	birch	no		Burchert et al. 1990

Continued on next page

Table III. *Continued*

component	substrate	toxicity demonstrated?	microorganism	references
1-syringylacetol	birch	no		Burchert et al. 1990
syringyl glycerol	birch	no		Burchert et al. 1990
aldehydes				
cinnamaldehyde	poplar	yes	*S. cerevisiae*	Ando et al. 1986
coniferyl aldehyde	birch	no		Burchert et al. 1990
furfural (2-furaldehyde)	red oak	yes	*P. stipitis*	Tran and Chambers 1986
	wood and ag. residues	no		Kostenko and Nemirovskii 1992
p-hydroxybenzaldehyde	poplar	yes	*S. cerevisiae*	Ando et al. 1986
	birch	no		Burchert et al. 1990
	wood and ag. residues	no		Kostenko and Nemirovskii 1992
p-hydroxycinnamaldehyde	poplar	no		Ando et al. 1986
5-hydroxymethylfurfural	birch	no		Burchert et al. 1990
	wood and ag. residues	no		Kostenko and Nemirovskii 1992
sinapaldehyde	red oak	no		Tran and Chambers 1986
sinapyl aldehyde	birch	no		Burchert et al. 1990
syringaldehyde	red oak	yes	*P. stipitis*	Tran and Chambers 1986
	poplar	yes	*S. cerevisiae*	Ando et al. 1986
	birch	no		Burchert et al. 1990
	wood and ag. residues	no		Kostenko and Nemirovskii 1992

Continued on next page

Table III. *Continued*

component	substrate	toxicity demonstrated?	microorganism	references
vanillin (4-hydroxy-3-methoxybenzaldehyde)	red oak	yes	*P. stipitis*	Tran and Chambers 1986
	poplar	yes	*S. cerevisiae*	Ando et al. 1986
	birch	no		Burchert et al. 1990
	wood and ag. residues	no		Kostenko and Nemirovskii 1992
other				
lignofuran substances	wood and ag. residues	no		Kostenko and Nemirovskii 1992
syringoyl methyl ketone	red oak	no		Tran and Chambers 1986

hydrolysis of acetylated xylan is the inhibitory component present at the highest concentrations in a variety of hemicellulose hydrolyzates, as shown in many of the studies cited above. The pK_a of acetic acid is 4.8 at 30°C (*49*). Because it is the protonated form of the acid that is inhibitory, the toxicity of acetic acid increases at lower pH, as shown in Figure 1. Conducting fermentation at a pH well above the pK_a reduces acetic acid inhibition, and enables improved performance to be achieved (*22,37,44,47*).

Only a few studies have investigated inhibition of growth or fermentation performance using pure forms of inhibitory components other than acetic acid. Studies have generally used different microorganisms as well as different levels of pure forms of suspected inhibitory components. For comparative purposes, the most comprehensive studies to date were analyzed to determine the relative toxicity of tested compounds. Inhibition was quantified in terms of performance relative to control fermentations in which no inhibitors were present. Relative performance was assessed on the basis of ethanol concentrations (*26*), ethanol yields (*50*), non-ethanol solvent concentrations (*51*), and the extent of growth (*51,52*). The relative degree of inhibitor toxicity was calculated for each of the compounds tested in terms of percent inhibition of performance per mmol of inhibitory compound. Relative toxicity levels were grouped into qualitative categories of weak, moderate, strong, and extreme, indicating increasing levels of specific toxicity. Results of this analysis are summarized in Table IV.

There are three noteworthy points that are evident upon examination of Table IV. First, although acetic acid is generally considered to be the most significant toxic component in hydrolyzates, its significance as an inhibitor derives from the fact that it is present at high concentrations. As Table IV shows, acetic acid is only weakly inhibitory on a percent inhibition per mmol basis. Second, numerous compounds show strong to extreme relative toxicity, most notably the aromatic aldehydes and acids and C_6 to C_9 straight-chain organic acids. The analysis of data from Tran and Chambers (*26*) shows that C_6 to C_9 organic acids are most inhibitory to *P. stipitis* on a per mmol basis (C_9 pelargonic > C_8 caprylic > C_6 caproic), followed by aromatic aldehydes (vanillin > syringaldehyde) and aromatic acids (vanillic > syringic). The relative toxicity of furfural and acetic acid was much lower than that of other compounds tested. No comprehensive studies on the inhibition of *E. coli* have been performed. A rough estimate can be obtained from the results of Nishikawa et al. (*51*) for *Klebsiella pneumoniae*, since this organism is physiologically similar to wildtype *E. coli*. Of the 7 monoaromatic components tested, the aromatic aldehydes, p-hydroxybenzaldehyde and syringaldehyde, were the most toxic, followed by p-hydroxybenzoic acid and then vanillin. Vanillic and syringic acids were moderately inhibitory. The third point evident from Nishikawa et al.'s (*51*) data is that sensitivity to the presence of particular inhibitory components is different for fermentation and growth.

Efforts have been made to correlate inhibition with compound structure, particularly for the aromatic species (*50*). Unfortunately, generalizations are difficult to make because organisms exhibit different sensitivities to inhibitory compounds. For example, syringic acid is moderately inhibitory to *P. stipitis* and *K. pneumoniae* (see Table IV) when present at levels of 0.4 mmol (0.08 g/L) (*26*) and 1.0-2.5 mmol (0.2-0.5 g/L), respectively, but is only weakly inhibitory to *Candida utilis* at 20 mmol

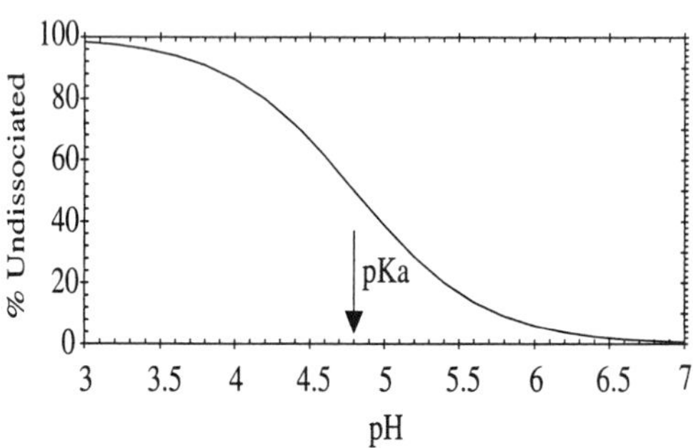

Figure 1. Percent Undissociated Acetic Acid as a Function of pH.

Table IV. Relative Toxicity of Inhibitory Compounds

Relative Degree of Inhibition Among Compounds Tested* (% inhibition/mMol)				References
weak	moderate	strong	extreme	
furfural acetic acid	palmitic acid syringic acid	vanillin caproic acid syringaldehyde vanillic acid	pelargonic acid caprylic acid	Tran and Chambers (*26*) (inhibition of fermentation)
m-hydroxybenzoic acid vanillic acid	p-hydroxybenzoic acid vanillin syringaldehyde	p-hydroxybenzaldehyde	cinnamaldehyde cinnamic acid	Ando et al. (*50*) (inhibition of fermentation)
vanillyl alcohol	vanillic acid syringic acid	p-hydroxybenzaldehyde syringaldehyde p-hydroxybenzoic acid vanillin		Nishikawa et al. (*51*) (inhibition of fermentation)
vanillin vanillyl alcohol	vanillic acid p-hydroxybenzoic acid syringic acid	p-hydroxybenzaldehyde syringaldehyde		Nishikawa et al. (*51*) (inhibition of growth)

Continued on next page

Table IV. *Continued*

Relative Degree of Inhibition Among Compounds Tested* (% inhibition/mMol)				References
weak	moderate	strong	extreme	
gallic acid p-hydroxybenzoic acid vanillyl alcohol syringic acid conipheryl alcohol	guaiacol p-hydroxybenzaldehyde vanillic acid 2,6-dimethoxyphenol phenol	acetovanillon pyrocatechol vanillin syringaldehyde	o-vanillin cinammic acid pyrogallol salicylic acid benzoic acid 3,4-dimethoxybenzaldehyde	Mikulášová et al. (*52*) (inhibition of growth)

*Tran and Chambers (*26*) investigated inhibition of ethanol fermentation by *P. stipitis* using concentrations of pure compounds similar to those detected in red oak hydrolyzates; Ando et al. (*50*) investigated inhibition of ethanol fermentation by *S. cerevisiae* using 1.0 g/L concentrations of pure compounds; Nishikawa et al. (*51*) investigated inhibition of growth and 2,3-butanediol fermentation by *Klebsiella pneumoniae* using a range of sub-1.0 g/L concentrations of pure compounds; Mikulášová et al. (*52*) determined LD_{50} concentrations for *C. utilitis* growth inhibition.

(4 g/L) (*52*). It is not inhibitory at all to *S. cerevisiae* at 5 mmol (1.0 g/L) (*50*). Similarly, *P. stipitis*, *K. pneumoniae*, and *C. utilis* exhibit higher sensitivity to vanillin than *S. cerevisiae* (Table IV).

Detoxification Treatments

Hydrolyzate composition varies significantly depending upon the type of lignocellulosic feedstock and the pretreatment method employed. In general, however, inhibitory compounds are present at sufficiently high levels that detoxification procedures beyond direct neutralization to fermentation pH are required to achieve favorable conversion performance. A variety of detoxification treatments have been used to increase the fermentability of acidic hemicellulose hydrolyzates, including neutralization, overliming, heating, steam stripping, rotoevaporation, ion exchange, extraction, and treatment with activated carbon and molecular sieves. Research reports investigating these different methods are listed in Table V.

Overliming acid hydrolyzates with calcium hydroxide (lime) to pH 10 or higher and then reacidifying to fermentation pH is the detoxification technique cited most frequently in recent reports, although additional treatments such as steam stripping or heating or addition of sulfite are often used in combination with overliming to achieve even greater detoxification than is possible using overliming alone (*53-58*). The mechanism of action of overliming remains unclear. Strickland and Beck (*33*) showed that the benefit of sulfite addition is reduced although not entirely eliminated when hydrolyzates are overlimed or heated (as opposed to just neutralized), suggesting that there may be similarities in the mechanisms by which these different approaches detoxify hydrolyzates. Calcium hydroxide has been shown to be superior to other bases such as potassium hydroxide and sodium hydroxide (*32,46*). Substantial precipitation of calcium sulfate (gypsum) occurs when calcium hydroxide is used to neutralize or overlime hydrolyzates generated using sulfuric acid. Researchers usually remove the precipitate by filtration or centrifugation to improve hydrolyzate fermentability. It is likely that some inhibitory components are precipitated by divalent calcium ions, or bind to or otherwise associate with precipitated solid gypsum (*46*). However, conflicting literature exists concerning the extent to which inhibitory components are removed by overliming. For example, Strickland and Beck (*32,33*) report that overliming treatment reduces the concentration of furfural and possibly of metal ions, and van Zyl et al. (*46*) report a 17% decrease in acetic acid concentration following neutralization of sugarcane bagasse hemicellulose hydrolyzate with lime. In contrast, Frazer and McCaskey (*59*) found that acetic acid and furfural concentrations in oak hydrolyzates were unchanged following overliming treatment. Rather, they observed that overliming resulted in a 25% decrease in the concentration of phenolic compounds.

After overliming, the most widely cited detoxification methods include heating, steam stripping, and vacuum distillation, which remove acetic acid and other volatile inhibitory components. Steam stripping in particular has been used successfully as a stand-alone method to increase hydrolyzate fermentability (*60,61*). Several other methods have been used to substantially reduce the concentration of phenolic inhibitory components (and in some cases the concentration of acetic acid as well), most notably ion exchange and solvent extraction. Frazer and McCaskey (*59*) used

Table V. Summary of Methods Used to Detoxify Biomass Hydrolyzates

Detoxification method	References
Neutralization (includes pH shifts such as overliming, as well as precipitation of lignin using acid)	Strickland and Beck (*32,33*) Fein et al. (*31*) Tran and Chambers (*25,26*) van Zyl et al. (*46*) Frazer and McCaskey (*59*) Perego et al. (*70*) Wenzhi et al. (*64*) Chahal et al. (*83*) Roberto et al. (*53*) Barbosa et al. (*54*) Beall et al. (*19*) Lawford and Rousseau (*56-58*)
Removal of volatiles (includes steam stripping, vacuum distillation, and rotary evaporation)	Slapack et al. (*44*) Fanta et al. (*84*) Strickland and Beck (*33*) Beck (*42*) Yu et al. (*61*) Parekh et al. (*60*) Wilson et al. (*28*) Gans et al. (*82*) Chahal et al. (*83*) Buchert et al. (*43*) Roberto et al. (*53*) Barbosa et al. (*54*)
Ion exchange (includes cation and anion exchange resins as well as ion exclusion chromatography)	Watson et al. (*41*) Fein et al. (*31*) Tran and Chambers (*25,26*) Frazer and McCaskey (*59*) van Zyl et al. (*47*) Wenzhi et al. (*64*) Buchert et al. (*43*) Ferrari et al. (*37*)
Extraction (includes washing with water and extraction with organic solvents)	Mes-Hartree and Saddler (*77*) Slapack et al. (*44*) Frazer and McCaskey (*59*) Wilson et al. (*28*)
Adsorption (includes treatment with activated carbon or molecular sieve)	Fein et al. (*31*) Tran and Chambers (*25,26*) Frazer and McCaskey (*59*) Roberto et al. (*53*)

ion exchange to detoxify hydrolyzate prior to fermentation by *K. pneumoniae*, for example, achieving greater than 50% removal of phenolic components. In general, however, stand-alone treatments are insufficient to achieve substantial detoxification. Rather, detoxification treatments that utilize a combination of methods must be used to produce highly fermentable hydrolyzates. Examples include using rotoevaporation followed by extraction with ethyl acetate to make aspen hemicellulose hydrolyzates readily fermentable by *P. stipitis* (*28*), and the use of heating and overliming to render pine hemicellulose hydrolyzates highly fermentable by recombinant *E. coli* (*54*). In each of these cases, the first technique principally removes volatile components such as acetic acid and furfural (rotoevaporation and heating, respectively), whereas the second removes phenolic and other organic components (extraction and overliming, respectively).

Other Approaches to Alleviating Inhibition

In addition to detoxification of hydrolyzates prior to fermentation, process operating conditions can be manipulated to minimize inhibition of fermentation performance. As discussed above, operation at or near neutral pH has been used successfully by a number of researchers to minimize inhibition by acetic acid. Another approach that has been successful is to inoculate at high initial cell concentrations (*27,31,55,61-64*). A somewhat related approach, suggested by a number of researchers, is to adapt strains to achieve greater tolerance to inhibitory compounds present in biomass hydrolyzates (*25,26,31,44,45,65*). This approach is particularly intriguing in the context of continuous processes provided cell recycle is possible. However, Wayman and co-workers are the only researchers to report substantial success using adaptation (*61,62,66*). Highly productive strains of xylose-fermenting yeasts were developed by repeated culture of yeasts on aspen hydrolyzates supplemented with urea and diammonium phosphate (*60*). A *P. stipitis* strain was reportedly adapted after only 12 passes, whereas adaptation of a *C. shehatae* strain required 72 passes.

Discussion

Feasibility of Incorporating Detoxification Treatments. A critical examination of research conducted on conversion of hemicellulose hydrolyzates suggests that NREL's performance goals for pentose fermentation are attainable, possibly by *P. stipitis* strains and certainly by recombinant *E. coli* constructs, provided hydrolyzates are sufficiently detoxified prior to fermentation. As the literature demonstrates, hydrolyzates can be substantially detoxified using a variety of methods. However, detoxification treatments generally remain chemical or energy intensive, or both. The incorporation of the detoxification procedures employed to achieve performance values reported in Table II into an overall process would increase process complexity and impose costs not currently considered in technoeconomic assessments of biomass-to-ethanol processes. Unfortunately, studies on detoxification treatments have not considered the economic impact of detoxification on overall production costs, and the economic feasibility of incorporating post-pretreatment detoxification remains uncertain. Sensitivity analyses are urgently needed to determine the extent to which the additional energy or chemical inputs required for hydrolyzate detoxification will affect ethanol production costs.

Compositional Similarities Among Biomass Hydrolyzates. Reported research demonstrates that hydrolyzate composition is highly dependent upon biomass species and pretreatment characteristics. Hydrolyzates from a variety of sources share some compositional features, because acids (e.g., acetic, glucuronic, and galacturonic) and carbohydrate breakdown products such as furfural and HMF are generally present. However, levels of these compounds vary widely depending upon pretreatment methodogy. The composition and concentration of components present at lower levels, in particular lignin-derived monoaromatic species, are extremely sensitive to pretreatment conditions as well as biomass type. In general, syringyl-based lignin breakdown products are found at significant levels in hardwood hydrolyzates. Buchert et al. (*43*) observed that although birch contains equal proportions of guaiacyl- and syringyl-based lignin, syringyl derivatives are more abundant in birch hydrolyzates. They hypothesized that this occurs because a high percentage of guaiacyl units are present in highly condensed regions of lignin. Other researchers, however, have stated that hardwoods are primarily composed of syringyl-based lignin (*67*), in which case the presence of guaiacyl-based inhibitors in hardwood hydrolyzates would not be expected. It remains necessary and critically important to more accurately determine the composition of hydrolyzates and to identify which components are the major causes of observed process inhibition.

Significantly, studies often do not report whether pretreated material is derived from whole tree (i.e., not debarked) or debarked material. It is likely that bark-derived extractives are highly inhibitory to microbial activity, owing to the fact that bark functions in living trees as a protective barrier to microbial attack. Unfortunately, bark-associated inhibitory compounds have not been investigated to any significant extent.

Significant compositional differences are anticipated between hardwoods and herbaceous species, both of which are considered as potential energy crops. Acetic acid levels in hemicellulose hydrolyzates of herbaceous species, although largely uninvestigated, should be approximately 50% lower than is typically observed in hardwood hydrolyzates owing to the lower degree of hemicellulose acetylation in herbaceous species relative to hardwoods (*2-4*). On the other hand, coumaric, ferulic, and glucuronic acid concentrations will probably be higher in hydrolyzates of herbaceous species than in those of hardwoods (*68,69*). Research being carried out at NREL will determine the extent to which these projections are correct and may also identify other significant differences in inhibitor composition that exist between hardwood and herbaceous hydrolyzates.

Opportunities to Improve Efficiency of Detoxification Treatments. Ethanol-from-biomass process economy and efficiency will be improved by reducing the cost and complexity of detoxification treatment(s). The possibility exists that phenolic components can be removed from hydrolyzates without significant additional chemical or energy (e.g., steam) inputs. Many monomeric and oligomeric lignin components released during pretreatment are highly unstable and can rapidly condense at room temperature into higher molecular weight species (*43*). This observation suggests that techniques might be developed to condition and then filter hydrolyzates to remove precipitated phenolic components prior to fermentation. Although there is limited definitive evidence supporting the efficacy of this type of approach, Perego et al. (*70*) report improved efficacy of overliming when the temperature of hydrolyzate is raised prior to final filtration. However, although it is possible to develop chemically and energetically efficient methods for reducing the toxicity caused by phenolic components, developing efficient processes for removal of components such as acetic acid and furfural, which are usually present at much higher (i.e., greater than 1 g/L) concentrations, will undoubtedly prove challenging.

Process Considerations Influencing Biocatalyst Selection.

Substrate Range. Several monomeric sugars other than D-xylose are typically present in hydrolyzates, including D-glucose, D-mannose, D-galactose, and most significantly, L-arabinose (*24*). In addition, hydrolyzates often contain appreciable levels of oligosaccharides. It is important that microorganisms selected for conversion of such hydrolyzates have the ability to ferment these sugars rapidly and at high yield to maximize conversion performance. Broad substrate range and reduced sensitivity to catabolite repression by glucose (or other sugars) are key attributes to be considered when selecting a biocatalyst.

Diauxic sugar utilization patterns in which glucose is preferentially utilized prior to other sugars have been observed for a number of ethanologenic fungi and yeast (*24,29,30*) and bacteria (*24*). The problem of reduced volumetric productivity as a result of lags in the onset of utilization of other sugars following glucose depletion may not be a significant problem, however, as glucose levels in hemicellulose hydrolyzates are often low. However, differences in substrate range could be important, particularly if economically significant levels of sugars other than glucose and xylose are present.

In terms of substrate range, bacteria appear superior to yeast (*24*). Some of the recombinant ethanologenic bacterial strains recently developed are capable of fermenting virtually all of the monomeric sugars present in lignocellulosic materials, as well as some oligomeric carbohydrates, to ethanol (*20,71,72*). Some xylose-fermenting yeasts (and many fungi) can ferment sugars other than xylose to ethanol, however. In addition to xylose, *P. stipitis, P. tannophilus, and C. shehatae* are capable of fermenting glucose, mannose, and galactose to ethanol (*73,74*). Some strains of *P. stipitis* can also ferment cellobiose and xylobiose (*21,28*). Significantly, however, arabinose cannot be fermented to ethanol by *P. stipitis* (*74*) or *C. shehatae* (*74,75*). These findings have been confirmed in research conducted at NREL, and it appears that the inability to ferment arabinose is a general characteristic of wild type xylose-fermenting yeasts (*24*). Thus there is a significant disadvantage to using yeasts such as *P. stipitis* for the conversion of hemicellulose hydrolyzates of agricultural residues and herbaceous crop species, because these feedstocks contain economically significant levels of arabinose (*2-4*).

Tolerance to Inhibitory Components. The concentrations of inhibitory components in hemicellulose hydrolyzates vary depending upon the type of lignocellulosic material used and the pretreatment and detoxification methods and conditions employed.

However, if process water is recycled for cell and enzyme recovery or nutrient reuse, or both, inhibitory components are likely to eventually accumulate to high levels regardless of their level following detoxification. Therefore, in addition to exhibiting the ability to hydrolyze and ferment a variety of sugars, microorganisms and enzymes selected or developed, or both, for converting pretreated biomass hydrolyzates need to be able to tolerate inhibitory components.

Tolerance to acetic acid, arguably the dominant inhibitory component in hardwood hydrolyzates, is strongly pH dependent. Carrying out fermentation at a pH well above the pK_a of acetic acid substantially reduces inhibition by acetic acid, but primarily limits the selection of microorganisms that can be considered to recombinant bacteria because wild type xylose-fermenting yeast perform poorly at or near neutral pH. Studies by du Preez et al. (*74*), for example, show that the performance of *P. stipitis* falls off at pH greater than 5.5. Therefore, the frequently touted advantage of yeast exhibiting good performance at low pH where the potential for contamination is lower may be overcome by the need to operate at higher pH (where performance is poor) to reduce inhibition by acetic acid. In contrast to wild type xylose-fermenting yeasts, recombinant bacteria perform well at or near neutral pH where acetic acid inhibition is minimized (*22,23,54-56*).

Tolerance to potentially inhibitory compounds other than acetic acid should also be considered when selecting biocatalysts for fermentation of hemicellulose hydrolyzates. However, only limited studies of the tolerance of microorganisms to inhibitory components have been performed, primarily to assess tolerance to furfural. Furfural completely inhibits growth of *P. stipitis*, *Pachysolen tannophilus*, and *S. cerevisiae* at concentrations above 2-3 g/L (*30,53,76*). Similarly, furfural concentrations above 1.1 g/L inhibit performance of recombinant *E. coli*, with complete inhibition occurring at 4.5 g/L (*55*). However, furfural and HMF are not inhibitory to cellulase enzyme activity at concentrations up to 5 g/L (*77*). Furfural is reportedly assimilated by *S. cerevisiae* (*76*).

Microorganisms often can assimilate or transform inhibitors, or both, to potentially less toxic compounds. Transformation of furfural to furfuryl alcohol has been reported for *P. stipitis* (*78*) and *P. tannophilus* (*27*). Weigert et al. (*78*) showed that furfuryl alcohol causes a linear decrease in growth of *P. stipitis* with increasing concentration, with 50% inhibition occurring at a concentration of 1.6 g/L and complete inhibition occurring at 3.2 g/L. Aromatic aldehydes such as vanillin and syringaldehyde are also transformed by some organisms. *K. pneumoniae* is reported to convert vanillin and syringaldehyde to vanillyl alcohol and syringyl alcohol, respectively, and to further metabolize these compounds to other unidentified products (*51*). Conversion of vanillin to vanillyl alcohol is also reportedly carried out by *S. cerevisiae* (*51,78*). Weigert et al. (*78*) hypothesized that transformation of furfural to furfuryl alcohol is catalyzed by alcohol dehydrogenase (ADH) enzyme, and it is likely that this is also the mechanism by which other aldehydes are reduced to their corresponding alcohols. Unfortunately, the literature in this area is sparse, and further research will be required to determine if bacteria or yeast are superior biocatalysts in their ability to tolerate and detoxify hydrolyzates.

Outlook

Xylose-fermenting yeasts such as *P. stipitis* show good potential for achieving high-conversion yields on detoxified hydrolyzates. However, aeration is required and there is a significant performance trade-off between yield and productivity (*79*). In contrast, recombinant enteric bacteria achieve high yield and high productivity under anaerobic conditions. Moreover, recombinant bacteria can ferment arabinose and also perform well at high pH where acetic acid inhibition is reduced. These are significant advantages. Broad substrate range enables process yields to be maximized. Ability to ferment at or near neutral pH eliminates, or at least decreases the need to reduce the concentration of acetic acid by steam stripping, ion exchange, or other methods. For these reasons, the recombinant *E. coli* and *K. oxytoca* strains recently developed by Ingram and co-workers are currently the best candidates for conversion of hydrolyzates. Given the pace of developments in metabolic engineering, however, it is unclear how long these organisms will remain the front-runners for use in large-scale conversion processes.

The development of even better recombinant strains is anticipated in the near future. For example, the development of recombinant xylose-fermenting *S. cerevisiae* yeast strains is likely to occur soon, because numerous laboratories around the world are pursuing this. *S. cerevisiae* strains are historically recognized as having higher ethanol tolerance than bacteria, and being more tolerant of acidic conditions. Although xylose-fermenting yeasts do not perform well at low pH in real hydrolyzates, *S. cerevisiae* ferments hexoses well at low pH and in non-detoxified substrates (*80*). Under comparable conditions, *S. cerevisiae* is able to ferment glucose at volumetric productivities more than threefold higher than those achieved on xylose by the best xylose-fermenting yeast (*81*). It will be interesting to examine how the poorer energetics of xylose fermentation affect the productivity of xylose fermentation by recombinant yeast.

In addition to the progress in metabolic engineering, the direction of future hemicellulose conversion research depends on the impact that detoxification treatment(s) is projected to have on overall process economics. If detoxification treatments are determined to be cost-prohibitive, development of xylose-fermenting microbes with much greater tolerance to inhibitory components will need to be pursued unless substantially improved pretreatment processes can be developed which produce highly fermentable hydrolyzates. On the other hand, if economically sound detoxification procedures can be developed, existing microbes should be sufficient for achieving technical performance objectives.

Conclusions

Detoxification of hydrolyzates increases pentose fermentation performance to the point that performance goals are achievable, although the economic feasibility of incorporating detoxification treatments into ethanol production processes is unknown and requires further exploration. A number of putatively inhibitory components have been identified in pretreated lignocellulosic materials. However, only a limited number of these compounds have actually been tested for toxicity to microbes currently favored for use in pentose fermentation processes. Acetic acid is the inhibitory component typically present at the highest levels, although its toxicity on a per mmol basis is considerably lower than that of other inhibitory components. Approaches for minimizing inhibition by acetic acid include removing acetic acid prior to fermentation and conducting fermentations at or near neutral pH. Alleviation of inhibition by acetic acid and other toxic components can also be achieved by using chemically intensive detoxification treatments and acclimatizing strains, or using large inocula and nutrient supplementation.

Acknowledgement

This work was supported by the Ethanol from Biomass project of the Biofuels Systems Division of the U.S. Department of Energy.

Literature Cited

1. Hinman, N.D.; Wright, J.D.; Hoagland, W.; Wyman, C.E. *Appl. Biochem. Biotechnol.* **1989,** *20/21*, 391-401.
2. Torget, R.; Werdene, P.; Himmel, M.; Grohmann, K. *Appl. Biochem. Biotech.* **1990**, *24/25*, 115-126.
3. Torget, R.; Walter, P.; Himmel, M.; Grohmann, K. *Appl. Biochem. Biotech.* **1991,** *28/29*, 75-86.
4. Torget, R.; Himmel, M.; Grohmann, K. *Appl. Biochem. Biotech.* **1992**, *34/35*, 115-123.
5. Barnett, J.A.; Payne, R.W.; Yarrow, D. *Yeasts: Characteristics and Identification*; Cambridge University Press: Cambridge, MA, 1983; pp 467-469.
6. Ingram, L.O.; Alterthum, F.; Ohta, K.; Beall, D.S. *Developments in Industrial Microbiol.* **1990,** *31*, 21-30.
7. Kosaric, N.; Wieczorek, A.; Cosentino, G.P.; Magee, R.J; Prenosil, J.E. *Ethanol Fermentation,* In *Biotechnology*; Rehm, H.J.; Reed, G., Eds.; Verlag Chemie: Weinheim, Germany, 1983, Vol. 3; pp 257-385.
8. Lindén, T.; Hahn-Hägerdal, B. *Enzyme Microb. Technol.* **1993**, *15*, 321-324.
9. Lastick, S.M.; Tucker, M.Y.; Beyette, J.R.; Noll, G.R.; Grohmann, K. *Appl. Microbiol. Biotechnol.* **1989,** *30*, 574-579.
10. Lastick, S.M.; Mohagheghi, A.; Tucker, M.P.; Grohmann, K. *Simultaneous Fermentation and Isomerization Xylose to Ethanol*; NREL Technical Report TP-231-3396, National Renewable Energy Laboratory: Golden, CO, 1990; pp 248-264.
11. Lastick, S.M.; Mohagheghi, A.; Tucker, M.P.; Grohmann, K. *Appl. Biochem. Biotechnol.* **1990,** *24/25*, 431-439.
12. Hahn-Hägerdal, B.; Lindén, T.; Senac, T.; Skoog, K. *Appl. Biochem. Biotechnol.* **1991,** *28/29*, 131-144.
13. Jeffries, T.W. *Trends Biotechnol.* **1985**, *3*, 208-212.
14. *CPD-5: The Conversion of C5 Sugars to Ethanol*; Manderson, G.J., Ed.; International Energy Agency/Forestry Energy Agreement, Cooperative Program D-5 Report No. 1; Massey University, Massey, New Zealand, 1985.
15. Skoog, K.; Hahn-Hägerdal, B. *Enzyme Microb. Technol.* **1988,** *10*, 66-80.
16. Schneider, H. *Crit. Rev. Biotechnol.* **1989,** *9(1)*, 1-40.

17. Jeffries, T.W. In *Yeast: Biotechnology and Biocatalysis*, Verachtert, H.; De Mot, R., Eds.; Marcel Dekker: New York, NY, 1990; pp 349-394.
18. McMillan, J.D. *Xylose Fermentation to Ethanol: A Review*; NREL TP-421-4944; National Renewable Energy Laboratory: Golden, CO, 1993; available from the National Technical Information Service, Springfield, VA 22161.
19. Beall, D.S.; Ohta, K.; Ingram, L.O. *Biotech. Bioeng.* **1991,** *38*, 296-303.
20. Burchhardt, G.; Ingram, L.O. *Appl. and Environ. Microbiol.* **1992**, *58*: 1128-1133.
21. Prior, B.A.; Kilian, S.G.; du Preez, J.C. *Process Biochem.* **1989**, *24*, 21-32.
22. Lawford, H.G.; Rousseau, J.D. *Appl. Biochem. Biotechnol.* **1992,** *34/35*, 185-204.
23. Lawford, H.G.; Rousseau, J.D. *Developments in Hemicellulose Bioconversion Technology*; In *Energy from Biomass & Wastes XVI*; Institute of Gas Technology: Chicago, IL, 1993; pp 559-598.
24. McMillan, J.D.; Boynton, B.L. *Appl. Biochem. Biotech.* **1994**, *45/46,* in press.
25. Tran, A.V.; Chambers, R.P. *Biotech. Lett.* **1985,** *7*, 841-846.
26. Tran, A.V.; Chambers, R.P. *Enzyme Microb. Technol.* **1986**, *8*, 439-444.
27. Chung, I-S.; Hahn, T-R. *J. Korean Ag. Chem. Soc.* **1987,** *30*: 311-314.
28. Wilson, J.J.; Deschatelets, L.; Nishikawa, N.K. *Appl. Microbiol. Biotechnol.* **1989,** *31*, 592-596.
29. Ueng, P.P.; Gong, C.S. *Enzyme Microb. Technol.* **1982,** *4*, 169-171.
30. Detroy, R.W.; Cunningham, R.L.; Herman, A.I. *Biotech. Bioeng. Symp. No. 12*, **1982,** 81-89.
31. Fein, J.E.; Tallim, S.R.; Lawford, G.R. *Can. J. Microbiol.* **1984**, *30*, 682-690.
32. Strickland, R.C.; Beck, M.J. *Effective Pretreatments and Neutralization Methods for Ethanol Production From Acid-Catalyzed Hardwood Hydrolyzates Using Pachysolen Tannophilus*. Proc. 6th Intl. Symp. Alcohol Fuels Technology: Ottawa, Canada, 1984, Vol. 2; pp 220-226.
33. Strickland, R.C.;Beck, M.J. *Effective Pretreatment Alternatives For the Production of Ethanol From Hemicellulosic Hardwood Hydrolyzates*; Proceedings of the 9th Symposium on Energy from Biomass and Wood Wastes: Lake Buena Vista, FL, 1985; pp 915-939.
34. Lawford, H.G.; Rousseau, J.D. *Appl. Biochem. Biotechnol.* **1991,** *28/29*, 221-236.
35. Lawford, H.G.; Rousseau, J.D. *Biotech. Lett.* **1992,** *14*, 421-426.
36. Qureshi, N.; Manderson, G.J. *J. Ind. Microbiol.* **1991,** *7*, 117-122.
37. Ferrari, M.D.; Neirotti, E.; Albornoz, C.; Saucedo, E. *Biotech. Bioeng.* **1992,** *40*, 753-759.
38. Lindén, T.; Hahn-Hägerdal, B. *Enzyme Microb. Technol.* **1993,** *11*, 583-589.
39. Leonard, R.H.; Hajny, G.J. *Ind. Eng. Chem.* **1945,** *37*, 390-395.
40. Stanek, D.A. *TAPPI* **1958**, *41*, 601-609.
41. Watson, N.E.; Prior, B.A.; Lategan, P.M.; Lussi, M. *Enzyme Microb. Technol.* **1984,** *6*, 451-456.
42. Beck, M.J. *Biotech. Bioeng. Symp. No. 17*, **1986,** 617-627.
43. Buchert, J.; Niemelä, K.; Puls, J.; Poutanen, K. *Process Biochem. International* **1990,** 176-180.

44. Slapack, G.; Edey, E.; Mahmourides, G.; Schneider, H. *Ethanol from Hexoses and Pentoses in Spent Sulfite Liquor, Biotechnology in the Pulp & Paper Industry*; Publication No. 21205, National Research Council Canada: London, Ontario, Canada, 1983.
45. Lee, Y.Y.; McCaskey, T.A. *Tappi J.* **1983,** *66*, 102-107.
46. van Zyl, C.; Prior, B.A.; du Preez, J.C. *Appl. Biochem. Biotech.* **1988,** *17/18*, 357-369.
47. van Zyl, C.; Prior, B.A.; du Preez, J.C. *Enzyme Microb. Technol.* **1991**, *13*, 82-86.
48. du Preez, J.C.; Meyer, P.S.; Kilian, S.G. *Biotech. Lett.* **1991**, *13*, 827-832.
49. *CRC Handbook of Chemistry and Physics*, 71st Edition; Lide, D.R. Ed.; CRC Press: Boca Raton, FL, 1990; p **8**-37.
50. Ando, S.; Arai, I.; Kiyoto, K.; Hanai, S. *J. Ferment. Technol.* **1986,** *64*, 567-570.
51. Nishikawa, N.K.; Sutcliffe, R.; Saddler, J.N. *Appl. Microbiol. Biotechnol.* **1988,** *27*, 549-552.
52. Mikulášová, M.; Vodný, Š.; Pekarovičová, A. *Biomass* **1990,** *23*, 149-154.
53. Roberto, I.C.; Lacis, L.S.; Barbosa, M.F.S.; de Mancilha, I.M. *Process Biochem.* **1991,** *26*, 15-21.
54. Barbosa, M. de F.S.; Beck, M.J; Fein, J.E.; Potts, D.; Ingram, L.O. *Appl. Environ. Microbiol.* **1992,** *58*, 1382-1384.
55. Beall, D.S.; Ingram, L.O.; Ben-Bassat, A.; Doran, J.B.; Fowler, D.E.; Hall, R.G.; Wood, B.E. *Biotech. Lett.* **1992,** *14*, 857-862.
56. Lawford, H.G.; Rousseau, J.D. *Biotech. Lett.* **1991**, *13*, 191-196.
57. Lawford, H.G.; Rousseau, J.D. *Xylose to Ethanol: Enhanced Yield and Productivity Using Genetically Engineered Escherichia coli*; In *Energy from Biomass & Wastes XV*, Institute of Gas Technology: Chicago, IL, 1991.
58. Lawford, H.G.; Rousseau, J.D. *Appl. Biochem. Biotechnol.* **1993,** *39/40*, 667-685.
59. Frazer, F.R.; McCaskey, T.A. *Biomass* **1989**, *18*, 31-42.
60. Parekh, S.R.; Parekh, R.S.; Wayman, M. *Process Biochem.* **1987,** 85-91.
61. Yu, S.; Wayman, M.; Parekh, S.K. *Biotech. Bioeng.* **1987,** *24*, 1144-1150.
62. Parekh, S.R.; Yu, S; Wayman, M. *Appl. Microbiol. Biotechnol.* **1986,** *25*, 300-304.
63. Wayman, M; Parekh, S.; Chornet, E.; Overend, R. *Biotech. Lett.* **1986,** *25*, 749-752.
64. Wenzhi, L.; Li-Ming, Z.; Shikai, Z.; Ho, W.Z. *Fermentation of Xylose in Sulfite Spent Liquor*, In *Biotechnology in Pulp and Paper Manufacture*; Kirk, T.K.; Chang, H.M., Eds.; Butterworth-Heinemann: Boston, MA, 1990; pp 323-328.
65. Johnson, M.C.; Harris, E.E. *J. Am. Chem. Soc.* **1948,** *70*, 2961-2963.
66. Wayman, M.; Doan, K. *Biotech. Lett.* **1992,** *14*, 335-338.
67. Chum, H.L.; Douglas, L.J.; Feinberg, D.A.; Schroeder, H.A. *Evaluation of Pretreatments of Biomass for Enzymatic Hydrolysis of Cellulose*; SERI/TP-231-2183; Solar Energy Research Institute: Golden, CO, 1985.
68. Smith, D.C.; Bhat, K.M.; Wood, T.M. *World J. Microbiol. Biotechnol.* **1991,** *7*, 475-484.
69. Kostenko, V.G.; Nemirovskii, V.D. *Composition of Hydrolyzates Produced During Acid Hydrolysis of Plant Raw Material* (translated from *Gidroliznaya i Lesokhimicheskaya Promyshiennost* [USSR]); **1992,** *1*, 7-8.

70. Perego, P.; Converti, A.; Palazzi, E.; Del Borghi, M.; Ferraiolo, G. *J. Ind. Microbiol.* **1990,** *6*, 157-164.
71. Alterthum, F.; Ingram, L.O. *Appl. Environ. Microbiol.* **1989,** *55*, 1943-1948.
72. Wood, B.E.; Ingram, L.O. *Appl. Environ. Microbiol.* **1992,** *58*, 2103-2110.
73. Maleszka, R.; Wang, P.Y.; Schneider, H. *Enzyme Microb. Technol.* **1982,** *4*, 349-352.
74. du Preez, J.C.; Bosch, M.; Prior, B.A. *Appl. Microbiol. Biotechnol.* **1986**, *23*, 228-233.
75. Jeffries, T.W.; Sreenath, H.K. *Biotechnol. Bioeng.* **1988,** *31*, 502-506.
76. Sanchez, B.; Bautista, J. *Enzyme Microb. Technol.* **1988,** *10*, 315-318.
77. Mes-Hartree, M.; Saddler, J.N. *Biotech. Lett.* **1983,** *5*, 531-536.
78. Weigert, B.; Klein, C.; Rizzi, M.; Lauterbach, C.; Dellweg, H. *Biotech. Lett.* **1988,** *10*, 895-900.
79. Boynton, B.L.; McMillan, J.D. *Appl. Biochem. Biotech.* **1994,** *45/46,* in press.
80. Olsson, L.; Lindén, T.; Hahn-Hägerdal, B. *Appl. Biochem. Biotechnol.* **1992,** *34/35*, 359-368.
81. Ligthelm, M.E.; Prior, B.A.; du Preez, J.C. *Appl. Microbiol. Biotechnol.* **1988,** *28*, 63-68.
82. Gans, I.; Potts, D.; Matsuo, A.; Tse, T.; Holysh, M.; Assarsson, P. *Process Development For Plug Flow Acid Hydrolysis and Conversion of Lignocellulosics to Ethanol;* Bioenergy Proc. 7th Can. Bioenergy R&D Seminar; Hogan, E., Ed.; National Research Council Canada: London, Ontario, Canada, 1989; pp 419-423.
83. Chahal, D.S.; Ouellet, A.; Hachey, J.M. In *Biotechnology in Pulp and Paper Manufacture*; Kirk, T.K.; Chang, H.M., Eds.; Butterworth-Heinemann: Boston, MA, 1990; pp 303-310.
84. Fanta, G.F.; Abbott, T.P.; Herman, A.I.; Burr, R.C.; Doane, W.M. *Biotech. Bioeng.* **1984,** *26*, 1122-1125.

RECEIVED April 8, 1994

Chapter 22

Anaerobic Digestion of Municipal Solid Waste

Enhanced Cellulolytic Capacity Through High-Solids Operation Compared to Conventional Low-Solids Systems

Christopher J. Rivard[1], Nicholas J. Nagle[1], Rafael A. Nieves[1], Abolghasem Shahbazi[2], and Michael E. Himmel[1]

[1]Applied Biological Sciences Branch, National Renewable Energy Laboratory, 1617 Cole Boulevard, Golden, CO 80401–3393
[2]Plant Science and Agricultural Engineering Department, North Carolina A&T State University, Greensboro, NC 27412

Anaerobic digestion of municipal solid waste (MSW) represents a waste disposal option that results in the production of a gaseous fuel (methane) and an organic residue suitable for use as a soil amendment. The rate-limiting step in this process is the hydrolysis of polymeric substrates, such as cellulose. Increasing the digester sludge total solids approximately 4-5 fold (high-solids systems) allows for enhanced process organic loading rates (6.6-fold) over conventional low-solids digester systems. Cellulase enzyme activities were quantitatively recovered from the sludge solids of laboratory-scale anaerobic digester systems fed a processed MSW feedstock using a detergent extraction protocol. Levels of β-D-glucosidase and exoglucanase activity were similar for both low- and high-solids systems on a sludge solids basis, while endoglucanase activity was 3-fold greater in the high-solids system. Additionally, the high-solids system functions with 82% fewer microbes per gram of sludge as compared to the low-solids system. Further analysis of discrete cellulase activities was performed using non-denaturing (native) gel electrophoresis and zymogram activity staining for endoglucanases. Partial purification of discrete cellulase activities from anaerobic digester sludge was carried out by recycling free-flow isoelectric focussing using the Rainin RF-3 system. Preliminary data indicates both cationic and anionic forms of both β-D-glucosidase and endoglucanase while exoglucanase activity was determined only in fractions with a pH value greater than 4.

The anaerobic digestion process takes place through the synergistic action of a consortium of microorganisms with biocatalytic activities including polymer hydrolysis, fermentation, acidogenesis, and methanogenesis. Numerous reviews have been published detailing the level of understanding of the anaerobic digestion process

0097–6156/94/0566–0438$08.00/0

and the terminal steps of methane production (*1-10*). The hydrolytic bacteria in these consortia use cellulase enzymes to depolymerize cellulose to simple sugars. The anaerobic digestion of lignocellulosic materials, such as municipal solid waste (MSW) and biomass, is limited by the rate of hydrolysis (*9,11*). The primary biodegradable polymer in biomass and MSW is cellulose. Extracellular hydrolytic enzymes, such as cellulases, amylases, lipases, and proteases, have been shown to be present in anaerobic digester effluents (*12*). Also, preliminary experiments indicate that hydrolytic enzymes added to the anaerobic digestion process *in situ* increase rates of conversion (*13*). Yet the types, activities, stabilities, and relative concentrations of these enzymes have not been examined rigorously. Although similar in some characteristics to the rumen systems, little information is available specifically on the characteristics of most of the cellulolytic enzymes produced by anaerobic bacteria found in digesters fed high-cellulose feedstocks, such as MSW. Dramatic improvements in digester performance may eventually be possible using knowledge of the types and levels of enzyme activities, both naturally occurring and from augmentation (*13*), in the anaerobic digestion system.

This work serves to further evaluate the relative levels of cellulase enzymes present in both conventional low-solids, continuously stirred tank reactor (CSTR) systems and novel high-solids systems. Additionally, the level of hydrolytic enzyme is correlated with preliminary evaluations of total active microbial populations resident in these systems to obtain a picture of the relative cellulolytic capacity of these systems.

Materials and Methods

Feedstocks. The MSW feedstock used in this study was obtained from Future Fuels, Inc., Thief River Falls, Minnesota. The MSW was processed using a combination of mechanical and manual separation. The MSW feedstock was obtained in two fractions: the food/yard waste fraction and the paper and paperboard materials (also referred to as refuse-derived fuel [RDF] in the form of densified pellets). The food/yard waste fraction was stored at 4°C until it was blended with the RDF-MSW fraction. Most of the mixed MSW was stored at -20°C until it was used. The mixed MSW feedstock was determined to be 72.7% total solids (TS) and 12.5% ash; 87.5% of the total solids (TS) was present as volatile solids (VS).

Previous investigations of the anaerobic bioconversion of MSW feedstocks identified the need for nutrient supplementation (*14*). Therefore, a nutrient solution as previously described (*14*), was added to adjust the moisture content of the feedstock, as well as to ensure sufficient nutrients for robust biological activity.

Anaerobic Digester Operation. The laboratory-scale high-solids reactors used in this preliminary study were described previously (*15*). Each consists of a cylindrical glass vessel positioned with a horizontal axis and capped at each end. The agitator shaft runs horizontally along the axis of the cylinder and mixing is achieved with a rod-type agitator (tines) attached to the shaft at 90-deg angles and in opposing orientation. Shaft rotation is provided by a low-speed, high-torque, hydraulic motor (Staffa, Inc.,

England). The glass vessel was modified with several ports including two 3/4-in. ports for liquid introduction and gas removal and a 2-in. ball valve (Harrington Plastics, Denver, Colorado) used for dry feed introduction and effluent removal.

The four high-solids reactors used in this study were maintained at 37°C in a temperature-controlled warm room. The reactors were batch fed daily by adding the relatively dry MSW feedstock and a liquid nutrient solution. Sludge was removed from the reactors on a semi-weekly basis and stored at 4°C until it was analyzed.

Four low-solids anaerobic digesters with 3.5-L working volumes and semi-continuous stirring (15 m of each 1/2 h) were constructed and operated as previously described (*16,17*). The digesters were operated in a temperature-controlled warm room to maintain the 37°C constant temperature. The anaerobic reactors were batch fed daily as described above for the high-solids systems with the exception of the addition of water to maintain a 14-day retention time. In the batch feeding protocol, a volume of effluent equivalent to the volume of feed added was removed daily to maintain the reactor sludge volume at 3.5 L. In the operation of the reactors, the solids retention time was equivalent to the hydraulic retention time (i.e., 14 days).

Feedstock/Digester Effluent Analysis. The solids concentrations of both feedstocks and digester effluent samples were determined using 1-g aluminum weigh tins. A 20- to 30-g sample was loaded into preweighed tins and dried for 48 h at 45°-50°C. The dried sample was then cooled to room temperature in a laboratory desiccator and weighed using a Sartorius balance (Model 1684MB). The percent TS was calculated on a weight/weight basis, and the percentages of VS and ash were determined by combustion of the dried samples at 550°C for 3 h in a laboratory-scale furnace.

Feedstock materials were analyzed for levels of chemical oxygen demand (COD) as previously described (*18*). The COD assay employed the micro-determination method with commercially available "twist tube" assay vials (Bioscience, Inc., Bethlehem, Pennsylvania).

Levels of volatile organic acids (C_2-C_5 iso- and normal-acids) were determined by gas-liquid chromatography (GLC) using a Hewlett-Packard Model 5840A gas chromatograph equipped with a flame ionization detector, a Model 7672A autosampler, and a Model 5840A integrator (all from Hewlett-Packard). The chromatograph was equipped with a glass column packed with Supelco 60/80, Carbopack C/0.3%, Carbowax 20M/0.1% H_3PO_4 for separations.

The feedstock was also analyzed with respect to specific polymer content as determined by the standard forage fiber analyses of acid detergent fiber (ADF) and neutral detergent fiber (NDF) as previously described (*19*).

Gas Analysis. Total biogas production in high-solids reactor systems was measured daily using pre-calibrated wet tip gas meters (Rebel Point Wet Tip Gas Meter Co., Nashville, Tennessee). Total biogas production in low-solids CSTR systems was determined from calibrated water displacement reservoirs. The composition of the biogas produced was determined by gas chromatography as previously described (*20*). For this analysis, a Gow-Mac (Model 550) gas chromatograph equipped with a Porapak Q column and a thermal conductivity detector was used.

Theoretical Methane Yield. The theoretical methane yield for the MSW feedstock was calculated as previously described (*21*) from the feedstock COD value. The ratio of actual methane yield for a given anaerobic fermentation system to the theoretical methane yield calculated from the feedstock COD value is a measure of the extent of the organic carbon conversion of the substrate added.

Cellulase Enzyme Assay Methodology. Digester samples from both low-solids and high-solids systems fed the MSW feedstock were sampled on a weekly basis over a 4-week period for analysis. Digester sludge samples were first diluted to 1%-2% TS. The diluted samples were then split into two equal 30-mL samples. One of the samples was used for analysis of TS content. The second sample was used for enzyme assessment. For enzymatic assays, the diluted digester sludge sample was subjected to centrifugation at 15,000x*g* for 20 m at room temperature to concentrate the particulate fraction. The supernatant was discarded and the pellet was resuspended in 15 mL of 100 mM Tris buffer at pH 7.0. Triton X-155 was then added to obtain a final concentration of 0.1% (vol/vol) and the sample was gently mixed for 16-20 h at 4°C. The samples were then centrifuged at 15,000x*g* for 20 m at 4°C. The supernatant was removed and filtered using a 0.45-μm disposable Acrodisk syringe filter (Gelman Sciences, Ann Arbor, Michigan). The filtered supernatant was then assayed for various cellulase enzyme activities. As a control, the supernatant from the initial particulate concentration was assayed for enzyme activity and no activity above the assay background was determined.

The determination of ß-D-glucosidase (EC 3.2.1.21), endoglucanase (EC 3.2.1.4), and exoglucanase (EC 3.2.1.74) in detergent extracts of digester sludge samples was determined as previously described (*22*). One modification to the assay protocol for "apparent" exoglucanase activity was the substitution of phosphate-swollen cellulose in 100 mM Tris, pH 7.0, for Whatman #1 filter paper strips.

Large-Scale Enzyme Extraction Protocol. Large amounts of active high-solids anaerobic digester sludge (1-2 kg) were extracted by dilution (1:10) with 100 mM Tris, pH 7.0. The Triton X-155 was added to a final concentration of 0.1% (vol/vol). The slurry was mixed for 24 h at 4°C using a reciprocal shaker to extract active hydrolytic enzymes from the sludge solids. Following incubation, the preparation was centrifuged at 5000x*g* for 15 m and 4°C, the pellet was discarded and the supernatant again centrifuged at 17,000x*g* for 5 m. The pellet was again discarded and the supernatant filtered through progressive Capsule filters including 3 μm, 0.45 μm, and 0.2 μm (Gelman Sciences, polyurethane membrane). The filtered supernatant was then concentrated 50-fold using an Amicon stirred cell with a polysulfone 10,000-MW cutoff (PM10) membrane. The concentrated sample was stored at 4°C until it was analyzed.

Electrophoresis, Zymogram Activity Staining, and Preparative Isoelectric Focusing. To verify carboxymethylcellulose degrading activity (CMCase) and to visualize the location of the endoglucanases, the concentrated digester extract was electrophoresed and visualized by zymogram staining (*23*). For this procedure, the sample was electrophoresed on Pharmacia native gradient Phastgels and immediately placed on agar media containing CMC. The agar plates were incubated at 37°C for 60 m and endoglucanase activity was visualized as cleared zones after staining with Congo red dye. Identical gels were prepared and stained for total protein using Coomassie blue.

Isoelectric focusing employing continuous recycle was performed using the RF-3 instrument (Rainin, Woburn, Massachusetts). The instrument was operated as recommended by the manufacturer. Briefly, the instrument was filled with a 3% solution containing ampholytes (Protein Technologies Inc., Tucson, Arizona) in a 10% glycerol (vol/vol) solution. The ampholyte solution contained a stipulated pH range of 3-10. The ampholyte solution was prefocused for 15 m and approximately 8 mL of concentrated digester extract was applied in the middle of the gradient. The proteins were focused for 1 h at 1500 V, followed by focusing at 500 V for 30 m. Fractions were collected and the specific sample pH was determined. Prior to the determination of specific cellulase activities, the fractions were adjusted to pH 7 with either 0.1 M HCl or 0.1 M NaOH.

Total Microbial Enumerations. As a measure of the total viable microbial numbers from sludge samples of both high- and low-solids reactor systems, a rich plating medium was utilized to obtain growth of a wide spectrum of microorganisms. The growth medium contained the following components per liter of distilled water: K_2HPO_4, 0.7 g; KH_2PO_4, 0.54 g; $MgSO_4 \cdot 7H_2O$, 1.0 g; $CaCL_2 \cdot 2H_2O$, 0.2 g; NH_4Cl, 0.5 g; yeast extract (Difco), 7.5 g; bacto-peptone (Difco), 7.5 g; glucose, 5.0 g; trace vitamin solution (*24*), 10 mL; trace mineral solution (*24*), 10 mL; cysteine HCl, 0.3 g; resazurin, 0.001 g; and agar, 15.0 g. The medium components were prepared under anaerobic conditions as previously described (*25*) using 500-mL serum bottles. Following sterilization of the agar growth medium, the agar was tempered in a 60°C water bath. The tempered agar medium was transferred to an anaerobic chamber (Coy Laboratory Products, Inc., Ann Arbor, Michigan). Sterile disposable petri dishes were previously transferred to the chamber to allow outgassing of oxygen from the plates. The agar medium was transferred to petri dishes and allowed to solidify. Poured plates were maintained within the chamber for 5 days to reduce moisture in advance of spread plating. Representative digester sludge samples were serially diluted in 9-mL blanks containing 50 mM K_2HPO_4 buffer (pH 7.4). During dilution, the sludge samples were repetitively vortexed to dislodge microbial cells that might be associated with the particulate fraction of the sample. A 0.1-mL aliquot of each dilution within the series was pipetted onto individual plates and spread plated. The inoculated plates were allowed to incubate for 7 days within the chamber at room temperature (20°C). Following incubation, the plates were removed from the chamber and colonies were counted using a darkfield counter.

Results

The anaerobic bioconversion of an actual MSW feedstock was compared using laboratory-scale digestion systems. In this study, a conventional low-solids CSTR system is compared to a novel high-solids completely mixed system. Both systems were fed an identical MSW feedstock. Various fermentation parameters, such as pH, VS, and polymer conversion demonstrated that both systems were stable and operating at steady state (data not shown). The data shown in Table 1 indicate that while similar levels of overall bioconversion occurred in the two systems, the high-solids system was functioning at 6.6 times the organic loading rate and with 4.4 times more sludge solids (i.e., corresponds to less process water in industrial applications).

The hydrolysis of cellulose, the major polymer of most biomass and MSW feedstocks, has been demonstrated to represent the rate-limiting step in anaerobic bioconversion for these materials (*9,11*). We have developed protocols for quantitatively recovering active digester-resident enzymes for study in order to generate informed strategies for further system enhancement. The comparison of discrete cellulase enzymes extracted from the low- and high-solids digester systems is shown in Table 2. In general, similar levels of β-glucosidase and exoglucanase are present when compared on a per gram of sludge solids basis. In contrast, however, levels of endoglucanase activity were approximately 3-fold greater in the high solids system. It should be noted that the level of sludge solids is significantly greater in the high-solids system and, thus, the overall level of hydrolytic enzyme activity per unit volume is greater (and far greater in the case of endoglucanase activity), as compared to the low-solids system (*26*).

The low- and high-solids digester systems were also examined for total active microbial numbers utilizing a complex growth medium. Table 3 indicates that based upon sludge total solids, the high-solids digester system functions with 82% fewer microbes per gram sludge solids as compared to the conventional low-solids system.

The ratios of discrete cellulase activities to total microbial numbers may be used to compare the cellulolytic productivity of the digester system. The data shown in Figure 1 indicate that the high-solids system exhibits increased cellulolytic productivity (5-19 fold) over the conventional low-solids system. Additionally, the concept of enzyme retention may be especially important here as the high-solids sludge may play a role in stabilizing and storing active enzymes.

However, little is known about the digester-resident cellulase enzymes present especially in high-solids anaerobic digester systems. Hampering this work is the complex nature of sludge, which presents a significant problem to the purification of active enzyme for detailed study. Protocols were developed to extract and concentrate active cellulase enzymes from the high-solids system. Initial analysis of concentrated digester extract by native gradient-gel electrophoresis is shown in Figure 2. Identical gels were examined for endoglucanase activity and total protein, both of which are not well separated and appear as a smear. This inability to separate proteins in this system is most likely due to the presence of detergents and microbe-derived lipids, protein, and nucleic acids.

A recent technological advancement is the use of recycling free-flow isoelectric focusing for preparative purification of proteins from crude mixtures. Free-flow isoelectric focusing allows for the large-scale separation of proteins by isoelectric pH using ampholytes. The system is also relatively unaffected by precipitates (often prevalent in crude protein extracts) because of the unobstructed "free-flow" design

Table 1.

Comparison of Anaerobic Bioconversion of MSW Feedstock for Low- and High-Solids Digestion Systems

Digester	Organic Loading Rate g COD/L•d	Biogas Productivity mL/L•d	Methane content %	Methane Productivity mL/L•d	Effluent Total Solids %	% Bioconversion (based on COD)
Low Solids	2.3	918 ± 129	59.8	549 ± 77	4.9 ± 0.3	68.2
High Solids	15.3	6225 ± 880	60.1	3741 ± 529	21.5 ± 0.3	69.9

Table 2.

Comparison of Hydrolytic Enzyme Levels for Low- and High-Solids Digestion Systems

Cellulase Activity	Low Solids	High Solids
ß-D-glucosidase (μmol pNP released/m)		
(activity units/g digester sludge solids)	0.24 ± 0.04	0.34 ± 0.06
endoglucanase (μmol glucose released/m)		
(activity units/g digester sludge solids)	0.33 ± 0.12	1.11 ± 0.18
exoglucanase (μmol glucose released/m)		
(activity units/g digester sludge solids)	0.29 ± 0.04	0.27 ± 0.01

Table 3.

Comparison of Microbial Enumerations for Low- and High-Solids Digestion Systems

Total Cell Number (colony forming units, [CFU])	Low Solids	High Solids
(CFU/g digester sludge solids)	$9.6 \pm 2.2 \times 10^8$	$1.7 \pm 0.3 \times 10^8$

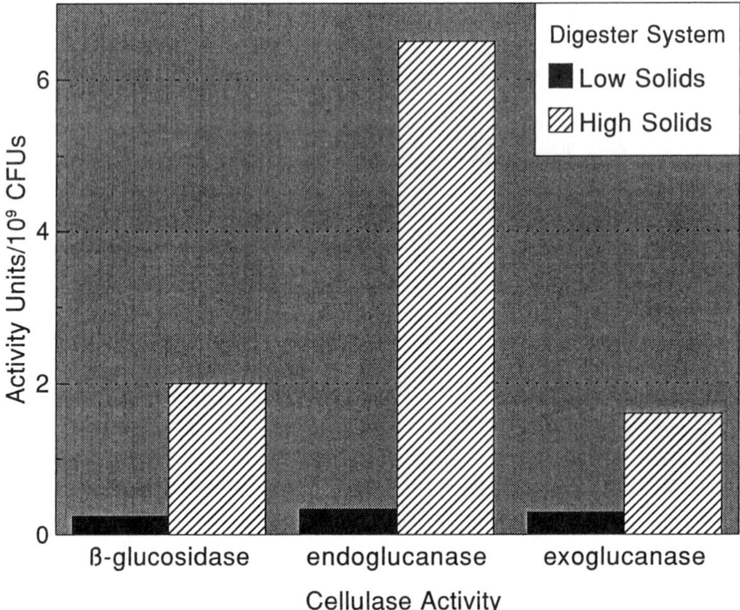

Figure 1. Comparison of cellulolytic productivity as determined by the ratio of discrete cellulase activity units to total microbial number for low-and high-solids digester systems described in Tables 1-3.

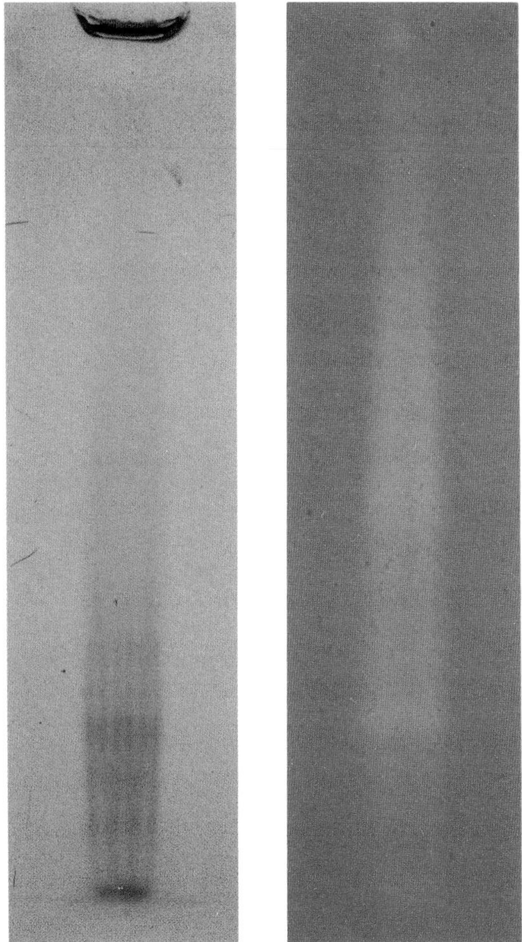

Figure 2. Analysis of concentrated digester extracted proteins from the high-solids system by native gradient gel electrophoresis. Identical gels were stained for total protein using Coomassie blue and endoglucanase activity (zymogram) was determined by CMC staining with Congo red.

(i.e., this system does not utilize screens, gels, or gradients to control convection). The RF-3 system was used to resolve the concentrated digester extract sample from the high-solids system. These data, shown in Figure 3, demonstrate the presence of both anionic and cationic proteins determined to possess β-glucosidase and endoglucanase activity. Dissimilar was the discovery of exoglucanase activity only in fractions with pH values in excess of 4.

Discussion

The economics of anaerobic digestion as a solid waste disposal option may be enhanced through improvements in process rates, yields, and by reducing system size. Enhancing the anaerobic digestion process by reducing reactor size requirements has been demonstrated using high-solids technology. This process is also referred to as anaerobic composting. A variety of high-solids process designs have been evaluated, including non-mixed batch processes, and mixed systems utilizing a variety of agitation systems (*27*). Most of these systems demonstrate operation at solids levels approaching 35%-40% TS. Furthermore, using a well-controlled, completely mixed design, we have demonstrated that the rates of conversion for the high-solids design are superior to conventional low-solids systems when treating a high cellulose content feedstock such as MSW (*28*). Combining the effects of reduced water content and enhanced organic loading rates, the high-solids system represents a reduction in the required reactor size of 90%-95% compared to conventional low-solids systems (*27*). This study further demonstrates the improvement in cellulolytic capacity afforded by operation of the anaerobic digestion system at high-solids levels. If it is assumed that the proportion of cellulolytic microbes is similar in low- and high-solids systems, the data indicate that the high-solids environment stimulates cellulase enzyme production by the hydrolytic microbial population. This conclusion is not totally unexpected, as cellulose has been demonstrated to be an effective inducer of cellulase production (*29-33*). By virtue of lower available water, the high-solids system maintains an environment in which the microorganisms are in constant contact with the solid cellulosic matter, and therefore, are more consistently induced as compared to low-solids systems. These solids may also play a role in "archiving" enzymes in an active state.

However, if strategies are to be developed for further enhancing the cellulolytic capacity of the digestion system, a thorough understanding of the digester-resident enzymes is necessary. To this end, protocols have been developed for effectively extracting and concentrating active hydrolytic enzymes from the complex sludge material. Additionally, in this preliminary work, recycling free-flow isoelectric focusing was determined to be effective in developing partially purified cellulase enzymes. Initial focusing patterns for the crude concentrated digester extract sample indicate similarities to fungal cellulases in that the digester contains β-glucosidase and endoglucanase enzymes that are both anionic and cationic (*34*). Potentially quite different from fungal systems, digester exoglucanase activities were demonstrated only for proteins focusing to pH 4.0 or greater. Studies evaluating the nature of anionic and cationic cellulases and optimization of separation protocols are in progress.

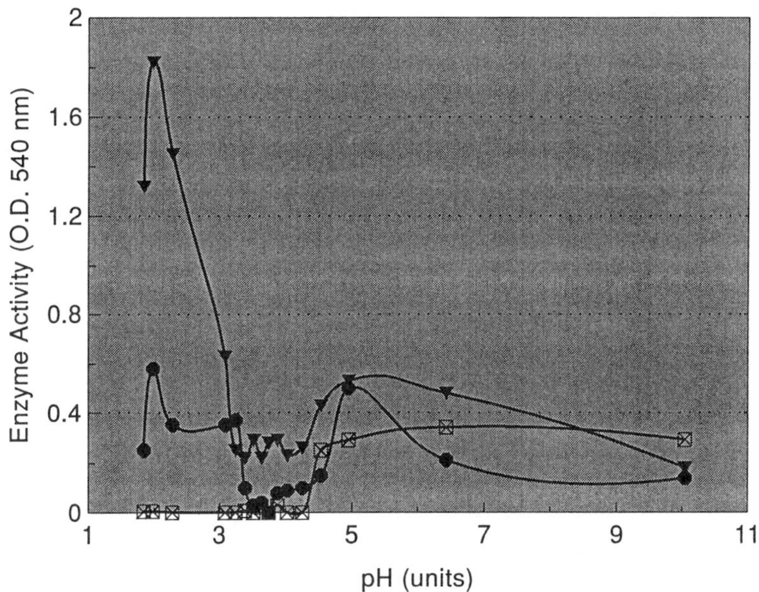

Figure 3. Separation of discrete cellulase activities from concentrated digester extract using the recycling free-flow isoelectric focusing (RF-3) system.

Acknowledgments

This work was funded by the Energy from Municipal Waste Program of the U.S. Department of Energy and managed by the National Renewable Energy Laboratory.

Literature Cited

1. Buswell, A. M.; Hatfield, W.D. *Anaerobic Fermentations*, State of Illinois Department of Registration and Education Bulletin No. 32; State of Illinois: Chicago, IL, 1936.
2. McCarty, P.L. *Public Works* **1964**, *95*, 107.
3. Wolfe, R.S. *Adv. Microbial Physiol.* **1971**, *6*, 107.
4. Mah, R.A.; Ward, D.M.; Baresi, L.; Glass, T.L. *Ann. Rev. Microbiol.* **1977**, *41*, 309.
5. Zeikus, J.G. *Bacteriol. Rev.* **1977**, *41*, 514.
6. Balch, W.E.; Fox, G.E.; Magrum, L.J.; Woese, C.R.; Wolfe, R.S. *Microbiol. Rev.* **1979**, *43*, 260.
7. Bryant, M.P., *J. Animal Sci.* **1979**, *48*, 193.
8. Clausen, E.C.; Sitton, O.C.; Gaddy, J.L., *Biotech. Bioeng.* **1979**, *21*, 1209.
9. Boone, D.R. *Appl. Environ. Microbiol.* **1982**, *43*, 57.
10. Daniels, L.; Sparling, R.; Sprott, G.D. *Biochim. et Biophys. Acta* **1984**, *768*, 113.
11. Noike, T.; Endo, G.; Chang, J-E.; Yaguchi, J-I.; Matsumoto, J-I. *Biotechnol. Bioeng.* **1985**, *27*, 1482.
12. Adney, W.S.; Rivard, C.J.; Grohmann, K.; Himmel, M.E. *Biotech. Lett.* **1989**, *11*, 207.
13. Himmel, M.E.; Adney, W.S.; Rivard, C.J.; and Grohmann, K; *Detection of Extra-Cellular Hydrolytic Enzymes in Anaerobic Digestion of MSW*; SERI/SP-231-3520; National Renewable Energy Laboratory: Golden, CO, 1989; pp 9-23.
14. Rivard, C.J.; Vinzant, T.B.; Adney, W.S.; Grohmann, K.; Himmel, M.E. *Biomass* **1990**, *23*, 201.
15. Rivard, C.J.; Himmel, M.E.; Vinzant, T.B.; Adney, W.S.; Wyman, C.E.; Grohmann, K. *Appl. Biochem. and Biotech.* **1989**, *20/21*, 461.
16. Rivard, C.J.; Bordeaux, F.M.; Henson, J.M.: Smith, P.H. *Appl. Biochem. and Biotech.* **1987**, *17*, 245.
17. Henson, J.M.; Bordeaux, F.M.; Rivard, C.J.; Smith, P.H. *Appl. Environ. Microbiol.* **1986**, *51*, 288.
18. *Standard Methods for the Examination of Water and Wastewater*; Greenberg, A.E.; Conners, J.J.; Jenkins, D., Eds.; American Public Health Association: Washington, DC, 1981.
19. Goering, H.K.; Van Soest, P.J. *U.S. Dept. of Agriculture Handbook #379*, U.S. Department of Agriculture: Washington, DC, 1970.
20. Rivard, C.J.; Himmel, M.E.; Grohmann, K; *Biotech. Bioeng. Symp.* **1985**, *15*, 375.
21. Owen, W.F.; Stuckey, D.C.; Healy, J.B.; Young, L.Y.; McCarty, P.L. *Water Res.* **1979**, *13*, 485.

22. Rivard, C.J.; Adney, W.S.; Himmel, M.E. In *Enzymes in Biomass Conversion*; Leatham, G.; Himmel, M.E., Eds.; American Chemical Society: Washington, DC, 1991, Vol. 460; pp 22-35.
23. Biely, P.; Markovic, O.; Mislovicova, D. *Anal. Biochem.* **1985**, *144*, 147.
24. Balch, W.E.; Fox, G.E.; Magrum, L.J.; Woese, C.R.; Wolfe, R.S. *Microbiol. Rev.* **1979**, *43*, 260.
25. Hungate, R.E. In *Methods in Microbiology*; Norris, J.R.; Ribbons, D.W., Eds.; Academic Press: New York, NY, 1969.
26. Rivard, C.J.; Nieves, R.A.; Nagle, N.J.; Himmel, M.E. *Appl. Biochem. Biotech.* **1994**, *45/46*, (in press).
27. Overend, R.P.; Rivard, C.J. In *Proceedings from the First Biomass Conference of the Americas: Energy, Environment, Agriculture, and Industry*; NREL/CP-200-5768; National Renewable Energy Institute: Golden, CO, 1993, Vol. 1; pp 470.
28. Rivard, C.J. *Appl. Biochem. Biotech.* **1993**, *39/40*, 71.
29. Kubicek, C.P.; Messner, R.; Gruber, F.; Mach, R.L.; Kubicek-Pranz, E.M. *Enzyme Microb. Technol.* **1993**, *15*, 90.
30. Gong, C.-S.; Tsao, G.T. *Annu. Rep. Ferment. Processes* **1979**, *3*, 111.
31. Breuil, C.; Kushner, D.J. *Can. J. Microbiol.* **1976**, *22*, 1776.
32. Montenecourt, B.S.; Nhlapo, S.D.; Trimino-Vazquez, H.; Cuskey, S.; Schamhart, D.H.J.; Eveleigh, D.E.; In *Trends In the Biology of Fermentations for Fuels and Chemicals*; Hollaender, A., Ed.; Plenum Press: New York, NY, 1981; pp 33-53.
33. Montenecourt, B.S.; Schamhart, D.H.J.; Eveleigh, D.E. *Spec. Publ. Soc. Gen. Microbiol.* **1979**, *3*, 327.
34. Coughlan, M.P. *Biotech. Genet. Eng. Rev.* **1985**, *31*, 39.

RECEIVED March 17, 1994

Chapter 23

Role of Acetyl Esterase in Biomass Conversion

James C. Linden, Stephen R. Decker, and Meropi Samara

Departments of Microbiology and of Agricultural and Chemical Engineering, Colorado State University, Fort Collins, CO 80523

Acetyl esterases that deacetylate native xylan occur in cellulolytic systems of several fungi. The biochemical and kinetic properties of acetyl xylan esterases (AXE) that have been purified are reviewed. One of the major resistances of lignocellulosic materials to hydrolysis is the high degree of acetylation of the hemicellulose fraction. This is especially true for hardwood biomass. Enzymatic deacetylation may be a prerequisite for the technological utilization of biomass feedstocks for ethanol production.

Currently, the main conversion of plant biomass to fuels is from cellulose to ethanol. This process is incomplete as it does not utilize the entire sugar content of the plant biomass. In addition to cellulose, two other carbohydrate polymers are abundant in plant biomass. These other polymers are hemicellulose and pectin. Utilization of these carbohydrates, in addition to cellulose, may greatly improve the productivity and yield of fuels from biomass. The main problem with the utilization of hemicellulose is the high degree of substitution found on the hemicellulose backbone. These substituents interfere with the enzymatic degradation of the hemicellulose to monomeric sugars through steric hindrance. Removal of these side-chain groups would greatly facilitate the conversion of hemicellulose sugars to ethanol.

Hemicellulose in Plant Cell Wall Structure

The most abundant source of carbohydrates in the world is plant biomass, which represents the lignocellulosic materials that comprise the cell walls of all plant cells. Plant cell walls are divided into two sections, the primary and the secondary cell walls. The primary cell wall, which functions to provide structure for expanding cells (and hence changes as the cell grows) is composed of three major polysaccharides, and one group of glycoproteins. The main polysaccharide, the most abundant source of carbohydrates, is cellulose. The second most important sugar source is

0097–6156/94/0566–0452$08.00/0

hemicellulose and the third polysaccharide is pectin. The compounds contained in the strongly basic glycoprotein group are referred to as extensins. The secondary cell wall, which is produced after the cell has completed growing, also contains polysaccharides and is strengthened by lignin covalently cross-linked to the carbohydrate polymers, cellulose and hemicellulose.

Cellulose is a linear β-(1→4)-D-glucan and comprises 20 to 30 percent of the primary cell wall. Pectins are "chelating-agent-extractable polysaccharide plus chemically similiar inextractable polysaccharides"*(1)*. Pectins are rich in D-galacturonic acid and also contain, in decreasing order, arabinose, galactose and rhamnose. Hemicellulose is a general term used to refer to cell wall polysaccharides that are not celluloses or pectins. Hemicelluloses include a variety of compounds including xylans, xyloglucans, arabinoxylans, mannans, and others. Hemicelluloses are always branched with a wide spectrum of substituents along the backbone polysaccharide. Hemicellulose is hydrogen bonded to cellulose as well as to itself, and can be isolated by extraction with alkali or dimethyl sulfoxide. The hydrogen bonding is thought to help stabilize the cell wall matrix and render the matrix insoluble in water. Hemicellulose is also bonded to lignin through ferulic acid residues which covalently link lignin to the arabinosyl and 4-*O*-methyl glucuronyl residues of hemicellulose, further enhancing the stability of the cell wall.

Xylan Structure

Hemicelluloses that are based on a xylose backbone, regardless of source, are termed xylans. Xylans occur widely and may be present in all higher land plants; they have been isolated from grasses, legumes, ferns, mosses, softwoods and hardwoods *(2)*. Xylan is a linear β-(1→4)-D-xylopyranose polymer with substituents attached at the number 2 and 3 hydroxyl. These substituents include α-L-arabinose and α-D-glucuronic acid residues, the 4-O- methyl ester derivative of α-D-glucuronic acid, cinnamate-based esters, some short oligosaccharides, and acetyl esters (Figure 1). Examination of xylans isolated from leaves and stems of several grasses and cereals show that these xylans carry low proportions of L-arabinofuranose units on the *O*-3 position in the main chain, and the same or lower proportions of D-glucuronic acid and/or its 4-*O*-methyl ester on the *O*-2 position in the main chain. Xylans from *Zea mays* include in their backbone D- and more rarely L-galactose units. Some xylans isolated from a few grasses carry high proportions of arabinose units in their backbone. The xylans from immature grasses are also esterified with phenolic acids *(3)*. Coumaroyl and feruloyl moieties are thought to anchor the xylan chains to the lignin matrix by crosslinking the two polymers *(2, 4)*.

The substituent chemical groups found on the xylan chain may be attached to the main chain as single units or occasionally as complex branches *(5)*. The role of these groups is not clear. They may play a major role in the resistance to hydrolyzing agents, such as enzymes. The degree of substitution of carbohydrate side chains varies greatly and has a major influence on the solubility and functionality of the xylan. The more substituted xylans cannot self-hydrogen bond, and become more water soluble. Xylans substituted to a low degree hydrogen bond not only to themselves, rendering them insoluble, but also to cellulose, screening it from degradation. Highly substituted xylans do not hydrogen bond to cellulose very well and may have only a limited protective effect against cellulose degradation. A high degree of xylan acetylation also increases the solubility of xylan by preventing interchain hydrogen bonding *(6)*. According to Poutanen, *et al.* *(7)* acetate esters impacted xylan hydrolysis in two ways; acetyl substituents increase the solubility of polymeric xylan but also hinder complete enzymatic hydrolysis of oligosaccharides to xylose and xylobiose. By removing acetyl groups and allowing interchain hydrogen bonding, AXE opens up binding sites for endoxylanase. Too much deacetylation without concurrent fragmentation of the polymer by endoxylanase, on the other hand, causes precipitation of the xylan, rendering it unavailable to the xylanases.

Acetylated Xylan. Acetylated xylan is mainly a component of higher land plants e.g. trees *(8, 3, 9)* and grasses *(3, 10)* etc. The high content of acetic acid found in many wood and cereal hydrolysates suggests that xylan is generally acetylated. The acetyl contents of some hardwoods are shown in Table I. The degree of acetylation is different for xylans of different origin. Usually, when isolated under milder conditions than alkali extraction, such as using DMSO, seven out of ten xylose units are found to be acetylated *(6, 11)*. However, lower and higher degrees of substitution have been reported. In hardwoods and mature grasses every second xylose residue is acetylated *(12)*, while in mature beech leaves, almost all xylose residues carry acetyl groups *(10)*. In a study of a hardwood hemicellulose preparation, the acetyl substituents were located at the hydroxyl groups of carbons 2 or 3 or both positions (2:3:2+3) of the xylopyranose ring in the ratio 24:22:10 *(13)*. It is assumed by analogy that the acetyl groups found in other xylans can be assigned to similar positions *(10)*. It should be noted, however, that during extraction of xylan, transesterification can occur readily so these reported ratios may not reflect the true situation in native xylan.

Enzymes Relevant to Xylan Digestion

Xylanase. Xylanase is a general term given to the enzymatic system that degrades xylan. The system is based on two core enzymes; endoxylanase and β-xylosidase. Endoxylanase acts on the long chain, low substituted xylan polymers, cutting the long chains into short oligomers. A high degree of substitution on the chain prevents efficient hydrolysis, probably due to steric hindrance induced by the side chains. The short oligomers produced are degraded by β-xylosidase, yielding monomeric xylose and preventing end-product inhibition of the endoxylanase. β-xylosidase is most active on xylobiose, with decreasing activity on increasing chain length *(14, 15)*. β-xylosidase is also hindered by substituents on the xylan oligomer.

Coumaroyl and Feruloyl Esterases. In addition to the two core depolymerizing enzymes, other accessory enzymes are present in the xylanase system. Two of these are coumaroyl esterase and feruloyl esterase, which remove coumaric and ferulic acid esters (respectively) from the xylan chain. Coumaric acid and ferulic acid are cinnamate-based acids that occur in an ester linkage to α-L-arabinofuranose units attached to the xylan chain. They function to protect the xylan chain from degradation by preventing xylanase access to the polymer. They also form cross-links to the lignin matrix, giving structural support to the cell wall *(16)*. Ferulic acid has also been shown to form diferulic acid bridges between different xylan chains, giving additional strength through cross-linking of the xylan chains. Coumaroyl and feruloyl esterase removes these moieties, opening up the cell wall structure and allowing xylanase enzymes easier access to the xylan polymer. Although most reports indicate that there is a separate enzyme for removing each acid, a couple of studies have presented evidence of a single enzyme capable of removing both coumaroyl and feruloyl acid from the xylan chain *(16, 17)*.

α-L-arabinofuranosidase and α-D-glucuronidase. Three other substituents are common on xylan; α-L-arabinofuranose, α-D-glucuronic acid, and 4-*O*-methyl glucuronic acid. These also need to be removed to prevent steric hindrance of the xylanase enzymes. The arabinose residues act as attachment points for ferulic and coumaric acid, as well as other sugars and are removed by α-L-arabinofuranosidase. α-D-glucuronic acid and its 4-*O*-methyl ester are removed by α-D-glucuronidase.

Acetyl Esterase. Acetyl esterase is a general term for an enzyme that cleaves an esterified acetic acid residue from a compound. In this paper, acetyl esterase refers to the enzyme that cleaves acetic acid from a short acetylated xylan oligomer and is inactive against acetylated long-chain xylan polymers. This enzyme acts on the short acetylated end-products of xylan degradation, removing any remaining acetyl groups and allowing β-xylosidase access to the xylose oligomer.

Acetyl esterases are distinguished from acetyl xylan esterases by the specificity of acetyl esterase for acetylated xylan oligomers and certain chromophoric substrates which mimic these short oligomers. Acetyl esterase is not active against intact long-chain acetylated xylan. The activities of acetyl esterases on these chromophoric substrates indicates that the enzymes are probably active against small degradation products such as acetylated xylose, xylobiose, and xylotriose. Acetyl esterases have also been shown to be restricted to deacetylating xylobiose only at the C-3 position *(18)*.

Acetyl xylan esterase (AXE). In contrast to acetyl esterase, AXE is active against long-chain intact acetylated xylan. By removing acetyl groups from the long chain xylose polymer, AXE opens up binding sites for endoxylanase, allowing production of short xylan fragments. Acetyl esterases active on acetylated xylan have been detected in enzyme systems from fungi, streptomycetes, plants and animals. Acetyl xylan esterases from fungal systems have been found to exhibit higher activities against acetylated xylan than any acetyl esterase of non-fungal origin *(28)*.

Assay of Xylanase, Acetyl Xylan Esterase, and Acetyl Esterases

Xylanase. Most studies on xylan digestion have employed deacetylated xylan as the test substrate. Native acetylated xylan is difficult to obtain. Xylans in native cell walls are insoluble in all known unreactive solvents, because of their covalent crosslinking to lignin, which prevents their isolation in an unmodified state *(19)*. The procedures most commonly used for xylan isolation are harsh and usually utilize strong solutions (4-15%) of sodium or potassium hydroxide *(9, 12)*. Treatment with alkaline solutions saponifies most or all ester groups and frees ester crosslinks *(20)*. Therefore, these xylan preparations are extensively modified. This xylan preparation is useful for studying the action of endoxylanase or β-xylosidase, but it is not appropriate for the study of other xylanolytic accessory enzymes or for the study of the entire xylan-degrading system.

Acetyl xylan esterases. One of the largest drawbacks to studying the activity of acetyl xylan esterases directly is the lack of a readily available substrate. There are two general types of substrates currently used to assay for AXE. One type contains an acetyl group in an ester linkage with some easily detectable molecule. When the acetate is cleaved off, the indicator is freed and can be easily detected. Examples of this type include *p*-nitrophenyl acetate, α-napthyl acetate, and 4-methylumbelliferyl acetate. These substrates have the advantage of being readily available and quickly detected by simple means, such as colorimetric or fluorescent assays. The disadvantages include instability and nonspecificity. Many organisms produce a multitude of acetyl esterase enzymes, many of which will cleave these simple molecules. Because of this, activity of an enzyme against these molecules is not a reliable indication of acetyl xylan esterase activity.

The other type of substrate is acetylated xylan, either naturally derived or chemically acetylated. Isolation of "native" xylan is accomplished in two ways, either by steam extraction or by thermomechanical extraction. In steam extraction, wood fibers are steamed at 200°C and the water soluble fraction is dialyzed against water, providing "native" hemicellulose *(21)*. The thermomechanical method involves treating wood fibers in an aqueous phase thermomechanical treatment loop for 103 s at 220°C with a shear value of 27 MPa. This treatment solubilizes about 60% of the starting hemicellulose. After centrifugation and dialysis, the hemicellulose is freeze dried for storage *(22)*. Alternatively, DMSO can be used to extract intact xylan from delignified biomass *(23)*.

In order to obtain a substrate comparable to the native acetyl xylan, methods like chemical acetylation of commercially available xylans have been used in recent studies of xylan digestibility *(24, 25)*. Xylan purified from wood fibers is acetylated by mixing with DMSO at room temperature, then heating slowly to 55°C over a twenty minute period to solublize the xylan. Potassium borate is added to a final concentration of 0.8% (w/v) and stirred for 10 minutes at 55°C. Acetic anhydride (200 mL, 60°C) is added over five minutes. The mixture is then dialyzed for about five days against running tap water and then 24 hours against distilled water, after which it is lyophilized *(25)*.

It is difficult to determine what type of xylan best imitates native xylan. In a study comparing the action of different AXEs on different types of acetylated xylan, Kahn *et al.* examined the AXE activities of a number of *Aspergillus* species *(22)*. Nine different strains of *Aspergillus* were studied, including five strains of *A. niger* and four other species. Two different xylans were studied. Native xylan in this study was obtained by thermomechanical pulping of aspen wood. Analysis of this product showed that it was about 70% sugar and 16% acetic acid, with about 85% of the sugar being xylan, the rest being mannose, glucose and cellobiose. The second xylan was a chemically acetylated larchwood xylan. Two thirds of the xylose residues in the native hemicellulose were acetylated, compared to every xylose residue being acetylated in the synthetic acetyl xylan. Results of the study showed that most of the species were able to deacetylate native hemicellulose more efficiently than synthetic acetyl xylan, indicating that AXE acts differently on chemically acetylated xylan than on native xylan, possibly due to different acetylation patterns.

Acetyl Esterases. Acetyl esterases are easily assayed by using the previously mentioned colorimetric and fluorescent substrates. Although some AXEs may also cleave these substrates, the inactivity of acetyl esterase on intact acetylated xylan allows non-AXE acetyl esterase to be readily distinguished from AXE.

Specific Examples of Investigated Enzymes (History)

Acetyl esterases were documented in plant cell wall degradation a few decades ago. In 1963, Frohwein *et al.* *(26)* demonstrated the existence of acetyl esterases in unpurified enzyme preparations. With these preparations various acetylated substrates, such as acetyl-mannose, acetyl-glucose, acetyl-maltose and acetyl-cellobiose, were hydrolyzed. Almost twenty years later in 1981, the role of acetylated polysaccharides was clarified when Williams and Withers reported the presence of acetyl esterases again *(27)*. These workers isolated more than 100 bacterial cultures from bovine rumen contents based on their ability to degrade plant cell wall structural polysaccharides. Thirty of the more active xylan degrading isolates, which were used to study the production and activity of the principal enzymes involved in the hydrolysis of plant cell wall polysaccharides, demonstrated esterase activity.

Esterases proved to act on acetylated xylan were not described until 1985. Acetyl xylan esterase [E.C. 3.1.1.6] was first reported as a new, distinguishable enzyme by Biely *et al.* *(28)*. The presence of AXE was demonstrated using a nondialyzable fraction of birch hemicellulose as a substrate and the release of acetic acid as an indication of enzyme activity. The reaction was proven to be enzymatic by using solutions of different protein content. Esterases of fungal, animal and plant systems were compared; fungal AXE showed the highest activity (Table II). The purified AXE from *A. niger* studied by Kormelink, *et al.* *(29)* was tested against acetylated pectins and was shown to have no activity against them, suggesting the presence of another class of acetyl esterases specific for acetylated pectin.

Synergism Studies

AXE and Xylanase. Acetyl ester groups protect the xylan backbone from enzymatic attack and contribute to the resistance of plant cell wall to degradation. Until recently, the effect of acetylation on the digestibility of native xylans has been overlooked. Acetylation greatly impedes enzymatic hydrolysis of xylan. From studies of the digestibility of dried grasses carried out in ruminants, *(30, 31)* the presence of acetylated xylan was shown to be an important factor influencing digestibility of the plant cell wall. Higher degrees of the acetylation were related to lower degrees of digestion *(19)*. Further research on the acetylated xylan itself showed that the xylan was more digestible by esterase-free xylanase preparations after it had been chemically deacetylated *(24)*. The same xylan was found to be more susceptible to enzymatic hydrolysis when the xylanase was supplemented with acetyl xylan esterase enzymes *(32)*. Puls *et al.* *(13)* showed that acetyl-4-*O*-methylglucuronoxylan from beechwood, treated first with acetyl xylan esterase and then xylanase, showed a high

Table I. Acetyl content of various hardwoods[*]

Hardwood	Acetyl content (% w/w)
Aspen	3.4
Balsa	4.2
Basswood	4.2
Beech	3.9
Yellow Birch	3.3
White Birch	3.1
Paper Birch	4.4
American Elm	3.9
Shellbark Hickory	1.8
Red Maple	3.8
Sugar Maple	3.2
Mesquite	1.5
Overcup Oak	2.8
Southern Red Oak	3.3
Tan Oak	3.8

[*] Data from ref. 46.

Table II. AXE activity in fungal and bacterial preparations*

Organism	Acetyl xylan esterase (nkat/mL)
Trichoderma reesei	366
F. oxysporum	53
Streptomyces olivochromogenes	28
Aspergillus awamori	3
Bacillus subtilis	1

*Data from ref. 40.

degree of hydrolysis. Much higher amounts of xylose, xylobiose and xylotriose were produced if AXE was added prior to the xylanase, than when xylanase was used alone or with xylanase followed by acetyl xylan esterase. Another study of an AXE from *A. niger* by Kormelink *et al.* showed that deacetylation of acetyl xylan was required before xylanases from *A. awamori* could depolymerize the xylan *(29)*.

In a study of AXE from *Trichoderma reesei*, xylanases and esterases were shown to have synergistic action in the breakdown of glycosidic bonds *(28)*. This cooperative effect of xylanases on the deesterification of acetyl xylan by esterases was attributed to a preference of esterases for deacetylation of shorter fragments of the substrate. The results of this study were later supported by Poutanen *et al. (33)*. Complete degradation of hemicellulose requires a combination of enzymes (Figure 1).

It was also demonstrated by Bachmann and McCarthy that AXE from *T. fusca* BD21 operated in a cooperative fashion with other enzymes of the xylanolytic system *(34)*. Liberation of acetic acid from acetyl xylan by AXE increased approximately two-fold when endoxylanase was present, compared to use of AXE alone.

Acetylation of xylan can greatly impede enzymatic hydrolysis by microorganisms, and may even be an important factor influencing digestibility of lignocellulose in the rumen. As a consequence, the enzymatic deacetylation of xylan may be a prerequisite for its breakdown. Presumably, deacetylation enhances the ability of the microorganisms to digest hemicellulose. McDermid *et al.* studied an AXE from *Fibrobacter succinogenes*, which colonizes digestion resistant plant material such as vascular tissue in the rumen *(35)*. The extracellular AXE produced by this organism was shown to deacetylate intact hemicellulose without the presence of xylanases, suggesting that the AXE helps prepare the hemicellulose for degradation. Two acetyl xylan esterases were produced by this organism; one working on intact xylan, the other working on fragmented xylan. The latter enzyme was lost in the purification step that removed most of the xylanase activity. This finding is supported by other studies. Biely *et al.* isolated a *T. reesei* AXE that worked very well on birchwood acetylated xylan, yet Poutanen and Sundberg isolated an AXE from the same species that deacetylated acetyl xylan only in the presence of xylanase, working only on short oligomers of acetylated xylan *(6, 15, 32, 36)*. Synergism between the xylanase enzymes and the acetyl xylan esterase enzyme was demonstrated in both studies. Biely *et al.* showed that the xylanase enzymes were much more efficient in degrading xylan after it had been deacetylated. Poutanen and Sundberg, however, showed that the esterase was much more effective in deacetylation after the xylan had been partially degraded. Biely first demonstrated synergism among these enzymes in 1986 *(32)*. These and other studies have suggested that the action of xylanases, acetyl esterases and AXE are all required to get high efficiency of xylan degradation.

Ester groups play an important role in the mechanisms of plant cell wall resistance to enzyme hydrolysis. Mitchell *et al.* found in both aspen wood and wheat straw that the degree of acetylation of xylan was inversely related to the rate of enzymatic hydrolysis of the hemicellulose *(24)*. Aqueous hydroxylamine solutions were used under very specific conditions to deacetylate the two plant materials to values up to 90 percent deacetylation. With increasing degrees of deacetylation, the xylan fraction became five to seven times more digestible. The cellulose fraction also became more accessible and two to three times more digestible. The effect leveled

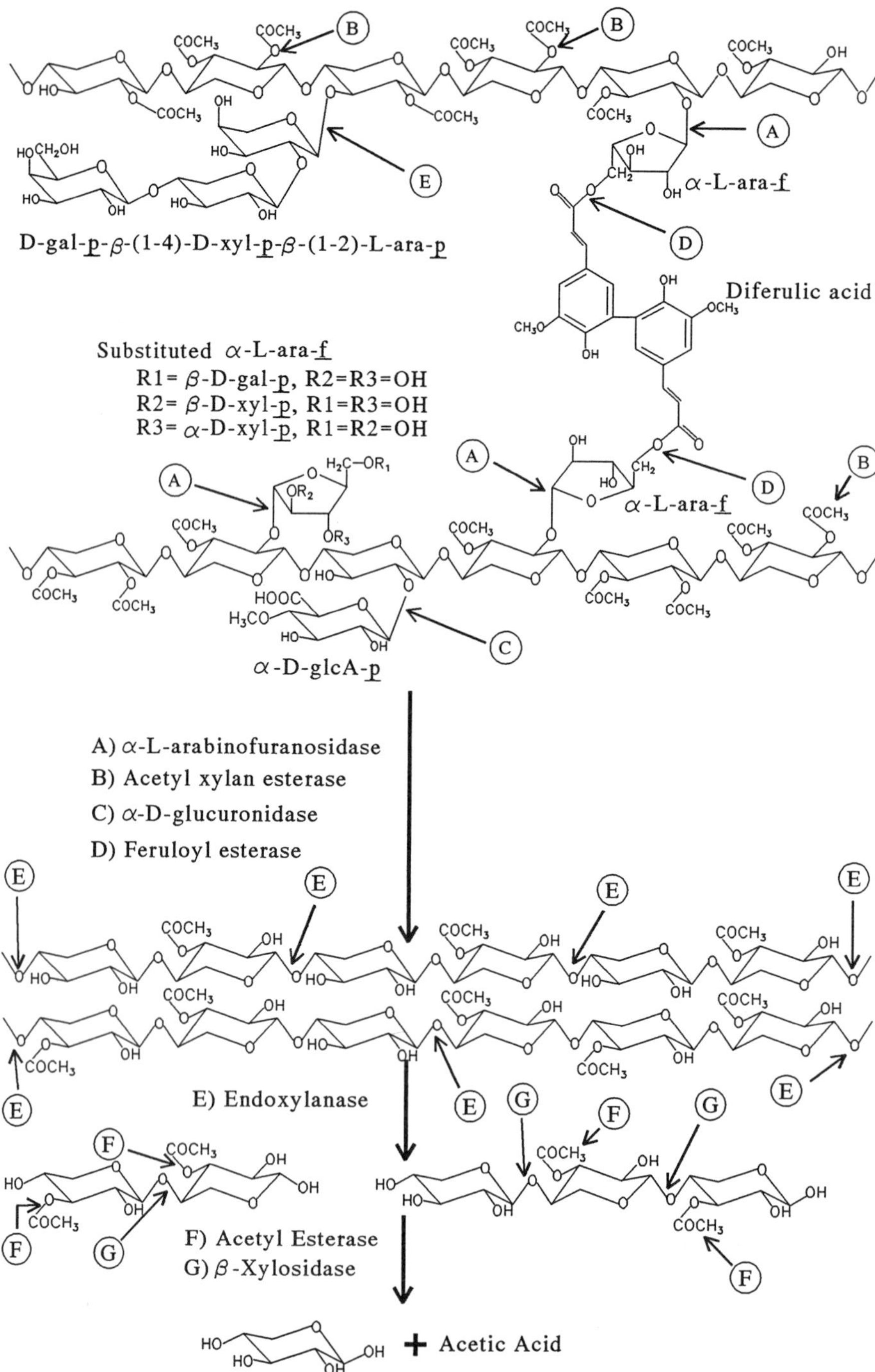

Figure 1. The structure and enzymology of native xylan degradation.

off at approximately 75 percent deacetylation, and other mechanisms such as lignin shielding came into play for providing resistance to enzyme hydrolysis.

Overend and Johnson found that lignin-carbohydrate complexes from *Populus deltroides* contained *O*-acetyl residues *(37)*. Studies were conducted to characterize three distinct lignin-carbohydrate complexes (LCC) isolated from steam exploded poplar. There was no detectable release of reducing groups when these were treated with xylanase in the absence of AXE. However, AXE isolated from *Schizopyllum commune* in addition to xylanase was sufficient for removal of monosaccharides from the LCC fragments.

Due to the high degree of acetylation in native xylan, xylanases have only limited access to the polymer backbone in the absence of esterases. Poutanen *et al.* showed that accessory enzymes, such as AXE, a feruloyl esterase and an acetyl esterase, were necessary enzymes involved in the hydrolysis of arabinoxylans from wheat *(15)*. *A. oryzae* was reported as a good source of these esterase enzymes. In another study, AXE purified from *T. reesei* was used separately, simultaneously and sequentially with an *A. oryzae* xylanase to quantitate the effects of acetate esters on the yield and nature of the products of hydrolysis *(7)*. Acting alone, endoxylanases were able to depolymerize acetyl-4-*O*-methyl glucuronoxylan from beechwood. Because of steric hinderance of acetyl groups, only small amounts of xylose, xylobiose and xylotriose were recovered. Simultaneous hydrolysis with xylanase and AXE produced xylobiose and xylotriose as the main hydrolysis products. Xylanase followed by AXE produced mainly xylotriose; the opposite sequence, AXE followed by xylanase, produced mainly xylose and xylobiose. Removal of acetyl groups prior to xylanase action resulted in smaller degradation products than any other mode of degradation.

Kong *et al.* found that aspen wood substrates with varying degrees of deacetylation yielded proportional degrees of enzymatic hydrolysis *(38)*. Deacetylation was controlled in a stoichiometric manner by the ratio of alkali to wood that was used in the preparation of the hemicellulose. The greater the degree of deacetylation, the greater the yield of sugars produced after enzymatic hydrolysis. The conclusion reached from this work was that for aspen wood, both acetyl groups and lignin were important barriers to enzyme hydrolysis, whereas the xylan backbone itself was not.

Xylanase/Cellulase Synergism. In enzymatic hydrolysis of plant cell walls, cellulose digestion is highly dependent on hemicellulose digestion. Although other factors such as crystallinity and lignin content have been suggested as barriers to the enzymatic attack on lignocellulose *(38)*, the key to increasing lignocellulose digestibility depends on the increase of the cellulose surface that is accessible to the enzymes. Hemicellulose surrounds the cellulose fibrils, protecting them from any biological attack. This makes it a necessity to hydrolyze hemicellulose first. Digestion of hemicellulose loosens the rigid, complex structure of the cell wall and exposes the cellulose surface to cellulase attack.

Physicochemical Studies

Comparison of AXE from intra- and extracellular fractions of *T. fusca* by SDS-PAGE indicated that the intracellular AXE was an 80 kDa dimer. When released from the cell, the dimer separated into two 40 kDa subunits which were the predominant extracellular form of the enzyme *(34)*. The AXE from A. niger studied by Kormelink, *et al.* had a pI of 3.0-3.2, an optimal pH of 5.5-6.0 with a stability range of 3.0-8.0 and an optimal temperature of 50°C; activity was rapidly lost above this temperature *(29)*.

Using *T. reesei* RUT C-30 as a source, Sundberg and Poutanen isolated two different AXE enzymes *(39)*. The two enzymes had exactly the same mobility on SDS-PAGE gels and were distinguished by differing isoelectric points.

Some properties of these and other acetyl xylan esterases are summarized in Table III.

Enzyme Production

Poutanen *et al.* compared the xylanolytic enzymes produced by fungal and bacterial strains *(40)*. *Trichoderma reesei* produced the most active xylanolytic enzymes (Table 2). This study was valuable in that it also compared different substrates used to grow the microorganisms. Of all the actinomycetes studied, only two mesophilic species have been the objects of close AXE study, *Streptomyces flavogriseus* and *S. lividans* *(25, 41)*. These streptomycetes are generally well suited to production of AXE. Johnson *et al.* *(25,41)* studied the response of different *Streptomyces* strains to different lignocellulosic materials. Some of the results are shown in Table IV. Conditions have been optimized for production of AXE from *A. niger* ATCC 10864. Using 5 L fermentation vessels AXE (assayed by deacetylation of chemically acetylated oat spelt xylan) volumetric productivity is greatest at 33°C, 1.5 vvm aeration and 400 rpm agitation with no pH control (Linden, J.C., *et al. Appl. Biochem. Biotechnol.*, in press.).

A number of other studies have been carried out on production of AXEs from the fungal and bacterial sources listed in Table III, with the studies consisting basically of comparisons between different microorganisms, different strains of the same species, various substrates and different assay conditions *(21, 22, 35, 36, 42-44)*.

Bachmann and McCarthy isolated the individual xylanolytic enzymes of the thermophilic actinomycete *Thermomonospora fusca* BD21 *(34)*. The growth yield was not affected to any great extent using different carbon sources but xylanase production was impacted. Xylanase production was 100 times greater when xylan was used as the main carbon source, compared to the use of glucose or arabinose and 5 times greater than levels obtained when induced by xylose alone. Extracellular acetyl esterase activity was very low (<0.1 U/mL) when grown on xylan and not detectable when the monosaccharides were used as the main carbon source.

Lee identified 350 yeast strains of different genera that produced the acetyl esterase activity and characterized the properties of the esterase activity produced extracellularly by the yeast *Rhodotorula mucilaginosa* *(45)*. The most interesting result was the lack of xylanolytic enzymes in the *R. mucilaginosa* cultures. In most, if not all, other systems studied, AXE is coproduced with endoglucanases and

Table III. Some properties of acetyl xylan esterases from various sources

Species		pH	T°	Substrate	Units	SA	Fraction	MW	Reference
Thermomonospora fusca	BD 21	5.7	RT	Acetylated	0.44	nd	5X lysate	80 KDa[1]	34
				oat spelt xylan	0.39	nd	20X filtrate	(40 KDa dimer)	
Trichoderma reesei	QM 9414	6.5	30°C	Birchwood	0.146	0.26		nd	21
T. reesei	RUT C-30			acetylated xylan	0.37	0.37		nd	
Schizophyllum commune					0.35	0.38		nd	
Aspergillus niger	06275	6.5	37°C	Larchwood	0.43	nd	culture	nd	22
A. niger	10575			acetylated xylan	0.95	1.5	filtrate	nd	
A. niger	10864				0.91	2.09		nd	
A. niger	20107				0.07	0.30		nd	
A. niger	42418				1.42	1.40		nd	
A. versicolor	11730				2.92	nd		nd	
A. clavatus	18214				0.03	nd		nd	
A. japonicus	20236				1.65	5.0		nd	
A. nidulans	24919				0.43	0.51		nd	
Fibrobacter succinogenes	S85	6.5	39°C	Birchwood	nd	0.18	20X filtrate	55 KDa	35
				acetylated xylan		8.63	Q-sepharose		
A. niger	DS 16813	5.0	30°C	Birchwood	55	1.8	filtrate	30.48 KDa	29
				acetylated xylan	180	32.3	DEAE column		
Streptomyces flavogriseus	45-CD	6.0	50°C	Chemically	3.11	8.64	purified and	nd	25
S. olivochromogenes				acetylated	4.72	12.27	concentrated	nd	
Streptomyces	C248			larchwood	34.91	129.30	culture	nd	
Streptomyces	C254			xylan	38.8	110.35	filtrate	nd	
Streptomyces	R-39				0	-		nd	
Trichoderma reesei	RUT C30	5.0	50°C	Steamed				34[1], 20[2] KDa	37
AXE I				acetylated	21	1230	53X filtrate		
AXE II				xylose	20	1250	54X filtrate		
Raw filtrate				oligomers	444	23	1X filtrate		

1- determined by SDS PAGE
2- determined by gel filtration

Table IV. Effect of lignocellulosic materials on levels of AXE*

	Lignocellulosic Material							
	None		Wheat Bran		Oat spelt xylan		Sugar cane bagasse	
Organism	U/mL	S.A.	U/mL	S.A.	U/mL	S.A.	U/mL	S.A.
S. flavogriseus	0.13	0.35	3.11	8.64	4.42	12.3	0.06	0.24
S. olivochromogenes	0.05	0.09	4.72	9.25	5.86	17.2	3.22	7.22
Strain C248	---	---	34.91	129.3	---	---	---	---
Strain C254	---	---	38.8	110.4	---	---	---	---

S.A.- Specific Activity

*Data from Johnson *et al. (41).*

endoxylanases, except, apparently, in the case of the yeast *Rhodotorula mucilaginosa* where it is produced free from any xylanolytic enzyme.

Biely *et al.* purified and characterized AXE from two strains of *Trichoderma reesei* and a strain of *Schizophyllum commune (21)*. The *T. reesei* strains produced low amounts of AXE when grown on glucose, xylose or cellobiose, although using cellobiose gave slightly greater AXE yields than glucose or xylose. Growth on larchwood xylan gave approximately 9-fold and 4-fold increases in the production of AXE in strains QM 9414 and RUT C-30, respectively, compared to the use of simple sugars. Using acetyl xylan or cellulose as a carbon source increased the AXE production by about 14-fold in strain QM 9414, relative to use of simple sugars. The highest amounts of AXE were obtained from a medium containing 0.9% (w/v) larchwood xylan and 0.1% (w/v) cellulose, yielding a 40-fold increase in AXE production over the monosaccharides alone for strain QM 9414 but only a 5-fold increase in strain RUT C-30. When compared to the basal expression when grown on glucose, xylose and cellobiose, *S. commune* showed a similar pattern of AXE induction. AXE production increased 5-fold, 33-fold and 88-fold for acetyl xylan, larchwood xylan plus cellulose, and cellulose alone respectively. Larchwood xylan alone seemed to have little affect on AXE production in *S. commune* and cellulose seemed to be the best inducer in this organism. These differences suggest different regulatory control in *T. reesei* and *S. commune.*

All of the *Aspergillus* strains tested by Kahn *et al.* also gave the highest AXE yields when grown on xylan *(22)*. Sucrose did not induce AXE expression in any of the strains tested. Although cellulose did induce AXE production, it was always less than that induced using xylan.

Summary

Hemicellulose in most plant biomass is highly acetylated. Removal of the acetyl residues and other substituents either by enzymatic action or by conditions of biomass pretreatment enhances hemicellulose and cellulose accessibility to hydrolytic enzymes. Enzyme systems have evolved in degradative organisms which utilize a host of debranching enzymes to remove substituents from the xylan backbone and allow access by the xylanase enzymes. Acetyl xylan esterase is an important accessory enzyme in this enzymatic system.

Acetyl xylan esterase works better on polymeric xylan than on short xylan oligomers. Xylanases cannot effectively degrade completely acetylated xylan, due to steric hindrances. It has been suggested that as AXE starts to deacetylate xylan, it opens up binding sites for endoxylanase. Continuous deacetylation of the xylan chain opens up more endoxylanse binding sites, resulting in partial degradation of the xylose backbone. The short xylose oligomers produced by endoxylanase action are still partially acetylated. Acetyl esterase completes the deacetylation of these short xylan oligomers, which are too small to be effectively bound by AXE. Xylobiose is the preferred substrate for acetyl esterase, although is it active on other short acetylated xylose oligomers. Deacetylation by acetyl esterase is limited to the C3 position of the xylopyranose structure. AXE is also limited in deacetylation of acetyl xylan by the presence of attached glucuronic acid and its 4-*O*-methyl ester, as well as other side chains which sterically prevent deacetylation of the immediately adjacent acetyl groups. The completely debranched xylan oligomers are then hydrolyzed by β-xylosidase, yielding free xylose as the final end product.

Literature Cited

1. Fry, S.C.; *The growing plant cell wall: Chemical and metabolic analysis;* John Wiley and Sons, Inc.: New York, NY, 1988; pp 103-119.
2. Fry, S.C.; *Annual Rev. Plant Physiol.* **1986,** *37,* 165-186.
3. Chesson, A.; Gordon, A.H.; Lomax, J.A. *J. Sci. Food Agric.* **1983,** *34,* 1330-1340.
4. Mueller-Harvey, I.; Hartley, R.D.; Harris, P.J.; Curzon, E. *Carbohydr. Res.* **1986,** *148,* 71-85.
5. Aspinall, G. O. *Adv. Carbohyd. Chem.* **1959,** *14,* 429.
6. Poutanen, K.; Sundberg, M.; Korte, H.; Puls, J. *Appl. Microbiol. Biotechnol.* **1990,** *33,* 506-510.
7. Poutanen, K.; Tankanen, M.; Korte, H.; Puls, J. *Enzyme Microb. Technol.* **1991,** *13,* 483-486.
8. Wilkie, K.C.B.; *Chem. Tech.* **1983,** *13,* 306-319.
9. Timell, T.E.; *Wood Sci. Technol.* **1967,** *1,* 45-70.
10. Bacon, J.S.; Gordon, A.H.; Morris, E.A. *Biochem. J.* **1975,** *149,* 485-487.
11. Preston, R.D.; *The physical biology of plant cell walls;* Chapman & Hall: London, 1974.
12. Timell, T. E.; *Adv. Carbohydr. Chem.* **1964,** *19,* 247-302.
13. Puls, J.; Tenkanen, M.; Korte, H.E.; Poutanen, K. *Enzyme Microb. Technol.* **1991,** *13,* 483-486.
14. Kormelink, F.J.M. *Ph.D. Thesis, Agricultural University Wageningen.* **1992,** 3-44.
15. Poutanen, K.; Tenkanen, M.; Korte, H.; Puls, J. In *Enzymes in Biomass Conversion;* Leatham, G.F.; Himmel M.E., Eds.; ACS Symposium Series 460; American Chemical Society: Washington, DC, 1991, 426-436.
16. Tenkanen, M.; Schuseil, J.; Puls, J.; Poutanen, K. *J. Biotechnol.* **1991,** *18,* 69-83.
17. Castanares, A.; McCrae, S.I.; Wood, T.M. *Enzyme Microb. Technol.* **1992,** *14,* 875-884.
18. Christov, L.P.; Prior, B.A. *Enzyme Microb. Technol.* **1993,** *15,* 460-475.
19. Grohmann, K.; Mitchell, D.J.; Himmel, M.E.; Dale, B.E.; Schroeder, H.A. *Appl. Biochem. Biotechnol.* **1989,** *20/21,* 45-61.
20. Coughlan, M.P.; Hazlewood, G.P. *Biotechnol. Appl. Biochem.* **1993,** *17,* 259-289.
21. Biely, P.; MacKenzie, C.R.; Schneider, H. *Can. J. Microbiol.* **1988,** *34,* 767-772.
22. Khan, A.W.; Lamb, K.A.; Overend, R.P. *Enzyme Microb. Technol.* **1990,** *12,* 127-131.
23. Hägglund, E.; Lindberg, B.; McPherson, J. *Acta Chem. Scand.* **1956,** *10,* 1160-1164.
24. Mitchell, D.J.; Grohmann, K.; Himmel, M.E.; Dale, B.E.; Schroeder, H.A. *J. Wood Chem. Technol.,* **1990,** *10,* 111-121.

25. Johnson, K.G.; Fontana, J.D.; MacKenzie, C.R. *Methods Enzymol.* **1988,** *160,* 551-560.
26. Frohwein, Y.Z.; Zori, U.; Leibowitz, J. *Enzymologia* **1963,** *26,* 193-200.
27. Williams, A.G.; Withers, S.E. *J. Appl. Bacteriol.* **1981,** *51,* 375-383.
28. Biely, P.; Puls, J.; Schneider, H. *FEBS Letters* **1985,** *186,* 80-84.
29. Kormelink, F.J.M.; Lefebvre, B.; Strozyk, F.; Voragen, A.G.J. *J. Biotechnol.* **1993,** *27,* 267-282.
30. Morris, E.J.; Bacon, J.S.D. *J. Agric. Sci. Camb.* **1977,** *85,* 327-340.
31. Wood, T.M.; McCrae, S.I. *Phytochemistry* **1986,** *25,* 1053-1055.
32. Biely, P.; MacKenzie, C.R.; Puls, J.; Schneider, H. *Bio/Technol.* **1986,** *4,* 731-733.
33. Poutanen, K.; Puls, J.; Linko, M. *Appl. Microbiol. Biotechnol.* **1986,** *23,* 487-490.
34. Bachmann, S.L.; McCarthy, A.J. *Appl. Environ. Microbiol.* **1991,** *57,* 2121-2130.
35. McDermid, K.P.; Forsberg, C.W.; MacKenzie, C.R. *Appl. Environ. Microbiol.* **1990,** *56,* 3805-3810.
36. Poutanen, K.; Sundberg, M. *Appl. Microbiol. Biotechnol.* **1988,** *28,* 419-424.
37. Overend, R.P.; Johnson, K.G. In *Enzymes in Biomass Conversion;* Leatham G.F.; Himmel, M.E., Eds.; ACS Symposium Series 460; American Chemical Society: Washington, DC, 1991, pp 270-287.
38. Kong, F., Engler, C.R. and Soltes, E.J. *Appl. Biochem. Biotechnol.* **1992,** *34/35* 23-35.
39. Sunberg, M.; Poutanen, K. *Biotech. Appl. Biochem.* **1991,** *13,* 1-11.
40. Poutanen, K.; Ratto, M.; Puls, J.; Viikari, L. *J. Biotechnol.* **1987,** *6,* 49-60.
41. Johnson, K.G.; Harrison, B.A.; Schneider, H.; MacKenzie, C.R.; Fontana, J.D. *Enzyme Microb. Technol.* **1988,** *10,* 403-409.
42. Johnson, K.G.; Silva, M.C.; Mackenzie, C.R.; Schneider, H.; Fontana, J.D. *Appl. Biochem. Biotechnol.* **1989,** *20/21,* 245-258.
43. Biely, P.; MacKenzie, C.R.; Schneider, H. *Methods in Enzymology.* **1988,** *160,* 700-707.
44. Poutanen, K.; Puls, J. *Appl. Microbiol. Biotechnol.* **1988,** *28,* 425-432.
45. Lee, H.; To, R.J.B.; Latta, R.K.; Biely, P.; Schneider, H. *Appl. Environ. Microbiol.* **1987,** *53,* 2831-2834.
46. Samara, M. *M.S. Thesis*, Colorado State University, 1992.

RECEIVED March 28, 1994

Chapter 24

Metabolism of Xylose and Xylitol by *Pachysolen tannophilus*

Jie Xu and K. B. Taylor

Department of Biochemistry, Fermentation Facility, University of Alabama at Birmingham, University Station, Birmingham, AL 35294

The conversion of both xylose and xylitol to ethanol has been demonstrated in cell-free extracts of *Pachysolen tannophilus*. The facts that xylitol is metabolized in the extracts but not in whole cells in the absence of nystatin demonstrate that transport across the cell membrane limits its metabolism in whole cells. The metabolism of both xylose and xylitol requires NAD and ADP, but the metabolism of xylose requires and NADPH-generating system in addition. Xylose isomerase increases the rate of ethanol formation from xylose, but some xylitol accumulates nevertheless. Therefore, the cell-free system metabolizes xylose as two independent, sequential pathways, one to synthesize xylitol and one to convert it to ethanol. The consequences for process-improvement strategies are discussed.

With the increased awareness of the limitations in the world supplies of petroleum came an increased interest in gasoline extenders and additives. The oxygenated additives tert-butyl ether, methanol and ethanol were found to have the additional advantages of octane enhancement and lower rates of air pollution. The use of ethanol on a national scale is associated with further advantages (*1,2*): the security of fuel sources; a more favorable foreign trade balance; and the renewability of the raw materials for production. Although ethanol can be produced either from a petroleum component or by fermentation, the latter source is the only one associated with all of the advantages above. The most attractive raw materials for ethanol fermentation are cane sugar, corn sugar, and sugar derived from wood and crop residues (lignocellulosic materials). In this country cane sugar is too expensive to be a source of fuel ethanol and corn ($98-$104 per ton) (*2*), can be competitive only in special circumstances, such as government subsidy. However, lignocellulosic material ($20-$70 per ton) could be competitive, if the technical limitations associated with processing and fermentation could be overcome. Although the approximate price of

0097–6156/94/0566–0468$08.00/0

corn-derived ethanol is currently $1.20 per gallon, it is estimated that the cost of fuel ethanol from lignocellulosic materials should be $0.60-$0.80 per gallon (*3*).

The most productive and cost effective processing of lignocellulosic material currently consists of mild acid hydrolysis to release a fraction containing primarily xylose (15%-40% of dry weight) followed by strong acid treatment to release the glucose (30%-60%), (*4,5,6*). The sugars are then fermented, usually separately, to ethanol. However, there are two steps in particular that limit the productivity and increase the cost of this process: the hydrolysis of cellulose and the fermentation of xylose. The most promising solution to the former problem is enzymatic hydrolysis, but currently the enzymes remain too expensive and the hydrolysis is incomplete. The problem with xylose fermentation arises from a poor overall yield, which compromises the productivity and increases the cost. It is estimated that the efficient conversion of xylose would reduce the cost of ethanol by about $0.40 (*7*).

Organisms. A number of both yeasts and bacteria will convert xylose to ethanol and other products (*8,9,10*), but most of the attention has been given to three yeast strains: *Pachysolen tannophilus, Candida shehatae,* and *Pichia stipitis.* Yeast has the advantages of growing in a simple, inexpensive medium, growing at a low pH to minimize contamination, and being the subject of much past experience. Although both bacteria and yeast generally use transketolase and transaldolase to convert D-xylulose to hexose (as the phosphate esters) and convert the hexose to ethanol by the glycolytic enzymes, they differ with respect to the methods used to convert D-xylose to D-xylulose. Most bacteria carry out the conversion with a single enzyme, xylose isomerase, whereas most yeast reduce the xylose to xylitol and subsequently reoxidize it to xylulose.

The principal limitation to the ethanol yield in the yeast fermentation is the accumulation of xylitol in molar amounts equal to or exceeding that of ethanol (*11*). Furthermore, the accumulated xylitol is not further utilized. Fermentation under micro-aerobic conditions reduces the amount of xylitol (*12,13*) accumulated, but the oxygen also promotes the further oxidation of ethanol.

Pathways. Besides the accumulation of xylitol the evidence for the pathway converting xylose to xylulose consists of the identification and isolation of both xylose reductase (*14,15,16,17*) and xylitol dehydrogenase (*18,19*) as well as the demonstration that xylose induces both of them (*20,21*). The gene for each of the two enzymes from *Pichia stipitis* has been identified and sequenced (*22,23*), and part of the amino acid sequence of xylose reductase from *Pachysolen tannophilus* is known (*14*). Workers in several laboratories have described two forms of xylose reductase, one of which is NADH-specific and the other of which uses NADH or NADPH (*21*), but other results suggest that there is a single protein with dual specificity (*24*). In any case the enzyme that requires NADPH predominates in *Pachysolen tannophilus.*

The evidence for the pathway for the pentose-to-hexose conversion consists of the demonstration of transaldolase and transketolase in *Pachysolen tannophilus.* However, Ratledge and coworkers (*25,26*) demonstrated the enzyme, pentulose-5-phosphate phosphoketolase in a number of yeast strains including some *Candida,* some *Pichia*, and *Pachysolen tannophilus.* Although the activity was lower in these three strains than in some others, the results bring up the possibility of an alternative pathway that may be associated with a higher theoretical molar yield, 2.0 instead of 1.67.

Process Improvements. The ethanol yield has been improved by the addition to the medium of fatty acids (*27*), polyethylene glycol (*28*), azide (*29*), or acetone (*30*). In addition mutants have been reported (*31,32,33*). However, none of these strategies seems to have become widely used in practice. Xylose isomerase has been utilized in conjunction with the fermentation to facilitate the conversion of xylose to xylulose (*34*), but neither the equilibrium of the reaction nor the pH optimum of the enzyme is favorable (*35*). Several workers have described strains of *Saccharomyces cerevisiae* (*36,37*) and *Schizosaccharomyces pombe* (*38*) into which the xylose isomerase gene has been cloned, but the productivity was low presumably because of low levels of transport systems, xylulokinase, transketolase, and transaldolase.

Questions. Several biochemical questions remain, listed below, whose answers should generate additional process improvements.

1. What is the pathway for xylose metabolism in *Pachysolen tannophilus*?
2. What are the small-molecular-weight requirements for xylose conversion?
3. What are the causes of xylitol accumulation?
4. What are the small-molecular-weight requirements for xylitol conversion?
5. Why is xylitol not utilized by normal cells?
6. Why is oxygen required for optimum ethanol production?

Although the evidence for the transaldolase-transketolase pathway is rather convincing, some evaluation should be made of the flux through the phosphoketolase enzyme, because this might influence the theoretical yield.

The most important questions deal with the causes of xylitol accumulation and its non-utilization. Several investigators have attributed xylitol accumulation to the fact that the majority of the xylose reductase activity requires NADPH, whereas the xylitol dehydrogenase requires NAD (*39,40*). However, if this nucleotide imbalance were the only cause, the utilization of xylose would quickly stop, because of the low concentration of intracellular nucleotides. Since the overall conversion of xylose to ethanol is isoelectronic, the question becomes one of the identification of the primary electron source for the reduction of xylose and the identification of an additional electron sink for NAD regeneration. Since no partially oxidized compounds accumulate in significant concentration (*11*), the electron source must be associated with complete oxidation to carbon dioxide. In addition it is attractive to hypothesize that oxygen stimulates ethanol production and reduces xylitol accumulation because it acts as an electron sink to regenerate NAD. However, results are needed to test the hypothesis.

The poor xylitol utilization was inconsistent with its status as an intermediate. Although the utility of oxygen and acetone in xylitol conversion, as well as its redox status, indicates that an electron acceptor is necessary, xylitol still accumulates to a very significant extent and the utilization of extracellular xylitol remains very poor in the presence of oxygen.

The present work was undertaken to provide answers to these and other questions. However, the order in which they are answered is dictated by the logic of the experiments instead of the logic of the pathway or the logic of the questions.

Results

Xylitol Utilization. Whole, resting cells of *Pachysolen tannophilus* converted xylose to ethanol (yield, 1.11 mol/mol) but did not convert xylitol (*41*). However, a cell-free extract, prepared by French pressure cell and low-speed centrifugation, produced ethanol from both xylose (yield, 1.1 mol/mol) and xylitol (yield, 1.64 mol/mol). Therefore, the cell-free extract converts xylitol to ethanol with the theoretical yield for the transaldolase-transketolase pathway. In order to test the hypothesis that the permeability of the cell membrane to xylitol limits its metabolism experiments were done with whole cells in the presence of nystatin, amphotericin B, and filipin; all agents that increase the cell-membrane permeability of yeast and fungi (*42,43*). The rate of xylitol conversion to ethanol by resting cells of *Pachysolen tannophilus* increased 40-fold in the presence of nystatin (100 µg/mL), and the other two agents, amphotericin B and filipin produced similar results (*44*). Nystatin produced little or no effect on the metabolism of either xylitol or xylose in cell-free extracts. Therefore, the rate of metabolism of extracellular xylitol by this yeast is limited by its transport rate across the cellular membrane. Acetone, apparently, is effective in the stimulation of xylitol metabolism because it too increases the permeability of the cell membrane. This action of acetone on cell membranes is well known in a variety of organisms.

Requirements for Xylitol Metabolism. Although the aerobic conversion of xylitol to ethanol by the cell-free extract was quantitative, the anaerobic rate was immeasurably small (*41*). Therefore, we have demonstrated that oxygen is required for the metabolism of xylitol. The oxygen requirement is predicted from the fact that xylitol is at an overall lower oxidation state than ethanol plus carbon dioxide. In addition the supernatant, soluble fraction of the cell-free extract, after centrifugation for 100,000 x g for 1 h, failed to metabolize xylitol aerobically in the absence of the pellet, membrane fraction; which was shown to have oxygen uptake activity in the presence of succinate.

According to the transaldolase-transketolase pathway both ADP and NAD should be required in catalytic amounts for xylitol metabolism. The ADP is required for those enzymes that generate ATP and for those that require ATP. The NAD is converted to NADH in the oxidation of xylitol to xylulose and of glyceraldehyde phosphate to diphosphoglycerate. Part of the NAD is regenerated in the reduction of acetaldehyde to ethanol and the rest in the reduction of molecular oxygen. Confirmation of these requirements was provided by incubation with xylitol of a fraction of the cell-free extract isolated by gel filtration to remove the small-molecular-weight compounds. Both ADP (100 μmolar) and NAD (50 μmolar) were necessary for the conversion of xylitol to ethanol (*41*).

The requirement for both NAD and oxygen is difficult, if not impossible, to explain if significant metabolic flux is through the phosphoketolase pathway as described above. The generation of ethanol from pentose by the latter pathway requires two moles of reduced nucleotide per ethanol to reduce the acetylphosphate. This putative requirement for reduced nucleotide is inconsistent with the demonstrated requirement of oxygen, presumably to oxidize reduced nucleotide. Therefore, the experimental results support the hypothesis that little metabolic flux utilizes the phosphoketolase pathway.

Requirements for Xylose Metabolism. In order to simplify the investigation the putative pathway for xylose metabolism was separated into two parts by the inclusion of xylose isomerase (Spezyme GI, Finnsugar Biochemicals, Inc.) in the incubations with the fractions of the cell-free extract. In this way the pathway from xylulose to ethanol can be studied separate from that responsible for the conversion of xylose to xylulose. Preliminary experiments showed that ethanol formation from xylose by the crude cell extract was stimulated 9-fold by added xylose isomerase.

In the Presence of Xylose Isomerase. The activity of the HMW fraction (Fraction A, Table 1) was restored by the addition of ADP. Although the addition of NAD^+ stimulated ethanol formation in this preparation, it was not absolutely required (fraction A). However, treatment three times with gel filtration and once with charcoal (fraction B, Table 1) accomplished the complete removal of NAD^+, and the requirement could be demonstrated. The optimum concentration of NAD^+ is 50 μM, whereas that of ADP or ATP is 200 μM respectively under the experimental conditions (Fig. 1).

In contrast to the experience with xylitol, xylose is metabolized by soluble fractions of the cell-free extract from which the membranes and organelles have been removed by centrifugation (Table 1), and xylose is metabolized in the absence of oxygen.

In the Absence of Xylose Isomerase. Although the addition of NAD^+ and ADP restored the production of ethanol from xylose by the HMW fraction, in the presence of xylose isomerase, their addition failed to restore ethanol formation from xylose by the same fraction in the absence of xylose isomerase. In addition, the inclusion of $NADP^+$, NADPH, or ATP in addition to NAD^+ and ADP, failed to restore ethanol formation from xylose (Table 2). Furthermore the inclusion of a variety of other coenzymes and metal ions in the presence of NAD^+, $NADP^+$ and ADP also failed to restore activity.

The fact that added NADPH was rapidly consumed by the HMW fraction in the absence of xylose (Fig. 2) suggested that other reactions compete for NADPH and that experiments would require a high initial concentration. However, the failure of the attempts to demonstrate activity by the initial addition of NADPH is consistent with the fact that, at higher concentration (>100 μM), both NADPH and $NADP^+$ inhibited ethanol formation from xylose by the unfractionated cell extract (Table 3). In fact NADPH at concentrations as high as 12 mM failed to restore ethanol formation from

Table 1.
Effects of Sephadex Column and Charcoal Treatment of the High Speed Supernatant Fraction (HSSF)[a].

Fraction	Additional Components		EtOH (mg/mL)
HMW	-	-	<0.01
HMW	-	NAD^+	<0.01
Fraction A[b]	ADP	-	0.12
Fraction A	ADP	NAD^+	0.64
Fraction B[c]	ADP	-	<0.01
Fraction B	ADP	NAD^+	0.62

[a]The reaction mixture contained 2 mg protein/mL, xylose isomerase (53 U/mL), and xylose (50 mM) plus ADP (0.5 mM) and NAD^+ (0.1 mM) as indicated.

[b]Fraction A, the HMW fraction of the high speed supernatant fraction, was subjected to two consecutive gel filtration treatments.

[c]Fraction B (Fraction A after treatment with charcoal).

Table 2.
Effect of Various Cofactors on Ethanol Formation from Xylose by the HMW Fraction of the Crude Cell Extract[a].

Protein fraction	Substrate	Additional component	EtOH (mg/mL)
Crude extract	xylose	-	1.01
HMW	xylitol	-	2.09
HMW	xylose	-	<0.01
HMW	xylose	NADH	<0.01
HMW	xylose	FAD	<0.01
HMW	xylose	PQQ[b]	<0.01
HMW	xylose	TPP[b]	<0.01
HMW	xylose	Mn^{2+}	<0.01
HMW	xylose	Zn^{2+}	<0.01

[a]The reaction mixture contained either the HMW fraction or the crude extract (8.8 mg protein/ml), ATP (0.1 mM), ADP (1.0 mM), NAD^+ (0.2 mM), NADPH (0.1 mM) and xylose (71 mM). The concentration of the additional components was 0.1 mM.

[b]TPP, thiamine pyrophosphate; PQQ, pyrroloquinoline quinone.

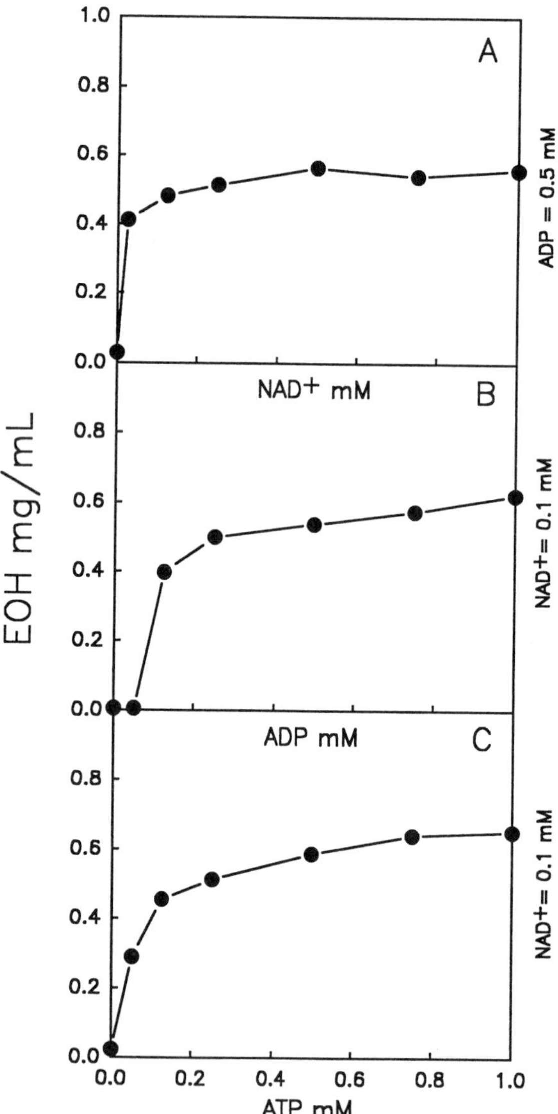

Figure 1. Effect of concentrations of NAD^+, ADP or ATP on ethanol formation from xylose in the presence of xylose isomerase by the HMW fraction of high speed supernatant fraction (fraction B). The reaction mixture contained the same amount of protein, xylose, xylose isomerase as described in Table 1. A: varying NAD^+ concentration; B: varying ADP concentration; C: varying ATP concentration.

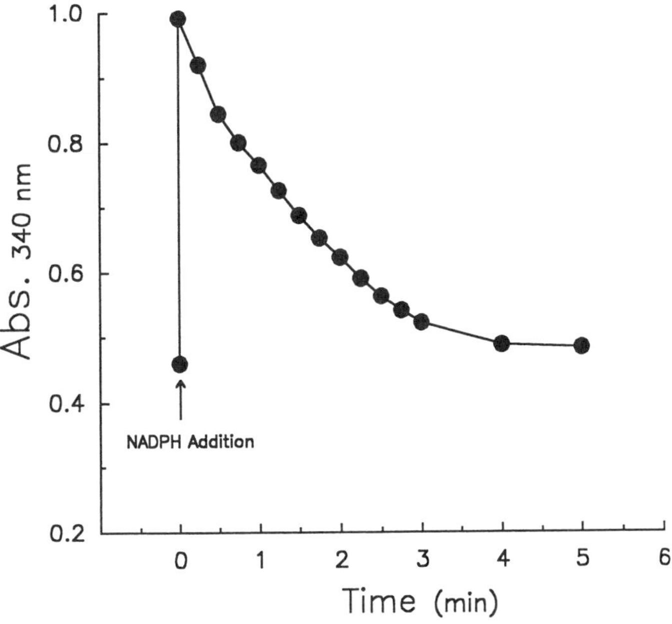

Figure 2. Endogenous consumption of NADPH by the HMW fraction of the crude cell extract. The reaction mixture contained 0.2 mM NADPH and the HMW sample (2 mg/mL). The reaction was started by adding NADPH (0.2 mM) and the absorbance was monitored at 340 nm.

Table 3.
Effect of NADPH and $NADP^+$ on Ethanol Formation from Xylose by the Crude Cell Extract[a].

Additional Coenzyme	Concentration (mM)	EtOH (mg/mL)	Inhibition %
None	-	1.58	0
NADPH	0.1	1.39	12
NADPH	0.5	0.46	71
NADPH	1.0	0.21	87
$NADP^+$	0.5	1.02	35
$NADP^+$	1.0	0.50	68

[a]The reaction mixture contained the crude cell extract (15.5 mg protein/mL) and xylose (62 mM).

xylose by the HMW fraction in the presence of NAD^+, ADP, and ATP. Therefore, three NADPH generation systems were tested to maintain a constant supply of the reduced coenzyme at low concentration: partially purified transhydrogenase from *E. coli K-12*, NADP-linked isocitrate dehydrogenase from porcine heart, and NADP-linked alcohol dehydrogenase from *Thermoanaerobium brockii*. The amount of each enzyme used was calculated on the basis of its specific activity and the rate of xylitol to ethanol conversion, such that at least 0.5 mg ethanol/ml should be formed during a 2-hour incubation period. Although adequate transhydrogenase activity was demonstrated in the first preparation, it was ineffective in the restoration of ethanol production. In contrast, both the NADP-linked isocitrate dehydrogenase and the alcohol dehydrogenase stimulated the formation of ethanol from xylose by about 60 and 80-fold respectively (Table 4). The controls, containing no xylose, show that little ethanol was derived from the components of the NADPH generation systems themselves, although the ethanol from isocitrate was higher than that from cyclopentanol.

Although the reasons for the requirement for the NADPH-generating system will be discussed in detail later, the requirement indicates that the NADPH-requiring xylose reductase is responsible for practically all of the xylose metabolism in these extracts.

Xylitol Accumulation. A number of factors contribute to the causes of xylitol accumulation. Certainly the fact that xylose reductase requires NADP whereas xylitol dehydrogenase requires NAD (nucleotide imbalance) is one of the factors. Another is certainly the fact that extracellular xylitol is not normally utilized by whole cells.

Xylitol was produced during xylose metabolism in the presence of xylose isomerase (Table 5). Furthermore, in the absence of added $NADP^+$, the gas chromatographic assay for ethanol contained an additional peak that had the same retention time as acetaldehyde. Therefore, some of the NADH generated by glyceraldehyde phosphate dehydrogenase is reoxidized in the generation of xylitol, presumably from xylulose by xylitol dehydrogenase. Although the results show that $NADP^+$ had little effect on the amount of ethanol made from xylose, NADP increased the xylitol accumulation by 2.5-fold, increased the xylose consumption by 2-fold, and eliminated the acetaldehyde peak (Table 5). The ethanol to xylitol ratios in the absence and presence of $NADP^+$ are 1.39 and 0.49 respectively. Therefore, most of the xylitol is generated by xylose reductase in these extracts.

Since the source of the electrons (NADPH) for xylitol formation must oxidize at least part of the substrate completely to carbon dioxide, there are only two possibilities in the context of the known pathways: the tricarboxylic acid cycle oxidizing ethanol (or acetate) or the pentose-hexose cycle. In the former isocitrate dehydrogenase would be the principle catalyst for the generation of NADPH, whereas in the latter pentose is converted to hexose by the transaldolase-transketolase pathway and then glucose-6-phosphate is oxidized back to pentose (ribulose-5-phosphate) by glucose-6-phosphate dehydrogenase and 6-phosphogluconate dehydrogenase with the generation of two moles of NADPH. The fact that inhibition of the tricarboxylic acid cycle with fluorocitrate produced no difference in the amount of xylitol accumulated in extracts provides good support to the hypothesis that the primary electron source for xylitol accumulation is the pentose-hexose cycle (*41*).

Table 4.
Restoration of Ethanol Formation from Xylose in the HMW Fraction of the Crude Cell Extract in the Presence of an NADPH Generation System[a].

Reaction mixture	Substrate	EtOH produced (mg/mL)
None	Xylose	<0.01
IC[b] + ICDH[b]	none	0.057
IC + ICDH	Xylose	0.710
cPen-OH[b] + ADH[b]	none	0.016
cPen-OH + ADH	Xylose	0.925
TH[b] + NADPH (0.1 mM)	Xylose	<0.01

[a]The reaction mixture contained the HMW fraction (10.5 mg protein/mL), ATP (0.1 mM), ADP (1.0 mM), NAD^+ (0.2 mM), $NADP^+$ (0.2 mM) plus additional components as indicated. The indicated reaction mixture contained xylose (50 mM), isocitrate (50 mM), isocitrate dehydrogenase (5 unit/mL), cyclopentanol (55 mM), alcohol dehydrogenase (0.5 unit/mL), and transhydrogenase (1 mg/mL).

[b]IC, sodium isocitrate; ICDH, $NADP^+$-linked isocitrate dehydrogenase; cPen-OH, cyclopentanol; ADH, $NADP^+$-linked alcohol dehydrogenase; TH, transhydrogenase preparation.

Table 5.
Comparison of the Ratio of Formation of Ethanol to Xylitol by Different Systems[a].

Protein fraction	xylose used (μmol)	xylitol produced (μmol)	EtOH produced (μmol)	EtOH/Xylitol ratio (mol/mol)
Whole Cell				5.75[b]
Crude Extract				3.63[b]
HSSF				0.63[b]
Fraction B[c] + XI[c] + ADP + NAD^+	15.4	7.9	11.0	1.39
Fraction B+ XI + ADP + NAD^+ + $NADP^+$	29.1	19.9	9.7	0.49

[a]The reaction mixture contained ADP (0.5 mM), NAD^+ (0.1 mM), and $NADP^+$ (0.1 mM). The concentrations of protein, xylose, and xylose isomerase are the same as in Table 1.
[b]From Xu and Taylor (ref. *44*).
[c]Same as in Table 1; XI, xylose isomerase.

In normal xylose metabolism NADPH is generated from the metabolism of residual pentoses and hexoses. In the experiments above with purified fractions it was necessary to add an NADPH-generating system to generate some pentoses and hexoses. Xylose is reduced to xylitol, but the poor coupling of the rate with the rate of xylitol dehydrogenase causes xylitol to accumulate faster than it is utilized. The intracellular xylitol leaks out of the cell and becomes diluted in the medium, from which it is not reutilized because the transport back into the cell is limiting. The intracellular xylitol can be metabolized, if oxygen is present. Thus oxygen is required for optimum xylose metabolism.

Strategies for Process Improvement

The most easily implemented strategy for process improvement would include in the medium inhibitors of the source of electrons for xylitol formation, either before, during or after biomass formation. Ideally in the interest of selectivity phosphoglucose isomerase, glucose-6-phosphate dehydrogenase, or 6-phosphogluconate dehydrogenase would be inhibited, and only partially, because some xylitol formation must be maintained. However, known selective inhibitors of these enzymes, particularly ones that cross the cell membrane, are very difficult to identify. In addition the timing and the exact degree of inhibition may be difficult to control. Furthermore, the utilization of carbon for the generation of NADPH and the requirement for oxygen would persist.

The transformation and expression of the gene for transhydrogenase in one of the xylose-fermenting yeasts would relieve the nucleotide imbalance. However, the effectiveness of this enzyme could not be demonstrated in the experiments above. Moreover, it would also be necessary to inhibit completely the electron source for xylitol generation. If these two conditions could be accomplished, the organism should convert xylose to ethanol anaerobically with the theoretical yield.

According to the results above the transformation and expression of the gene for xylose isomerase in one of the xylose-fermenting yeasts should facilitate the conversion of xylose to ethanol, but it would also be necessary to inhibit both the xylose reductase and xylitol dehydrogenase, in order to prevent xylitol accumulation. However, the transformants should be able to ferment xylose to ethanol anaerobically.

Literature Cited

1. Lynd, L.R., *Appl. Biochem. Biotechnol.* **1990**, *24/25*, 695-719.
2. Wyman, C.E.; Hinman, N.D., *Appl. Biochem. Biotechnol.* **1990,** *24/25*, 735-753.
3. Lynd, L.R.; Cushman, J.H.; Nicholas, R.J.; Wyman, C.E. *Science* **1991,** *259,* 1318-1323.
4. Whistler, R.L.; Bachrach, J.; Bowman, D.R. *Arch. Biochem.* **1948,** *19,* 25-33.
5. Weihe, H.D.; Phillips, M.J. *Agr. Res.* **1940**, *60,* 781-786.
6. Rosenberg, S.L. *Enz. Microb. Technol.* **1980,** *2,* 185-193.
7. Hinman, N.D.; Wright, J.D.; Hoagland, W.; Wyman, C.E. *Appl. Biochem. Biotechnol.* **1989**, *20/21,* 391-401.
8. Skoog, K.; Hahn-Hagerdal, B. *Enz. Microb. Technol.* **1988,** *10,* 66-80.

9. Toivola, A.; Yarrow, D.; van den Bosch, E.; van Duken, J.; Scheffers, W.A. *Appl. Environ, Microbiol.* **1984,** *47,* 1221-1223.
10. Nigram, J.N.; Ireland, R.S.; Margaritis, A.; Lachance, M.A. *Appl. Environ. Microbiol.* **1985**, *50,* 1486-1489.
11. Taylor, K.B.; Beck, M.J.; Huang, D.H.; Sakai, T.T. *J. Ind. Microbiol.* **1990,** *6,* 29-41.
12. du Preez, J.C.; Prior, B.A.; Monteiro, A.M.T. *Appl. microbiol. Biotechnol.* **1984,** *19,* 261-266.
13. Debus, D.; Methner, H.; Schulze, D.; Dellweg, H. *Appl. Microbiol. Biotechnol.* **1983,** *17,* 287-291.
14. Bolen, P.L.; Bietz, J.A.; Detroy, R.W. *Biotechnol Bioeng. Symp.* **1985,** *15,* 129-148.
15. Rizzi, M.; Erlemann, P.; Anh, B.T.N.; Dellwig, H. *Appl. Microbiol. Biotechnol.* **1988,** *29,* 148-154.
16. Ditzelmuller, G.; Kubicek, C.P.; Wohrer, W.; Roeher, M. *Can. J. Microbiol.* **1984**, *30,* 1330-1336.
17. Verduyn, C.; Van Kleef, R.; Frank, J.; Schreuder, H.; Van Dijken, J.P.; Scheffers, W.A. *Biochem. J.* **1985,** *226,* 669-677.
18. Ditzelmuller, G.; Kubicek, C.P.; Wohrer, W.; Rohr, M. *FEMS Microbiol. Lett.* **1984**, *25,* 195-198.
19. Morimoto, S.; Matsuo, M.; Azuma, K.; Sinskey, A.J. *J. Ferment. Technol.* **1986,** *64,* 219-225.
20. Bolen, P.L.; Detroy, R.W. *Biotechnol. Bioeng.* **1985,** *27,* 302-307.
21. Verduyn, C.; Frank, J.; van Dijken, J.P.; Sheffers, W.A. *FEMS Microbiol. Lett.* **1985,** *30,* 313-317.
22. Takuma, S.; Nakashima, N.; Tantirungkij, M.; Kinoshita, S.; Okada, H.; Seki, T.; Yoshida, T. *Appl. Biochem. Biotechnol.* **1991,** *28,* 327-40.
23. Kotter, P.; Amore, R.; Hollenberg, C.P.; Ciriacy, M. *Current Genet.* **1990**, *18,* 493-500.
24. Bolen, P.; Roth, K.A.; Freer, S.N. *Appl. Environ. Microbiol.* **1986,** *52,* 660-664.
25. Evans, C.T.; Ratledge, C. *Arch. Microbiol.* **1984,** *139,* 48-52.
26. Ratledge, C.; Holdsworth, J.E. *Appl. Microbiol. Biotechnol.* **1985,** *22,* 217-221.
27. Dekker, R.F.H. *Biotechnol. Bioeng.* **1986,** *6,* 605.
28. Hahn-Hagerdal, B.; Joensson, B.; Lohmeier-Vogel, E. *Appl. Microbiol. Biotechnol.* **1985,** *21,* 173-175.
29. Hahn-Hagerdal, B.; Chapman, T.W.; Jeffries, T.W. *Appl. Microbiol. Biotechnol.* **1986,** *24,* 287.
30. Alexander, N. *Appl. Microbiol. Biotechnol.* **1986**, *25*, 203-207.
31. Jeffries, T.W. *Enz. Microb. Technol.* **1984**, *6*, 254-258.
32. Lochke, A.H.; Jeffries, T.W. *Enz. Microb. Technol.* **1986**, *8*, 353-359.
33. Lee, H.; James, A.P.; Zahab, D.M.; Mahmourides, G.; Maleszka, R.; Schneider, H. *Appl. Environ. Microb.* **1986**, *51*, 252-258.
34. Gong, C.S.; Chen, L.F.; Flickinger, M.C.; Chiang, L.C.; Tsao, G.T. *Appl. Environ. Microbiol.* **1980**, *41*, 430-436.
35. Olivier, S.P.; du Toit, P.J. *Biotechnol. Bioeng.* **1986**, *28*, 684-699.
36. Sarthy, A.V. *Appl. Environ. Microbiol.* **1987**, *53*, 1996.

37. Batt, C.A.; Carvallo, S.; Easson, D.D., Jr.; Akedo, M.; Sinskey, A.J. *Biotechnol. Bioeng.* **1986**, *28*, 549.
38. Chan, E.C.; Ueng, P.P.; Chen, L. *Biotechnol. Lett.* **1986**, *8*, 231.
39. Bruinenberg, P.M.; DeBot, P.H.M.; Van Dijken, J.P.; Scheffers, W.A. *Eur. J. Appl. Microbiol. Biotechnol* **1983**, *18*, 287-292.
40. Bruinenberg, P.M.; DeBot, P.H.M.; Van Dijken, J.P.; Scheffers, W.A. *Appl. Microbiol. Biotechnol* **1984**, *19*, 256-260.
41. Xu, J.; Taylor, K.B. *Appl. Environ. Microbiol.* **1993**, *59*, 231-235.
42. Bolard, J. *Biochim. Biophys. Acta* **1986**, *864*, 226-234.
43. Sutton, D.D.; Arnow, P.M.; Lampen, J.O. *Proc. Soc. Exp. Biol. Med.* **1961**, *108*, 170-175.
44. Xu, J.; Taylor, K.B. *Appl. Environ. Microbiol.* **1993**, *59*, 1049-1053.

RECEIVED March 1, 1994

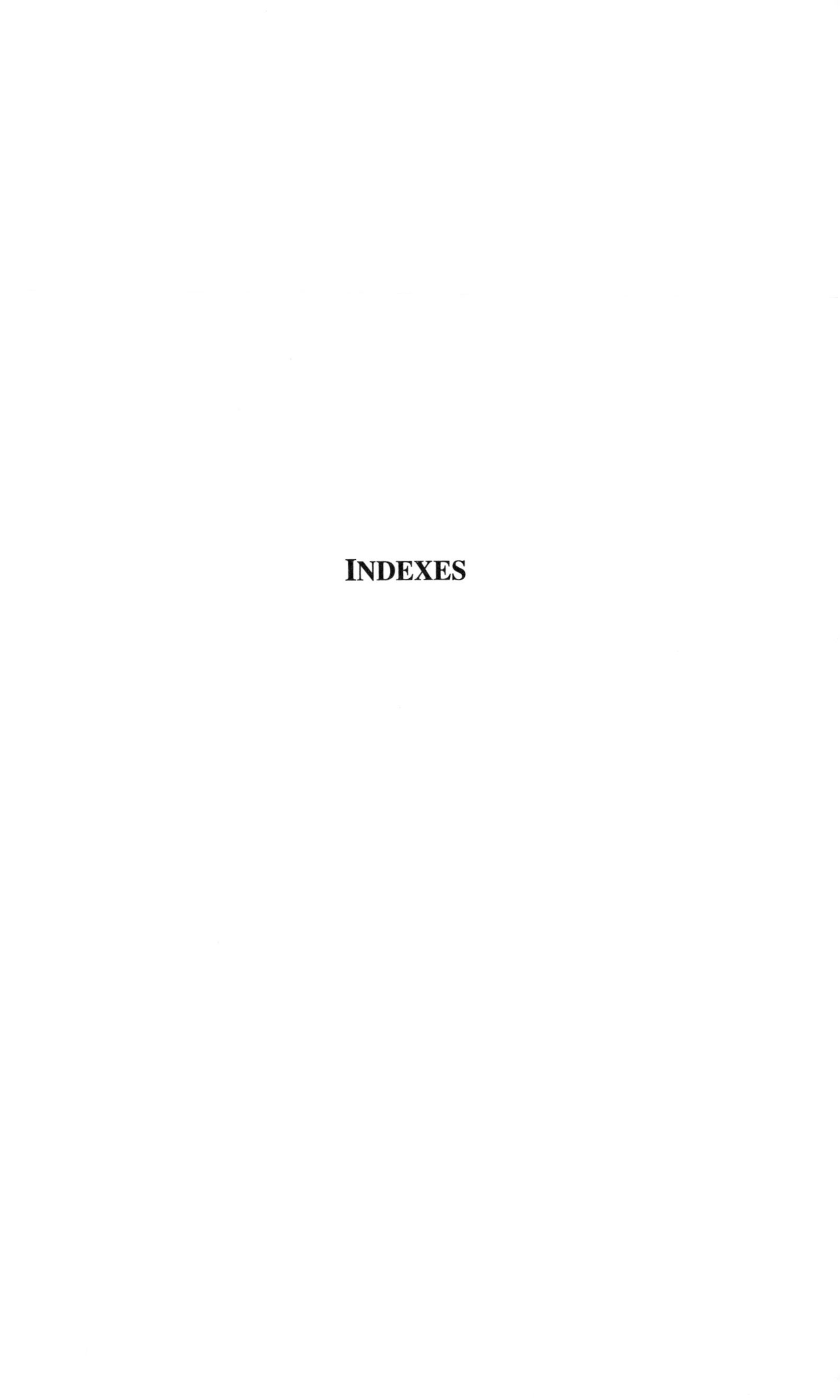

INDEXES

Author Index

Affiliation Index

Subject Index

E

O

P

Production: Beth Harder & Charlotte McNaughton
Indexing: Deborah H. Steiner
Acquisition: Barbara Pralle & Rhonda Bitterli
Cover design: Amy Hayes

Printed and bound by Maple Press, York, PA